D1662265

F. Müller M. Leupelt Eco Targets, Goal Functions, and Orientors

Springer

Berlin
Heidelberg
New York
Barcelona
Budapest
Hong Kong
London
Milan
Paris
Santa Clara
Singapore
Tokyo

SVP

Felix Müller   Maren Leupelt (Eds.)

Eco Targets, Goal Functions, and Orientors

With 77 Figures and 55 Tables

Springer

Dr. Felix Müller
Dipl. Ing. Maren Leupelt

Universität Kiel
Ökologie-Zentrum
Schauenburger Straße 112
D-24118 Kiel
Germany

ISBN 3-540-63679-X Springer-Verlag Berlin Heidelberg New York

Die Deutsche Bibliothek - CIP Einheitsaufnahme

Ecc targets, goal functions and orientors : with 55 tables / Felix Müller , Maren Leupelt (ed.). - Berlin ; Heidelberg ; New York ; Barcelona ; Budapest ; Hong Kong ; London ; Milan ; Paris ; Santa Clara ; Singapore ; Tokyo : Springer, 1998
ISBN 3-540-63679-X

Typesetting: Camera ready by editors
Cover Design: Design & Production, Heidelberg
SPIN 10568571 31/3137 - 5 4 3 2 1 0 - Printed on acid-free paper

Foreword

This volume comprises the proceedings of the International Workshop on Ecological Goal Functions, held at the Schleswig-Holstein Cultural Center of Salzau, August 30 - September 4, 1996. The conference – first in a series – intended to be convened at Salzau at 1 - 2 year intervals to address various aspects of theoretical and application-oriented ecology, was initiated, organized and carried out under the auspices of the Ecology Center of the Kiel University. It featured keynote addresses, invited lectures, submitted papers, and posters. 32 contributions written by authors from eight countries, were selected to be presented in this book.

From the very rich discussions of the workshop, some general characteristics emerged which might become important for a deeper understanding of the nature of evolving systems or, in other words, systems with a history, described by variables with a high degree of interdependence. These characteristics include the following: Speaking of 'goal functions' is a convenient 'façon de parler', since a logical analysis of the formal structure of teleological and causal explanations shows that both are analogous with regard to the inherent structural typology and the basic mode of explanation. Teleological interpretations introduce motives or objectives of actors into the set of 'antecedens' conditions relevant for system evolution, and are consequently a subset of causal interpretations. Hence it ensues that, for the sake of terminological clarity, the use of the term "goal" should be limited to the modeling context, while in descriptions of the real-world situation terms like "teleonomy" or "finality" appear more appropriate.

Considering the latter terms, a further distinction is indicated. Static teleonomy or fitness means that a certain arrangement, for instance poisonous substances or thorns protecting plants against grazing cattle, or imitative colorations and mimicries, are useful for certain 'purposes'. By way of contrast dynamic teleonomy indicates a directiveness of processes, and here again different phenomena can be distinguished which are often confused. Each system which attains a time-independent condition can be described as if its present behavior were dependent on that final state. Contrary to this 'direction of events' is the directiveness based upon structure, meaning that a specific arrangement of structures leads the process in such a way that a certain result is achieved. This clearly applies to the function of man-made machines yielding products desired or performances foreseen but is also true in living systems which are characterized by a structural order of processes widely surpassing all machines in complication and evolved in order to maintain the system itself. An important part of these processes is represented by homeostasis, i.e. those processes through which the material and energetical situation of the organism is maintained constant. These regulations are governed, in a wide extent, by feedback mechanisms, as emphasized by cybernetics.

There is yet another basis for organic regulations, namely equifinality, i.e. the fact that the same final state can be reached from different initial conditions and in different ways. This is the case in open systems, insofar as they attain a steady state. Thus equifinality is responsible for the primary regulability of living systems and supersystems which cannot be based upon predetermined structures or mechanisms, but on the contrary, exclude such mechanisms. They were in former times regarded as arguments for vitalism. Nowadays it is already classical wisdom of thermodynamically oriented system-ecology that a state of sufficient nonequilibrium is maintained within living systems and their relations with the environment. Thus state variables like entropy or negentropy and the energy parameter exergy are very useful descriptors and integral measures of ecosystem organization.

Finally, there is true finality, meaning that the actual behavior is determined by the foresight of the goal. As such, true purposiveness is, in particular, characteristic of human behavior, and it is fundamentally connected with the evolution of the symbolism of language and concepts.

Equating the phase space of ecological systems with a geographic landscape, there exist multiple stable points and regions, or dynamic regimes, for a system; in switching between them, the system has the capability of undergoing qualitative change. Evolution implies an ordered succession of such transitions; at many levels autocatalysis seems to be an important "driving mechanism". Resilience is a measure of the ability of a system to absorb changes of state variables, driving variables, and parameters. It is high near the boundaries of a stable regime (i.e. near the maximum sustainable nonequilibrium), where fluctuations abound and stability (qua return to an equilibrium state) is low; inversely, high stability generally implies low resilience - a system geared to short-term efficiency and productivity. However, evolution seems to further flexibility of systems at all levels; this implies that long-term viability through the development of a capability to deal with the unexpected is favored over short-term efficiency and productivity. Popper's propensity concept with its conditional probabilities is a very promising tool to further and more precisely elucidate the underlying mechanisms.

A convenient, if not comprehensive, taxonomy of views regarding sustainability evolved in the course of discussions. First, the "input-output" view assumes that the internal dynamics of the ecosystem are more or less in a steady state. Second, the related "state" definition requires simply that a sustainable system be one in which a state can be maintained indefinitely. The "capital" or "stock" view requires the maintenance of natural capital at or above current levels and, thus, that the products of the ecosystem be used at a rate within that system's capacity for renewal. Sustainability is thereby ensured by living off the income rather than the ecological capital. This involves the perpetuation of the character and natural processes of the ecosystem and indefinite maintenance of its integrity without degrading the integrity of other ecosystems. Finally, there is the "potential throughput" view, emphasizing the use of resources within the capacity of those resources to renew themselves. According to this view, sustainability is defined on the basis of maintenance of potential, so that ecosystems can provide the same quantity and quality of goods and services as in the past. Potential is emphasized rather than stocks, biomass, or energy levels.

In looking for an organizing principle for measuring sustainability, three levels of activity can be identified. The first level involves determining measurements of sustainability. By defining the basis for measuring sustainability at the ecosystem or landscape level, the relevant measurements can be aggregated into indicators at the second level. Therefore, it is important to assemble a generic list of indicators of sustainablity to serve as a checklist for determining the correct ground for each of the ecosystems. Specific measurements must then be identified for each indicator. However, it is important to define the status of this generic list of indicators at the regional and global levels, and this requires a composite index, which is the third level.

Finally, sustainable development is not an ecological problem nor a social problem nor an economic problem, but an integrated feature at all three. This means that effective investments in sustainable development simultaneously retain and encourage the adaptive capabilities of people, of business, and of nature (Holling v. Schumpeter). The effectiveness of those capabilities can turn the same unexpected event into an opportunity for one system or a crisis for another. Adaptive capacities are dependent on processes that permit renewal in society, economies, and ecosystems. For nature, it is the structure of the biosphere; for business and people, it is usable and useful knowledge; and for society as a whole, it is trust.

This is what this book is all about. It is divided into four main sections, addressing various problems of, and suggested solutions to, the complex problems of an integrated, systemary environmental management. Chapter 2 is devoted to ecosystem development; Chapter 3 deals with the philosophical basis of the goal function and orientors concepts; Chapter 4 presents problems of combining natural and human orientors; and, finally, Chapter 5 covers eco targets as goal functions in environmental management. May this broad coverage help focus the interest of environmentalists, planners and policy makers alike on the all-important environmental issues facing modern life.

Kiel, November 1997 Prof. Dr. Otto Fränzle

Preface

Goal Functions, Orientors and Eco Targets may be considered as elements of an interdisciplinary effort that attempts to utilize and couple principles of ecosystem research and ecological systems theory with human systems and environmental management. The Salzau workshop on ecological goal functions, in autumn 1996, was devoted to the fields of ecosystem research, general and applied ecology, economy, philosophy and sociology. It was a stimulating opportunity to discuss ecosystem theoretical concepts with an interdisciplinary group of scientists. Four work groups have elaborated important components of the workshop schedule. Their results have been integrated into the introductions and conclusions of the single chapters.

It was generally agreed that a joint meeting of different disciplines is of great importance for the development of a unifying concept that combines the holistic ecological theory with environmental practice. Furthermore, the discussion showed clearly that it is an urgent necessity to start an intensive mediation process between these potential partners.

The general opinion substantiates that we are ready to use the data and results of ecosystem research in a suitable and sound way. However, the corresponding basic ideas can only develop and cope with reality if there is a constant and intense dialogue between different theories and between theory and planning-practice. Publishing the workshop results, we hope that the reader will profit from an initial inter- and intra-scientific exchange.

We wish to thank the Federal Ministry for Education, Research, Science and Technology for the kind and efficient support of the workshop and this volume.

We also thank all the persons who were involved in the Salzau-meeting, those contributing with a paper, those taking part at the vivid discussions, and those supporting the organization of the meeting and this book: Guiseppe Bendoricchio, Mario Catizzone, Villy Christensen, Uta Eser, Irene Gabriel, Ursula Gaedke, Folke Günther, Benno Hain, Stefanie Hari, Eugene Krasnov, Bai-Lian Li, Sabine Ludwigshausen, Alessandro Marani, Kati Mattern, Georg Schulze-

Ballhorn, Tim Uhlenkamp, Wiebke Wewer, Wilhelm Windhorst and Hartmut Zwölfer.

Finally, we like to thank all the persons who did take part in the review-process of this book: Robert Alexy, Peter Allen, Jan Barkmann, Guiseppe Bendoricchio, Dieter Birnbacher, Reinhard Bornkamm, Hartmut Bossel, Broder Breckling, Wolfgang Cramer, Klaus Dierssen, Sabine Dittmann, Wolfgang Dombrowski, Uta Eser, Brian Fath, Otto Fränzle, Marc Gessner, Giulio Genoni, Ulrich Haber, Susan Haffmanns, Ulrich Hampicke, Robert A. Herendeen, Rainer Hingst, Georg Hörmann, Rainer Horn, Sven E. Jørgensen, Ludger Kappen, Hans Kastenholz, Roman Lenz, Dmitrii O. Logofet, Sievert Lorenzen, Sylvia Opitz, Claudia Pahl-Wostl, Bernhard C. Patten, Harald Plachter, Wolfgang Rath, Heiner Reck, Michael Rühs, Claus Schimming, Joachim Schrautzer, Wolf Steinborn, Milan Straskraba, Robert E. Ulanowicz, Christian Wissel, Wilhelm Windhorst and Roland von Ziehlberg.

Kiel, Januar 1998 Maren Leupelt and Felix Müller

Contents

List of Contributors

Badura, Jens, M.A.
Environmental Ethics, Center for Ethics in Sciences and Humanities, University of Tuebingen, Keplerstrasse 17, 72074 Tübingen, Germany

Bass, Brad, Dr.
Institute for Environmental Studies, 33 Willcocks St., University of Toronto, Toronto, Ontario M5S 3E8, Canada

Baird, Dan, Prof. Dr.
Department of Zoology, University of Port Elizabeth, P.O. Box 1600, Port Elizabeth 6000, South Africa

Barkmann, Jan, Dipl.-Biol.
Ecology Center, Schauenburger Strasse 112, University of Kiel, 24118 Kiel, Germany

Berg, Astrid, Dipl.-Geogr.
Institute for Landscape Planning and Landscape Ecology, University of Rostock, Justus-von-Liebig-Weg 6, 18051 Rostock, Germany

Bork, Hans-Rudolf, Prof. Dr.
Center for Agricultural Landscape and Land Use Research, (ZALF), Wilhelm-Pieck-Strasse 72, 15374 Müncheberg, Germany

Bossel, Hartmut, Prof. Dr.
Am Galgenköppel 6, 34289 Zierenberg, Germany

Breckling, Broder, Dr.
Ecology Center, Schauenburger Strasse 112, University of Kiel, 24118 Kiel, Germany

Bröring, Udo, Dr.
Institute for Ecology, Technical University of Cottbus, P.O. Box 10 13 44, 03013 Cottbus, Germany

Cousins, Steven, Dr.
International Ecotechnology Research Centre, Cranfield, Bedford MK43 OAL, UK

Deppert, Wolfgang, Prof. Dr.
Department of Philosophy, University of Kiel, Leibnizstrasse 6, 24118 Kiel, Germany

Dierßen, Klaus, Prof. Dr.
Institute of Botany, University of Kiel, 24098 Kiel, Germany

Dilly, Oliver, Dr.
Ecology Center, University of Kiel, Schauenburger Strasse 112,
24118 Kiel, Germany

Fath, Brian, Ph.D., M.A.
Institute of Ecology, University of Georgia, Athens, GA 3060, USA

Gnauck, Albrecht, Prof. Dr.
Institute of Ecosystems and Environmental Informatics, Technical University Cottbus,
P.O. Box 10 13 44, 03013 Cottbus, Germany

Herrmann, Sylvia, Dr.
Institute of Landscape Planning and Ecology, University of Stuttgart,
Keplerstrasse 11, 70174 Stuttgart, Germany

Jeltsch, Florian, Dr.
Departement of Ecosystem Analysis, Center for Environmental Research, Leipzig-Halle, PF 2, 04301Leipzig, Germany

Jong de, Folkert, Dr.
Common Secretariat for the Cooperation on the Protection of the Wadden Sea,
Virchowstrasse 1, 26382 Wilhelmshaven, Germany

Jørgensen, Sven E., Prof. Dr.
Royal Danish School for Pharmacy, Department of Environmental Chemistry,
University Park 2, 2100 Copenhagen Ø, Denmark

Jüdes, Ulrich, Dr.
Institute of Pedagogics of the Natural Sciences, University of Kiel,
Olshausenstrasse 62, 24098 Kiel, Germany

Krotscheck, Christian, Dr.
Institute of Chemical Engineering, Technical University of Graz, Inffeldgasse 25,
8010 Graz, Austria

Kutsch, Werner, Dr.
Ecology Center, University of Kiel, Schauenburger Strasse 112, 24118 Kiel, Germany

Leupelt, Maren, Dipl.-Ing. agr.
Ecology Center, University of Kiel, Schauenburger Strasse 112, 24118 Kiel, Germany

Marques, João C., Dr.
Institute of Marine Research (IMAR), Department of Zoology, University of Coimbra,
3049 Coimbra Codex, Portugal

Marticke, Hans-Ulrich, Dr.
Independent Experts Committee for the Environmental Code, Kurfürstendamm 200,
10719 Berlin, Germany

Metzner, Andreas, Dr.
Institute for Environmental Issues of Social Sciences, Technical University of Cottbus, P.O. Box 10 13 44, 03013 Cottbus, Germany

Müller, Felix, Dr.
Ecology Center, University of Kiel, Schauenburger Strasse 112, 24118 Kiel, Germany

Nielsen, Søren N., Dr.
Royal Danish School for Pharmacy, Department of Environmental Chemistry, University Park 2, 2100 Copenhagen Ø, Denmark

Pardal, Miguel Ã., Dr.
Institute of Marine Research (IMAR), Department of Zoology, University of Coimbra, 3049 Coimbra Codex, Portugal

Patten, Bernhard C., Prof. Dr.
Department of Zoology and Institute for Ecology, University of Georgia, Athens, GA 30602, USA

Potthast, Thomas, Dipl.-Biol.
Environmental Ethics, Center for Ethics in Sciences and Humanities, University of Tuebingen, Keplerstrasse 17, 72074 Tübingen, Germany

Radermacher, Walter, Dipl.-Oec.
Work Group Environmental Economic Accounting System, German Federal Statistical Office, Gustav-Stresemann-Ring 4, 65180 Wiesbaden, Germany

Reiche, Ernst-Walter, Dr.
Ecology Center, University of Kiel, Schauenburger Strasse 112, 24118 Kiel, Germany

Riedel, Wolfgang, Prof. Dr.
Institute for Landscape Planning and Landscape Ecology, University of Rostock, Justus-von-Liebig-Weg 6, 18051 Rostock, Germany

Roch, Isolde, Dr.
Institute for Ecological Regional Development e.V., Weberplatz 1, 01217 Dresden, Germany

Rounsevell, Marc, Dr.
Soil Survey and Land Research Centre, School of Agriculture, Food and Environment, Cranfield University, Silsoe, Bedfordshire MK45 4DT, UK

Roweck, Hartmut, Prof. Dr.
Institute for Water Management and Landscape Ecology, University of Kiel Hermann-Rodewald-Strasse 9, 24118 Kiel, Germany

Ruth, Matthias, Dr.
Department of Geography, Center for Energy and Environmental Studies, Boston University, 675 Commonwealth Avenue, Boston, MA 02215-1401, USA

Schwarz, Astrid E., Dipl.-Biol.
Institute for Landscape Ecology, Technical University of München-Weihenstephan, 85350 Freising, Germany

Svirezhev, Yuri, Prof. Dr.
PIK – Potsdam Institute for Climate Impact Research, P.O. Box 60 12 03, 14412 Potsdam, Germany

Theobald, Werner, Dr.
Department of Philosophy, University of Kiel, Leibnizstrasse 6, 24118 Kiel, Germany

Trepl, Ludwig, Prof. Dr.
Institute for Landscape Ecology, Technical University of München-Weihenstephan, 85350 Freising, Germany

Ulanowicz Robert E., Prof. Dr.
Chesapeake Biological Laboratory, Unversity of Maryland, P.O. Box 38, Solomons, Maryland 20688, USA

Werner, Armin, Dr.
Center for Agricultural Landscape and Land Use Research, (ZALF), Wilhelm-Pieck-Strasse 72, 15374 Müncheberg, Germany

Wiegleb, Gerhard, Prof. Dr.
Institute for Ecology, Technical University of Cottbus, P.O. Box 10 13 44, 03013 Cottbus, Germany

Zölitz-Möller, Reinhard G., Dr.
Ministry of Environment, Nature and Forestry of the State Schleswig-Holstein, P.O. Box 6209, 24123 Kiel, Germany

Chapter 1

Introduction: Targets, Goals, and Orientors

1 Targets, Goals and Orientors

Felix Müller, Maren Leupelt, Ernst-W. Reiche and Broder Breckling

1.1 The Variety of Targets

This volume attempts to combine ecological theory with environmental practice, thermodynamics with environmental planning, network theory with agricultural strategies, theoretical aspects such as self-organization and emergence with environmental economy, jurisdiction, philosophy and sociology.

This volume exemplifies the enormous potential of interdisciplinarity, demonstrating the transfer of concepts across the scientific levels, from empirical measurements to theory construction, from theory testing to practical tool development, and from validated methods to real-world decision making processes.

This volume deals with a broad range of problems and attitudes concerning the identification, definition, realization, and the dynamics of targets in an interdisciplinary environmental context. The basic idea is that any modification of the human relation to nature is inevitably connected with a change of targets and motivations. Thus, the environmental crisis is a crisis of conflicting goals, attractions and intentions which cover a wide range between human exploitation of natural resources on the one hand and undisturbed, natural self-organization of ecosystems on the other. All human input into natural systems thus is a result of considerations mediating the different aims on the extreme ends of this range.

We will ask whether scientists can derive strategies for sustainable development by reflecting basic ecological principles. The fundamentals of these reflections will be ecosystem science and ecosystem theory. Within this framework the central questions of the book will be: What are the general principles of ecosystem development? How are the arising principles of ecological self-organization and emergence exemplified on different scales? Can we define certain regularly appearing attractors and system-based orientors throughout the development? What can we learn from these ecosystem properties? Can we use those orientor principles in environmental management? Could orientors be applied to define new sustainable strategies for landscape management?

Analyzing the general problems and questions, a variety of arguments and features concerning the semantic field "target", which covers a wide field including ecological attractors and orientors as well as individual and social human goals, will be analyzed in detail and integrated on the following pages. Because

F. Müller M. Leupelt (Eds.)
Eco Targets, Goal Functions, and Orientors

the notions "target", "goal", and "aim" will very often be at the center of the discussions, they will be featured in the next chapters from different points-of-view.

1.2 Human Aims: The Central Elements of Environmental Conflicts

All the intentional human actions are directed towards a goal. Even if we seem to be roaming about aimlessly, our inner life is determined by a complex hierarchy of objectives. In this sense, the psychologist Alfred Adler postulated in 1927 that "nobody would be able to think, feel, will, or even dream if this was not determined, stipulated, limited, and directed by an inherent goal" (Adler 1980, p. 31). Biologically it may easily be declared that this inherent orientation of the individual life is survival, a fact which should rationally imply the survival of the species as an emergent property of individual instincts and their sexual drives. But human actions are additionally filtered, driven and regulated psychologically. These actions can be summarized as the individualistic endeavor to be an integrated, valuable, and acknowledged member of the human society and its subsystems (Dreikurs 1969). As a consequence, our behavior is strongly influenced by the changing priorities among the physiological and psychological goals, as well as the outcome of the various goal conflicts, which in the end form our personal characters. As we all know, whenever human beings with different characters start interacting, diverging and competing goals easily become evident. In many of such cases the best strategy to solve the problems has turned out to be a reflection of the goals that the individuals involved are heading for, consciously or unconsciously. Therefore, a critical analysis of the dominating hierarchy of orientations, aims and targets is an approved method for the attempt to adjust human behavior to the requirements of circumstances.

Of course, the human goal hierarchies also determine our attitudes towards nature and towards the relative importance of the natural systems' health and integrity (see Constanza et al. 1992 or Woodley et al. 1993). Generally, human life depends on natural resources and on the activities and influences of other life forms: Nature provides essential prerequisites for human welfare (see de Groot 1992), such as production functions (e.g. oxygen or food supply), carrier functions (e.g. recovery or disposal), information functions (esthetical and cultural information), or regulation functions (e.g. climatic or hydrological regulations). However, looking back to the early 17th century, man did claim power upon nature. Since Bacon's concepts, which we could today designate as elements of a rather outsized human self-confidence, mankind has striven for the mastery of nature by developing science and technology. The subsequent submission and exploitation of nature did not only bring power and wealth, but did also cause social, economic and at last ecological problems. Since we still are and always will be dependent upon nature the key to our actual environmental, social and economic problems is a close cooperation with nature. Therefore, our targets have to be changed, on an individualistic level as well as on the level of society. And, as to cooperate with another "party" means partnership and requires knowledge about the partner, we have to study nature from that modified point-of-view. The results of these studies should be the basis for the imperative adaptation of our targets, because this is a basic supposition for the necessary change of our environmental behavior.

1.3 Environmental Targets: The Goals of Sustainability, Integrity and Health

An initialized modification of targets has been visible in environmental policy since the sustainability debate started (WCED 1987; Ekins 1992; Brand 1997; Teichert et al. 1997). One aspect that was introduced with the sustainability concept originates in its long-term character: The temporal extents of the objectives are not restricted to the typical four years election period but to generations. With this attitude the developmental capacities of ecosystems, i.e. their potentials for future self-organization, succession and evolution, have become emphasized targets of environmental policy. Consequently, a sustainable landscape management has to practice a holistic strategy, it has to argue about integrated ecosystems instead of structural units of the community alone, and it has to integrate ecological, social and economical goals. Therefore, the sustainability concept is an interdisciplinary strategy, which brings together aspects of ecological qualities as well as the many different interpretations of human life quality. Also, the spatial extent of the debate has changed: The area that sustainability evaluations should refer to, has to include neighboring systems as well as indirectly linked zones, which may be far away from the locality of direct action. And finally, sustainable management strategies have to include indirect effects, they have to take into account non-linearities, chronic stress effects, decouplings of processual networks and combinations of these factors. In other words, they have to be based on an ecosystemic approach. Thus, the sustainability concept demands new measures, new methods, and – last but not least – new goals. Trendsetting directions and objectives have been proposed, e.g. with the concepts of ecosystem health (Costanza et al. 1992) or ecological integrity (Woodley et al. 1993). Both strategies are based on an approach that integrates ecosystem structures and functions. Both approaches are oriented at the complexity of ecological systems and at the idea to support the systems' long-term developments. Constanza et al. (1992) define "health" as an ecosystem feature which refers to the basic characteristics: vigor (metabolic activity), organization, and resilience. "Integrity" has been described by Kay (1993) as the ability of a system to maintain its organization and to develop in sequences of self-organized processes, thus integrity comprises health, buffer capacity, and the self-organization capacity.

In summary, recently a paradigmatic change of fundamental orientations is taking place in ecology and environmental practice. The basic ideas, models and leading principles are scrutinized, and thus they are become more flexible in a phase of change. Also the methods of environmental evaluation and decision support are modified. This improvement can be characterized by an increased consciousness of the significance of goals and constructions of goal hierarchies, reaching from leading environmental ideas to eco targets, environmental quality objectives or environmental standards. This development is accompanied by a prolonged search for suitable indicators which satisfactory represent the degree of success with respect to the environmental aims that have been and will be set.

1.4 Ecological Orientors: Attractors and their Consequences

Both, the setting of a goal and the ensuing reflections in order to reach it are intellectual activities. Natural systems can neither define a goal nor reflect on it. Nevertheless, for biologists and evolutionary theoreticians it seems to be extremely tempting to impute a certain teleological purpose to natural features: "This anatomic structure has developed to enable the organism to hunt its prey more efficiently", "Aerenchyma is present in certain water plants to enable a better gaseous exchange even below water". The critical point about these and similar statements is that they provoke the impression of an organism to be heading for a 'Lamarckian' evolutionary self-optimization. In contrast, the Darwinian theory has shown that the causality is based on other fundamental processes: mutation, competition, and selection, which may lead to new, perhaps even more efficient strategies of resource utilization. These procedures are continuously proceeding without any conscious intentions involved.

On the ecological scale, a similar teleological temptation can be found, especially in the context of succession theory. Succeeding systems are understood as aiming at certain climax states which potentially are stable for long periods of time because the potentials of the sites are utilized perfectly when a climax stage – a stage of an optimal internal adaptation of the bioceonotic community – is attained. Thus, we have to formulate scientific hypotheses very carefully in this context: Of course, ecosystems, unlike human systems, do not have a goal. There is no intellectually fixed destination that natural development would lead to. In addition, there is no chance to formulate or realize the wish to grow into a specific direction in these systems. From a scientific perspective, there are no religious, mystic, or vital forces such as Driesch's 'vis vitalis' or Theilhard de Chardin's final purpose 'omega'. All there is are coexisting and competing organisms which constantly interact with their environments and which therefore continuously change their abiotic life conditions. Vice versa, with the non-living ecosystem components the biotopes of the organisms are steadily modified, and, thus, there is a constant change, caused by internal as well as external actors. Throughout that change, the basic selection processes take place as described above. They are also observable on a small scale, and their local consequences of course are not as striking as global evolutionary radiations. However, the basic interorganismic processes are the same in both contexts.

Consequently, also the scientific strategies to cope with these processes can be paralleled: evolutionary biologists observe the anatomy and physiology of living organisms and compare them with historical forms or with related species. Then they put their objects into an evident temporal order and in a third step they draw functional and causal conclusions and posit new hypotheses. Ecologists also analyze the structures and the physiology of ecological systems, they compare them with neighboring, related or historical systems, and they build hypotheses on the basic developmental trends which connect the different stages. Following this procedure in many cases, ecologists have been able to draw some abstract conclusions and generalizations: Succession attributes could be found particularly on a functional scale while the exact prognosis of the developing ecosystem structure – the equipment with different species in different abundance – has turned out to be very problematic. In the second chapter of this volume, some of these characteristics will be reported, and it will be tested whether the proposed tendencies really can be taken as valid generalizations.

In the last years additional aspects have been developed in ecosystem theory. The concepts of self-organization and the fundamental laws of thermodynamics, results from network theory, information theory, and cybernetics, have elucidated succession from a new ecosystemic and theoretical perspective. Certain system features which can be denoted as emergent or collective ecosystem properties, are changed regularly by self-organized ecological development. A central hypothesis of this volume refers to the idea that certain states of these attributes can be taken as orientors or attractors, which means as stages that ecological systems usually develop towards.

Orientors are aspects, notions, properties, or dimensions of systems which can be used as criteria to describe and evaluate the system's developmental stage (Bossel 1992a). The degree of orientor satisfaction which is represented by the distance of an observed state from an optimum point, can be taken as a respective indicator. The combination of degrees of satisfaction from different orientors leads to multiple indications of systemic properties. One focal question will be whether the degree of naturalness and the degree of disturbance in an ecological system can be characterized by such orientors.

The orientation principle and orientor theory (Bossel 1992b) are closely connected with the idea of optimization: A certain feature or a group of elements increases up to a maximum level which is a consequence of the constraints that regulate the respective functional unit. The optimization process implies a high conceptual vicinity with technical procedures and machinery. Thus, it is not easy to deal with the highly individual and hardly predictable self-organization potential of ecological systems. Furthermore, the optimization aspect introduces an originally economic attitude into ecosystem analysis: One of its focal points is the economy of the system's energy budget, including the idea of a minimization of efforts, and thus an increasing efficiency of the processes (Kull 1995). Another conceptual problem arises because evolutionary optimization is not oriented towards future structures, but only effective in the present situation.

Considering evolutionary processes from an ecological perspective leads to the idea of co-evolution. At this point optimization means that the single elements of a system – the organisms – become more and more adjusted to each other. This mutual adaptation causes many ecosystem properties, such as efficient reductions of losses or optimization of efficiencies on the ecosystem level (Weber et al. 1989). Many of the corresponding properties will be discussed in this book, and several of them are proposed to be introduced into environmental evaluation procedures.

If we utilize these orientation concepts in ecology, many applications are possible. Most experience has been gained in ecological modeling. There, certain state variables have been taken as goal functions, which are observed under different model constellations, e.g. different species compositions. Assuming that the system will always develop towards a state with a higher value of the respective goal function, optimizational calculations have been carried out very successfully, e.g. in applying the thermodynamic variable exergy as a goal function.

1.5 A Variety of Questions - and a Variety of Positions

What we can conclude from the introductory discussion of human aims, environmental targets, and ecological orientors, is the enormous significance of target setting processes as well as a strong need for interdisciplinarity, if we want to integrate these different target approaches within one concept. In addition, there is a high demand for discussions which could lead to an ecosystemic environmental evaluation system. These highly relevant items will be discussed and described in the following chapters. The pervasive questions are the following:

- Question A: What are the potential consequences of ecosystem evolution and development for the definition of goals for environmental management?
- Question B: Which are the suppositions for a valid derivation of ecological orientors in self-organized systems?
- Question C: Will it be possible to develop a hierarchy of system-oriented, holistic goals for the realization of principles like sustainability, health, and integrity, based upon an integration of eco-centric and anthropo-centric orientors? Which are the corresponding indicators and how can they be quantified, implemented, and evaluated?
- Question D: How can natural trends of ecosystem development be coupled with the goals of society? Which are the orientors of human systems, how can they be quantified, and how can they be combined with the ecological orientors?

- Question E: How are ecological goal functions correlated with the general features of self-organization, emergent properties, thermodynamics, gradient degradation, and general ecosystems dynamics?

The theoretical ecological and ecosystem fundamentals for potential answers to these mentioned questions are summarized in the following 6 theses:

- Thesis A: During the development of ecosystems, important measurable properties are regularly optimized.
- Thesis B: Ecological orientors can be used to distinguish systems states and to characterize different systems.
- Thesis C: One set of ecological orientors defines both, the structural and the functional features of the investigated system. Thus, orientors can be used for a holistic ecosystems characterization.
- Thesis D: Ecological orientors are based on thermodynamic principles. They indicate general properties of dissipative living systems. Therefore, they represent the potential for self-organization.
- Thesis E: Ecological orientors indicate the degree of naturalness in ecosystems.
- Thesis F: Ecological orientors are a good basis for finding usable indicators for ecosystem health, ecological integrity, or sustainablity.

The questions and hypotheses will be analyzed in the following chapters. In chapters two to five we will move from theory to practice, while weaving a net among the participating scientific disciplines.

The second Chapter will deal with the theoretical and the ecological background of the orientor and goal function concept: The six hypotheses mentioned above, form one pillar of this volume, and thus will be the guiding principles for this chapter. Central questions will be the definition of ecological orientors, their interrelationships with the concepts of self-organization and emergence, and their potentials to be used for ecosystem protection purposes. These discussions are based on papers that include thermodynamic approaches, papers from network theory, information theory, utility theory, succession theory and self-organization. In addition, the role of orientors in the Holling-model of ecosystem dynamics will be discussed, along with a number of case studies describing empirical and modeling experience with orientors and goal functions.

In Chap. 3 the discussion will continue with philosophical aspects of the goal functions-concept. What is the epistemological status of the orientor concept? Is it possible to avoid an invalid teleology when describing goal functions or orientors? How far is the concept rooted in socially preconceived prejudices that influence the perception of nature? How are orientors coupled with ethical values? We will address the orientor concept with arguments from economics and the social sciences in Chap. 4: There exists a diversity in ecosystems, a diversity of orientors, and a diversity of targets. This Chapter will introduce ecological and anthropogenic targets and their parameters based on the concept of conflicting utilization interests. The general questions deal with the interface of ecological and human goal functions. How can both functions be integrated and how can their balance be used in environmental management? Is it useful to integrate the orientor concept into the sustainability discussion? If so, on what temporal and spatial scales should it be done?

Chap. 5 will deal with practical problems of landscape management. This chapter is very important for the claim of unifying different and diverging goal functions. Ecological targets must be differentiated according to their feasibility, and the conclusions should provide usable proposals for ecosystem management as well as for the implementation of the theoretical principles. In the last chapter, the results will be summarized and the following questions will be discussed: What is the indication for future ecosystem-protection-strategies? How can this concept be implemented into future ecosystem research and environmental planning?

Each Chapter will include an introduction, which presents the focal questions and the main points. In addition, each Chapter will conclude with a discussion of the papers contained therein. The general questions raised in this volume and the results of the workshop discussions will be addressed in a final general conclusion.

References

Adler A (1980) Menschenkenntnis. Fischer, Frankfurt/Main, p 255

Bossel H (1992a) Modellbildung und Simulation - Konzepte, Verfahren und Modelle zum Verhalten dynamischer Systeme. Vieweg-Verlag, Braunschweig, p 400

Bossel H (1992b) Real structure process description as the basis of understanding ecosystems and their development. Ecological Modelling 63:261-276

Brand R (1997) Begriffsdschungel: Chronologie der Entwicklung und Diskussion zur Agenda 21. Politische Ökologie 52:25

Costanza R, Norton BG and Haskell BD (eds) (1992) Ecosystem Health - New goals for environmental management. Island Press, Washington DC, p 269

De Groot R S (1992) Functions of nature. Wolters - Noorhoff

Dreikurs R (1969) Grundbegriffe der Individualpsychologie. Stuttgart

Ekins P (1992) 'Limits to growth' and 'sustainable development': grappling with ecological realities. Ecolog Economics 8:269-288

Kay JJ (1993) On the nature of ecological integrity: some closing remarks. In: Woodley S, Kay JJ and Francis G (eds) Ecological integrity and the management of ecosystems. University Waterloo and Canadian Park Serv, Heritage Res Center, Ottawa, pp 201-214

Krohn W, Krug H-J, Küppers G (eds) (1992) Selbstorganisation: Jahrbuch für Selbstorganisation in der Geschichte der Wissenschaften. Band 3, Duncker & Humbolt, Berlin

Kull U (1995) Evolution, Evolutionstheorien und Optimierung. In: Kull U, Ramm E and Reiner R (eds) Evolution und Optimierung. Stuttgart, pp 11-62

Teichert V, Stahmer C, Karcher H, Diefenbacher H (1997) Quadratur des Kreises. Politische Ökologie 52:55-57

WCED (Brundtland SH) (1987) Our Common Future. Oxford University Press, Oxford

Weber BH, Depew DJ, Dyke C, Salthe SN, Schneider EO, Ulanowicz RE and Wicken JS (1989) Evolution from a thermodynamic perspective: An ecological approach. Biology and Philosophy 4:373-405

Woodley S, Kay JJ and Francis G (1993) Ecological integrity and the management of ecosystems – University Waterloo and Canadian Park Serv, Heritage Res Center, Ottawa, p 220

Chapter 2

The Theoretical Approach: Tendencies of Ecosystem Development

2.1 The Physical Basis of Ecological Goal Functions – Fundamentals, Problems and Questions

Felix Müller and Brian Fath

The following Sections (2.2 to 2.15) serve as a general foundation for the Chapters 3 to 5 of this book. The papers impart the essential ecological information and the basic (eco)system theoretical background of the orientor and goal function concepts. Thus, this Chapter is a base for the discussions on the role of goals and attractors, for the setting of targets, for the significance of repeatedly occurring tendencies in ecological development and for the application of these principles to empirical case studies. Fourteen contributions from ecology, mathematics, physics, systems analysis, modeling and agricultural science will explain these phenomena from different scientific points-of-view. The basic theoretical ideas will be presented in the first part of the section (Sect.s 2.2 to 2.11), while a second group of papers (Sect.s 2.12 to 2.15) —including case studies, applications and progress reports— discusses their significance and their empirical evidence. These presentations will be used to prove the general hypotheses of the orientor approach, which have been summarized in Chap. 1 and which will be formulated with more detail in the following paragraphs. The essential results of the individual papers will be integrated in the concluding paper of this section (Sect. 2.16).

The central idea of the orientor approach, which most of the following hypotheses are focused on, originates in modern systems analysis, complexity science and synergetic. It refers to the idea of self-organizing processes, that are able to build up gradients and macroscopic structures from the microscopic "disorder" of non-structured, homogeneous element distributions in open systems, without receiving directing regulations from the outside. In such dissipative structures the self-organizing process sequences in principle generate comparable series of constellations that can be observed by certain emergent or collective systems features. Thus, similar changes of certain attributes can be observed in different environments. Utilizing these attributes, the development of the systems seems to be oriented toward specific points or areas in the state space (see Fig. 2.1.1.). The respective state variables which are used to elucidate these dynamics, are termed orient*ors*. Their technical counterparts in modeling are called *goal functions*. They are variables or objects of optimization procedures, which are guiding the systems' functions as well as their structural devel-

F. Müller M. Leupelt (Eds.)
Eco Targets, Goal Functions, and Orientors

papers of Bossel in Sect. 2.2, Jørgensen et al. in Sect. 2.4, and Nielsen et al. in Sect. 2.14). The specific points in the state space that are approached asymptotically, may be described as *attractors*. Their positions are determined by the potentials of the subsystems, as well as the supersystem of constraints which are limiting the focal system's developmental degrees of freedom. These tendencies and attractors can be found in very different systems, but in particular they are abundant in living systems, which can be characterized by an enormous creativity referring to the evolution and succession of forms, actors, and interactions in a comprehensive hierarchy of structures and processes. The central question the following Chapters will focus on, is whether – in spite of nature's enormous structural creativity and in spite of all the very special circumstances of the unique, "non-comparable" single ecological cases – such oriented, self-organizing tendencies can also be found in ecological systems. And if they can, we have to ask for the general principles that are governing those phenomena.

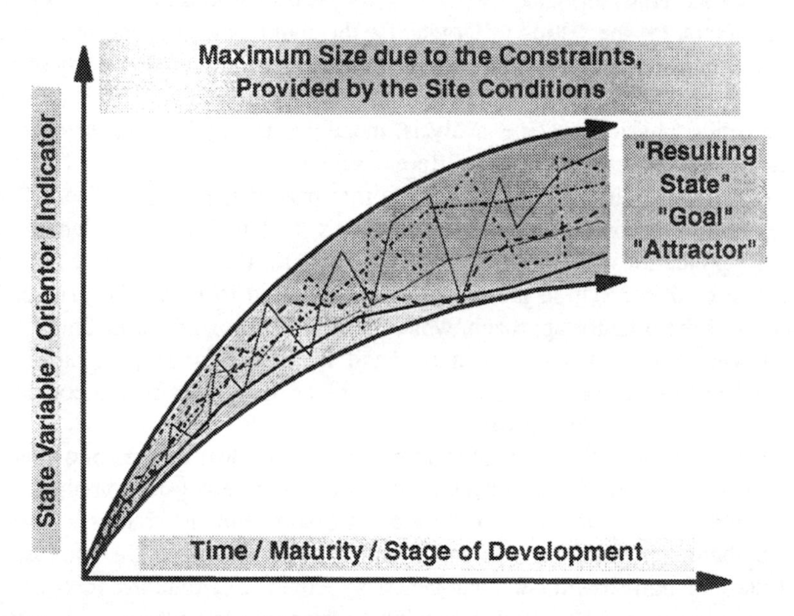

Fig. 2.1.1. A generalized model of developmental trends of ecological state variables, indicators or orientors. To illustrate the conceptual uncertainty the tendency is directed toward a range of potential states.

To illustrate these questions and to formulate the respective concepts in more detail, seven hypotheses will be used as a framework of the forthcoming contributions. These hypotheses address the existence, observability, scale, function, and controllability of ecological orientors. The first hypothesis suggests nothing more than the existence of orientors: *During the development of ecosystems, important measurable properties are regularly optimized.* The corresponding question is this: Are ecological orientors, goal functions, and ecological attractors valid models and evident concepts from an ecological and systems-theoretical point-of-view? All authors in this Chapter will discuss this question, and as a result the space of validity for this hypothesis will be enclosed, especially from a methodological point-of-view. The aspects of this discussion and the derivation of different orientors will reach from the conventional succession theory (see Bröring and Wiegleb Sect. 2.3) through thermodynamics (Jørgensen et al. Sect.s 2.4 and 2.7; Marques et al. Sect. 2.5; and Svirezhev Sect. 2.6), network theory (Patten Sect. 2.8; Fath and Patten Sect. 2.9; and Ulanowicz Sect. 2.10) up to Holling's theory of cycling ecological dynamics (Bass Sect. 2.11). Furthermore, in a number of case studies of rather different origins, such as terrestrial ecosystems (Kutsch et al. Sect. 2.12), marine ecosystems (Baird in Sect. 2.13), coastal ecosystems (Nielsen et al. Sect. 2.14), and soils (Cousins and Rounsevell Sect. 2.15), the orientor principles will be demonstrated and experience about their application will be reported.

The second hypothesis claims the applicability of orientors: *Ecological orientors can be used to distinguish systems states and to characterize different systems.* If this idea is correct, orientors could be well used to describe ecological entities and to classify them from a developmental point-of-view. The appropriate questions are these: Are the quantitative differences concerning the orienting state variables high enough to distinguish maturity stages unambiguously? Are orientors suitable to serve as ecological or ecosystem indicators? Is there a clear connection between the indicator and its indicandum? Is the resulting distinction useful as a classification? Concerning these questions, especially orientor theory, which will be discussed in Bossel's paper (Sect. 2.2), can be used as a helpful tool. Further information can be extracted from the case studies and from the examples that are included in most of the relevant papers. The third hypothesis was formulated in the following manner: *One set of ecological orientors defines both, the structural and the functional features of the investigated system. Thus, orientors are holistic ecological characteristics.* Only if this prerequisite is fulfilled, the orientors can be used to describe ecosystems as inseparable units in the demanded holistic manner. The corresponding strategic questions are: How many of such indicators do we need to describe the whole system satisfactory? How can the single orientor data be aggregated into an applicable indication scheme? Can all the different orientors be treated as equivalent elements? Or are there differences in scale, significance, or expressiveness? If there are, how can these indicators be integrated into a comprehensive model? Which are the interrelationships between the orientors? Again, all papers will provide some ideas concerning these questions, whereby Jørgensen's parameter comparisons (Sect. 2.7) and Patten's derivation of multiple ecosystem properties (Sect. 2.8) will be used as focal sources for answers as well as the quantitative case studies from Baird (Sect. 2.13) and Kutsch et al. (Sect. 2.12).

Hypothesis D that has been discussed with emphasis during the workshop refers to the theoretical foundation as well as the potential applicability of goal functions and orientors: *Ecological orientors are based on thermodynamic principles. They indicate general properties of dissipative living systems. Therefore, they represent a potential for self-organization.* These points suggest a very high generality of the orientor approach on the one hand. On the other hand, they refer to the idea that orientors can be used for the evaluation of ecosystem states. Modern concepts of ecosystem protection such as health or integrity include the idea that a central goal of environmental management must be the enhancement of the capacity of ecosystems to continue the self-organized development in presence of environmental perturbations. This is a support of adaptability and flexibility and thus an enhancement of the self-organization capacity. Will the results of the papers be suitable to propose a further concept development in that direction? How can the ideas be implemented into the young concepts of ecosystem protection? Very close to this question lies the hypothesis E. It suggests that *ecological orientors indicate the degree of naturalness in ecosystems.* If this is true, we can come back to the initial question of this book: As nature has developed continuously and successfully, proceeding several maturing periods of orientor optimization, we can ask what humans can learn from the underlying principles and their multiple realizations. Furthermore, it is the "balance of nature" (Naturhaushalt) which has to be protected with the highest priority of many environmental laws. Therefore, it should be discussed within which framework the processes that guarantee the preservation of this pictorial equilibrium can be analyzed and utilized in conservation and management contexts. The final hypothesis suggests a very high potential in this context: *Ecological orientors are a good basis for finding usable indicators for ecosystem health, ecological integrity, or sustainability.* The papers in this Chapter identify many different orientors and goal functions each which gives different information about the system. The list herein is not exhaustive, but it represents the current development in this area. It is possible that some orientors will prove to be more useful than others particularly for certain applications. However, the objective is not to prune the tree of orientors, but to showcase and explain it. The following papers will give many answers for these questions and demonstrate support for the general hypotheses. In Sect. 2.16 we will try to find some summarizing answers, concepts and ideas on the base of the next 14 studies.

2.2 Ecological Orientors: Emergence of Basic Orientors in Evolutionary Self-Organization

Hartmut Bossel

Abstract

According to orientation theory, multidimensional value orientation is a basic emergent feature of evolutionary adaptation of systems to environments characterized by particular physical conditions, sparse resources, variety, variability, change, and other systems. The need for balanced satisfaction of essential system values, i.e. basic orientors (existence, effectiveness, freedom of action, security, adaptability, coexistence) emerges in response to these environmental challenges. This process of emerging value orientation can also be demonstrated in computer experiments with artificial organisms (animats) using genetic algorithms to study knowledge growth and organization. Basic orientor emphasis emerges differently among individuals of the animat population, resulting in different lifestyles. Pathological behavior and system failure result if there is insufficient attention to any of the basic orientors. The basic orientor approach allows comprehensive assessments of system fitness and performance, in particular also of the feasibility and viability of future development paths. Goal functions are context-specific and system-specific expressions of basic orientor requirements.

2.2.1 Basic Orientors of System Viability and Sustainability

System viability is functionally related to both the system and its properties, and to the system's environment and its properties. A system can only exist and prosper in its environment if its structure and functions are adapted to that environment. If a system is to be successful in its environment, the particular features of that environment must be reflected in the system's structure and functions.

There is obviously an immense variety of system environments, just as there is an immense variety of systems. But all of these environments have some common general properties. These properties will be reflected in systems. These reflections, or basic orientors, orient not just structure and function of systems, but also their behavior in the environment (Bossel 1977, 1987, 1994). The term

F. Müller M. Leupelt (Eds.)
Eco Targets, Goal Functions, and Orientors

'orientor' is used to denote (explicit or implicit) normative concepts that direct behavior and development of systems in general. In the social context, values and norms, objectives and goals are important orientors. Ecosystems and organisms pursue certain goal functions as orientors (Müller 1996). Orientors exist at different levels of specifity within an 'orientor hierarchy'. The most fundamental orientors, the 'basic orientors', are identical for all complex adaptive systems. Orientors are 'dimensions of concern'; they are not specific goals.

System environments that can be found on earth are characterized by six fundamental **environmental properties**: *1. normal environmental state, 2. scarce resources, 3. variety, 4. variability, 5. change, 6. other systems.*

Normal environmental state: The actual environmental state can vary around this state in a certain range.

Scarce resources: Resources required for a system's survival are not immediately available when and where needed.

Variety: Many qualitatively very different processes and patterns of environmental variables occur and appear in the environment constantly or intermittently.

Variability: The normal environmental state fluctuates in random and unpredictable ways, and the fluctuations may occasionally take the system far from the normal state.

Change: In the course of time, the normal environmental state may gradually or abruptly change to a permanently different normal environmental state.

Other systems: The behavior of other systems introduces changes into the environment of a given system.

These fundamental properties of any environment are each unique, i.e., each property cannot be expressed by any combination of other fundamental properties. If we want to describe a system environment *fully*, we have to say something about *each* of these properties. Their content is system-specific, however. The same physical environment presents different environmental characteristics to different systems existing in it. Systems have to develop adequate responses to these basic properties of their environment. Their system structures and their behaviors will be shaped by corresponding fundamental criteria, or basic orientors.

The **basic system orientors** corresponding to the six basic properties of the system environment are (Bossel 1977, 1992, 1994) *1. existence, 2. effectiveness, 3. freedom of action, 4. security, 5. adaptability, 6. coexistence.* The basic system orientors are direct counterparts of the fundamental environmental properties:

Existence: Attention to this orientor is necessary to insure the immediate survival and subsistence of the system in the *normal environmental state*.

Effectiveness: The system should on balance (over the long-term) be effective (not necessarily efficient) in its efforts to secure *scarce resources* from, and to exert influence on its environment.

Freedom of action: The system must have the ability to cope in various ways with the challenges posed by *environmental variety.*

Security: The system must be able to protect itself from the detrimental effects of *environmental variability*, i.e. variable, fluctuating, and unpredictable conditions outside of the normal environmental state.

Adaptability: The system should be able to change its parameters and/or structure in order to generate more appropriate responses to challenges posed by *environmental change.*

Coexistence: The system must be able to modify its behavior to account for behavior and orientors of *other* (actor) *systems* in its environment.

Psychological Needs: For sentient beings, we must add psychological needs as an additional orientor.

Obviously, the system equipped for securing better overall orientor satisfaction will have better fitness, and will therefore have a better chance for long-term survival and sustainability (Krebs and Bossel 1997). Quantification of orientor satisfaction therefore provides a measure for system fitness in different environments. This can be done by identifying indicators that tell us how well each of the orientors is being fulfilled at a given time. In other words, the basic orientors provide a checklist for asking suitable questions and defining the proper indicators for finding out how well a particular system is doing in its environment, and how it is contributing to the development of the total system of which it is a part.

Each of the orientors stands for a unique requirement. That means that a minimum of attention must be paid to each of them, and that compensation of deficits of one orientor by over-fulfillment of other orientors is not possible. For example, a deficit of 'freedom of action' in a society cannot be compensated by a surplus of 'security'. Note, however, that uniqueness of each of the orientor *dimensions* does not imply independence of individual orientor *satisfactions*: For example, better satisfaction of the security orientor may require a sacrifice in freedom of action because financial resources are needed for the former, and are then unavailable for the latter. But that does not mean that 'freedom' can be used as a substitute for 'security'.

Viability, i.e. health and fitness of a system therefore require adequate satisfaction of each of the system's basic orientors. In ecosystems, we can (normally) expect balanced attention to basic orientors at the level of organisms, populations, and whole ecosystems. In societal systems, simultaneous attention to basic orientors (or derived criteria) should be applied in all planning processes, decisions, and actions. Comprehensive assessments of system behavior and development must therefore be multi-criteria assessments. In analogy to Liebig's Principle of the Minimum, the system's development will be constrained by the orientor that is currently 'in the minimum'. Particular attention will therefore have to focus on those orientors that are currently deficient.

In the orientation of system behavior, we deal with a two-phase assessment process where each phase is different from the other. *Phase 1:* First, a certain minimum satisfaction must be obtained separately for each of the basic orientors. A deficit in even one of the orientors threatens long-term survival. The system will have to focus its attention on this deficit. *Phase 2:* Only if the required minimum satisfaction of *all* basic orientors is guaranteed is it permissible to try to raise system satisfaction by improving satisfaction of *individual* orientors further – if conditions, in particular other systems, will allow this.

The basic orientor proposition has three important implications:

1. If a system evolves in a normal environment, then that environment forces it to implicitly or explicitly ensure minimum and balanced satisfaction of each of the basic orientors.
2. If a system has successfully evolved in a normal environment, its behavior will exhibit balanced satisfaction of each of the basic orientors.
3. If a system is to be designed for a given environment, proper and balanced attention must be paid to satisfaction of each of the basic orientors.

The third implication has particular relevance for the creation of programs, institutions, and organizations in the socio-political sphere (Bossel 1996, 1998).

Note that for a specific system in a specific environment, each orientor will have a specific meaning. I.e. 'security' of a nation is a multi-facetted objective set with very different content from the 'security' of an individual. However, the systems-theoretical background for satisfaction of the 'security' orientor is the same in both cases.

In many systems, in particular ecosystems, 'goal functions' are often more immediately obvious than the basic orientors that cause the emergence of these goal functions in the first place. Goal functions can be viewed as appearing on a level below the basic orientors in the hierarchical orientation system. They translate the fundamental system needs expressed in the basic orientors into concrete goals linking system response to environmental properties.

With regard to finding a comprehensive indicator set for ecosystem integrity, such an approach, based on ecosystem 'goal functions', has been outlined in considerable detail (Müller 1996). Ecosystem goal functions emerge as general ecosystem properties in the coevolution of ecosystem and environment. They can be viewed as ecosystem-specific responses to the need to satisfy the basic orientors. Major ecosystem goal functions are (according to Müller 1996): Optimization of: use of solar radiation, material and energy flow intensities (networks), matter and energy cycling (cycling index), storage capacity (biomass accumulation), nutrient conservation, respiration and transpiration, diversity (organization), hierarchy (signal filtering).

The emergence of basic orientors in response to the general properties of environments can be deduced from general systems theory, as has been done here, but supporting empirical evidence and related theoretical concepts can also be found in such fields as psychology, sociology, and the study of artificial life.

If basic orientors are indeed the consequence of adaptation to general environmental properties, and therefore of fundamental importance to the viability of individuals, then we can expect them to be reflected in our emotions. This is indeed the case (Bossel 1978, 1997): Each of the basic orientors has a characteristic counterpart in our emotions. Also, we find that all societies have developed methods of punishment by selective basic orientor deprivation (Bossel 1978, 1997). Depending on kind and severity of the offense, the delinquent is denied full satisfaction of one of the basic orientors. In his work on 'human scale development', Manfred A. Max-Neef has classified human needs according to several categories which can be mapped one-to-one on the basic orientors (Max-Neef 1991; Bossel 1997).

Characteristic differences in the behavior ('life styles', 'life strategies') of organisms, or of humans or human systems (organizations, political or cultural groups) can be explained by differences in the relative importance attached to different orientors (i.e. emphasis on 'freedom', or 'security', or 'effectiveness', or 'adaptability') in phase 2 (i.e. after minimum requirements for all basic orientors have been satisfied in phase 1) (Krebs and Bossel 1997).

Cultural theory (Thompson et al. 1990) identifies five types of individuals in the social world, each having characteristic and distinct value orientation and lifestyle: 'egalitarians', 'hierarchists', 'individualists', 'fatalists', and 'hermits'. Orientation theory explains these different lifestyles in terms of different basic orientor emphasis, and furthermore fills two obvious gaps in cultural theory: 'innovators' and 'organizers'. The egalitarian stresses partnership in *coexistence* with others, the hierarchist tries to gain *security* by regulation and institutionalized authority, the individualist tries to keep his *freedom* by staying free from control by others and the 'system', and the fatalist just tries to secure his *existence* in whatever circumstances he finds himself in. The autonomous hermit is of no practical relevance to the social system. The innovator stresses the basic orientor *adaptability*, while the organizer concentrates on *effectiveness*.

One can find solid evidence of the basic orientors even in computer experiments with 'animats' simulating the evolution of intelligence in artificial life (Krebs and Bossel 1997). This artificial intelligence evolves differently in different animats, resulting in different life-styles. The differences can be traced back to different emphases on the basic orientors. These value dimensions emerge in the animat's cognitive system as it gradually learns to cope with its environment. These experiments in artificial life show that values are not subjective inventions of the human mind, but are basic system requirements emerging from a system's interaction with its environment. Results of these experiments are reported in the following.

2.2.2 Simulation of Cognitive Self-Organization

The genetic algorithms of Holland and coworkers (Holland 1975, 1992) are models of biological adaptive processes that have now been widely and successfully applied to a wide spectrum of adaptation and optimization problems (Grefenstette 1985, 1987; Schaffer 1989; Belew and Booker 1991). In particular, these algorithms have been used to simulate learning and adaptation of artificial animals ('animats') in a simulated environment containing 'food' and obstacles (Wilson 1985).

In Krebs and Bossel (1997), Wilson's (modified) animat was used to demonstrate the emergence of basic orientors in self-organizing systems having to cope with complex environments. The animat model incorporates essential features of a simple animal in a diverse environment. Being an open system, an animal depends on a flow of energy from the environment. In the course of its (species) evolution, it has to learn to associate certain signals from the environment with 'reward' or 'pain' and to either seek or avoid their respective sources (energy gain or energy loss). This learning phase (of populations) will eventually lead to the establishment of cognitive structure and behavioral rules which are approximately optimal in the particular environment (with respect to maximization of reward, minimization of pain, and securing survival). These behavioral rules incorporate knowledge which enable intelligent behavior.

The animat is designed to simulate this process. It can pick up sensory signals from its environment (containing 'food' and 'obstacles'), and classify them with available rules to determine an appropriate action (direction of movement). After a successful move, the strength of rules leading up to it is increased by sharing in the reward (i.e. energy gain). New rules are occasionally generated by either random creation, or by genetic operations (crossing-over and recombination). They are added to the existing rule set, and compete with the other rules for reward. Unsuccessful rules are not reinforced and lose strength and influence in the rule set.

The training process consists of placing the animat at a random 'empty' location in an environment with specific environmental properties, and allowing it to move around searching for food. A collision with an obstacle causes a loss of energy and throws the animat back to its previous position. Rules leading to success are rewarded. A genetic event of rule generation may occur with a prescribed probability. Random rules are created in unknown situations. The process is repeated for a large number of steps (up to $85*10^5$ in our experiments). Eventually, a set of behavioral rules develops which allows 'optimal' behavior under the given set of conditions.

Note that this 'optimal' behavior has not been defined in terms of an objective function guiding the evolution of the set of behavioral rules. The rule set develops solely from the reinforcement of rules which lead to food or avoid collisions. An explicit energy balance accounts for all energy losses associated with movement, collisions with obstacles, and rule generation, and energy gains due to uptake of food. The development of the rule set is then driven by the requirement to optimize energy pickup in the given environment (with specific resource availability), while allowing for environmental variety, variability, and change specific for that environment. Neglect of these properties is penalized by lack of fitness, and threat to survival, and causes disappearance of deficient rules. Other criteria besides efficiency will therefore be reflected in the set of behavioral rules. Since these were not expressly introduced, we must recognize them as emergent value orientations.

The animat can perceive objects in its immediate vicinity (i.e. in the eight surrounding cells). This information is translated into a 'sense vector', and converted to its internal representation. This message is compared to situation classifiers already available in memory, where each stored classifier specifies a corresponding action rule, and carries a 'strength' related to its previous success. Matching classifiers are collected in a match list. If the match list is empty, a new classifier is created. All classifiers on the match list bid in proportion to their strength (and may also pay a corresponding 'tax'). The highest bidding classifier pays its bid and determines the next action. Bid (and taxes) are redistributed to the previous winner. This ensures that chains of classifiers leading up to successful action are rewarded ('bucket brigade'). The specified action is executed, changing the animat's position in its environment. The winning classifier receives reward or punishment (if any) corresponding to the action. After the payoff event, the genetic rule discovery process may be triggered. The process employs the genetic operations of crossover and mutation. New rules are added to the classifier population and assigned the average strength of their parents. The cycle starts again by detecting the new external situation.

2.2.3 Environmental Properties and Animat Orientors

In a stable environment where food is distributed in a completely regular pattern, the genetic algorithm would eventually lead to optimization of the animat's movements in a regular grazing pattern, with a single objective: maximum food uptake. The regular grazing pattern reflects the complete certainty of the next step, which the animat learns through the bucket brigade despite its one-step vision.

In more complex and diverse environments the animat, because of its limited 'vision', may not know for several steps which situation it will encounter next. It will therefore have to develop decision rules which have greater generality and are applicable to (and will be reinforced by) different motion sequences with different outcomes. Another objective is now implicitly added: to secure food under the constraint of incomplete information, i.e. a security objective. Note that this is an emergent property which is not explicit in the reward system (which still rewards only food uptake). It becomes 'obvious' to the system only through the 'pain' (energy loss) associated with a neglect of the security dimension. Failure to heed this implicit security objective will reduce food uptake and may endanger survival. On the other hand, the pressure to 'play it safe' will occasionally mean giving up relatively certain reward. With other words, efficiency is traded for more security.

Orientation theory (Bossel 1977, 1994) deals in a more general way with the emergence of behavioral objectives (orientors) in self-organizing systems in general environments. The proposition is that if a system is to survive in a 'normal' environment characterized by the six fundamental properties above, it must be able to (1) physically exist in this environment (existence), (2) effectively harvest necessary resources (effectiveness), (3) freely respond to environmental variety (freedom of action), (4) protect itself from unpredictable threats (security), (5) adapt to changes in the environment (adaptability), and (6) interact productively with other systems (coexistence). In other words, it must learn to pay attention to each of the six basic orientors.

These concepts were applied to analyze behavior and performance of the animat in different environments. The animat experiment contains all components necessary for a study in the basic orientor framework. It is assumed that the system is physically able to exist in this environment; the orientor 'existence' is therefore satisfied and will not be considered further. Also, we deal with a single animat only and do not consider interactions with other individuals, i.e. the 'coexistence' orientor plays no role in this study. This means that we only have to deal with the four orientors 'effectiveness', 'freedom of action', 'security', and 'adaptability', corresponding to the four environmental properties 'resource availability', 'variety', 'variability', and 'change'. For these properties as well as for the orientors, quantitative indicators have to be defined which reflect the particular problem setting.

Animat fitness depends on its ability to maintain a positive energy balance in the long term. This energy balance is therefore at the core of the orientor satisfaction assessment. At each step, energy uptake (by food consumption) and energy losses (by collisions with obstacles, motion, and learning of rules) are recorded and used to compute the momentary energy balance. Attention to all orientors is mandatory to ensure a positive energy balance even under adverse environmental conditions.

Quantitative measures must be defined for characterizing the different properties of the environments used in the animat experiments (Krebs and Bossel 1997). *Resource availability* is defined by relating the amount of resources to the (minimum) travel distance required for harvesting all resources. *Environmental variety* is defined by relating the number of diverse sense vectors distinguishable in the environment to the number of spatial positions occupied by 'food' or 'obstacles'. *Environmental variability* is defined by the probability that a 'food' object is actually a collision object. *Environmental change* is expressed by relating the time constant of environmental change to the time constant of system change.

Animat performance in different environments is compared by using measures of orientor satisfaction. These have to be defined using relevant parameters of animat performance. The orientor measures are defined such that values < 1 indicate a threat to survival, while values ≥ 1 show adequate orientor satisfaction.

Effectiveness is expressed by subtracting the energy consumption rate (for motion, collision, and learning) from the energy uptake rate (food harvesting) and relating this net gain to the rate at which energy is (theoretically) available (= resource availability × speed of motion). If the measure of *effectiveness* drops below 1, the net energy balance of the organism is negative, and survival is at stake. The better the ratio of net energy gain vs. energy availability, the better is the satisfaction of the effectiveness orientor. Effectiveness is therefore crucial for the survival of the organism, and the remaining orientors are defined in terms of it.

Freedom of action reflects the system's ability to cope with environmental variety by employing the behavioral variety embodied in its cognitive structure. The experimental procedure for determining *freedom of action* consists of increasing the environmental variety by adding randomly generated new patterns and observing the animat's resulting performance. Eventually, the *effectiveness* measure drops below 1 when environmental variety reaches a critical level. The animat can then no longer cope with environmental variety, cannot sustain a positive energy balance, and hence cannot survive. The *freedom of action* measure in another environment is defined by relating its variety to that of the critical environment.

Security reflects the system's ability to cope with environmental uncertainty. In this set of experiments, environmental variability is gradually increased by introducing random uncertainy into the recognition of (food) objects. When environmental variability is increased to a critical value, the animat can no longer cope with this environment, its *effectiveness* drops below 1, and it cannot survive. The *security* measure of a given system in a given environment is then found by relating its variability to that of the critical environment (and taking the inverse).

Adaptability reflects the ability to adapt to a changing environment. By definition, this must capture the performance of the learning process. This means that – in contrast to the measurement of the freedom of action and security orientors – the learning process must be part of the experiment. We can define a nondimensional measure of *adaptability* by relating the time required for adaptation to the time constant of environmental change (and taking the inverse). If either the environment changes too quickly, or the time required for adaptation (to this particular change) is too long (or both), the system can no longer adapt completely, and survival is threatened.

2.2.4 Summary of Experimental Results and Conclusions

Since the animat's training depends on a number of random factors, each animat develops a different cognitive system (classifier set), even though final performance may be similar. In order to show general tendencies despite these individual differences, mean values over large populations (50 individuals) were obtained. These dealt with (1) results of the training process in two (otherwise identical) environments having different variety and variability, and with (2) performance of animats after transfer from their training environment to environments challenging them with more variety, or variability, or change. In the final part of the study, (3) results of individual simulations were compared, interpreting individual differences in terms of different emphases of basic orientors. Full details are given in Krebs and Bossel (1997).

2.2.4.1 Emergence of Basic Value Orientations

The animat experiments confirm the basic proposition of orientation theory: In a self-organizing adaptive system evolving in a complex environment, multi-dimensional value orientation emerges as a result of fitness selection. I.e., the system learns to secure balanced orientor satisfaction. The time records of the learning process show very clearly how the measures for each of the basic orientors gradually rise to values above the survival value (=1), where they saturate as the learning process stagnates. The margin above this critical value is a measure of fitness of the system.

In the animat experiments (and similarly, in 'real life'), balanced multi-dimensional attention to the basic orientors emerges from the simple one-dimensional mechanism of rewarding success in the given environment. Thus, in the course of its evolutionary development in interaction with its environment, the system evolves a complex multi-dimensional behavioral objective function from the very unspecific requirement of 'fitness'. Conversely, this also means that balanced attention to these emergent basic orientors is necessary for system viability and survival – they would not have emerged unless important for the system.

2.2.4.2 Emergence of Individual Differences in Value Orientation

'Balanced attention' still leaves room for individual differences in the relative emphasis given to the different orientors. Individuals belonging to the populations used in the animat experiments evolve significant differences in value emphasis. Three different 'lifestyles' in particular became obvious: 'specialists', 'generalists', and 'cautious types' (Fig. 2.2.1). 'Specialists' stress the *effectiveness* orientor. They perform best in the environment in which they were trained, but lose their fitness quickly as system variety or variability is increased. 'Generalists' stress the *freedom of action* orientors. They have adequate but mediocre performance in their training environment, cannot deal well with increasing variability, but can maintain adequate performance even for very significant increases of variety. 'Cautious types' stress the *security* orientor. Their performance even in their training environment is low, and they cannot tolerate much environmental variety. But they have learned to tolerate even very significant increases in variability.

These individual variations, while not significantly reducing performance in the standard training environment, provide comparative advantage and enhanced fitness when resource availability, variety, or reliability of the environment change. They also result in distinctly different behavioral styles. However, pathological behavior will follow if orientor attention becomes unbalanced (e.g. dominant emphasis on effectiveness, or freedom of action, or security, or adaptability), even if the system should still manage to secure the resources for survival.

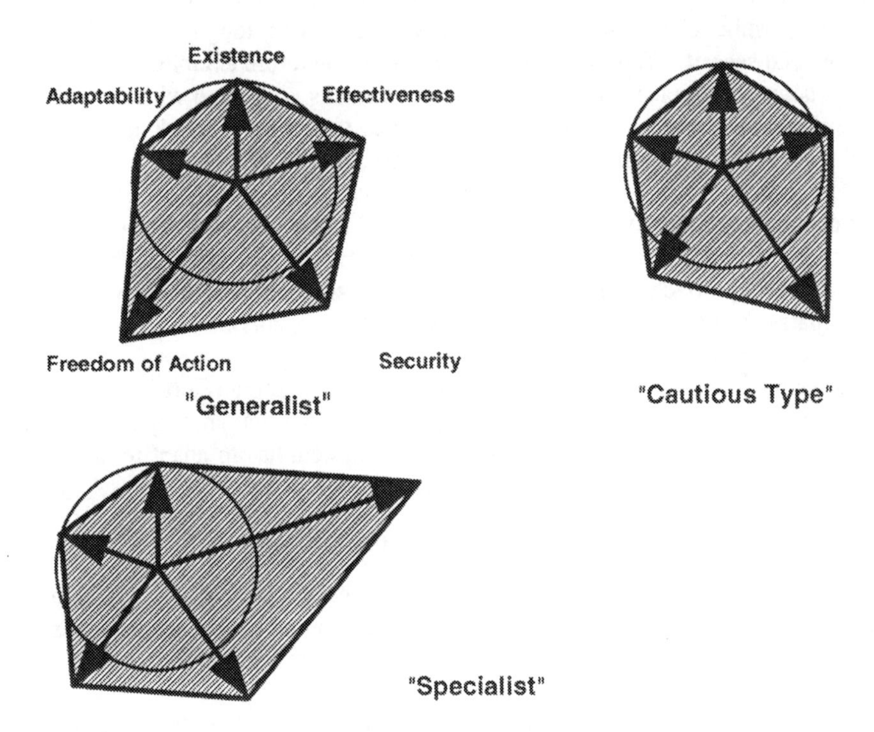

Fig. 2.2.1. Orientor stars showing different basic orientor emphasis of individual animats

2.2.4.3 General Conclusions Concerning Systems Science

In terms of van Heerden's (1968) definition of intelligence („Intelligent behavior is: to be repeatedly successful in satisfying one's (psychological) needs in diverse, observably different, situations on the basis of past experience"), animat individuals develop intelligent behavior in an even wider sense than suspected by Wilson (1985): They also develop a complex objective function (balanced attention to basic orientors), or value orientation. Serious attention to basic values (existence, effectiveness, freedom, security, adaptability, coexistence) is therefore an objective requirement emerging in, and characterizing self-organizing adaptive systems in complex environments.

Self-organizing systems are dissipative systems by their very nature: The processes of maintenance and organization of structure require that the system must be able to maintain a resource throughput (energy, materials, information). The system's energy balance therefore plays a central role, which is also reflected in the orientors: Given satisfaction of the existence orientor, satisfaction of the effec-tiveness orientor is essential for survival. Satisfaction of the freedom of action and security orientors are preconditions for longer-term effectiveness (and survival), while adaptability in turn ensures effectiveness, long-term freedom of action, and security. The system of basic orientors therefore exhibits a hierarchical structure of urgency, with different time constants associated with the different orientors (existence: immediate, effectiveness: short term, freedom of action and security: longer term, adaptability: long term).

The animat's rule reinforcement method ('bucket brigade') very effectively builds up a cognitive model which enables anticipatory behavior: Since rewards flow back to earlier rules leading to later pay-off, the activation of the initial rules in a pay-off chain means that the system suspects possible pay-off and anticipates the near future, i.e. it has a model of the results of its actions under the given circumstances.

Since the cognitive system with its inherent value structure is a function of the environment to which the system has adapted, it follows that an adaptive system will also undergo value change if the cognitive system has to adapt to an environment with different, or changing environmental properties.

2.2.4.4 General Conclusions Concerning System Behavior

Training of animats in different environments, and the performance of animat individuals in environments that differ from their training environments, leads to some general conclusions which are in full agreement with everyday observations and general systems knowledge:

'Generalists' have a better survival chance than 'specialists' if moved to an environment of greater variety.

'Cautious types' have a better survival chance than others if moved to a less reliable environment.

Training in more unreliable and/or more diverse environments increases performance with respect to the security and/or freedom of action orientors at the cost of the effectiveness orientor.

Training in an uncertain environment teaches caution and improves fitness in a different environment.

Learning caution (better satisfaction of the security orientor) takes time and decreases effectiveness.

Investment in learning (i.e., energy cost of learning in the animat) pays off in better fitness: the learning investment is (usually) much smaller than the pay-off gain.

2.2.5 Implications and Applications

Orientation theory in conjunction with a simulated organism (animat) has been found to be useful in analyzing and understanding processes of self-organization and system evolution, and in particular the evolution of cognitive structure and resultant behavior. The concepts are also applicable and useful to the design of (complex) systems which have to cope with a given environment. Such systems must be designed with a cognitive unit which allows balanced attention to the full (multidimensional) set of basic orientors. One-dimensional criteria (e.g. cost-effectiveness, maximization of gross national product, etc.) will normally be insufficient and lead to pathological behavior. The approach therefore has particular significance for path studies of potential future development of human systems. In such studies, the time development of relevant system indicators (state variables) must be mapped on the basic orientor dimensions in order to determine the potential contribution to system development, and to its potential for survival and development.

Quantification of a system's orientor satisfactions and comparison of the orientor stars (Bossel 1987, 1997) for competing systems provides a more comprehensive picture and fuller understanding of system performance than an aggregate measure of fitness (such as net energy accumulation over a given time period). In particular, the classifier sets developing during the learning phase in adaptive artificial intelligence systems (such as Wilson's animat) cannot be evaluated and compared directly with respect to their performance. However, the orientor mappings offer a means of objective performance evaluation and comparison.

Numerical assessment and comparison of system performance requires quantification of (1) measures of environmental properties, and (2) measures of the orientor dimensions. Both measures are system-dependent. The same environment means different things to different organisms, depending on their sensors and their system structure: A meadow has a different composition of environmental properties for a cow than for a honey-bee. Similar distinctions are required with respect to orientors: The security of a nation requires a different objective function than that of an individual. However, definition and quantification of suitable criteria and measures is usually possible, even in complex societal contexts (e.g. Bossel and Strobel 1978; Hornung 1988; Bossel 1994).

Ecological goal functions and basic orientors are intimately related: Ecological goal functions are translations of the (completely general) basic orientor needs into functional and structural requirements of specific systems in earthly environments. Basic orientors and goal functions are both part of the orientor hierarchy defining a system's normative structure at different levels of detail. Putting 'viability' (fitness) as the 'ultimate goal' of a system at the 'top' of the orientor hierarchy, the basic orientors are at the next level. Attaining the ultimate goal requires adequate contributions in terms of all basic orientors. Adequate fulfillment of the different basic orientor needs requires – on the next level of the orientor hierarchy – pursuit of concrete goal functions such as efficient use of resources, cycling, storage, diversity, hierarchy etc. The basic orientor concept acknowledges a common set of fundamental traits in all complex adaptive systems and recognizes the various goal functions and their diverse implementations in different systems as different expressions of identical essential system requirements.

References

Belew RK, Booker LB (eds) (1991) Proceedings of the Fourth International Conference on Genetic Algorithms. Morgan Kaufmann, San Mateo

Bossel H (1977) Orientors of nonroutine behavior. In: Bossel H (ed) Concepts and Tools of Computer-assisted Policy Analysis. Birkhäuser, Basel, pp 227-265

Bossel H (1978) Bürgerinitiativen entwerfen die Zukunft – Neue Leitbilder, neue Werte, 30 Szenarien. Fischer, Frankfurt/M

Bossel H (1987) Viability and sustainability: Matching development goals to resource constraints. Futures 19-2:114-128

Bossel H (1992) Real-structure process description as the basis of understanding ecosystems and their development. Ecological Modelling 63:261-276

Bossel H (1994) Modeling and Simulation. A K Peters, Wellesley MA and Vieweg, Wiesbaden

Bossel H (1996) Ecosystems and society: Implications for sustainable development. World Futures 47:143-213

Bossel H (1997) Deriving Indicators of Sustainable Development. Environmental Modeling and Assessment 1:193-218

Bossel H (1998) Paths to a Sustainable Future. Cambridge University Press, Cambridge, UK

Bossel H, Strobel M (1978) Experiments with an 'intelligent' world model. Futures 10:191-212

Grefenstette JJ (ed) (1985) Proceedings of the First International Conference on Genetic Algorithms and Their Applications. Lawrence Erlbaum, Pittsburgh PA

Grefenstette JJ (ed) (1987) Proceedings of the Second International Conference on Genetic Algorithms and Their Applications. Morgan Kaufmann, San Mateo CA

Holland JH (1975) Adaptation in Natural and Artificial Systems. University of Michigan Press, Ann Arbor

Holland JH (1992) Adaptation in Natural and Artificial Systems. MIT Press, Cambridge MA (revised edn of 1st edn, University of Michigan 1975)

Hornung BR (1988) Grundlagen einer problemfunktionalistischen Systemtheorie gesellschaftlicher Entwicklung – Sozialwissenschaftliche Theoriekonstruktion mit qualitativen, computergestützten Verfahren. Europäische Hochschulschriften, Reihe XXII, Soziologie, Bd. 157, Peter Lang, Frankfurt/M Bern New York

Krebs F, Bossel H (1997) Emergent value orientation in self-organization of an animat. Ecological Modelling 96:143-164

Max-Neef MA (1991) Human Scale Development. Apex Press, New York London

Müller F (1996) Ableitung von integrativen Indikatoren zur Bewertung von Ökosystem-Zuständen für die Umweltökonomische Gesamtrechnung. Ökologiezentrum, University Kiel

Schaffer JD (ed) (1989) Proceedings of the Third International Conference on Genetic Algorithms. Morgan Kaufmann, San Mateo CA

Thompson M, Ellis R, Wildavsky A (1990) Cultural Theory. Westview Press, Boulder San Francisco Oxford

van Heerden PJ (1968) The Fountain of Empirical Knowledge. Wistik, Wassenaar, The Netherlands

Wilson SW (1985) Knowledge growth in an artificial animal. In: Grefenstette JJ (ed) Proceedings of the First International Conference on Genetic Algorithms and Their Applications. Lawrence Earlbaum, Pittsburgh PA, pp 16-23

2.3 Ecological Orientors: Pattern and Process of Succession in Relation to Ecological Orientors

Udo Bröring and Gerhard Wiegleb

Abstract

A theoretical analysis reveals that the concept of "succession" is generally applied both to community and ecosystem phenomena, which must be kept strictly apart. "Succession" is also applied to phenomena which we call "dynamics", "change" and "development". Additionally, any satisfactory theory on the temporal behavior of ecological systems must account for both succession and its logical counterpart, constancy. Furthermore, it is shown that the community level is difficult to distinguish from both the population and the ecosystem level. Important community characteristics which enable us to delimit the community from other observation levels can be drawn from the analysis of semiautonomous behavior of hierarchically ordered levels. A list of aspects relating to semiautonomy distinguishing the community both from the ecosystem and the population is presented. The confusion relating to key concepts of spatial and temporal relations in ecological systems is outlined. Different terminologies are necessary to speak about observational and theoretical entities. "Pattern" and "process" are adequate terms of an observational terminology, while "structure" and "function" are terms of theoretical reasoning. Also, discreteness or continuity of phenomena must be described in different terms. The term "process" accounts for continuous phenomena, while the term "event" is necessary to describe discrete ones.

An analysis of special problems of succession theory shows that descriptive succession theory must take into account three classes of change: changes in species composition, changes in important functional characters, and changes in type. An exhaustive causal repertoire of successional mechanisms is listed. The same mechanisms lead to either change or constancy, or to past, present, and future states of a system. Special emphasis is put on the deduction of zero-force laws. An inquiry of the reduction of community phenomena to population phenomena shows that invasion, maintenance, and extinction require different models of causality. While a causal theory for invasion is elaborated, maintenance and extinction are not yet well understood. Predictive succession theory is also not well developed. In an overview 15 successional theories are listed which

F. Müller M. Leupelt (Eds.)
Eco Targets, Goal Functions, and Orientors

claim to explain temporal community phenomena. These approaches use a wide range of mechanisms. Two theories (resource ratio hypothesis, inhibition model) explain different kinds of temporal behavior and are therefore regarded as the most advanced ones. Two examples of succession are presented in order to test the explanatory power of successional theories. In both cases a narrative explanation of succession on the level of species composition is possible a posteriori. The two cases include a great deal of surprise, and both cases do not comprise any indication of maturation, retrogression, or orderly serial development.

An analysis of the relation of successional trends to ecologically based "goal functions" in conservation and planning shows that goals of a democratic society are diverse and pluralistic. This is also the case with respect to goals of nature conservation. By analyzing an empirical sequence of primary succession in the former coal mining area of East Germany we investigate the question whether environmental goals can directly be deduced from successional trends of ecological communities. All the distinguished and described stages have a certain value in the context of conservation. Each stage is characterized by special features of contradicting value concerning erosion potential, nutrient losses, water balance, diversity of different plant and animal groups. Thus no goal functions derived from successional trends can be adopted as goals for nature conservation.

2.3.1 Introduction

Many ecologists maintain that "succession" is the second most important ecological concept (see e.g. Cherrett 1989) after "the ecosystem". This estimation is in fact supported by two reasons. First, the concept of succession appeared at the same time as ecology as a self-confident discipline at the end of the 19th century (Burrows 1990; McIntosh 1976; 1981). Second, the development of successional theories repeatedly influenced the progress of general ecological theory (see McIntosh 1981; Wiegleb 1989). We regard these facts to be more than mere coincidence.

Founders of ecology, like E. Warming, strongly emphasized the dynamic nature of their research object. Early ecologists agreed with one important tendency of the philosophy of science during the 19th century. Nature and society were mainly seen as processes ("Prozeßcharakter der Natur" in German idealism). Exponents of this way of thinking were F. Schelling, G. Hegel, and A. Schopenhauer, later K. Marx and A. N. Whitehead. Also Ch. Darwin´s theory of evolution, of which both E. Haeckel and E. Warming were close adherents, fits into the general scheme.

However, after one century of dispute on empirical and theoretical questions of succession the subject is far from being settled. Many competing and contradictory theories have been presented in recently published textbooks (e. g. Burrows 1990; Begon et al. 1990). Terminology in succession theory is often confusing.

Not any two key note papers or textbooks use the same terminology. Furthermore, while dealing with a complex dynamic system, recent succession theory is not yet connected properly to the terminology of modern systems theory (cf. concepts like "emergence", "evolution", "hierarchy", "causality", "chaos", "fractality", "noise" and "complexity", see Ulanowicz 1979, 1990; O'Neill et al. 1986; Kolasa and Pickett 1989; Wiegleb and Bröring 1996). This can be regarded as a major shortcoming for the future development of a comprehensive succession theory.

We therefore decided to widen the scope of our study beyond a state-of-the-art review on succession theory and its applicability in the context of ecologically derived goals. We will try to clarify some basic elements of current theory with special reference to

- the observational level to which succession and related concepts refer, and the necessary requirements for recognizing something like a community level "between" the population and ecosystem level,
- the philosophical implications of descriptional terms of succession theory with relation to space and time,
- the descriptive, explanatory and predictive aspects of succession theory in general, and the explanatory value of some detailed succession theories,
- the regularities of empirical successional sequences and their implications for the definition of social goals.

2.3.2 Some General Considerations on Ecological Theory

2.3.2.1 Concepts of Change and Constancy

We started our investigation with the question: What kind of behavior of an ecological entity on which observational level should be called "succession"? By collecting different definitions of the term "succession" from textbooks, we found that it is used with quite different connotations:

- "Succession" may be defined in the context of a climax concept, or in modern words, referring to a system having an "attractor". This is a strong delimitation to a certain class of systems, to which ecological systems may not even belong.
- Any change of a community or ecosystem may be regarded as "succession". In this case succession is ubiquitous and thus meaningless as a scientific concept (similar to the use of the term "evolution", see Potthast 1996).
- The process of change "through time as represented by a series of observed states" (Allen and Starr 1982: 278) may be stressed; here the temporal but not necessarily regular "series of states" that are "observed" are constitutive for "succession".

- Only non-directional change may be classified as "succession". We had some difficulties to understand the meaning of the term "non-directional" in the English-language literature. We question whether it is possible to think of succession only in terms of past states of the system without referring to future states.

Furthermore we found that it is by no means clear to which observational level the concept of succession applies. We use the term "observational level" as observer-defined entity in the sense of hierarchy theory (Allen and Hoekstra 1992; Müller 1992; Wiegleb 1996a). The following four observational levels are dealt with, namely the population, the community, the ecosystem, and the landscape, which are usually regarded as important levels of ecological theory formation (O'Neill 1989; Begon et al. 1990; Allen and Hoekstra 1992). As the levels are related to numerous connotations, the definitions we use for the purpose of this paper are found in Table 2.3.1.

Observational levels can provide "ecological units" (Jax 1996; Jax et al. 1997) or "entities", which are either concrete manifestations of a given observational level in the "real" world, or theoretical constructs (models) derived from the observation of a unit. We deliberately distinguish between the more general concept of the "observational level" (with its hierarchical connotation) and the specific concepts of "units" and "entities", which need not necessarily be seen in a hierarchical context.

Table 2.3.1. Definitions of observational levels

Observational level	Definition
Population	Any set of at least two countable entities (individuals) of a species in a given space at a given time
Community (species assemblage, taxocoenosis)	Any set of at least two populations of different species in a given space and time regardless of further statistical, structural or functional relations
Ecosystem	The community of a given space and time and its abiotic environment
Landscape (biome)	Any large scale arrangement of communities or ecosystems in a concrete space at a given time

Since "succession" as an ecological concept is approximately 40 years older than the concept of the "ecosystem", its early development was closely connected with the development of the "community" concept. However, later successional ideas were also applied to the ecosystem. Populations and landscapes were generally not treated in a successional context, and changes of populations and landscapes were referred to as "dynamics". Nevertheless it must be decided whether the term "succession" should refer to community processes only, or to both community and ecosystem processes, or be abandoned in favor of a more general terminological framework.

As a potential solution of the problem, we advance the following general scheme (Table 2.3.2) for the distinction between the basic concepts of "dynamics", "change" and "development". We summarize the contents of the Table as follows:

- "Dynamics" is ubiquitous, and taken for granted. The stochastic "causes" of dynamic behavior of a system can be studied (see below, causal repertoire).
- "Change" is an observable net effect of dynamics. As the observable net effect of dynamics can also be "non-change", both change (succession) and non-change (constancy) have to be explained with the same set of stochastic causes. The change itself is an event-classification (Johnstone 1986).
- "Development" can only be assumed in the case of an orderly sequence of change. "Orderly" may refer to serial, cyclical or u-shaped sequences. "Non-orderly" may refer to any other sequence that is more than a fluctuation around a stable point. We suggest that it might be more fruitful to distinguish between "predictable" and "unpredictable" change (surprise, emergence) rather than between "directional" and "non-directional" change in the context of a model. Models may refer to climax states, mosaic cycles, transition probabilities between states, etc. (the conventional wisdom of succession theory).

Table 2.3.2. Distinction between dynamics, change and development

Term	Definition	Remark
Community dynamics	A set of (stochastic) processes performed by *populations and individuals* belonging to the *community*	Dynamics is ubiquitous
Ecosystem dynamics	A set of (stochastic) processes performed by *biotic and abiotic agents* belonging to the *ecosystem*	
Community change	The net effect of these processes *on the community level*	As a net effect also "no change" = "constancy" may occur
Ecosystem change	The net effect of these processes on *the ecosystem level*	
Community development	Any orderly sequence of change	Also "non orderly change" = no development may occur
Ecosystem development		

"Succession theory" has to deal with all the phenomena listed. Succession can be studied with emphasis on both community and ecosystem phenomena. The present paper focusses on the problems resulting from the study of dynamic community phenomena. It can be concluded from Table 2.3.2 that both change and constancy can occur as net effects of dynamics. Thus any general scientific ecological theory has to account for both. This dichotomy can be observed throughout history of ecology. A large body of theories has been developed for constancy, "stability" and related concepts.

Table 2.3.3. Attitudes towards change and constancy based on various ecological authors

Author	Constancy	Change
Linné	Balance-of-Nature and Constancy of Species as general laws	Struggle and catastrophies serve for balance, anecdotal observations on succession
Darwin	Selection as an agent of change (species formation), but also constancy (species maintenance)	Population growth and intraspecific variability as prerequisites for evolution
Warming	The community as a typological constant (on a small scale)	Succession based on local competition for resources, selective forces of the habitat
Clements	Climax state of the local community (small scale) and biome (large scale) as functional constants	Successional series based on disturbance, invasion, facilitation and various dynamic processes, aiming for constancy and equilibrium
MacArthur	Community stability (constancy) via species diversity (species interaction)	Invasion and extinction on islands
Odum	Ecosystem constancy via self-regulation, trophic-dynamic complexity, etc.	Matter and energy flow, successional trends, maturation
Neo-Gleasonian	Potential of community constancy via population maintenance, species interaction and habitat selection	Potential of community change via disturbance, safe sites and various other mechanisms
Enlightened Environmentalist	Long-term sustainability of ecosystems and landscapes, conservation of biodiversity	Complexity, dynamics and hierarchy as dominant scientific paradigms, short-term goals derived from potential ecosystem development

A general overview of attitudes to constancy and change during the last two centuries is given in Table 2.3.3. In the writings of Linné, "constancy" as a scientific concept was a dominant idea, even though it was associated mainly with species. However, since the times of Darwin, Warming and the development of self-conscious ecology, both constancy and change have been dealt with equal intensity. Most of the prominent authors of ecological textbooks or key note papers have developed important theories for both kinds of phenomena. It may be that at the peak of the diversity-stability dispute more ecological thought was devoted to constancy and stability than to succession and change. In the list of

Cherrett (1989) ecological concepts related to constancy and stability were still ranked highly. The dichotomy of constancy and change has also found its way into the writings of authors dealing with conservation (conservation ecologists, environmentalists) whose goals are mainly related to aspects of constancy (sustainablity, biodiversity) in a changing world.

2.3.2.2 Conclusions on the Nature of the Community

Despite the empirical studies of Cherrett (1989) and Shrader-Frechette and McCoy (1993) we maintain that the "community" is the central level of obser-vation and research in ecology (see Trepl 1988; Egerton 1968, 1979; McIntosh 1985). E. Warming, Ch. Elton and R. MacArthur, who advanced successful research programs in ecology (see Warming 1895; Elton 1927; MacArthur 1972), regarded themselves as community ecologists. Important pioneers of succession research (Warming, Clements 1916 and Gleason 1926) were likewise community ecologists.

The community is obviously a level which is difficult to define (see Wilson 1991; Shrader-Frechette and McCoy 1993). Ultimately, community and population, as introduced above, are only distinguishable by using a non-ecological taxonomic criterion (Allen and Hoekstra 1990). This cannot be avoided despite Harper's criticism (1982). The introduction of observational levels like the "food chain", "trophic level", "compartment", or "guild" does not solve this controversy. These levels can be regarded as special cases of the community based on another assembly criterion. Therefore, they reflect another degree of resolution. By neglecting the taxonomic component, these concepts are unable to reflect major elements of succession, like invasion and extinction.

"Community" and "ecosystem" can only be distinguished on a formal basis for the purpose of feasibility of practical research. Living organisms and their assemblies without physico-chemical environments are simply unthinkable (Begon et al. 1990). Allen and Hoekstra (1990) distinguished both "community" and "ecosystem" as different kinds of interference patterns of differently scaled processes (see also Wiens 1989; Allen and Hoekstra 1991; Wiegleb 1991). Despite formal similarities, the ecosystem concept cannot be replaced by concepts like "habitat", "niche" and "environ". All of these are purely relational concepts (interfaces, see below) without any spatial correlate. Even though we agree with Allen and Hoekstra (1990) that populations, communities, and ecosystems cannot be mapped exactly onto a landscape, all of these concepts have the connotation of a spatio-temporal component.

Even though the local population can not only be assigned to a "higher" observational level (the community) but also to a higher scale defined level (the metapopulation), population and community "consist" of individual organisms (Wiegleb 1996a). Even though the scaling of members of the same species is more similar than the scaling of members of different species (making it easier to recognize the population, Allen and Hoekstra 1990), populations and communities can be studied with similar biological methods (Urban et al. 1987; Wiegleb 1991). Communities, like populations, can exhaustively be described on the basis of growth forms, strategy types, individual organisms etc. The relation between the ecosystem and the community is more delicate. Logically, the ecosystem "contains" the community. However, besides the community itself, no further basic biological entity common to both the community and the ecosystem can be found. The only parts, common to both, biotic and abiotic components of the ecosystem are only chemical elements, atoms or some energy or "information" units.

This necessarily leads to a large methodological and epistemological gap between community and ecosystem ecology. Recently, a complete book has been devoted to the question of how to link biological and non-biological aspects of ecology (Brown 1995; Jones and Lawton 1995). As already pointed out, the search for interfaces between biotic and abiotic components of the ecosystem is older than the ecosystem concept itself. Three kinds of interfaces can be distinguished:

1. "Formal" interfaces, described in a theoretical or model terminology. Many key concepts of community ecology, like the habitat, the niche, the ecotope, and also the environ belong to this group (see Leser 1984; Patten 1982, 1990).
2. "Epistemic" interfaces, described in an observational language (Allen et al. 1984): scaling theory and scanning of grain and extent in relation to organismic activities.
3. "Realistic" interfaces, described without reference to formal or epistemic interfaces: Organism-centered approaches to heterogeneity in soils (Cousins 1997) are classified. Even the landscape can be regarded as an interface rather than an observational level (Forman and Godron 1986).

The above mentioned difficulties with a central concept of ecology have not only lead to the well-known divergence within community ecology (cf. organismic, individualistic and classificatory research program, Reise 1980). We think that the division of ecology has gone further. We postulate that other ecological research programs have tried to reduce community ecology to either physics or biology (Wiegleb 1996a). Strictly speaking these programs have tried to destroy community ecology or make it superfluous as a scientific discipline in its own right. In this sense the "two ecologies" may be more than just research programs in the sense of Lakatos (1978), even though the development of a paradigm in the sense of Kuhn (1976) did not occur.

The "physicist research program" was introduced by R. Lindeman (1942), and successfully implemented by E. P. Odum (1959, 1971) and his brother H. T. Odum (1983; see Golley 1993; Hagen 1992). It developed into "systems ecology" of the modern type (including thermodynamical, network analytical, and control theoretical approaches, Jörgensen 1992). We are not sure whether all of the adherents of the physicist program deliberately wanted to destroy any specific theory of ecology. However, a certain disregard of living beings and their special emergent properties is abundant in the writings of this school. Organisms are seen as "processors", "engineers" etc. in an otherwise physico-chemical world.

A side effect of the physicist research program is the confusion between the concepts of the "community" and the "ecosystem". It probably began with Odum (1959, 2nd ed.), who equaled "community system" and "ecosystem". A typical example of this confusion will be described. Ecosystem theories are often based on food webs as an integrative, organizational force (Cousins 1990, 1997). However, we maintain that it is almost impossible to base an ecosystem theory on food web theory, because ecosystems cannot be described exhaustively by food webs (for the concept of "exhaustive description", see Beckner 1974; Wiegleb and Bröring 1996). Only if the exclusive fate of primary production or biomass in an ecosystem was to be consumed in a living state, a description in terms of food webs would be adequate. A wider concept may also include the "detritic food chains". However, the processes of abiotic mineralization, immobilization, and export of organic matter may be important in various types of ecosystems to which a strict food web approach does not apply.

The "biologist research program" (Wiegleb 1996a) has several independent roots in population ecology (Andrewartha and Birch 1954), "individual ecology" (Harper 1967, 1975, 1982) and evolutionary ecology (stemming from various sources, Pianka 1992). Most adherents of this research program believe that life history traits or vital attributes of species are responsible for their behavior within communities. These traits or attributes of interest are believed to be evolutionary acquired. As natural selection (a population mechanism) and competition for resources (a community property) were seen in close connection (Trepl 1994), a reconciliation between community ecologists and population biologists was possible in various ways.

Physicists and biologists had very different attitudes to ecological systems. Ecosystem ecologists often believed that ecosystems are equilibrated, homeostatic, functioning, self-regulated, homogenous, stationary and ahistoric. Biologists, in particular British and American community and population ecologists, regarded communities as having a non-equilibrium status (Gleason 1926), being dependent on initial conditions (Egler 1954), and being spatially aggregated (Greig-Smith 1979).

2.3.2.3 On Ecological Hierarchies

If we try to conceptualize communities and ecosystems as "evolving hierarchical systems" (Allen and Starr 1982; Salthe 1985; Pickett et al. 1987; Wiegleb 1996a), we should be aware of the epistemological and practical consequences. The term may suggest that succession can be paralleled to biological evolution, cultural evolution and other dynamic processes in which living beings are involved.

However, expressions like "ecosystem evolution", or "community evolution" can be misleading to a certain extent. Only in a very general sense it is possible to speak of "evolving hierarchical systems" as a metaphor. At present only loose analogies can be discovered among various system types. There is no "general systems theory" (see Wiegleb and Bröring 1996). All "dynamic" systems must have a potential to change and a potential to remain stable (an old idea referring to Schelling, who additionally required the ability of permanent self-reproduction of natural kinds; for discussion see e.g. Schmied-Kowarzik 1989). They differ considerably in the initialization of change (mutation), the means of selection, and the storage and processing of information (Wiegleb and Bröring 1996). In evolution semi-autonomous entities (e. g. species, see below) have developed mechanisms to keep themselves independent, and therefore, show a long persistence. In succession only few of such mechanisms exist (see below) and stages of a series usually do not persist considerably long. The choice of different spatial and temporal scales alone inevitably leads to discrepancies among theories of different domains.

The recognition of a hierarchy of observational levels in ecology is dependent on either tangibility (Allen et al. 1984), scaling (Allen and Hoekstra 1990, 1992; Jax and Zauke 1991) or semi-autonomy. None of the observational units of ecology, except for some the individual organism, have a tangible surface. This finding led Allen and Hoekstra (1990) to introduce scaling operations for the purpose of an extensional definition of ecological units. According to this view, populations can be distinguished from communities because the population processes are differently and more uniformly scaled than community processes. Ecosystems can be distinguished from communities because of the different scaling of abiotic processes in contrast to biotic ones. This may be true, but we think that further means of distinction are necessary. We therefore introduce the concept of semi-autonomy in the context of level distinction.

Semi-autonomy means that mechanisms which lead to dependence, and mechanisms which lead to independence can be recognized. In the case of complete dependence of a part upon its whole, the description of the part is tautological. In the case of complete independence (autonomy), there are simply two different objects. The interesting relations lie in between. An example is shown in Table 2.3.4.

On the one hand organisms cannot be completely independent of their physical environment, on the other hand they are not completely determined by their environment. To a certain extent, they are dependent on the input of matter and energy, or, of the abiotic habitat conditions like moisture, light, temperature, and nutrients. Organisms have developed a wide variety of strategies which allow them to escape temporarily from their dependence upon the habitat. Even though the degree of physical dependence is strong, the recognition of mechanisms of independence (= semi-autonomy) makes biology and ecology a worth while object of study.

Table 2.3.4. Aspects of dependence and autonomy of single organisms in relation to their abiotic environment

Ecosystem dependence of organisms	Type of organism	Autonomy (by tolerance or avoidance)
disturbance	all organisms	behavior, r-strategy
energy input	animals	migration, homothermy
nutrient uptake	plant	modular growth, recycling
photosynthesis	green plants	storage of carbohydrates
water uptake and release	terrestrial organisms	water storage, efficient use
wind pollination	flowering plants	selfing

We will now use the same method to distinguish between community and population, two biologically defined levels (Table 2.3.5). There are synergistic and antagonistic mechanisms of dependence. As in the preceding case, both tolerance and avoidance mechanisms guarantee a certain degree of semi-autonomy. Both tolerance and avoidance can be related to synergistic and antagonistic mechanisms. It follows that if autonomy were perfect, the concept of community would no longer hold.

Table 2.3.5. Aspects of dependence and autonomy of populations in relation to their biotic environment

Community dependence of populations	*Autonomy (by avoidance or tolerance)*
herbivory, carnivory (eat)	omnivory
monophagy	polyphagy
predation (being eaten)	predation avoidance (mimicry, chemical warfare, behavior)
parasitism and disease	resistance, immune system
mutualism, facilitation	competitive dominance
symbiosis, coevolution	autonomous evolution
insect pollination	selfing, clonal growth
competition	competition avoidance (allelopathy, stress tolerance)

2.3.2.4 Terminology of Spatial and Temporal Concepts

So far, we have used terms like "pattern" and "process" as primitive terms to elucidate our concepts of dynamics, change and development. Do the terms "pattern" and "process" refer to observable entities or to theoretical constructs (models)? Scientific reasoning is divided into rationality of research, requiring an empirical terminology describing observable entities, and rationality of representation, requiring a terminology of theoretical entities, or models (Feyerabend 1981; Toulmin 1983; Mittelstraß 1992). Pattern can be regarded as a three-dimensional spatial observational entity. What is the temporal equivalent to pattern? Often "process" is regarded as the temporal observational equivalent to pattern, while terms like "structure and function" are obviously theoretical and based on inference (Table 2.3.6, see Wiegleb 1989).

Table 2.3.6. The concepts of "pattern" and "process" in theoretical and empirical language and in relation to continuity and discreteness

	theoretical	*empirical*
spatial	structure	**pattern**
temporal	function	**process**
	continuous	*discrete*
spatial	gradient	**pattern**
temporal	**process**	event

However, the assumption of a process is also based on causal inference. We propose to use the term "*event*" as a unidimensional temporal equivalent to pattern with respect to the distinction of continuity and discreteness.

In modern philosophy (neopositivism) "event" is used as an equivalent to "atomic fact" (Russel 1918; Wittgenstein 1922, but for the problem of individuation of events see Unwin 1996). We only perceive "events", not "substances". By preferring the term *"event"* the discreteness of our observation and modelling efforts is stressed. The relation between *"event"* and *"emergence"* needs to be studied.

We are not sure whether we can escape the paradox relation between "pattern" and its temporal equivalent. Whether "pattern determines process" or "process is a requisite for pattern" is often discussed (e. g. Wiegleb 1989; Leps 1990; Wiens 1995). Even pattern formation without reference to "any specific process" is advanced (With and King 1997). We assume that an event can be recognized without a change of pattern. Succession can be regarded as an ordered set of events.

There are several concepts referring to observable spatial entities, to temporal behavior of observable entities, and to inferred theoretical entities, and there are concepts relating to causal inference, or to teleological inference (see Table 2.3.7). Table 2.3.7 is empirically based on the abstracts of this symposium and on some textbooks. It is not meant to fix a certain terminology in a normative way. The terms in one row are not exactly synonymous, but form a cluster of concepts with a partial similarity of meaning. It can be shown that ecologists are intuitively well aware of the necessity of different terminologies for descriptive and inferential purposes. However, the use of various terms is not at all consistent. For example, what is called "pattern and process" by one author is called "structure and function" by the next. Particularly the terms "process", "structure" and "function" are used in different ways, and may represent quite contrasting concepts. This situation is very dissatisfying for the future development of a general ecological theory. More terminological exactness and more awareness of the philosophical implications of certain terms are required both in theoretical and practical ecology.

Table 2.3.7. Classification of theoretical and empirical concepts in ecology

Concepts referring to spatial distribution of observable entities
Architecture, parameter, pattern, spatial structure, spatial process, structure, state, state variable, texture
Concepts relating to temporal behavior of observable entities
Behavior, change, development dynamics, effect, event, flux variable, input-output, pattern formation, process, temporal pattern, time series, transition, transformation
Concepts referring to inferred theoretical entities
Agent, assemblage, compartment, component, entity, element, ganzheit, gestalt, hierarchy, model, network, node, subsystem, subunit, part, processor, system, system structure, subnetwork, supersystem, whole
Concepts relating to causal inference
Causal determinant, cause, context, function, interaction, mechanism, pathway, potential, process, propensity, proximate causation, relation, self organization, transfer, trigger, ultimate causation
Concepts relating to teleological inference
Adaptation, attractor, Balance-of-Nature, feed back, cycles, function, goal function, homeostasis, indirect effects, integration, macrodetermination, organization, regulation, self regulation, stability relation

2.3.5 Special Problems of Succession Theory

Practical succession research is usually carried out in three different fields (see also Pickett et al. 1987):

1. *Description of phenomena*: Definition of stages, constancy, resilience and duration of stages, transition probabilities between stages, floristic and faunistic characteristics of stages, biological characteristics of stages with references to growth forms, strategy types etc. by means of observation of permanent plots, transects, chronosequences, and subsequent explorative data analysis.
2. *Explanation of phenomena*: Investigation of the mechanisms of invasion, maintenance and decline of species by means of experiments, case studies, neutral models, correlation models, conceptual models etc.

3. *Prediction of phenomena*: Extrapolation of future developments from data by means of e.g. matrix models, time series, simulation techniques etc. and narrative presentation of scientific wisdom.

Each of these fields must have a satisfying body of theories. We will investigate the fields separately and then try to find some general conclusions. We will focus on descriptive and explanatory theory because these are the necessary requirements for prediction.

Table 2.3.8. Observable changes in communities

Observable change	*Event classification*
More species (invasion)	Gleasonian succession
Fewer species (extinction, emigration)	Gleasonian succession
Same number of species	No Gleasonian succession (= community constancy 1)
More biomass (more K-strategists, larger body size, longer life span, more biotic interaction, more patchiness)	Community maturation
Less biomass etc.	Community "retrogression"
Same biomass	No community maturation or retrogression (= community constancy 2)
More nutrient retention, exergy storage, indirect effects etc.	Ecosystem maturation
Less nutrient retention etc.	Ecosystem retrogression
Same nutrient retention etc.	No ecosystem maturation or retrogression (= ecosystem constancy)
Different type of community or ecosystem	Dependent on external classificatory systems
Same type of community or ecosystem	Dependent on external classificatory systems

2.3.3.1 Descriptive theory

We see that the observational concept of "change" of a "pattern", as introduced above, is still too abstract because it comprises different things in different research contexts (Table 2.3.8).

We can distinguish three classes of different successional phenomena:

1. *Change in species number*: We can observe change or constancy in the number of species of a community. The change in number of species has been termed "Gleasonian succession" by van der Valk (1981). This kind of succes-

sional phenomenon can be reduced to population phenomena like invasion, extinction, or migration. It can only be observed on the community level, not on the ecosystem level as defined above. "Species composition" may be regarded as a "structural characteristic" of a community, however, because of the necessity to introduce taxonomic information into the definition of the community, it is a privileged characteristic.

2. *Change in structural or functional characteristics*: Even in the absence of change in species composition different kinds of change over time can be observed, e. g. increase or decrease of biomass, average body size, age class distribution etc. These are genuine community phenomena. For these kinds of changes we can use the terms "maturation" and "retrogression", respectively. Ecosystem changes like nutrient retention, exergy storage etc. and community changes (see above, different scaling and degree of semi-autonomy) must not necessarily occur simultaneously and can be regarded as a separate class of observable changes. Therefore, maturation and retrogression can also be observed on the level of the ecosystem.
3. *Change of type*: A change of type can be observed both on the level of the community and the ecosystem. The recognition of this type of change includes an assessment of which characteristic of a community or ecosystem is regarded as so important that it is defined as another type of system (self-identity, Jax 1996; Jax et al. 1997).

2.3.3.2 Explanatory Theory

Causal Repertoires. Causal repertoires (Pickett and McDonnell 1989) are necessary requirements for any scientific explanation. In textbooks mechanisms like disturbance, habitat factors, predation, dispersal, competition, or facilitation are often regarded as "proximate causes" of community composition. They are either classified according to their (+, -, neutral) influence on a focal organism, or with reference to their discreteness and inside/outside-relation to the community (see Table 2.3.9). In population ecology some of these mechanisms are used as coefficients and correction factors in Lotka-Volterra-type models.

Lower level (population) mechanisms are regarded as "spontaneous" or stochastic actions. Focal level (community) mechanisms are "internal" and by definition biotic. Both classes of mechanisms constitute the requirements for semi-autonomy (see above). Upper level (ecosystem) mechanisms are by definition "external" and abiotic, and necessarily selective as not all external mechanisms change the community.

The present classification leads to some difficulties if used to characterize an ecosystem. Because, by definition, on the ecosystem level everything is inside, it is much more difficult to define "spontaneous action" (beyond spontaneous actions of living beings), and to define "external interactions". Also, the distinction between discrete and continuous processes is not at all clear on the level of the ecosystem. Many ecosystem ecologists may find these statements unusual, because they are used to speaking about a community as if it was an ecosystem. If we strictly distinguish between both levels, the statement will become immediately clear.

Table 2.3.9. Causal repertoire of community ecology according to external, internal, and spontaneous mechanisms with reference to their discreteness

	External (abiotic) interaction	*Internal (biotic) interaction*	*Spontaneous action of the parts*
Discrete	Disturbance	Predation	Dispersal
Continuous	Habitat factors (productivity, stress)	Competition, facilitation	Growth

If a repertoire can explain the current state it can also explain the sequence of states. Thus, the causal repertoire of general community ecology can be regarded as the same as that of succession theory. Likewise the causal repertoire must be sufficient to explain non-change or constancy. For example: Disturbance is not only a trigger of succession, but also an agent of non-succession, see e.g. "permanent pioneer communities".

Zero-Force Laws and Equilibrium Hypotheses. Our next question is: Which of the observable community phenomena need explanation? For example: Is there any necessity to explain the changing number of species on a site? If we regard succession as a gradient-in-time (Johnstone 1986; Peet and Christensen 1980) then there is no necessity to do so. Change is a kind of neutral hypothesis, while non-change needs explanation. A gradient in time hypothesis is non-explanatory. The stage S_1 does not explain the stage S_2. Neither does the species composition $N_{(t)}$ explain the species composition $N_{(t+1)}$. Our prediction that change may occur is based on the a priori knowledge of "safe sites" which must occur sometimes after disturbance, habitat factors which are in most combinations favorable to colonization with organisms, and the regional flora and fauna, which will contain species suitable to colonize the above mentioned sites. However, habitat factors or species characteristics alone do not explain everything. This assumption would lead to a "thinking in common factors" (Sattler 1986). What we need are system-dependent causes for both change and non-change.

The necessity for change can also be derived from the often discarded equilibrium theories of ecology. "Cooking-pot ecology" (Colwell 1984) is still a useful tool in ecology. Equilibrium theories are more or less identical with neutral hypotheses for change. Equilibrium theories are available for populations (logistic curve, law of constant yield), metapopulations (extinction risk), and communities (MacArthur-type island biogeography, Lotka-Volterra models on competition and predation, Tilman's resource ratio hypothesis, gap dynamics etc.). In all cases the theories predict that change must occur as long as an equilibrium is not reached, and non-change occurs near equilibrium.

We doubt whether similar equilibrium theories are available on the ecosystem level. We think that thermodynamic theories, particularly the concept of "exergy" might provide such a theory. However, at present, these theories are in a very premature state. It does not meet the scientific standards of community ecologists to regard life itself, information content of molecules, diffusion gradients, gradients in pressure and sea level as equivalent "forces" for the explanation of ecological phenomena.

Causality of Population Processes. Succession as a population process has been discussed by Peet and Christensen (1980) and van der Valk (1981) from different points of view. How can we relate our causal repertoire to the observable population processes? We examined the relative importance of causal agents for the observable events of invasion (immigration) and extinction and found a certain asymmetry with regard to possible causes of both classes of events.

Extinction is an event, which happens to organisms that are already within a community. Thus disturbance, predation, nutrient shortage, or competition can lead to direct causal interference. In this case we can use the trigger model of causality (Sattler 1986). A certain trigger (event) will cause the organism to change its observable state from being present to being absent.

Invasion is an event, that happens to objects that are not present before arrival. Causal interference in the above mentioned sense (exchange of matter, energy, or information) can only happen after arrival, and may decide about establishment and growth. Before arrival, only a potential for invasion can be recognized (a "safe site" or "enemy free space", sufficient nutrient supply etc.). A potential model of causality applies to this situation (Sattler 1986).

Maintenance (mostly fluctuation around a stable point) is a net effect of invasion and extinction, or triggers and potentials. Further potentials, which are not directly related to invasion or extinction events may be included (falling short of the minimum viable population size, facilitation) in the consideration.

As for causes, a disturbance can likewise be a direct trigger for extinction, and may lead to a potential for invasion. Almost all members of our causal repertoire can be related in one way or another to invasion, maintenance and extinction. Thus a satisfactory theory of succession must take into account this complex relation between different kinds of causes and different kinds of observable effects.

These problems have been dealt with in different subdisciplines of ecology. The theory of extinctions has been the domain of conservation ecology dealing with minimum viable population models (Soulé 1987), die-back phenomena of the dominants (life span), nutrient depletion (sustainable use), but it has also been the domain of disturbance ecology, auto-allelopathy and others. As yet, there is no full-fledged theory of invasion. Invasion phenomena have been studied from the viewpoints of theoretical population theory, island biogeography, and applied ecology (e.g. invasion of foreign plants), all treating the problem from different angles.

Table 2.3.10. Explanation of succession phenomena by theories and mechanisms advanced in key note papers and a recently published textbook (Drury and Nisbet 1973; Connell and Slatyer 1977; Noble and Slatyer 1980; Johnstone 1986; van der Maarel 1988; Jax 1994; Burrows 1990)

Phenomenon	Theory	Mechanisms proposed
No change, stability	Resource ratio-hypothesis	Nutrient gradient in time (positive correlation)
	Inhibition model	Competitive power of the dominants, environmental resistance
	Initial floristic composition hypothesis	Longevity, competitive dominance
	Gap replacement (direct replacement) hypothesis	Colonization ability of the dominants
	Climax hypothesis	Self-regulation (mature state only)
Fluctuation	Stochastic replacement hypothesis	Equal chance lottery of dispersal
	Hierarchic replacement hypothesis	Differential site availability, differential species availability (dispersal), differential species performance
	Tolerance model	Stress tolerance, safe sites with our without other community members
Cyclic or quasi-cyclic behavior	Mosaic cycle hypothesis	Disturbance, population phenomena of the dominants (age structure), colonization, large scale equilibrium
	Patch dynamics hypothesis	Disturbance, colonization, scaling
	Retrogression model	Deterioration of habitat conditions by the dominants
Serial change	Facilitation model	Disturbance, facilitation, dispersal competition
	Resource ratio-hypothesis	Nutrient gradient in time, inverse ratio of resources
	Distance-area-time-hypothesis	Equilibrium
	Inhibition model	Differential longevity, timing of invasion
	Unstable monoculture hypothesis	Empty niches, diversity-stability relations

A theory of maintenance is mainly based on empirical studies on populations (seed bank, vegetative propagation, seed germination, life cycle variation, anti-herbivore defense, intraspecific competition, and coexistence), despite a large body of theoretical thought on the ecosystem level (on "stability" relations, feed back loops, regulation, ecological function etc.).

Succession Theory and General Ecological Theory. Nevertheless, we can analyze successional theories with regard to their ability to explain observable phenomena. No less than 15 different genuine successional theories can be distinguished (see Table 2.3.10). They are able, or at least they promise, to explain different temporal features of the behavior of communities.

Two aspects can be pointed out to summarize the described relations among phenomena, theories, and mechanisms. First, the same hypothesis (e.g. the resource ratio hypothesis) may account for different types of change, or both change and constancy, according to the respective circumstances. Note that it is a priori not clear whether this is an advantage or a disadvantage. But we think that a hypothesis that explains different kinds of behavior is more general than one that explains only either change or stability. Second, it is obvious, that all hypotheses use more or less the same set of causal agents, but reach different conclusions. This is not surprising, as each ecological situation is different and the same mechanism may cause different effects in other circumstances.

2.3.3.3 Predictive Theory

The most problematic part of succession research is prediction of future species compositions, relevant community characteristics, or stages (types) of the community. The question is: Under which circumstances can succession be regarded as a non-trivial chain of surprising events, worthwhile of being studied with scientific methods? Here arise contradictions because surprise or emergence are usually not amenable to scientific study (Cariani 1992; Faber et al. 1992; Emmeche et al. 1994; Wiegleb and Bröring 1996).

At present we cannot propose any epistemologically satisfactory theory which bridges the conceptual gap between temporal emergence (the coming into being of the new) and model emergence (recognition of something new by an observer; see Wiegleb and Bröring 1996). We can only list some necessary ingredients of future predictive succession research:

- Application of Markov processes for model simulations of ecological successions. It is not clear whether real processes can be described successfully by means of Markov chains without running to equilibrium states and without including the aspect of the surprise. New models are available (Logofet 1996) and should be tested.
- Definition of bifurcations of ecological processes. The surprising element would be, which of the described ways the succession would turn. Maybe Petri nets which are more and more used for event based modelling (Sonnenschein and Gronewold 1995; Gnauck 1995) or the theory of switches (Wilson and Agnew 1996) can be used for this purpose.

2.3.3.4 Succession Examples 1 and 2

Some data sets from our own work were analyzed with regard to the question whether any succession theory can be applied to these special cases. First, we reanalyzed the dynamics of macrophyte vegetation in lowland rivers (succession example 1). We tried to elucidate the descriptive means which were used to illustrate the vegetation state and change. Furthermore we looked at the explanatory and predictive power of theories applied to the respective cases. The situation was analyzed in three steps:

Step 1: Between 1978 and 1982 a survey of the macrophytic vegetation of rivers in northern Germany – including more than 1500 sites – was carried out (Herr et al. 1989b). We found that appr. 80 % of the sites investigated could be classified into two community types:

- Community dominated by broad-leaved floating-leaved species like *Sparganium, Sagittaria, Nuphar* and broad-leaved *Potamgeton,* mostly occurring in lower courses with a current velocity of less than 0.3 m/s.
- Community dominated by small-leaved submersed species like *Elodea, Callitriche, Ranunculus,* and narrow-leaved *Potamogeton,* mainly restricted to smaller upper courses with current velocities of more than 0.3 m/s.

On a large scale a great regularity was visible, with two states of vegetation which may be regarded as some kind of "climax" for the respective ecological situation. An explanation was possible with respect to the dominant growth forms which are clearly related to the physical conditions of the habitats. The actual floristic composition within the two community types was quite diverse, and could not really be explained in detail.

Step 2: In 295 sites the species composition of 1985 was compared to that of 1946 (Herr et al. 1989a; Wiegleb et al. 1989). The individual sites were quite diversified with respect to species composition. According to statistical analyses only one site can be regarded as stable, 176 had completely changed their species composition. Greater regularities were found, however, on the species level with respect to decrease or increase in frequency of species.

- *Potamogeton gramineus* was widespread and frequent before, now it is completely extinct.
- *Potamogeton alpinus* was a rare species before, nowadays it is very frequently found in different kinds of waters.

These finding were not in accordance with the conventional wisdom of central European phytosociology, which judged both species as "oligotrophic" species of stagnant waters (cf. Oberdorfer 1990). We found that P. gramineus cannot maintain the oxygen balance in its big long-lived rhizomes under eutrophic conditions leading to anoxic sediments. *P. alpinus* has several very effective means of vegetative reproduction allowing the recolonization after anthropogenic disturbance (cutting, dredging) regardless of the trophic status (Wiegleb et al. 1989). Thus an explanation of the seemingly strange behavior was possible by reference to vital attributes (demography, life history, ecophysiology) of the respective species. The results also allow for a conditional prediction of vegetation development after eutrophication or intensifying of weed cutting.

Step 3: Six plots in two adjacents streams were chosen as permanent plots and have been investigated in more detail since 1977 (Wiegleb et al. 1989; Jax et al. 1996). In addition to species composition data, data on biomass, disturbance, water chemistry, hydrology, and climate are available. Results vary dramatically according to different spatial (within plot, among plots, among rivers) and temporal scales (seasonal, year to year, 2-19 years). On a given scale both explanation and surprise occurs:

- Three sites remained more or less unchanged with respect to growth form and general species composition. There is a scientific theory for explaining the oscillation around a stable point (incl. vegetative growth, recolonization processes after short-term disturbance etc., see Wiegleb et al. 1989).
- One site is now completely devoid of macrophytes because of ongoing strong anthropogenic disturbance. There is a scientific theory to explain this behavior as well.
- Two sites have changed in species composition, biomass and vegetation structure without any obvious change in habitat conditions or any obvious major disturbance event. In this case we find a surprising behavior of the system.

Noise and rare events make extrapolation of the observed sequences almost impossible. We cannot say how the sites will look like in 20 years – if they still exist – because surprise may also occur in "regular" cases. What we might call "scientific success" depends on the degree of resolution and the choice of scale in these cases. We were able to find narrative explanations in most cases. We were not aiming at an "understanding of the functioning of the system" nor at a complete mathematical formalization.

In succession example 2, species assemblages and spatial as well as temporal distribution of aquatic bugs (especially Corixidae) were studied in different ponds and ditches situated on the Wadden Sea Islands (Bröring and Niedringhaus 1988; Niedringhaus and Bröring 1988; Bröring, unpublished data). Furthermore the development of species composition was analyzed under different initial conditions and different environmental developments. It was shown that:

- Two species were always abundant (*Sigara lateralis*, *S. stagnalis*). Three further species showed high frequencies (*Sigara striata*, *Corixa punctata*, *Paracorixa concinna*), but reached abundance only in case of special combinations of habitat factors.
- According to the environmental conditions or the stages of environmental development (especially change in salinity and content of organic matter), several more specialized species were associated.
- The main external events were arrival of a new species able to reproduce, direct human impact (pollination, pumping fresh ground water), precipitation-regime during the summer season, invasion of brackish water into island areas with freshwater, increase or decrease of salinity and, due to pollination by birds or human impact increase or decrease of organic matter (note the different time scales of the events).

- An overlay of temporal and spatial patterns of species composition occurred (for examples see Niedringhaus and Bröring 1988).

With respect to the description of change, and the explanation and prediction of patterns and processes ("events") the following aspects can be summarized:

- The necessity to differentiate between short-term and long-term became evident. Time delay was observed by analyzing short-term change of species compositions.
- Some kind of network-causality arises between processes like precipitation-regime, pumping off of fresh groundwater during summer season, invasion of brackish water (which leads to fluctuation of the water levels) and decrease or increase of organic matter (which leads to increase or decrease of oxygen content). All these processes have effects on the migration activities of the species and their reproduction rate.
- Great variances in the data sets (species abundances and species compositions) remain inexplicable; this is partly due to the generally extremely high migration activities of Corixids, the extreme tolerance of the species towards change of environmental conditions, as well as some methodological error. The distinction between methodological error and systematic noise is nearly impossible.
- Concepts like "climax" or "maturation" seem quite inadequate in this case. Concepts of "noise", "emergence" and "fractality" seem to be applicable. However, a satisfactory mathematical description in these terms cannot be provided.

Similar to example 1, noise and rare events limit the possibility of forecasting species assemblages in time and space, particularly species abundance relations and species compositions in detail. Nevertheless it is substantiated that the abundant species will always be found. Narrative explanations of most species compositions and of the detailed development of community structure were possible in most cases a posteriori.

2.3.4 Conclusions on the Application of Successional "Goal Functions"

2.3.4.1 Goals in Society and Nature Conservancy

In the following we avoid the term "goal functions" in relation to social goals. "Goal functions" can only be found in models of succession or in any other scientific concept. If we analyze goals in modern society we find two contradicting types. First, social goals express the wish of semi-autonomy of the individual, which means both independence of man from the powers of nature, and independence of man from the power of man. We may well doubt that goals of the first type may ever be reached. Goals concerning the independence of man from the power of man are implemented in the constitution of any democratic society, and do not necessarily imply overexaggerated individualism.

Second, goals can be found which try to completely eliminate the autonomy of the individual and focus its interest on a "higher" whole or entity, e.g. nationalism, religious fundamentalism, eco-dictatorship. We reject these goals for many political and ethical reasons.

There is no such clear dichotomy of goals in nature conservancy. We find an extreme diversity of environmental goals or quality aims ("Leitbilder") in current German nature conservancy and planning, generating from a variety of world views and social backgrounds. The main groups of goals are (Wiegleb 1994, 1996b; Roweck 1995; Bastian 1996):

- *Biotic goals*: Biodiversity including species and genetic diversity, protection of typical, rare, or endangered species, protection of endangered habitats, creation of habitat networks etc.;
- *Abiotic goals*: Sustainability, ecological efficiency, resource protection, geotope protection, protection of geomorphological processes, critical loads, quality aims of soil, water and air pollution etc.; and
- *Aesthetic goals*: Protection of cultural landscapes, landscape architecture, wilderness etc.

Some of these goals refer to the intrinsic values of nature, while others are more anthropocentric referring to the either instrumental, the eudaimonistic, or the moral value of nature (Krebs 1996). However, in order to avoid any "naturalistic fallacy", none of these goals can be derived from ecosystem and community properties without taking the step of social assessment. Empirically defined successional trends, or goal functions of a successional model can be used as "ecological orientors" which serve as an orientation aid for finding out what is possible in specific ecological situations. E.g. we may decide to prefer a stable and species-rich biotic community in a stream to an unstable community with few species. Succession theory states that this is impossible. Nevertheless, it may be a useful goal to prevent further river pollution.

2.3.4.2 Succession Example 3

Another succession example taken from our recent research (Bröring et al 1995; Wiegleb 1996b) may illustrate the outlined principle in more detail. The characteristics of some common successional stages of communities of formerly devastated brown coal mining areas and their potential value for nature conservation are shown in Table 2.3.11 (their biotic, abiotic and aesthetic characteristics).

Changes in biotic characteristics of the community are neither linear nor predic-table in this case. In particular, the duration of a given stage is unknown. In some cases, the second stage is only reached after 70 years, in other cases, however, the fifth stage already is reached after 20 years. Some stages may be skipped (stage 3), other may occur several times (stage 4). As to the assessment of these stages in the context of conservation, we already find in the very beginning of stage 1, many rare specialized inhabitants of open habitats, often in high densities and species number, which may occur hundreds of kilometers away, at their closest. Later on, in a short grass prairie stage 3, with the abundance of flowering dicots, many species of butterflies, bees and wasps occur.

Table 2.3.11. Characteristics of some common successional stages of former brown coal mining areas and their potential values for nature conservation

Stages of primary succession	Biotic characteristics	Abiotic characteristics	Aesthetic characteristics
1. Bare sand	habitat of specialized Orthoptera, Dermaptera and Hymenoptera	wind erosion, leaching of cations, acidification of groundwater	bizarre landscape forms
2. *Pioneer vegetation with ruderal herbs*	habitat of specialized Hymenoptera	decrease of processes under 1., increase of processes under 6.	view on open landscape
3. Short grass prairie with Corynephorus and xerophytic herbs	high importance as feeding habitat for Lepidoptera and Hymenoptera-Aculeata	see 2	view on open landscape, with many colorful flowers
4. Tall grass prairie with Calamagrostis canescens	lowest species diversity in plants, high importance as corridors for carabid beetles and mice	see 2	view on open landscape
5. Scrub, open woodland	high species number of endangered species of Orchidaceae and Pyrolaceae	see 2	high structural diversity, smooth ecotones
6. Mixed Pine-Birch-Oak forest	low diversity of plants, high diversity of passerine birds	high evapotranspiration, no groundwater formation	lacking view, lacking margin effects

The tall grass stage 4 with *Calamagrostis* shows a low diversity in higher plants, but serves as an important corridor of migration for small animals like carabid beetles and mice. In the scrub stage 5 some extremely rare plant species with specialized habitat requirements which are no longer found elsewhere in the cultural landscape regularly occur. They vanish with the same regularity, as soon as the woodland gets denser. In the closed woodlands there is a low botanical diversity again, but a high species number of passerine bird species. Management decisions of the type "free succession everywhere", "partial biotop management" or "quick afforestation" will always destroy or favor one of those stages.

The abiotic characteristics change more regularly in relation to the respective successional stage. However, assessment criteria are even more contradictory both among themselves and in relation to biotic characteristics.

In stage 1 of the successional sequence, there is a high potential for wind erosion. Furthermore, we find leaching of cations and protons leading to an extreme acidification of ground and surface water (to pH 2.3). In the course of primary succession vegetation gets denser, and wind erosion does no longer play any significant role. The increase in evapotranspiration simultaneously leads to a decrease in groundwater formation. Leaching stops, since there is no longer any vertical water transport. Under woodland, there may be no groundwater formation at all which is problematic in this case. Former mining activities have caused an enormous water deficit for the entire landscape. In the long run, this threatens adjacent wet land areas (Spreewald) and even the drinking water supplies of cities (like Berlin). Thus the question whether woodland is estimated high or low depends on a decision in favor of a water volume management over a water quality management. Also the aesthetic characteristics of the landscape change with vegetation development. Many people prefer open landscapes to dense forests as recreational areas. Thus a complete afforestation of former mining landscape would not be tolerated.

We can learn from this example, that nature does not tell us what to do with our environment (for further discussion see also Bröring and Wiegleb 1990; Wiegleb and Bröring 1991; Costanza et al. 1992; Shrader-Frechette 1994; Wiegleb 1997). We have to decide this for ourselves in the pluralistic democratic discourse. But natural scientists can learn from the study of social systems that semi-autonomy is indeed an important requirement for both system development and persistence. Only semi-autonomous groups or individuals create innovation which is necessary to cope with the inevitable changes of the general context of social systems (internal: technological innovation; external: new powers). Therefore, semi-autonomy both on the level of the community and on the level of the ecosystem is indeed a necessary requirement for any useful concept of succession.

Acknowledgments. We thank F. Schulz, H. Fromm, H. Blumrich, J. Vorwald, A. Gnauck (Cottbus), W. Herr, H. Brux, D. Todeskino, R. Niedringhaus, W. Schultz, G.P. Zauke (Oldenburg), M. Struck (Hitzacker), K. Jax (Jena), T. Potthast, U. Eser (Tübingen), L. Trepl und A. Schwarz (Freising) for discussion about the various aspects of this paper. The work on river vegetation (G. Wiegleb) was carried out in cooperation with the Niedersächsische Landesamt für Ökologie. The work on pond faunas (U. Bröring) was carried out partly in cooperation with Bezirksregierung Weser-Ems. The work on mining landscapes was carried out in the framework of a BMBF-research project (Fkz 0339648) at the BTU Cottbus.

References

Allen TFH, Hoekstra TW (1990) The Confusion Between Scale-defined Levels and Conventional Levels of Organisation. J Veget Sci 1:5-13

Allen TFH, Hoekstra TW (1991) Role of Heterogeneity in Scaling of Ecological Systems Under Analysis. In: Kolasa J, Pickett STA (eds) Ecological Heterogeneity. Ecol Studies 86, New York, pp 47-68

Allen TFH, Hoekstra TW (1992) Toward a Unified Ecology Columbia. Univ Press, Columbia

Allen TFH, Starr TB (1982) Hierarchy: Perspectives for Ecological Complexity. University of Chicago Press, Chicago, p 310

Allen TFH, O'Neill RV, Hoekstra TW (1984) Interlevel Relations in Ecological Research and Management: Some Working Principles from Hierarchy Theory. USDA, Forest Service, General Technical Report RM-110:11

Andrewartha HG, Birch LC (1954) The Distribution and Abundance of Animals. University of Chicago Press, Chicago

Bastian O (1996) Ökologische Leitbilder in der räumlichen Planung – Orientierungshilfen beim Schutz der biotischen Diversität. Arch Naturschutz u Landschaftsforschung 34:207-234

Beckner M (1974) Reduction, Hierarchies and Organisms. In: Ayala FJ, Dobzhansky T (eds) Studies in the Philosophy of Biology Reduction and Related Problems. Berkeley University Press, Berkeley CA, pp 163-177

Begon M, Harper JL, Townsend CR (1990) Ecology: Individuals, Populations and Communities, 2nd edn., Blackwell, Oxford

Bröring U, Niedringhaus R (1988) Zur Ökologie aquatischer Heteropteren [...] in Kleingewässern der ostfriesischen Insel Norderney. Arch Hydrobiol 111:559-574

Bröring, U, Schulz, F, Wiegleb, G (1995) Niederlausitzer Bergbaufolgelandschaft: Erarbeitung von Leitbildern und Handlungskonzepten für die verantwortliche Gestaltung und nachhaltige Entwicklung ihrer naturnahen Bereiche. Z Ökol Natursch 4:176-178

Bröring U, Wiegleb G (1990) Wissenschaftlicher Naturschutz oder ökologische Grundlagenforschung. Natur u Landschaft 65:283-292

Brown JH (1995) Organisms and Species as Complex Aadaptive Systems: Linking the Biology of Populations with the Physics of Ecosystems. In: Jones, CG, Lawton, JH (eds) Linking Species and Ecosystems, Chapman and Hall, New York, pp 16-24

Burrows CJ (1990) Processes of Vegetation Change. Unwin, London

Cariani P (1992) Emergence and Artificial Life. In: Langton CG, Taylor C, Farmer JD, Rasmussen S (eds) Artifical Life II, Santa Fe Studies in the Science of Complexity, Proc Vol 10, Redwood City, CA, pp 775-797

Cherrett JM (1989) Key Concepts: the Result of a Survey of our Members' Opinion. In: Cherrett, JM (ed) Ecological Concepts. Blackwell, Oxford, pp 1-16

Clements FE (1916) Plant Succession An Analysis of the Development of Vegetation. Carnegie Inst Wash Publ No 242

Colwell RK (1984) What's New? Community Ecology Discovers Biology. In: Price P, Slobodchikoff CN, Gaud WS (eds) A New Ecology Novel Approaches to Interactive Systems. Wiley, New York, pp 387-396

Connell JH, Slatyer RO (1977) Mechanisms of Succession in Natural Communities and their Role in Community Stabiliy and Organization. Am Nat 111:1119-1144

Costanza R, Norton BG, Haskell BD (1992) Ecosystem Health New Goals for Environmental Management. Island Press, Washington DC

Cousins SH (1990) Countable ecosystems deriving from a new food web entity. Oikos 57:270-275

Drury WH, Nisbet IC (1973) Succcession. J Arnold Arboretum 54: 331-368

Egerton FN (1968) Studies of animal populations from lamarck to darwin. Journal Hist Biol 1:225-259

Egerton FN (1979) Changing concepts of the balance of nature. Quart Rev Biol 48:322-350

Egler FE (1954) Vegetation Science Concepts: I Initial Floristic Composition – a Fctor in Old-field Vegetation Development. Vegetatio 4:412-417
Elton C (1927) Animal Ecology. London
Emmeche C, Köpp S, Stjernfelt F (1994) Emergence and the Ontology of Levels: Search of the Unexplainable. Arbejdspapier 11 Afdeling for literaturvidenskab. Dept of Comparative Literature, University of Copenhagen
Faber M, Manstetten R, Proops J (1992) Toward an Open Future: Ignorance, Novelty, and Evolution. In: Costanza R, Norton BB, Haskell BD (eds) Ecosystem Health - New Goals for Environmental Management. Island Press, Washington DC, pp 72- 96
Feyerabend PK (1981) Erklärung, Reduktion und Empirismus. In: Probleme des Empirismus, Vieweg, Braunschweig, pp 73-125
Forman RTT, Godron M (1986) Landscape Ecology. John Wiley & Sons, New York
Gleason HA (1926) The individualistic concept of the plant association. Bull Torrey Bot Club 44:463-481
Gnauck A (1995) Einführung: Systemtheorie, Ökosystemvergleiche und Umweltinformatik. In: Gnauck A, Frischmuth A, Kraft A (eds) Ökosysteme: Modellierung und Simulation. Blottner, Taunusstein, pp 11-27
Golley FB (1993) The history of the ecosystem concept in ecology. Yale Uni Press, New Haven
Greig-Smith P (1979) Pattern in Vegetation. J Evcol 67: 755-779
Hagen J (1992) An Entangled Bank The Origins of Ecosystems. Ecology Rutgers University Press, New Brunswick
Harper JL (1967) A Darwinian Approach to Plant Ecology. J Ecol 55:247-270
Harper JL (1975) Population Ecology of Plants. Academic Press, New York
Harper JL (1982) After Description. In: Newman EI (ed) The Plant Community as a Working Mechanism. Blackwell, Oxford , pp 11-25
Herr W, Todeskino D, Wiegleb G (1989a) Übersicht über Flora und Vegetation der niedersächsischen Fließgewässer unter besonderer Berücksichtigung von Naturschutz und Landschaftspflege. Naturschutz u Landschaftspflege in Niedersachsen 18:145-284
Herr W, Wiegleb G, Todeskino D (1989b) Veränderungen von Flora und Vegetation in ausgewählten Fließgewässern Niedersachsens nach 40 Jahren (1946/86). Naturschutz u Landschaftspflege in Niedersachsen 18:121-144
Jax K, Zauke GP (1991) Maßstäbe in der Ökologie – ein vernachlässigter Konzeptbereich. Verh Ges Ökol 21:23-30
Jax K (1994) Mosaik-Zyklus und Patch-dynamics: Synonyme oder verschiedene Konzepte? Eine Einladung zur Diskussion. Z Ökol Naturschutz 3:107-112
Jax K (1996) The Units of Ecology. Towards an Intersubjective Defintion of Concepts, in Press
Jax K, Potthast T, Wiegleb G (1996) Skalierung und Prognoseunsicherheit bei ökologischen Systemen. Verh Ges Ökol 26:527-535
Jax K, Jones, CG, Pickett, STA (1997) The Self-identity of Ecological Units, in press
Johnstone IM (1986) Plant invasion windows: a time-based classification of invasion potential. Biol Rev 61:369-394
Jones CG, Lawton JH (eds) (1995) Linking Species and Ecosystems. Chapman, Hall, New York
Jörgensen SE (1992) Integration of Ecosystem Theories: A Pattern. Kluwer, Dordrecht: 376 pp
Kolasa J, Pickett STA (1989) Ecological Systems and the Concept of Biological Organisation. Proc Natl Acad Sci 86:8837-8841
Krebs A (1996) "Ich würde gerne aus dem Hause tretend ein paar Bäume sehen" Philosophische Überlegungen zum Eigenwert der Natur In: Nutzinger HG (ed) Naturschutz-Ethik-Ökonomie: Theoretische Grundlagen und praktische Konsequenzen. Metropolis, Marburg, pp 31-48
Kuhn TS (1976) Die Struktur wissenschaftlicher Revolutionen, 2nd ed. Suhrkamp, Frankfurt/Main
Lakatos I (1978) Die Geschichte der Wissenschaft und ihre rationale Rekonstruktion. In: Diederich W (ed) Theorien der Wissenschaftsgeschichte. Suhrkamp, Frankfurt/M, pp 55-119

Leps J (1990) Can underlying mechanisms be deduced from observed pattern. In: Krahulec F, Agnew ADQ, Willems JH (eds) Spatial processes in plant communities. Academia, Prague, pp 1-11
Leser H (1984) Zum Ökologie-, Ökosystem- und Ökotopbegriff. Natur u Landsch 59:351-357
Lindeman RL (1942) The trophic-dynamic aspect of ecology. Ecology 23:399-418
Logofet DO (1996) Inhomogeneous Markov Chain Models of Plant Succession: New Perspectives of an Old Paradigm. Ecological Summit 96, Abstract 163, University of Copenhagen
MacArthur RH (1972) Geographical Ecology. Harper and Row, New York
McIntosh RP (1976) Ecology since 1900. In: Taylor BJ, White TJ (eds) Issues and Ideas in America. Norman, Oklahoma, pp 353-372
McIntosh RP (1981) Succession and Ecological Theory. In West DC, Shugart HH, Botkin DB (eds) Forest Succession, Concepts and Application. Springer, New York, pp 10-23
McIntosh RP (1985) The Background of Ecology: Concept and Theory. Cambridge University Press, Cambridge
Mittelstraß J (1992) Rationalität und Reproduzierbarkeit. In: Janich P (ed) Entwicklungen der methodischen Philosophie. Frankfurt, pp 54-67
Müller F (1992) Hierarchical Approaches to Ecosystem Theory. Ecolog Modelling 63:215-242
Niedringhaus R, Bröring U (1988) Die Wanzen und Käfer süßer und brackiger Gewässer der jungen Düneninseln Memmert und Mellum (Heteroptera, Coleoptera). Drosera 88:329-340
Noble IR, Slatyer RO (1980) The Use of Vital Attributes to Predict Successional Changes in Plant Communities Subject to Recurrent Disturbances. Vegetatio 43:5-21
Oberdorfer E (1990) Exkursionsflora von Süddeutschland. 5. Aufl., Ulmer, Stuttgart
Odum EP (1959) Fundamentals of Ecology, 2nd edn., Saunders, Philadelphia
Odum EP (1971) Fundamentals of Ecology, 3rd edn., Saunders, Philadelphia
Odum HT (1983) Systems Ecology. An Introduction. Wiley, New York
O'Neill RV (1989) Perspectives in Hierarchy and Scale. In: Roughgarden J, May RM, Levin SA (eds) Perspectives in Ecological Theory. Princeton Univ Press, Princeton NJ, pp 140-156
O'Neill R, DeAngelis DL, Waide JB, Allen TFH (1986) A Hierarchical Concept of Ecosystems. Princeton University Press, Princeton
Patten BC (1982) Environs: Relativistic Elementary Particles for Ecology. Am Nat 119:179-219
Patten BC (1990) Environ Theory and Indirect Effects: A reply to Loehle. Ecology 71:2386-2393
Peet RK, Christensen NL (1980) Succession: a Population Process. Vegetatio 43:131-140
Pianka ER (1992) Evolutionary Ecology, 5th edn. Harper Collins, New York
Pickett STA, Collins SL, Armesto JJ (1987) A Hierarchical Consideration of Causes and Mechanisms of Succession. Vegetatio, 69:109-114
Pickett STA, McDonnell MJ (1989) Changing Perspectives in Community Dynamics: a Theory of Successional Forces. TREE 4:241-245
Potthast T (1996) Transgenetic Organisms and Evolution: Ethical Implications. In: Tomiuk J, Wöhrmann K, Sentker A (eds) Transgenic Organisms – Biological and Social Implications. Birkhäuser, Basel, pp 227-240
Reise K (1980) Hundert Jahre Biozönose: Die Evolution eines ökologischen Begriffes. Naturwiss Rundschau 33:328-335
Roweck H (1995) Landschaftsentwicklung über Leitbilder? Kritische Gedanken zur Suche nach Leitbildern für die Kulturlandschaft von morgen. LÖBF-Mitteilungen 4:25-34
Russel B (1918) The Philosophy of Logical Atomism. The Monist 28 (1919). dtv wissenschaft 4327, dtv, München
Salthe SN (1985) Evolving Hierarchical Systems. Columbia University Press, New York
Sattler R (1986) Biophilosophy: Analytic and Holistic Perspectives. Springer, Berlin
Schmied-Kowarzik W (1989) Friedrich Wilhelm Joseph Schelling. In: Böhme G (ed) Klassiker der Naturphilosophie: Von den Vorsokratikern bis zur Kopenhagener Schule. Beck-Verlag, München, pp 158-262
Shrader-Frechette KS (1994) Ecosystem Health: a New Paradigm for Ecological Assessment. TREE 9:456-457

Shrader-Frechette KS, McCoy ED (1993) Method in Ecology. Cambridge University Press, Cambridge
Sonnenschein M, Gronewold A (1995) Diskrete Petrinetze für individuenbasierte Modelle. In: Gnauck A, Frischmuth A, Kraft A (eds) Ökosysteme: Modellierung und SimulationBlottner, Taunusstein, pp 109-130
Soulé ME ed (1987) Minimum Viable Populations for Conservation. Cambridge University Press, Cambridge
Toulmin S (1983) Kritik der kollektiven Vernunft (Human understanding, vol 1). Suhrkamp, Frankfurt
Trepl L (1988) Gibt es Ökosysteme? Landschaft + Stadt 20:176-185
Trepl L (1994) Competition and Coexistence: on the Historical Background in Ecology and the Influence of Economy and Social Sciences. Ecol Modelling, 75/76:99-110
Ulanowicz RE (1979) Prediction, Chaos, and Ecological Perspective. In: Halfon E (ed) Theoretical Systems Ecology. Academic Press, New York, pp 107-117
Ulanowicz RE (1990) Aristotelian Causalities in Ecosystem Development. Oikos 57:42-48
Unwin N (1996) The Individuation of Events. Mind 105:315-330
Urban DL, O'Neill RV, Shugart HH (1987) Landscape Ecology: A Hierarchical Perspective Can Help Scientists Understand Spatial Pattern. BioScience 37:119-127
Van der Maarel E (1988) Vegetation Dynamics: Patterns in Space and Time. Vegetatio 77:7-19
Van der Valk A (1981) Succession in Wetlands: a Gleasonian Approach. Ecology 62:688-696
Warming EB (1895) Plantesamefund. Grundtraek af den oekologiske Plantegeografi. Kopenhagen
Wiegleb G (1989) Explanation and Prediction in Vegetation Science. Vegetatio 83:17-34
Wiegleb G (1991) Explorative Datenanalyse und räumliche Skalierung - eine kritische Evalu-ation. Verh Ges Ökol 21:327-338
Wiegleb G (1994) Einführung in die Thematik des Workshops "Ökologische Leitbilder". TUC Aktuelle Reihe 6:7-13
Wiegleb G (1996a) Konzepte der Hierarchietheorie in der Ökologie. In: Mathes K, Breckling B, Eckschmitt K (eds) Systemtheorie in der Ökologie. ecomed, Marburg, pp 7-24
Wiegleb G (1996b) Leitbilder des Naturschutzes in der Bergbaufolgelandschaft. Verh Ges Ökol 25:309-319
Wiegleb G (1997) Leitbildmethode und naturschutzfachliche Bewertung. Z Ökol u Natursch 6:43-62
Wiegleb G, Bröring U (1991) Wissenschaftlicher Naturschutz – Grenzen und Möglichkeiten. Garten + Landschaft 2/91:18-23
Wiegleb G, Bröring U (1996) The Position of Epistemological Emergentism in Ecology. In: Albers B, Dittmann S, Krönicke I, Liebezeit G (eds) The Concept of Ecosystems. Senckenbergiana Maritima 27 (3/6):179-193
Wiegleb G, Herr W, Todeskino D (1989) Ten Years of Vegetation Dynamics in Two Rivulets in Lower Saxony (FRG). Vegetatio 82:163-178
Wiens JA, (1989) Spatial Scaling in Ecology. Funct Ecol 3:385-397
Wiens JA (1995) Landscape Mosaics and Ecological Theory. In Hansson L, Fahrig L, Merriam G (eds) Mosaic Landscapes and Ecological Processes. Chapman and Hall, London, pp 1-26
Wilson JB (1991) Does Vegetation Science Exist? J Veget Sci 2:289-290
Wilson JB, Agnew ADQ (1992) Positive-feedback Switches in Plant Communities. Adv Ecol Res 23:264-326
With KA, King AW (1997) The Use and Misuse of Neutral Landscape Models in Ecology. Oikos 79:219-229
Wittgenstein L (1922) Tractatus Logico-Philosophicus. London

2.4 Thermodynamic Orientors: Exergy as a Goal Function in Ecological Modeling and as an Ecological Indicator for the Description of Ecosystem Development

Sven E. Jørgensen and Søren N. Nielsen

Abstract

The thermodynamic concept exergy is introduced and proposed as a goal function in ecological models, where it has successfully been applied for the development of structural dynamic models. It is discussed that the concept can be used as a translation of Darwin's theory to thermodynamics which leads to the formulation of a fourth law of thermodynamic: as a system receives a throughflow of exergy, the system will utilize this exergy to move away from thermodynamic equilibrium. If there are offered more than one pathways to move away from thermodynamic equilibrium the one yielding most stored exergy, i.e., with the most ordered structure or the longest distance to thermodynamic equilibrium by the prevailing conditions, will be selected. It is shown how an approximate and relative value of exergy can be found for a model of an ecosystem, including the information embodied by the organisms in the genes.

2.4.1 Introduction

Ecosystems have a self-organizing ability to move away from thermodynamic equilibrium which can be described by the thermodynamic function exergy. This may be utilized to formulate a goal function which can be applied to improve our present ecological models to account for the self-organizing ability by ecosystems. The term goal function should solely be applied in modeling context, while the term ecological indicator is more appropriate to be used when we are discussing the propensities that characterize the development of ecosystems (see Ulanowicz Sect. 2.10). This paper will introduce the concept of exergy, while examples for its application as a goal function in modeling and as an ecological indicator will be covered in the paper of Nielsen and Jørgensen (Sect. 2.7).

F. Müller M. Leupelt (Eds.)
Eco Targets, Goal Functions, and Orientors

Natural ecosystems are inconceivably complex and, for most, it is impossible to produce a description of ecosystem properties that encompasses all the details. Therefore most models must remain simple representations of real ecosystems. To be meaningful, a model must, however, possess the basic (holistic) properties of the system it is supposed to imitate.

Ecosystems self-regulate their process rates according to feedback from both source and product. They are able to replace ineffective sources, producers and processes with more effective ones to achieve a higher utilization of the resource. Is it possible to account for the entire hierarchy of self-regulation by the introduction of additional constraints in a model? If we presume that the regulation takes place according to a goal function, we would thereby be able to capture the flexibility that characterizes ecosystems. Many present models have rigid structures and a fixed set of parameters, reflecting that no changes or replacements of the components are possible. They describe the ecosystem as a rigid physical system. Structural-dynamic models introduce parameters (properties) that can change according to changing forcing functions (for further details see Nielsen and Jørgensen Sect. 2.7).

The biological cells that form the basic units of ecosystems are a result of a long evolution from organic soup to protobiont, to protocells and further on to ever more complex cells with very complex anabolic (synthesizing) processes. A wide spectrum of biochemical compounds with specific functions is produced in the cells. This ability is preserved by use of a very sophisticated genetic function and code to assure that no significant information is lost. The first part of the evolution which characterized the earth some 3600 to 4000 million years ago, was based on randomly produced organic compounds. Repeated use of "trial and error" found new pathways to create organization and move further and further away from the thermodynamic equilibrium, corresponding to an inorganic soup of the same elements. The development of genes made it possible to "store" a good solution and build on the shoulders of what was already achieved. More and more sophisticated biochemical processes resulted from further development. Later in the evolution, multicellular organisms, where more cells cooperated very closely as a unity, emerged. All organisms struggled for survival and attempted currently to adapt themselves to the prevailing conditions, determined by the external factors (forcing functions) and the other present organisms. This resulted in development of ecosystems as a cooperative phenomenon where different elements act together in some concerted way to obtain the best survival of a wide range of the most fitted species.

Exergy is a central concept in this context and can be used as an ecological indicator, as it expresses energy with a built-in measure of quality. It measures the energy that can do work, for instance the chemical energy in biomass. Exergy accounts natural resources (Eriksson et al. 1976) and can be considered as "fuel" for any system that converts energy and matter in a metabolic process (Schrödinger 1944). Ecosystems consume exergy, and an exergy flow through the system is necessary to keep the system functioning. Exergy stored in the system expresses directly the distance from the "inorganic soup" in energy terms, as will be explained further below.

2.4.2 Exergy

Exergy is defined as the amount of work (entropy-free energy) a system can perform, when it is brought into thermodynamic equilibrium with its environment. Fig. 2.4.1 illustrates the definition. The considered system is characterized by the extensive state variables (= variables dependent on the size of the system) S, U, V, N_1, N_2, N_3......., where S is the entropy, U is the energy, V is the volume and N_1, N_2, N_3 ..... are moles of various chemical compounds, and by the intensive state variables (= variables independent on the size of the focal system), temperature, T, pressure, p, the chemical potential of the components 1 to n, μc_1, μc_2, μc_3....The system is coupled to a reservoir, a reference state, by a shaft. The system and the reservoir are forming a closed system. The reservoir (the environment) is characterized by the intensive state variables To, po, μc_{1o}, μc_{2o}, μc_{3o}....and as the system is small compared with the reservoir, the intensive state variables of the reservoir will not be changed by interactions between the system and the reservoir. The system develops toward thermodynamic equilibrium with the reservoir and is simultaneously able to release entropy-free energy to the reservoir. During this process the volume of the system is constant as the entropy-free energy must be transferred through the shaft only. If a boundary displacement against pressure of the reference environment should take place, it would not be available as useful work on the surroundings. The entropy is also constant as the process is an entropy-free energy transfer from the system to the reservoir, but the intensive state variables of the system become equal to the values for the reservoir. The total transfer of entropy-free energy in this case is the exergy of the system. It is seen from this definition that exergy is dependent on the state of the system *and* the reservoir and not dependent entirely on the state of the system. Exergy is therefore not a state variable. In accordance with the first law of thermodynamics, the increase of energy in the reservoir, ΔU, is:

$$\Delta U = U - Uo \qquad (2.4.1)$$

where Uo is the energy content of the system after the transfer of work to the reservoir has taken place. According to the definition of exergy, Ex, we have:

$$Ex = \Delta U = U - Uo \qquad (2.4.2)$$

As $U = TS - pV + \sum_c \mu c\, Ni$, (see any textbook in thermodynamics), and

$$Uo = ToS - poV + \sum_c \mu co\, Ni, \qquad (2.4.3)$$

we get the following expression for exergy :

$$Ex = S\,(T - To) - V\,(p - po) + \sum_{c} (\mu c - \mu co)\,Ni \qquad (2.4.4)$$

As reservoir, or reference state, we can select the same (eco)system but at thermodynamic equilibrium, i.e., that all components are inorganic and at the highest oxidation state, if oxygen is present (nitrogen as nitrate, sulfur as sulfate and so on). The reference state will in this case correspond to the ecosystem without life forms and with all chemical energy utilized or as an "inorganic soup". Usually, it implies that we consider T = To, and p = po, which means that the exergy becomes equal to the Gibb's free energy of the system, or the chemical energy content of the system relative to the reference state. Notice that equation (2.4.4) also emphasizes that exergy is dependent on the state of the environment (the reservoir = the reference state), as the exergy of the system is dependent on the intensive state variables of the reservoir. When dealing with transfer processes, exergy flows are associated with mass flows as well as with flows of energy either as heat or as work across the control surface. Thus, in each application, we can identify the exergy input rate, Ex_i, to the process and the resulting (useful) exergy production, Ex_p. When we are using the expression energy efficiency, we actually mean exergy efficiency, as the energy efficiency always will be 100% according to the first law of thermodynamics. The exergy efficiency, eff, is however always less than 100% for real processes according to the second law of thermodynamics:

$$eff = Ex_p \,/\, Ex_i < 100\% \qquad (2.4.5)$$

Notice that exergy is not conserved – only if entropy-free energy is transferred, which implies that the transfer is reversible. All processes in reality are, however, irreversible, which means that exergy is lost (and entropy is produced). Loss of exergy and production of entropy are two different descriptions of the same reality, namely that all processes are irreversible and we unfortunately always have some loss of energy forms which can do work to energy forms which cannot do work. The energy is of course conserved by all processes according to the first law of thermodynamics. An energy efficiency of an energy transfer will always be 100%, while the exergy efficiency is of interest.

Exergy seems more useful to apply than entropy to describe the irreversibility of real processes, as it has the same unit, J, as energy and is an energy form, while the definition of entropy (unit: J/K) is more difficult to relate to concepts associated to our usual description of the reality.

Notice that information contains exergy. Boltzmann (1905) showed that the free energy of the information that we actually possess (in contrast to the information we need to describe the system) is kT ln I, where I are the pieces of information we have about the state of the system and k is Boltzmann's constant ($1.3803 \cdot 10^{-23}$ J / molecules·deg). It implies that one bit of information has the exergy equal to k T ln2. Transformation of information from one system to another is often almost an entropy-free energy transfer.

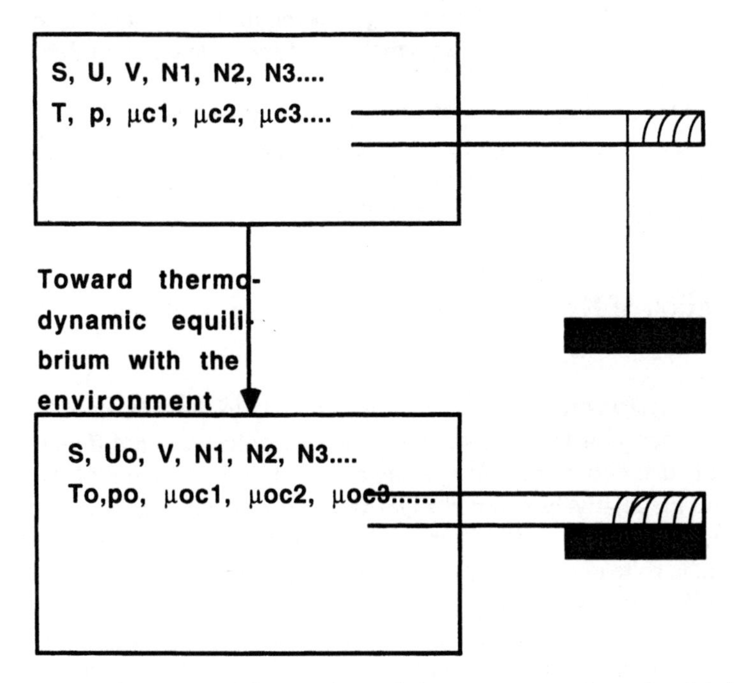

Fig. 2.4.1. Exergy is defined as the work the system can perform when it is brought into equilibrium with the reference state (its environment for instance). The work is on the figure is utilized by the shaft to lift a load

If the two systems have different temperatures, the entropy lost by one system is not equal to the entropy gained by the other system, while the exergy lost by the first system is equal to the exergy transferred and gained by the other system. In this case it obviously more convenient to apply exergy than entropy. The second law of thermodynamics can be expressed by the use of exergy as follows:

$$\Delta Ex_{\text{for any process}} \leq 0, \qquad (2.4.6)$$

which implies that exergy is always lost, i.e., work is lost in form of heat that cannot do work. These two formulations of the second law of thermodynamic by entropy and exergy are of course consistent.

As seen the exergy of the system measures the contrast - it is the difference in free energy if there is no difference in pressure, as may be assumed for an ecosystem and its environment - against the surrounding environment. If the system is in equilibrium with the surrounding environment the exergy is zero.

Since the only way to move systems away from equilibrium is to perform work on them, and since the available work in a system is a measure of the ability to do this, we have to distinguish between the system and its environment or thermodynamic equilibrium alias the inorganic soup. Therefore it is reasonable to use the available work, i.e., the exergy, as a measure of the distance from thermodynamic equilibrium.

As we know that the ecosystem due to the through-flow of energy has the tendency to develop away from thermodynamic equilibrium losing entropy or gaining negentropy and information we can put forward the following proposition: *Ecosystems develop toward a higher level of exergy.*

2.4.3 Application of Exergy in Ecosystem Theory and Darwin's Theory

Ecosystems are soft systems in the sense that they are able to meet changes in external factors or impacts with many varying regulation processes on different levels. The result is that only minor changes are observed in the function of the ecosystem, despite the relatively major changes in environmental conditions. It means that the state variables – but not necessarily the species – are maintained almost unchanged, in spite of changes in external factors.

It has been widely discussed during the last years, how it is possible to describe these regulation processes (Odum 1983; Straškraba 1980; Straškraba and Gnauck 1983), particularly those on the ecosystem level – i.e., the changes in ecological structure and the species composition.

The Neodarwinian theory expanded to include (1) coevolution, (2) the transfer of knowledge (information) from parents to children, (3) the ability of organisms to regulate their environment and thereby the selection pressure on them, (4) the D-genes. Darwin's theory states that the species, that are best fitted to the prevailing conditions in the ecosystem will survive. This formulation may be interpreted as a tautology. We should therefore prefer the following formulation: Life is a matter of survival and growth. Given the conditions, determined by the external and internal functions, the question is: which of the available organisms and species (and there are more available species than needed) have the combinations of properties to give the highest probability for survival and growth? Those species, or rather that combination of species, may be denoted the fittest and will be selected. Darwin's theory may in other words, be used to describe the changes in ecological structure and species composition, but can not be directly applied quantitatively with the present formulation, for instance in ecological modeling.

Ecosystems must be open or at least non-isolated. It is absolutely necessary for their existence. A flow of exergy through the systems is also sufficient to form an ordered structure (also named a dissipative structure by Prigogine and Nicolis (1988). Morowitz (1991) calls this latter formulation the fourth law of thermodynamics, but it would be more appropriate to expand this law to encompass a statement about *which* ordered structure among the possible ones the system will select, or which factors determine how an ecosystem will develop. This expanded version was formulated as a tentative fourth law of thermodynamics in Jørgensen (1992), but was already expressed without the pretentious name in Jørgensen and Mejer (1977), Mejer and Jørgensen (1979) and in Jørgensen (1982).

Growth may be defined as formation of ordered structure. In thermodynamics terms, growth means that the system is moving away from thermodynamic equilibrium. At thermodynamic equilibrium, the system cannot do any work, the components are inorganic and have the lowest possible free energy, and all gradients are eliminated. We use the expression growth of a crystal, growth of a society and growth of an economy to indicate that the structure in one way or another is getting larger. Biological systems have particularly many possibilities to grow or to move away from thermodynamic equilibrium. We use often the expression development to cover the many simultaneous directions of growth for more complex systems. It is crucial in ecology to know which pathways among the possible ones an ecosystem will select for development.

Darwin has given the answer to the question raised above, when one species is considered: the best fitted will survive. Survival means that the biomass of the species will be maintained or maybe even increased (growth). An organism or a population is exposed to many constraints, determined by the forcing functions on the ecosystems and the other organisms living in the ecosystem. The question is: who is winning the competition about the resources? Darwin gives the answer: the organism or the population with the properties, that are best coordinated (fitted) to the prevailing conditions. The winner's award is survival or even growth. In thermodynamic terms it means that the organisms that have their properties better coordinated to the prevailing conditions will be able to contribute most to the free energy or exergy of the system due to their biomass with the embodied information (which also in accordance with Boltzmann (1905) represents exergy; see Jørgensen et al., 1997). The exergy or chemical energy which can be used to do work in mineral oil is about 42 kJ/ g. For biomass with an average composition of proteins, carbohydrates and fat, it can be calculated to be approximately 18 kJ /g (the details of this calculations are given below and in Jørgensen et al., 1995).

Brown (1995) defines fitness as the rate at which resources in excess of those required for maintenance can be utilized for reproduction. He uses dW/dt, called reproductive power (W is the weight), to find the optimal body mass. So, he is asking the question: which size is best fitted? The answer is found by determination of the size with the highest growth potential, that is the size yielding the biggest increase of the biomass.

An ecosystem encompasses, however, many species. They cannot all obtain the biggest biomass independent of the other species - the species are interdependent. Darwin considered this complication, as the expression "prevailing conditions" is anticipated to include all the abiological and biological constraints imposed on the species, i.e., including the constraints originated from other species. The evolution and coevolution over a very long period have, however, implied that the species have been adapted to each other. They have by trial and error been able to find that they can move further away from thermodynamic equilibrium (get more growth) if they cooperate by adjusting their properties to each other. The effect of this cooperation is consistent with Patten (1991), where it is shown that the indirect effect often exceeds the direct one and for instance a predator-prey relationship also may be beneficial for the prey due to a number of factors including faster circulation of the nutrients.

On the other hand, a system cannot in principle and from a mathematical point of view be optimized with relation to two or more goals simultaneously. If two (or more) goals (for instance biomass of all the species) are at the maximum at the same time, it may just be a coincidence. Due to the coevolution and the adaptation of the species to each other, it will however not be surprising if the optimum biomass of for instance two or more species forming a food chain are almost coinciding.

The usual way to solve a multi-goal-optimization is by the use of weighting factors. In this context it would therefore be pertinent to use optimization of exergy indicating the distance from thermodynamic equilibrium, which as it will be shown below, can by approximations be computed as:

$$Ex = \Sigma \beta_i c_i \qquad (2.4.7)$$

where c_i is the biomass concentration of species i and β_i is the weighting factor expressing the information that the i'th species is carrying.

Darwin's Theory presumes, that populations consist of individuals, which:

1. on average produce more offspring than needed to replace them upon death – this is the property of high reproduction. Translated into thermodynamics: more possible pathways for the utilization of the energy flow are developed than the system and its energy can sustain. It implies that a competition among the pathways even among those, that are only slightly different, will be established.
2. have offspring, which resemble their parents more than they resemble randomly chosen individuals in the same population – this is the property of inheritance. Thermodynamically, it means that the properties, that have shown a better ability to utilize the energy flow to move as far away from thermodynamic equilibrium as possible by construction of more biomass to a high extent will be preserved. Genetics can explain, how this is possible.
3. vary in heritable traits influencing reproduction and survival due to differences in fitness to the prevailing conditions – this is the property of variation. The modernized Neo-Darwinism is able to give a long list of mechanisms, that can create new pathways. It implies that new possibilities are steadily created to meet the challenge of utilizing the energy flow. These possibilities

are tested under the prevailing conditions, and the successful ones are preserved according to 2. The evolution can therefore continue on the shoulders of the already found successful solutions and therefore steadily find new and better solutions, i.e., select the best genes, among all the present genes including the ones, that are continuously emerging by mutations and sexual recombinations.

It implies that the properties will be changed currently by the selection process to give the best possible survival at the prevailing conditions which for plants include the grazing and for the grazers the availability of food. The species cannot change the properties of the other species directly, but only because all species must consider all the other species in their effort to find a feasible combination of properties that is able to offer a higher probability of survival. This explains how the species get adapted to each other, and can cooperate on the joined goal to move as much as possible away from thermodynamic equilibrium. In principle, each of the species are striving toward their own goal: to get the highest possible growth for their own species. As these goals cannot be reached if the species don't adapt to the other species, because they are a part of the life conditions, the result will be that the species together move as much as possible away from thermodynamic equilibrium, i.e., give the system the highest possible exergy. It will in many cases coincide with the highest or closed to the highest biomass for most of the species.

These considerations are expressed in the following formulation of the tentative fourth law of thermodynamics: *If a system receives a through-flow of exergy, the system will utilize this exergy to move away from thermodynamic equilibrium. If there are offered more pathways than one to move away from thermodynamic equilibrium, the one yielding most stored exergy, i.e. with the most ordered structure or the longest distance to thermodynamic equilibrium by the prevailing conditions, will be selected.*

As it is not possible to prove the first three laws of thermodynamics by inductive methods, the tentative fourth law can at the best be proved by deductive methods. It implies that the fourth law should be investigated in as many cases as possible. Several modeling cases have been examined and they have all approved the tentative law; see Jørgensen (1986, 1988, 1990 and 1992a and b). The law has particularly been used successfully to develop models with dynamic structures, which will be mentioned in Nielsen and Jørgensen Sect. 2.7.

As seen from these contemplations the tentative fourth law of thermodynamics may be described as an extension of Darwin's theory from the species to the ecosystem level, applying the language of thermodynamics, which is useful in quantitative ecology and ecological modeling. The hypothesis has been supported by several modeling studies and has been a useful approach in the development of models with dynamic structures. Many more studies are needed to offer a full acceptance of the hypothesis or even better; experiments for instance in microcosms should be carried out to attempt to violate the hypothesis.

Ecosystems are open systems and receive an inflow of solar energy. It carries low entropy, while the radiation from the ecosystem carries high entropy.

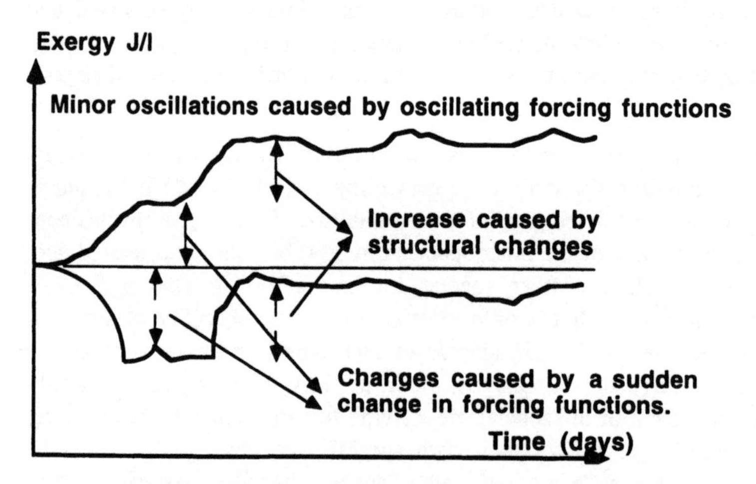

Fig. 2.4.2. Exergy response to increased and decreased nutrient concentration

If the power of the solar radiation is W and the average temperature of the system is T_1, then the exergy gain per unit of time, ΔEx is (Eriksson et al. 1976):

$$\Delta Ex = T_1 \cdot W\,(1/T_0 - 1/T_2), \qquad (2.4.8)$$

where T_0 is the temperature of the environment and T_2 is the temperature of the sun. This exergy flow can be used to construct and maintain structure far away from equilibrium.

Notice that the thermodynamic translation of Darwin's theory requires that populations have the above mentioned properties of reproduction, inheritance and variation. The selection of the species that contribute most to the exergy of the system under the prevailing conditions requires that there are enough individuals with different properties that a selection can take place – it means that the reproduction and the variation must be high and that once a change has taken place due to better fitness it can be conveyed to the next generation.

Notice furthermore that the change in exergy is not necessarily ≥ 0, it depends on the changes of the resources of the ecosystem. The proposition claims, however, that the ecosystem attempts to reach the highest possible exergy level under the given circumstances and with the available genetic pool ready for this attempt (Jørgensen and Mejer 1977 and 1979). Compare Fig. 2.4.2, where the reactions of exergy to an increase and a decrease in nutrient concentrations are shown.

2.4.4 Computation of Exergy

It is not possible to measure exergy directly – but it is possible to compute it according to equation (2.4.4), if the composition of the ecosystem is known.

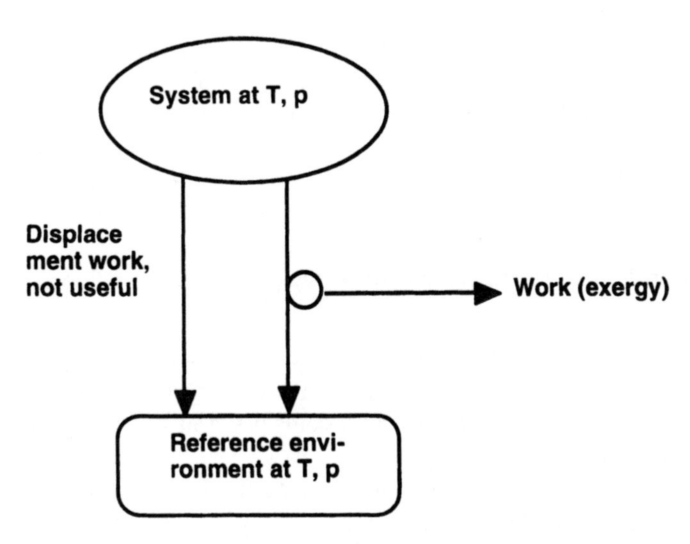

Fig. 2.4.3. The exergy content of the system is calculated in the text for the system relatively to a reference environment of the same system at the same temperature and pressure, but as an inorganic soup with no life, biological structure, information or organic molecules

If we presume a reference environment that represents the system (ecosystem) at thermodynamic equilibrium, which means that all the components are inorganic at the highest possible oxidation state (as much free energy as possible is utilized to do work) and homogeneously distributed in the system (no gradients), the situation illustrated in Fig. 2.4.3 is valid. As the chemical energy embodied in the organic components and the biological structure contributes by far most to the exergy content of the system, there seems to be no reason to assume a (minor) temperature and pressure difference between the system and the reference environment. Under these circumstances we can calculate the exergy content of the system as coming entirely from the chemical energy: $\Sigma_c (m_c - \mu_{co})$ Ni. We find by these calculations the exergy of the system compared with the same system at the same temperature and pressure but in form of an inorganic soup without any life, biological structure, information or organic molecules.

As (m_c - μ_{co}) can be found from the definition of the chemical potential replacing activities by concentrations, we get the following expressions for the exergy:

$$Ex = RT \sum_{i=0}^{i=n} C_i \ln C_i / C_{i,0} \tag{2.4.9}$$

where R is the gas constant, T is the temperature of the environment (and the system; see Fig. 2.4.3), while C_i is the concentration of the i'th component expressed in a suitable unit, e.g. for phytoplankton in a lake C_i could be expressed as mg/l or as mg/l of a focal nutrient. $C_{i,o}$ is the concentration of the i'th component at thermodynamic equilibrium and n is the number of components. $C_{i,o}$ is of course a very small concentration (except for i = 0, which is considered to cover the inorganic compounds), corresponding to a very low probability of forming complex organic compounds spontaneously in an inorganic soup at thermodynamic equilibrium.

It is important to underline that *all* computations of exergy have the following shortcomings:

- The computations will be based upon either a model or a limited number of measurements The results of the computations are therefore more appropriate for finding a *relative* difference in exergy by a comparison of an ecosystem under different conditions.
- The calculations – as all calculations in thermodynamics – are based upon approximations and assumptions. But as we draw conclusions on the basis of the differences in exergy rather than on the basis of absolute values, the results may be applicable in ecosystem theoretical context.

In addition, the application of equation (2.4.9) for the computation of exergy assumes the reference state shown in Fig. 2.4.3. It implies that the computed exergy will be entirely related to the chemical composition and the biological structure with its information. In most ecosystem cases the exergy found by this method is also the major contribution, as only minor temperature and pressure differences are realistic between an ecosystem and its environment.

The theory behind the application of exergy may be correct – but should of course be considered a hypothesis at this stage, but the practical application of the theory to real ecosystems will suffer from the above mentioned shortcomings.

The problem by application of equation (2.4.9) is related to the size of c_{io}. The problem related to the assessment of c_{io} has been discussed and a possible solution proposed in Jørgensen et al. (1995). For dead organic matter, detritus, which is given the index 1, it can be found from classical thermodynamics that:

$$\mu_1 = \mu_{1o} + RT \ln c_1 / c_{1o} \tag{2.4.10}$$

Where μ indicates the chemical potential. The difference $\mu_1 = \mu_{1o}$ is known for organic matter, e.g., detritus (Morowitz 1968), which is a mixture of carbohydrates, fats and proteins. If we use these figures, we get approximately 18 kJ/g detritus.

Generally, c_{io} can be calculated from the definition of the probability P_i to find component i at thermodynamic equilibrium:

$$P_i = c_{io} / \sum_{i=0}^{N} c_{io} \qquad (2.4.11)$$

If we can find the probability, P_i, to produce the considered component i at thermodynamic equilibrium, we have determined the ratio of c_{io} to the total concentration. As the inorganic component, c_0, is very dominant by the thermodynamic equilibrium, equation (2.4.11) may be rewritten as:

$$P_i \approx c_{io} / c_{0o} \qquad (2.4.12)$$

By a combination of equations (2.4.10) and (2.4.12), we get:

$$P_1 = [c_1 / c_{0o}] \exp[-(\mu_1 - \mu_{1o}) / RT] \qquad (2.4.13)$$

For the biological components, 2,3,4....N, the probability, P_i, consists of the probability for producing the organic matter (detritus), i.e., P_1, and the probability, $P_{i,a}$, to obtain the information embodied in the genes, which again determine the amino acid sequences. Living organisms use 20 different amino acids and each gene determines the sequence of about 700 amino acids. $P_{i,a}$, can be derived from the number of permutations among which the characteristic amino acid sequence for the considered organism has been selected. It means that we have the following two equations available to calculate P_i:

$$P_i = P_1 P_{i,a} \quad (i \geq 2;\ 0 \text{ covers inorganic compounds and 1 detritus}) \qquad (2.4.14)$$

and

$$P_{i,a} = 20^{-700g} \qquad (2.4.15)$$

where g is the number of genes. Equation (2.4.11) is reformulated to:

$$c_{io} = P_i c_{0o} \qquad (2.4.16)$$

Equations (2.4.16) and (2.4.9) are combined:

$$Ex \approx R \cdot T \cdot \sum_{i=0}^{N} [c_i \cdot \ln(c_i / (P_i c_{0o}))], \qquad (2.4.17)$$

This equation may be simplified by the use of the following good approximations for $i \geq 1$ (based upon $P_i << c_i$, $P_i << P_0$ and $1/P_i >> c_i$, $1/P_i >> c_{0o} / c_i$): $c_i / c_{0o} \approx 1$, and the inorganic component can be omitted. We obtain:

$$Ex \approx R \cdot T \cdot \sum_{i=1}^{N} c_i \cdot \ln(1 / P_i) \qquad (2.4.18)$$

By a combination of this equations with equations (2.4.13) and (2.4.14), we obtain the following expression for the calculation of exergy:

$$Ex/RT = (\mu_1 - \mu_{1o}) \sum_{i=1}^{N} c_i / RT - \sum_{i=2}^{N} c_i \ln P_{i,a} \qquad (2.4.19)$$

This equation can now be applied to calculate exergy for important ecosystem components, that are known from measurements or models. The free energy released per g of organic matter is about 18 kJ/g. R is 8.4 J/mol K and the average molecular weight of detritus is assumed to be 100,000. We get the following contribution of exergy by detritus per liter of water, when we use the unit g/l for the concentrations and T = 300K:

$$Ex_1 = 18 \cdot [c_i] \text{ kJ/l or } Ex_1/RT = 7.14 \cdot 10^5 \; c_i \text{ g/l} \qquad (2.4.20)$$

A typical uni-cell green algae has on average 850 genes. We use purposely the number of genes and not the amount of DNA per cell, which would include unstructured and nonsense DNA, which is different for different organisms. In addition, a clear correlation between the number of genes and the complexity has been shown (Li and Grauer, 1991). If it is assumed that an alga has 850 genes and the organic matter in algae contributes 5.8 kcal/g (Morowitz, 1968), the contribution of exergy per liter of water, using g/l as concentration unit would be:

$$Ex_{algae}/RT = 9.6 \cdot 10^5 \; c_i - c_i \ln 20^{-595000} = 27.5.0 \cdot 10^5 \; c_i \text{ g/l} \qquad (2.4.21)$$

The contribution to exergy from a simple prokaryotic cell which carries 470 genes can also be calculated :

$$Ex_{prokar}/RT = 9.6 \cdot 10^5 \; c_i + c_i \ln 20^{329\,000} = 19.5 \cdot 10^5 \; c_i \text{ g/l} \qquad (2.4.22)$$

Organisms with more than one cell will have DNA in all cells determined by the first cell.

The number of possible microstates becomes therefore proportional to the number of cells. Zooplankton has 100,000 cells approximately and 15 000 genes per cell. ln P_{zoo} can therefore be found as:

$$-\ln P_{zoo} = -\ln (20^{-15000*700} * 10^{-5}) \approx 315 * 10^5 \qquad (2.4.23)$$

As seen the contribution from the numbers of cells is insignificant. Similarly, $P_{fish,a}$ and the P-values for other organisms can be found (use the figures in Table 2.4.1).

The application of these values for a model consisting of inorganic material, IM, phytoplankton, P, zooplankton, Z, fish, F, and detritus, D and the energy content, would yield:

$$Ex/RT = 0\,IM + P(1.79*10^6) + Z(31.5*10^6) + F(2.52*10^8) + D(7.14 \cdot 10^5) + P(9.6 \cdot 10^5) + Z(10 \cdot 10^5) + F(9.6 \cdot 10^5) \; [g/l] \qquad (2.4.24)$$

The contributions from phytoplankton, zooplankton and fish to the exergy of the entire ecosystem are significant and far more than corresponding to the biomass. Notice that the unit of Ex/RT is g/l. Exergy can always be found in Joules per liter, provided that the right units for R and T are used. Equation (2.4.24) can be rewritten by converting g/l to g detritus/l by dividing by $(7.14 \cdot 10^5)$:

$$Ex / RT = P\,(3.9) + Z\,(45.5) + F\,(367) + (D)$$
$$\text{[exergy as g detritus exergy equivalents / l]} \qquad (2.4.25)$$

As can be seen from the equations, exergy is dominated by the contributions coming from information, originated from the genes of the organisms. The total exergy of an ecosystem cannot be calculated exactly, as we cannot measure the concentrations of all the components of an ecosystem, but we can calculate the contributions from the dominant components, for instance by the use of a model, that covers the components that are most essential for a focal problem.

Table 2.4.1. Approximate number of non repetitive genes

Organisms	Number of information genes	Conversion factor *)
Detritus	0	1
Minimal cell (Morowitz, 1992)	470	2.7
Bacteria	600	3.0
Algae	850	3.9
Yeast	2000	6.4
Fungus	3000	10.2
Sponges	9000	30.
Moulds	9,500	32
Plants, trees	10,000-30,000	30-87
Worms	10,500	35
Insects,	10,000-15,000	30-46
Jellyfish	10,000	30
Zooplankton	10,000-15,000	30-46
Fish	100,000-120,000	300-370
Birds	120,000	390
Amphibians	120,000	370
Reptiles	130,000	400
Mammals	140,000	430
Human	250,000	740

Sources: Cavalier-Smith (1985), Li and Grauer (1991) and Lewin (1994).

*) based on number of information genes and the exergy content of the organic matter in the various organisms, compared with the exergy contained in detritus. 1 g detritus has about 18 kJ exergy (=energy which can do work).

The exergy is found as the concentrations of the various components multiplied by weighting factors reflecting the exergy that the various components possess due to their chemical energy and to the information embodied in the genes. The calculation of exergy accounts for the chemical energy in the organic matter as well as for the information that is originated in the extremely small probability to form the living components, for instance algae, zooplankton, fish, mammals and so on spontaneously from inorganic matter. The weighting factors may also be considered as quality factors reflecting how developed the various groups are.

The calculations are also consistent with the classical application of thermodynamics on chemical equilibria. If we, for instance, consider the chemical reaction: zooplankton + oxygen <--> carbon dioxide + water + nutrients, the equilibrium constant, K, can in principle be defined in the usual way and the very low zooplankton concentration due to the low probability of the presence of zooplankton at thermodynamic equilibrium, will be reflected in a huge K-value, which can be translated to a high free energy = exergy in this case.

In this context information contributes considerably to the exergy of the system. This is, however, completely according to Boltzmann (1905), who gave the following relationship for the work, W, that is embodied in information:

$$W = R\,T \ln N \tag{2.4.26}$$

where N is the number of possible states, among which the information has been selected. N is the inverse of the probability to obtain the valid amino acid sequence spontaneously. It is furthermore consistent with Reeves's formulation (1991): "information appears in nature when a source of entropy becomes available but its (entire) entropy content is not emitted immediately."

The presented calculation does not include the information embodied in the structure of the ecosystem, i.e., the relationships between the various components, which is represented by the network

The information of the network encompasses the information of the components and the relationships of the components. The latter is calculated by Ulanowicz (1991) as a contribution to ascendancy. In principle, the information embodied in the network should be included in the calculation of the exergy of the entire ecosystem, but can often be omitted due to the following points:

1. The contributions from the network relationships (not from the components of the network, of course) are minor, compared with the contributions from the components. This is particularly true for models of networks, which are always extreme simplifications of the real network, although they attempt to account for the major flows of energy or mass.
2. In most cases a relative value of the exergy is sufficient to describe the direction of ecosystem development and growth.
3. In most cases the observed development is a change of the components not of the network structure. If the network is changing in addition to the compo-

nents (the knots of the network), it should be considered what this change would contribute to the exergy of the system.

4. The calculations of exergy will always be an approximation focusing on the most important components with respect to the changes taking place. A model (e.g. a conceptual model) is often used as a basis for these calculations, and a model is always a simplification of the real system, anyhow.

It may be concluded from these calculation that an approximate, relative measure of the exergy per unit of volume or per unit of area of the most important components can be assessed relatively easy as the sum of the concentrations multiplied by weighting factors that express the content of exergy in the organic matter and the information embodied in the various organisms. The contribution from inorganic components (component 0) is 0, and from dead organic matter (detritus) it is about 18 kJ/g. If the latter contribution is expressed as exergy equivalents of detritus, it is of course 1 g exergy equivalents/g. Exergy should always be considered a relative measure, in the sense that two different situations may be compared, as instance the exergy, resulting from two different sets of forcing functions. This is important for the use of these concepts in development of structural dynamic modelling (see Nielsen and Jørgensen Sect. 2.7) or for the assessment of ecosystem health.

We can distinguish two changes in exergy: a change caused by the external factors directly and a change caused by the response of the living organisms to the external factors. The former is related to the available resources in the ecosystem. If the phosphorus concentration is increased or decreased, the exergy will also increase or decrease. The latter change in exergy is caused by the effort of the organisms to survive and reproduce and will therefore reflect the many regulation mechanisms that an ecosystem and its organisms possess. Any change in the species composition or the ecological structure will therefore imply that the new structure and composition better fitted to the emerging conditions of the ecosystem.

In other words, whenever the external factors are changed, we observe a change in exergy, ΔEx, which can be expressed as:

$$\Delta Ex = \Delta Ex_E + \Delta Ex \qquad (2.4.27)$$

where the subscript E refers to the changes caused by the external factors directly, while the subscript I represents the effort of the organisms by adaptation to the new conditions (including those waiting in the wings) to get the best possible growth and reproduction out of the circumstances. ΔEx_E may be negative or positive, while ΔEx_I will always be ≥ 0 and the species giving the highest value will win (meaning it will be selected).

The idea with this formulation is that all externally caused changes may give any possible change in the exergy, but the effort of the ecosystem to "get the best" out of the situation including recovery after a stress situation always gives a contribution to exergy ≥ 0.

This effort can be expressed by calculation of the specific exergy which can be found from the weighting factors as follows:

$$Ex_{specific} / RT = \sum_{i=1}^{i=n} X_i \ln (X_i / X_{eq,i}) = \sum_{i=1}^{i=n} X_i \, \beta_i \tag{2.4.28}$$

$Ex_{specific}$ or also some times named $Ex_{structural}$ is as seen independent on the available resources, which is expressed by the total concentration C_{0o}. The specific exergy is therefore able to account for how well the ecosystem has been able to utilize the available resources, independent on the amount of resources.

2.4.5 The Tentative Fourth Law of Thermodynamics

It now seems feasible from the previous consideration to tentatively formulate a fourth or ecological law of thermodynamics, which may be considered a core law of ecosystem ecology and should therefore be repeated in this context: *If a system has a through-flow of exergy, it will attempt to utilize the flow to increase its exergy, i.e., to move farther away from thermodynamic equilibrium; if more combinations and processes are offered to utilize the exergy flow, the organization that is able to give the system the highest exergy under the prevailing conditions and perturbations will be selected.*

The first "exergy" in the formulation of the ecological law of thermodynamics shown above may be replaced by "low-entropy energy". Some may like to replace it by "negentropy", but as the classical thermodynamic expressions cannot be rigorously defined for systems far from equilibrium, preference should go toward using exergy. The second "*exergy*" (written in italics), may eventually be replaced by "order, information, maximum power, dissipation of energy or exergy, or dissipation of gradients".

A fundamental law cannot be proved but must be supported from many sides and be a workable "model" in many contexts. A fundamental law should facilitate the explanation of our observations, simplify and fit the overall scientific pattern and be consistent with all other laws. The tentative fourth law of thermodynamics seems to fulfill these requirements. In addition, it has been possible to show analytically in some specific ecosystem cases, that exergy follows the proposition given above (Jørgensen et al. 1992a, 1997). It furthermore bridges the gap between thermodynamics and Darwin's theory, which is considered an important feature of the law. It has therefore a strong support, but should be considered only tentatively anyhow. The coming years will reveal, if the hypothesis fits into the overall ecosystem-theoretical pattern, that will result from further research and development in the field. That is the real and needed test for a new fundamental law.

The energy flow through an ecosystem renders it possible to realize processes that require energy, as for instance construction of complex biochemical compounds: energy and simple inorganic compounds are converted to complex biochemical molecules. This is completely according to the first and second law of thermodynamics. However, nature offers many pathways for such processes that are competing and concerned with the selection process under the prevailing conditions, which are continuously changed.

Energy is in most cases not limiting, which may be deduced from the low energy efficiency of photosynthesis: only 2% of the solar radiation is converted to free energy of the produced organic matter. Shading may, of course, play a role and the shape of many plants and trees can be explained by an effective way to escape shading.

The question is which of the many biochemical pathways will win, or rather, which combination of pathways will win? The different pathways compete, however, in a very complex way, because the processes are many and complex and are therefore dependent on many factors such as:

1. at least 20 elements, of which some may be limiting,
2. competition from the other pathways,
3. temperature,
4. light (for photosynthetic pathways only),
5. ability to utilize the combined resources.

The conditions are furthermore varying in time and space. This implies that the history of the system also plays a role in selection of the organization that gives the highest exergy. Two systems with the same prevailing conditions will consequently not necessarily select the same species and food webs, because the two systems most probably have different histories.

Decomposition rates are of great importance, as they determine not only the ability of the different products (organisms) to maintain their concentrations, but also at which rate the inorganic compounds are recycled and can be reused. It is not surprising from this description that:

1. *There is room for "survival" of many pathways*, considering the heterogeneity in time and space, the many simultaneously determining factors and the many developed mechanisms to utilize different resources in the ecosystems.
2. *The competition is very complex and there are many possible pathways*. The description of the selection will therefore be very complex, too.
3. *As everything is linked to everything* in an ecosystem, it is necessary to look at the entire system. This implies that we have to find the combination of pathways - among the possible ones – that are able to move the entire system furthest away from the thermodynamic equilibrium – measured by the exergy of the total system. Because everything is linked to everything, it is obvious that every component in an ecosystem must consider the influence of all other components. The selection pressure comes from the forcing functions as well as from the other components. This explains the coevolution and the development of significant indirect effect.

4. It is not surprising that the highest exergy also means the highest ability to dissipate gradients and produce entropy, because the most developed system will require the most energy / exergy for maintenance, i. e., respiration (Kay and Schneider, 1991).
5. *Maintenance and development of biomass are extremely important for storage of the information level already achieved, i.e., to work on the "shoulders" of previous results. Without the ability to store information already gained it would be impossible to explain the rate of evolution* or, rather, there wouldn't have been evolution at all. The role of information storagemay be illustrated by a simple example. This book contains about 1,000,000 signs. If a chimpanzee should write the book by touching a keyboard (of, let us say, 50 different keys) randomly with a rate of let us say 1,000 000 signs per day, the probability that the book would be finished in one day would be $50^{(-1,000\ 000)}$. Even if the chimpanzee had worked since the Big Bang 15 billions years ago, the probability would still be less than 0.000 (more than 1.5 million zeros)...1. If on the other hand we preserve each time the chimpanzee has tried to type the book (1,000,000 signs) the signs correctly placed and next time let the chimpanzee only try to find randomly the incorrect signs and so on, there would be a probability close to one that the book would be finished in about 200 days!! The test on what is right and wrong in nature is carried out by the selection of the properties guaranteeing survival and growth and the genes preserve the results already achieved. The development of a mechanism to maintain information already gained has been crucial for the rate of evolution.

The interpretation of the proposed Fourth Law of Thermodynamics by use of models and reaction kinetics is a parallel to Dawkins' "Selfish Gene" (Dawkins 1989). The selfish gene produces survival machines to be able to protect the information stored in the gene and thereby maintains the level of information (exergy/biogeochemical energy) achieved.

Dawkins uses the expression "replicator" to underline the importance of the replication process. He talks about the gradual improvement by the replicators to ensure their own continuance in the world. There is no contradiction to the Gaia hypothesis. The existence of the selfish replicators is not surprising, when we consider that cooperation is generally more beneficial (selfish) than competition and it is probably true that the more complex the system becomes, the more beneficial the cooperation is.

The next obvious question may be: Which factors determine the ability of the system to move further away from the thermodynamic equilibrium, in order to obtain a higher level of information and more structural biogeochemical energy (exergy)? The evolution of ecosystems has created ever more possibilities to utilize the opportunity offered to the ecosystems by input of energy from solar radiation. The factors of importance in this context will be listed and commented on below.

The abilities of the ecosystems to utilize the through-flows of energy or exergy are rooted in the following properties:

1. *The variety of the gene pool.* The more genes, the more possibilities are given to find a better solution to obtain even more exergy. Furthermore, the more genes, the more mutations and sexual recombinations will occur and the more new possibilities the system will get to move further away from the equilibrium. It implies that the gene pools do not only determine the possibilities to get the highest possible exergy today but also determine how to get even better possibilities in the future.
2. *The chemical composition of the ecosystem.* The better the chemical composition of the ecosystem matches the needs of the biological components, the better the ecosystem will be able to utilize the energy flow. This may explain why the tropical rain forest has an enormous diversity. The chemical composition (and the temperature, see point 4) in tropical rain forests is almost ideal for growth, including the presence of the most important compound on earth: water.
3. *The temperature pattern.* The rate of utilization of the energy flow is dependent on the temperature. The closer the temperature pattern matches the optimum for growth and reproduction and vice versa, the higher the rate of utilization and the faster the system will be able to move away from thermodynamic equilibrium.
4. *Fluctuations and other changes of the forcing functions.* The changes of forcing functions will steadily pose new questions for the ecosystem: how to get the best survival and growth under the prevailing conditions just now? This challenge, provided that the system is not brought outside its framework of ability, will create new information. The natural fluctuations of the forcing functions have been governing (with some exceptions, of course) for billions of years. The genes have therefore been able to cope with these fluctuations and even understand to benefit from them by moving further away from the thermodynamic equilibrium - compare the role, that fluctuations may play on the overall biomass.

A dissipative structure is organized in a manner that it increases its internal exergy and dissipates more efficiently the flow of exergy, which traverses it. This implies a maximum accumulation of exergy in the system, expressed biologically in the law of growth of an organism and in population dynamics.

Population dynamics deals with the spatio-temporal evolution of living species in relation to each other. The concepts of ecological niche, territorially and of different species at the same site are examples of spatio-temporal populations. An example in point is afforded by the study of populations of plankton in the ocean (Dubois 1974). Plankton lives and develops in an environment which is submitted to currents and turbulent diffusion. Despite this random environment these populations develop according to a heterogeneous spatial structure over distances of 5 to 100 km with a life span of several weeks. This is the so-called patchiness effect, a spatial distribution of populations according to various geometrical figures. This order is remarkable, considering how water is constantly disturbed; a disturbance which will lead to a homogeneous spreading of the populations. Moreover, there is no correlation between this phenomenon and the chemical and physical properties of the environment such as nutrients, oxygen, salinity, light and temperature.

Dubois (1975) theoretically explains the emergence of patchiness. He suggests a competition between ecological interactions and the environment. In this approach a non-linear interaction of the prey and predator competes with transport phenomena by advection due to residual current and turbulent diffusion. The initiation of the patchiness effect is the result of the instabilities created by the advection.

If mankind on the other hand changes the forcing functions outside the natural limits the ecosystems, the challenge may be too difficult to meet for the ecosystem in spite of its well developed structure. It is what we experience today as the consequences of pollution problems.

Moving away from thermodynamic equilibrium may take the form of increasingly complicated chemical and physical structures, but a further antientropic movement requires that what has been already achieved is maintained, stored and built upon for further improvements. Therefore information and storages of information are so important. The great steps forward in evolution took place, when the genes were developed, when the coordination among more cells to form organisms was developed and when learning processes became an integrated part of the species properties. Pathways with the ability to accelerate the exchange of information among them were formed.

It is noticeable that information may create exergy and that through information the system will be able to move further away from the thermodynamic equilibrium. The more information the more pathways are available to increase exergy.

Information has a very low cost of energy. Developments of physical and chemical structures will increase the exergy, but how high an exergy level we are able to achieve is limited by energy and matter. Information has not these limitations, because information can be multiplied almost infinitely, since it does not or almost not consume energy and matter.

It is therefore not surprising that information has been used as the method of ecosystem and society to increase the level of exergy. Ecosystems by development of ever more species with an ever increasing ability made it possible to coordinate the functions of many cells, organs and organisms simultaneously.

This could be interpreted as the ecosystems' aim toward higher and higher complexity, but complexity may not be interchanged for information. A more complex ecosystem may be unable to cope with a given combination of forcing functions, which a simpler system can manage – for instance, a very eutrophic lake may have a simpler food web than an oligotrophic lake.

The crucial question in this context is not "Which structure is most complex?", but "Which structure gives the highest level of useful information storage?"

On the other hand it is important to emphasize that the diversity of the gene pool is important for the ability to create new possibilities for moving further away from the thermodynamic equilibrium and attain higher buffer capacities for the ecosystem. Therefore it is of great importance that we maintain the existing gene pool and maintain natural ecosystems for developments of new genes.

2.4.6 Concluding Remarks

The exergy principle has been presented as a tentative thermodynamic law, which may be used as a hypothesis in our effort to find a pattern of the presented ecosystem theories. The presented tentative law or hypothesis fits furthermore nicely into basic concepts of the other thermodynamic laws. The first law states the limitations in all possible processes by the conservation principles. The second law makes further limitations by introduction of the entropy concept and states that it is only possible to realize processes moving toward a higher entropy level in an isolated system. The tentative fourth law of thermodynamic asserts further limitations and indicates, which processes are biologically feasible, namely those (among many possible ones), which give the highest exergy under the prevailing conditions and perturbations in their widest sense. It will require a through-flow of exergy to realize these processes to combat the entropy production according to the second law of thermodynamics. The applicability of the law is therefore of particular interest for ecosystems, as these systems are characterized by many possible pathways and a through-flow of exergy.

References

Boltzmann L (1905) The Second Law of Thermodynamics. Populäre Schriften, Essay No. 3 (address to Imperial Academy of Science in 1886). Reprinted in English in Theoretical Physics and Philosophical Problems, Selected Writings of L Boltzmann. D. Reidel, Dordrecht

Cavalier-Smith T (1985) The Evolution of Genome Size. John Wiley and Sons, Chichester

Dawkins RD (1989) The Selfish Gene, 2nd Ed. Oxford University Press, Oxford

Eriksson B, Eriksson KE, Wall G (1976) Basic Thermodynamics of Energy Conversions and Energy Use. Institute of Theoretical Physics, Göteborg, Sweden

Jørgensen SE, Mejer HF (1977) Ecological buffer capacity. Ecological Modelling 3:39-61

Jørgensen SE, Mejer HF (1979) A holistic approach to ecological modelling. Ecol Modelling 7:169-189

Jørgensen SE (1982) A holistic approach to ecological modelling by application of thermodynamics. In: Mitsch W et al. (eds) Systems and Energy. Ann Arbor, Michigan, pp 61-72

Jørgensen SE (1986) Structural dynamic model. Ecological Modelling 31:1-9

Jørgensen SE (1988) Use of models as experimental tool to show that structural changes are accompanied by increased exergy. Ecological Modelling 41:117-126

Jørgensen SE (1990) Ecosystem theory, ecological buffer capacity, uncertainly and complexity. Ecological Modelling 52:125-133

Jørgensen SE (1992a) Integration of Ecosystem Theories: A Pattern: Kluwer Academic Publishers, Dordrecht, Boston, London

Jørgensen SE (1992b) Development of models able to account for changes in species composition. Ecological Modelling 62:195-209

Jørgensen SE (1994) Review and comparison of goal functions in system ecology. Vie Milieu 44 (1):11-20

Jørgensen SE (1997) Integration of Ecosystem Theories: A Pattern. 2nd Ed. Kluwer, Dordrecht

Lewin B (1994) Genes V. Oxford University Press, Oxford

Li W-H, Grauer D (1991) Fundamentals of Molecular Evolution. Sinauer Associates, Inc Publ Sunderland, Massachusetts

Mejer HF, Jørgensen SE (1979) Energy and ecological buffer capacity. In: Jørgensen SE (ed) State-of-the Art of Ecological Modelling. Environmental Sciences and Applications. 7. Proc Conf Ecological Modelling, 28th August 2nd September 1978, Copenhagen. International Society for Ecological Modelling, Copenhagen, pp 829-846

Morowitz HJ (1968) Energy Flow in Biology. Academic Press, New York

Odum HT (1983) Systems Ecology. Wiley Interscience, New York

Schrödinger E (1944) What is Life? Cambridge University Press, Cambridge

Straškraba M (1979) Natural control mechanisms in models of aquatic ecosystems. Ecological Modelling 6:305-322

Straškraba M (1980) The effects of physical variables on freshwater production: analyses based on models. In: Le Cren ED, McConnell RH (eds) The Functioning of Freshwater Ecosystems: International Biological Programme 22. Cambridge University Press, Cambridge, pp 13-31

Straškraba M, Gnauck A (1983) Aquatische Ökosysteme - Modellierung und Simulation: VEB Gustav Fischer, Jena. English Translation: Freshwater Ecosystems - Modelling and Simulation, Developments in Environmental Modelling 8. Elsevier, Amsterdam

Ulanowicz RE (1991) Formal agency in ecosystem development. In: Higashi M, Burns TP (eds) Theoretical Studies of Ecosystems, The Network - Perspective. Cambridge University Press, Cambridge, pp58-114

2.5 Thermodynamic Orientors: Exergy as a Holistic Ecosystem Indicator: A Case Study

João C. Marques, Miguel Ã. Pardal, Søren N. Nielsen and Sven E. Jørgensen

Abstract

Benthic eutrophication may cause qualitative changes in marine and estuarine ecosystems, for example the shift in primary producers. Subsequently, changes in species composition and trophic structure at other levels may often occur, and through time a new trophic structure might be selected. In structurally dynamic models such changes may be simulated using goal functions to guide ecosystem behavior and development. The selection of other species and other food webs may then be accounted by a continuous stepwise optimization of model parameters according to an ecological goal function.

Exergy has been applied as goal function in structurally dynamic models of shallow lakes. Hypothetically, exergy is assumed to become optimized during ecosystems development. Therefore, ecosystems are supposed to self-organize towards a state of an optimal exergy configuration. Exergy may then constitute a suitable system-oriented characteristic to express natural tendencies of ecosystems development, and simultaneously a good ecological indicator of ecosystems health.

Biodiversity, a powerful and traditional concept, is also an important characteristic of ecosystems structure. We found it suitable to test the intrinsic ecological significance of exergy. Therefore, we examined the properties of exergy (exergy and specific exergy) and biodiversity (species richness and heterogeneity) along an estuarine gradient of eutrophication, testing the hypothesis that they would follow the same trends in space and time. This hypothesis was only partially validated, since exergy, specific exergy and species richness decreased as a function of increasing eutrophication, but heterogeneity behaved mostly in the opposite way. Nevertheless, exergy and specific exergy behaved as hypothesized, providing useful information regarding the studied communities. They appeared therefore suitable to be used as goal functions in ecological models and as holistic ecological indicators of ecosystem integrity. Moreover, since exergy and specific exergy showed to respond differently to ecosystem seasonal dynamics, we recommend using both as complementary parameters.

F. Müller M. Leupelt (Eds.)
Eco Targets, Goal Functions, and Orientors

The method proposed by Jørgensen et al. (1995) to estimate exergy, which takes into account the biomass of organisms and the thermodynamic information due to genes, appeared to be operational, but more accurate (discrete) weighting factors to estimate exergy from organisms biomass need to be estimated. We propose to explore the assumption that the dimensions of active genomes, which are primarily a function of the required genetic information to build up an organism, are proportional to the relative contents of DNA in different organisms.

2.5.1 The Rationale

Changes in environmental factors often give origin to qualitative modifications in the ecosystems, such as species composition and biodiversity. For example, eutrophication processes may cause a shift in primary producers, which may also determine changes in species composition at other trophic levels, like the faunal composition. Through time, a sequence of modifications may give rise to the selection of a new structure of the trophic network.

To model these types of qualitative changes, which describe the development of ecosystems as a response to changes in external factors, the qualitative trophic alterations through time must be included in the models. The more conventional deterministic models of aquatic ecosystems, although efficient in casuistry terms, due to lack of generality are difficult to apply from one system to another. Nevertheless, it seems possible to incorporate the type of change described above through the development of structurally-dynamic models (Jørgensen 1993; Nielsen 1992, 1994, 1995), which has recently been started. The dynamic element of the model is needed to account for the adaptational and selective processes. This new generation of models may improve the existing ones, not only in the sense of increasing their predictive capability, but also by gaining a better understanding of ecosystems behavior, and consequently providing a better tool for environmental management. In this sense, besides the development of structurally dynamic models, it might be necessary to use soft parameter sets.

By applying the principles of optimization theory in ecology, structurally dynamic models can use goal functions to guide the model simulation of ecosystem behavior and development (Nielsen 1995). In such a case, in the simulation, the selection of other species and the selection of another food web is accounted for by a change of model parameters according to an ecological goal function. In structurally dynamic models, parameters may be introduced which change as functions of changing forcing functions and conditions of state variables. Therefore, the ability of the ecosystem for instance to move away from thermodynamic equilibrium may be optimized by a stepwise approach.

In ecological models goal functions are assumed to measure given properties or tendencies of ecosystems, emerging as a result of self-organization processes in their development. However, contrary to what happens in the run of structurally dynamic models, natural ecosystem development and adaptation does not pursue a goal, in the teleological sense. Nevertheless, from an environmentally static assessment point of view, the same mathematical expressions may constitute suitable measures of system oriented characteristics for natural tendencies of ecosystem development, and good ecological quality indicators. Such measures may then act as an adequate interface between modelling, where they are used in the scope of optimization theory, and empirical ecology, where they are utilized as environmental indicators.

In the last fifteen years several algorithms have been proposed as possible goal functions. Exergy, a holistic concept derived from thermodynamics (Jørgensen and Mejer 1979, 1981), which can be seen as energy with a built in measure of quality, appears to be a promising approach. Actually, in shallow lake models, exergy exhibited a possible role in expressing shifts in species composition and trophic structure (Jørgensen, 1988; Nielsen 1990, 1995).

Hypothetically, exergy is assumed to become optimized during ecosystem development. In other words, ecosystems are supposed to self-organize towards a state where this property is optimized (Jørgensen 1992). There are nevertheless theoretical and practical problems to be solved before this concept can be entirely accepted and used in models for the management of nature and as a holistic ecological quality indicator. Three major questions are empirically approached in this paper:

- What are the relationships between exergy and a more conventional ecological indicator like biodiversity?
- In practice, how can estimations of exergy values be made operational?
- To what extent are estimated exergy values able to add useful information regarding the state of an ecosystem?

The point in analyzing the relationships between exergy and biodiversity was that this latter concept, although somehow illusory, is powerful and intuitive, constituting a good available tool to test the ecological significance of goal functions when describing the ecosystem state. For instance, although there might be other factors involved, it is commonly accepted that, within a given ecosystem, polluted areas, e. g. eutrophied zones, will exhibit less complex communities, with a less complex trophic structure, and lower biodiversity than non polluted areas. Biodiversity may then be considered as an indicator of the quality state of the ecosystem. On the other hand, in what way could exergy, as a holistic ecological quality indicator, be effective in discerning between distinct states of the ecosystem when differences are relatively subtle? We empirically tested the following hypothesis: exergy and biodiversity will follow the same trends in space and time along an estuarine gradient of eutrophication. The chosen study site was the Mondego estuary, in the western Atlantic coast of Portugal (Fig. 2.5.1).

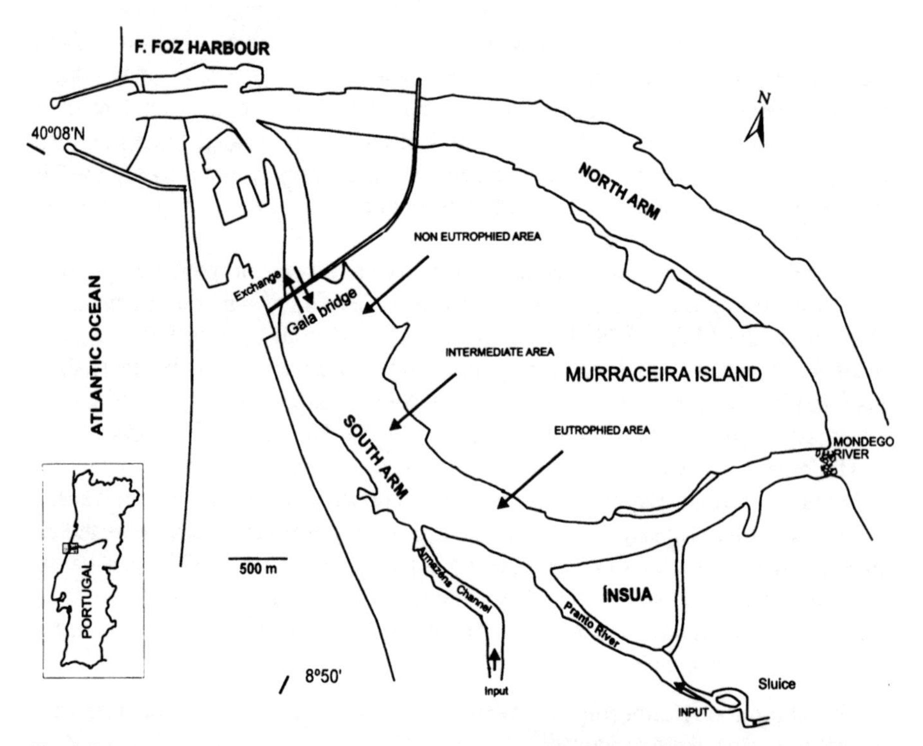

Fig. 2.5.1. The Mondego estuary: It consists of two arms, north and south, with very different hydrographic characteristics. The northern arm is deeper, while the southern one is almost silted up in the upstream areas. The water circulation in the southern arm is mostly due to tides and to the freshwater input of a tributary, the Pranto River, which is artificially controlled by a sluice. The southern arm of the estuary is eutrophied, and the gradient of eutrophication is indicated.

2.5.2 Materials and Methods

The benthic communities in the Mondego estuary were monitored fortnightly, from February 1993 to February 1994. Samples of macrophytes, macroalgae, and associated macrofauna were collected fortnightly at different sites, during low tide. The organisms were almost always identified to the species level, and their biomass was determined (g m^{-2} afdw). Corresponding to each biological sample, the following environmental factors were determined: salinity (g l^{-1}), temperature (^{o}C), pH, dissolved oxygen, ammonia, silica, chlorophyll a, nitrites, nitrates, and phosphates (mg l^{-1}), for the water, and organic matter contents (g m^{-2} afdw), for the sediments.

Sampling was performed along a gradient of eutrophication in the south arm of the estuary (Fig. 2.5.1), from a non eutrophied zone up to a heavily eutrophied zone, in the inner areas of the estuary. A *Zostera noltii* community is present in the non eutrophied zone and blooms of the green macroalgae *Enteromorpha* spp.

have been observed during the last decade in the heavily eutrophied zone. The pattern in the most eutrophied zone is that *Enteromorpha* spp. biomass normally increases from mid winter to early summer, when an algae crash usually occurs. A second but much less important algae biomass peak may be observed in late summer followed by a decrease in biomass through the fall season.

These measurements provided a comprehensive field data set on the spatial and temporal variation of benthic communities along the gradient of eutrophication, and was used to make exergy and biodiversity calculations.

To estimate exergy we used the method based upon the thermodynamic information due to genes proposed by Jørgensen *et al.* (1995). This method is considered the best candidate for exergy calculations of ecosystems, because it takes into account the organizational level of the organisms. Following this approach, an estimation of exergy may be given by

$$Ex = T \sum_{i=1}^{i=n} \beta_I C_i \qquad (2.5.1)$$

where T is the absolute temperature in Kelvin, Ci is the concentration of component i in the ecosystem (e. g. biomass of a given taxonomic or functional group), βi is a factor which roughly expresses the quantity of information embedded in the biomass. Detritus is chosen as the reference level, i.e. $\beta i = 1$, and exergy in the biomass of different types of organisms is expressed in detritus energy equivalents.

Consequently, the variation of exergy through time in an ecosystem may be caused from the variation of the biomass and information built in one unit of biomass (expressing the quality of the biomass)

$$\Delta Ex_{tot} = \Delta B_{iom}\, \beta i + \Delta \beta i\, B_{iom} \qquad (2.5.2)$$

If the total biomass ($Biom_{tot}$) in the system remains constant through time, then the variation of exergy (Ex_{tot}) will be a function of only the structural complexity of the biomass or, in other words, of the information embedded in the biomass. It may then be called specific exergy (SpEx), expressed as exergy per unit of biomass. For each instant, specific exergy is given by

$$SpEx = Ex_{tot} / Biom_{tot} \qquad (2.5.3)$$

Values of exergy and specific exergy were calculated from the biomass of the different organisms (g m^{-2} ash free dry weight – afdw) through the use of weighting factors that are able to discriminate different "qualities" of biomass (Table 2.5.1). For this purpose, taking into account the available set of weighting factors, data on organisms biomass was pooled as a function of higher taxonomic levels (e. g. Phylum or Class).

Table 2.5.1. The evolution of g DNA/cell, number of genes, and number of cell types (approximate figures are given) for different organisms. The concentration of each organism was multiplied by the proposed weighting factor to estimate exergy. The weighting factor accounts for the information embodied in the organism in addition to the simple biomass (g m^{-2} afdw). For this purpose, it is assumed that detritus (organic matter contents in sediments) does not contain relevant structural information. Sources: Li and Grauer (1991) in Jørgensen et al. (1995). All the values marked with * were not provided by any source but assumed as reasonable at the present state of knowledge, taking into account the evolutionary level of the groups concerned.

Organisms	10^{-12} g DNA/cell	Number of genes	Number of cell types	Weighing factor
Detritus	0	0	0	1
Bacteria	0.005	600	1 - 2	2.7
Algae	0.009	850	6 - 8	3.4
Yeast	0.02	2000	5 - 7	5.8
Fungus	0.03	3000	6 - 7	9.5
Sponges	0.1	9000	12 - 15	26.7
Plants, trees	-	10000 - 30000	-	30 to 90 *
Jellyfish	0.9	50000	23	144
Nemertineans	-	-	-	144*
Insects	-	-	-	144*
Crustaceans	-	-	-	144*
Annelid worms	20	100000	60	287
Molluscs	-	-	-	287*
Echinoderms	-	-	-	144*
Fish	20	100000 - 120000	70	287 - 344
Birds	-	120000	-	344
Amphibians	-	120000	-	344
Reptiles	-	120000	-	344
Mammals	50	140000	100	402
Human	90	250000	254	716

Regarding biodiversity, we took into consideration the species richness and the heterogeneity (species richness + evenness). For each date measurements were calculated using data on the species biomass (g m^{-2} afdw). For this purpose only macrofauna was taken into account. From the considerable assortment of indices designed by ecologists, we considered suitable the use of the Margalef index (I), to compute species richness, and of Shannon-Wienner's index (H') based on the information theory, to compute heterogeneity (Legendre and Legendre 1984; Magurran 1988).

The Margalef index is:

$$I = (n-1) / \log_e N \qquad (2.5.4)$$

where n is the number of species found and N is the total number of individuals, and the Shannon-Wienner index is given by:

$$H' = -\sum_{i=1}^{n} p_i \operatorname{Log} p_I \qquad (2.5.5)$$

where n is the number of species, and p_i is the proportion of the biomass of species i in a community were the species proportions are p_1, p_2, p_3, ...p_i, ...p_n.

Exergy and biodiversity values constitute static estimations of dynamic qualities of ecosystems. Therefore, to examine their properties as ecological indicators, a moving average (using the contiguously prior, present, and following dates) was applied to each value, in order to adjust potential bias due to sampling.

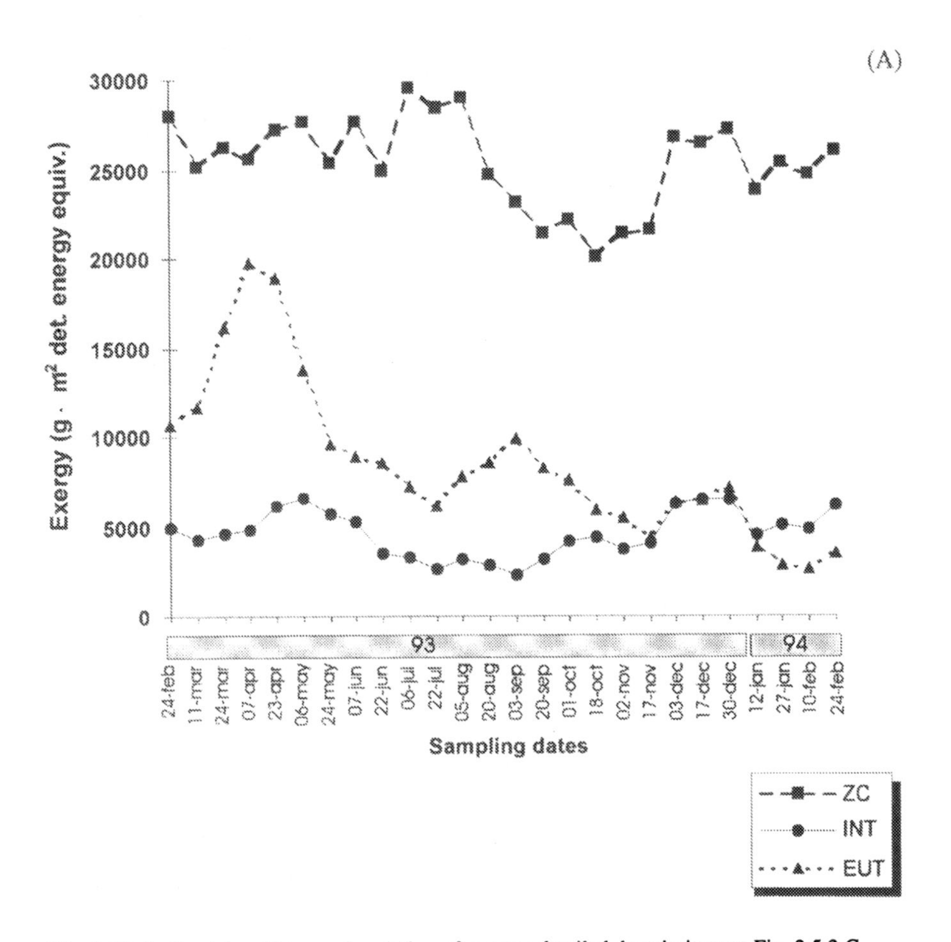

Fig. 2.5.2.A Spatial and temporal variation of exergy; detailed description see Fig. 2.5.2.C

2.5.3 Testing the Working Hypothesis

The spatial and temporal variation of exergy, specific exergy, species richness, and heterogeneity were analyzed, to test the hypothesis that these ecological indicators would capture changes in benthic communities in such a way that they would provide equivalent information about the ecosystem.

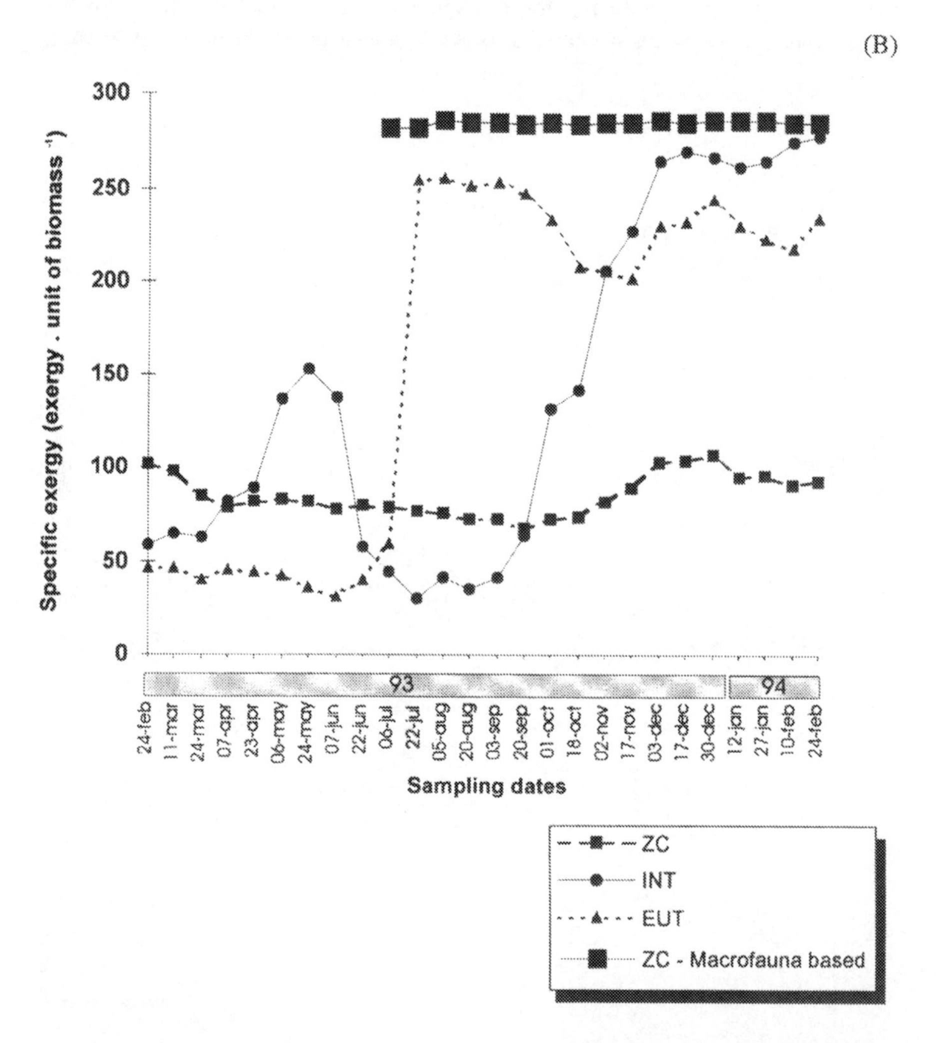

Fig. 2.5.2.B Spatial and temporal variation of specific exergy; detailed description see Fig. 2.5.2.C

Exergy (Fig. 2.5.2.A), as hypothesized, was found to be consistently higher in the *Zostera noltii* community than in the eutrophied areas. Additionally, during the spring and early summer of 1993, exergy values were higher in the most heavily eutrophied area when compared with the intermediate eutrophied area. This was obviously related with the intensity of the *Enteromorpha* bloom, determining much higher values for total biomass in the most eutrophied area (Fig. 2.5.2.C). Specific exergy was also consistently higher in the *Zostera noltii* community than in the eutrophied areas until late spring (Fig. 2.5.2.B), in accordance with the working hypothesis. However, the picture changed completely from early summer, when values became higher in the eutrophied areas.

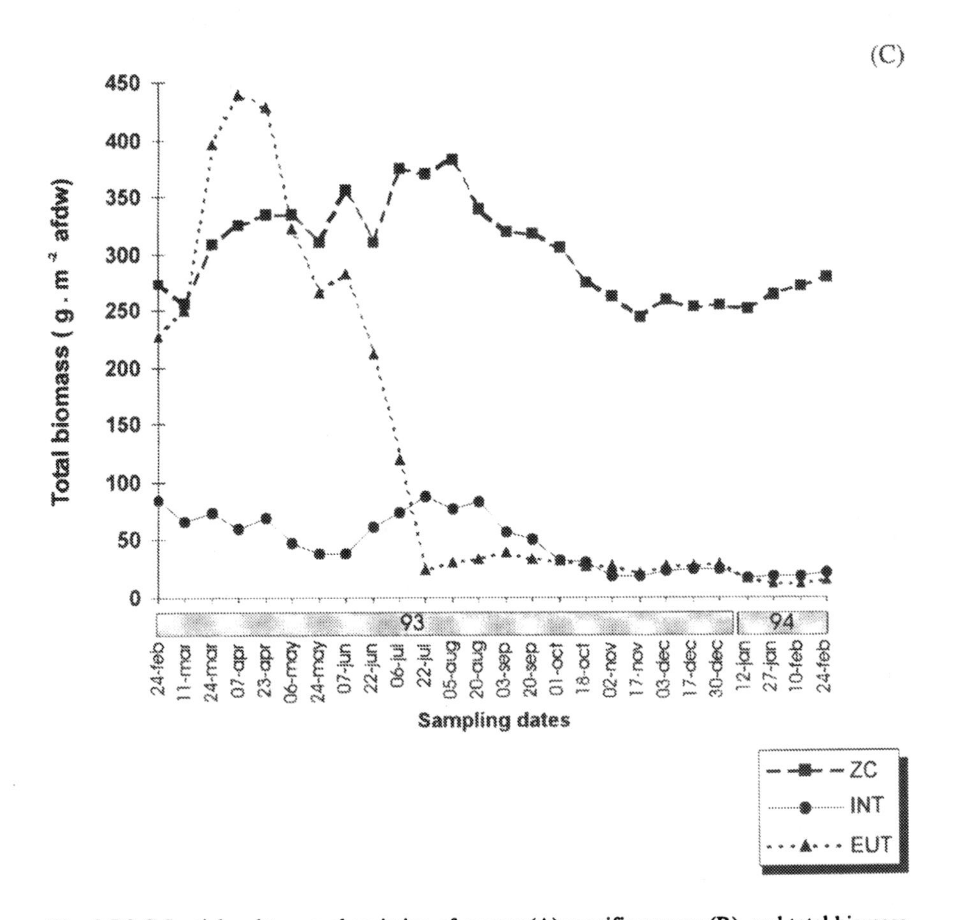

Fig. 2.5.2.C Spatial and temporal variation of exergy (A) specific exergy (B), and total biomass (C) in non eutrophied (*Zostera noltii* community) (ZC), intermediate eutrophied (INT) and eutrophied (EUT) areas. For specific exergy (B), we indicate the values estimated from the total biomass, taking into account the whole period, and the values estimated only from the macrofaunal biomass, for the period from 6 of July 1993 to 24 February 1994. In this last case, the aim was to allow the comparison of specific exergy between the *Zostera* community (non eutrophied area) and the eutrophied areas of the estuary, where after the algae crash the total biomass consisted basically of animals.

This was obviously a function of the macroalgae crash in the eutrophied areas (between the 22 of June and the 6 of July 1993), which resulted in not only a drastic reduction of the total biomass (Fig. 2.5.2.C) but also a change from a primary production based system toward a detritus based food web.

Since total biomass after the 6 of July consisted essentially of animals (consumers), primarily deposit feeders and detritic feeders (e.g., annelid worms and crustaceans), it is clear that the abrupt increase of specific exergy in the eutrophied areas after the algae crash does not reflect an augmentation of the structural complexity of the community, but simply the different quality of the biomass involved in the calculations. This becomes evident if we compare the specific exergy estimated for the non eutrophied area (*Zostera* community), taking only the macrofauna into account, with the values found for eutrophied areas (Fig. 2.5.2.B).

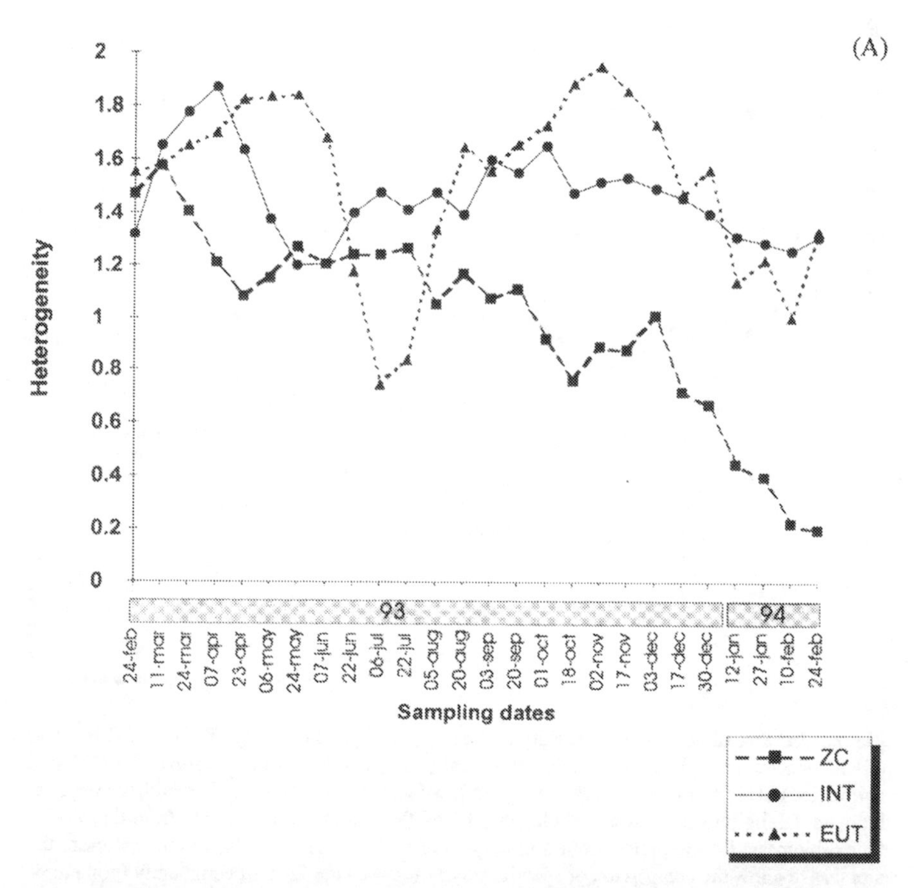

Fig. 2.5.3.A Spatial and temporal variation of heterogeneity; detailed description see Fig. 2.5.3.B

In fact, after the 6 of July, if we account for both primary producers and consumers, specific exergy is lower in the *Zostera* community than in the eutrophied areas. But if we account only for the consumers the specific exergy is clearly higher in the *Zostera* community, following the same pattern from before the algae crash, in agreement with the working hypothesis. We must conclude that the specific exergy may shift very drastically as a function of annual dynamics, like in communities dominated by r strategists, providing a spatial and temporal information that may not be related with the long term evolution and integrity of the system.

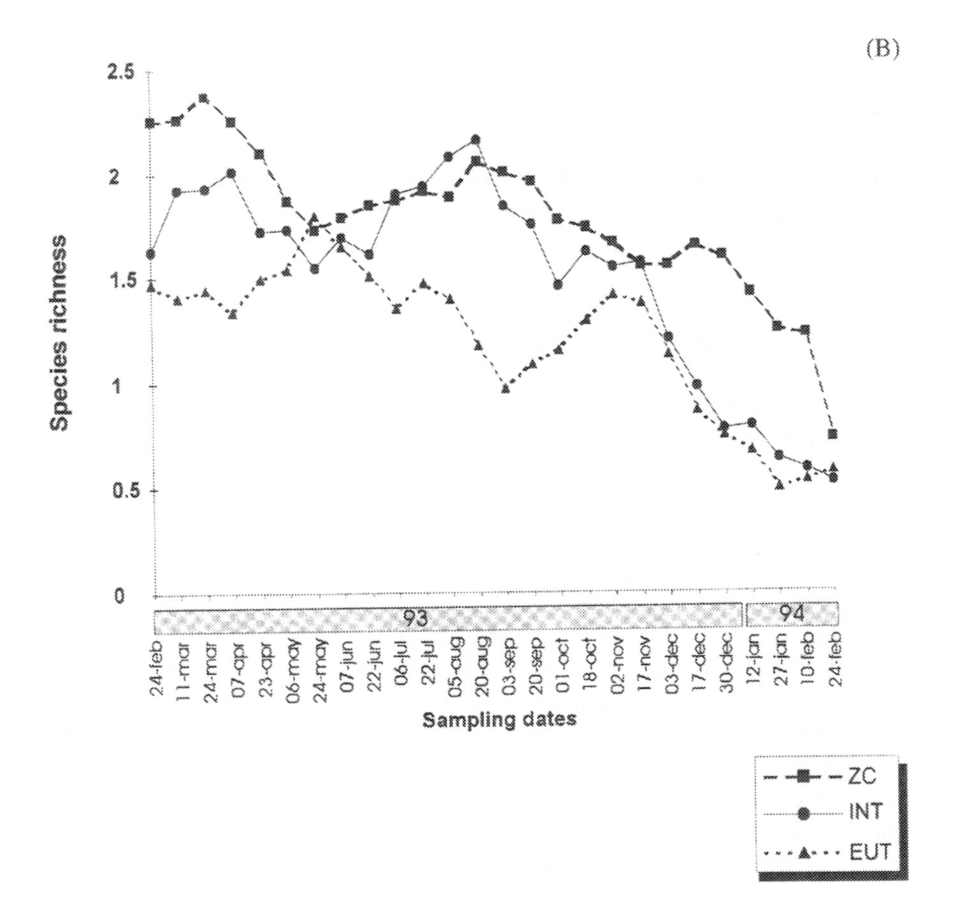

Fig. 2.5.3.B Spatial and temporal variation of heterogeneity (A), species richness (B) in non eutrophied (ZC), intermediate eutrophied (INC) and eutrophied (EUT) areas. Heterogeneity was computed using the Shannon-Wienner's index, and species richness using the Margalef's index.

The variation of species richness and heterogeneity (species richness + evenness) along the gradient of eutrophication provided quite different information (Fig. 2.5.3). Through time, as hypothesized, species richness was consistently higher in the non eutrophied area, corresponding to the *Zostera* community, decreasing along the gradient of eutrophication (Fig. 2.5.3.A). On the contrary, heterogeneity was always higher in the eutrophied areas (Fig. 2.5.3.B). The only exception was the decrease observed in the most heavily eutrophied area in early summer, which was related with the algae crash causing a drastic reduction of total biomass (Fig. 2.5.2.C).

The observed spatial variation of heterogeneity was not in agreement with the working hypothesis. This originates in the fact that the Shannon-Wienner's index integrates two components, the number of species (species richness) and their relative abundances (evenness). As expected, species richness decreased as a function of increasing eutrophication, but the extremely high concentration (dominance) of a few species in the *Zostera* community, namely *Hydrobia ulvae*, a detritic feeder and epiphytic grazer gastropod, and *Cerastoderma edule*, a filter feeder bivalve, decreased species evenness, and consequently heterogeneity values. In this case, lower values of heterogeneity must be interpreted as expressing higher biological activity of these species, probably due to the abundance of nutritional resources in the *Zostera* community, and not as a result of environmental stress (Legendre and Legendre 1984).

To what extent did exergy and biodiversity follow the same trends in space and time, as hypothesized, decreasing from non eutrophied to eutrophied areas? Taking into account the whole year data set, the variation of exergy and specific exergy along the eutrophication gradient (non eutrophied, intermediate eutrophied, and eutrophied areas) was significantly correlated ($P \leq 0.05$) (Fig. 2.5.4.A), providing an equivalent information from the system. Values were consistently higher and more stable in the non eutrophied area as compared to the eutrophied ones. Moreover, through the comparison of yearly exergy data series for each site (t test, $P \leq 0.05$) it was possible to distinguish between the three situations. However, for specific exergy, differences between the intermediate eutrophied and eutrophied areas were not significant, which suggests that exergy might be more sensitive to detect subtle differences.

Species richness and exergy appeared significantly correlated ($P \leq 0.05$), following a similar spatial pattern, both decreasing from non eutrophied to eutrophied areas (Fig. 2.5.4.B). On the contrary, heterogeneity and exergy appeared to be negatively, although not significantly correlated (Fig. 2.5.4.A), providing totally diverse information of the benthic communities along the eutrophication gradient. As explained above, this resulted from the properties of the heterogeneity measure.

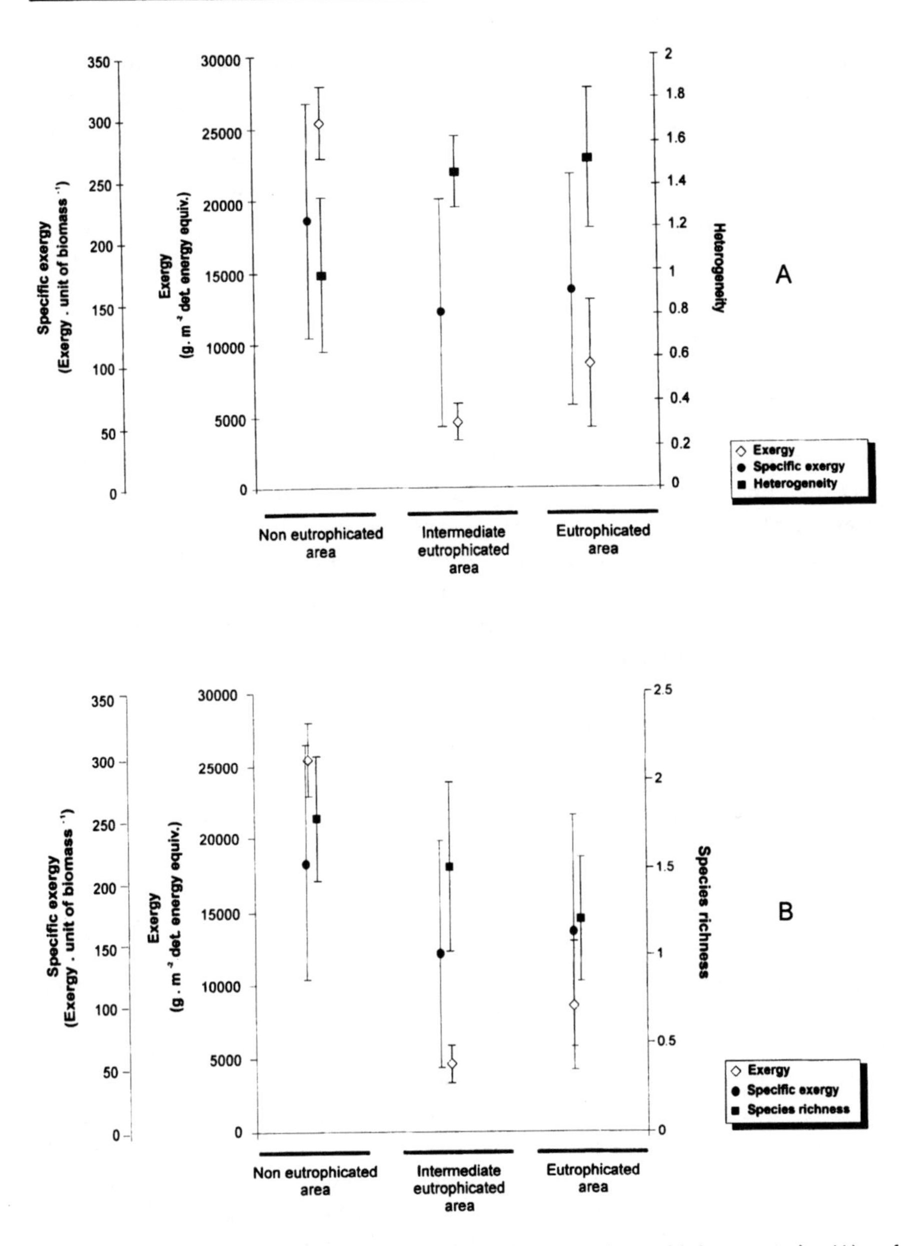

Fig. 2.5.4. Variation of exergy and specific exergy in comparison with heterogeneity (A) and species richness (B) along the gradient of eutrophication gradient. For each situation, respectively non eutrophied (ZC), intermediate eutrophied (INT), and eutrophied (EUT), we indicate the average values and the standard deviation, taking into account the entire yearly data set. The spatial variation of exergy and specific exergy was significantly correlated ($r = 0.59$; $P \leq 0.05$). The spatial variation of heterogeneity was not significantly correlated neither with exergy or specific exergy ($r = -0.48$ and $r = 0.38$ respectively; $P \leq 0.05$). The spatial variation of species richness was significantly correlated with both exergy and specific exergy ($r = 0.60$ and $r = 0.90$ respectively; $P \leq 0.05$).

Different results were obtained comparing the patterns of variation of species richness, heterogeneity and specific exergy. Species richness and specific exergy appeared clearly positively correlated ($P \leq 0.05$) (Fig. 2.5.4.B), while the patterns of variation of heterogeneity and specific exergy were distinct (Fig. 2.5.4.A). Moreover, from the comparison of yearly data series (t test, $P \leq 0.05$), heterogeneity values were not significantly different between the intermediate eutrophied and eutrophied areas, and therefore did not permit to discriminate the existing differences.

2.5.4 Discussion and Conclusions

The hypothesis that exergy and biodiversity would follow the same trends in space and time was validated with regard to species richness, but not for heterogeneity. In fact, exergy, specific exergy, and species richness responded as hypothesized, decreasing from non eutrophied to eutrophied areas, but heterogeneity responded in the opposite way, showing the lowest values in the non eutrophied area. On the other hand, exergy and species richness were able to grade situations presenting relatively subtle differences, but specific exergy and heterogeneity appeared to be less sensitive. Moreover, as ecological indicator, heterogeneity (measured using the Shannon-Wienner's index) appeared to be a more subjective interpretation. Exergy and specific exergy appeared to be able to provide useful information regarding the state of the benthic communities, and might therefore constitute suitable system-oriented characteristics, and may be good indicators of ecosystem integrity. Since specific exergy may shift drastically as a function of annual dynamics, it provides spatial and temporal information, which is different from static exergy measures. Therefore, we recommend using exergy and specific exergy as complementary parameters.

The method proposed by Jørgensen et al. (1995) to estimate exergy, which takes into account both biomass and the thermodynamic information due to genes, appears to be operational, but more accurate (discrete) weighting factors to estimate exergy from organisms biomass need to be determined (Table 2.5.1). Moreover, using the number of genes to express the thermodynamic information built in the biomass is not practical because genetic mapping available in published data is very scarce. Actually, most of the organisms in biotic systems have not been characterized owing to the long time procedures and high costs involved in the molecular work of gene analysis. We propose therefore to undertake a more practical approach.

The dimension of the active genome is primarily a function of the required genetic information to build up an organism. Since this genetic information is contained in DNA, it is reasonable to assume that the dimension of each active genome is roughly proportional to the contents of DNA in the nucleus of diploid cells of each organism. These contents may be determined through the isolation, purification, and analysis of cell nucleus from representative organisms. Obviously, in a certain extent, data produced will not be so accurate as data provided by genetic analysis, but this will not be a problem since the aim is to generate conceptual exergistic relationships between different kinds of organisms. In this case, their complexity will be accounted as an indirect measure of the quality (thermodynamic information) of the biomass, proportional to the distance of that matter to thermodynamic equilibrium.

References

Jørgensen SE (1988) Use of models to account for changes in species composition. Ecol Model 41:117-126

Jørgensen SE (1992) Integration of ecosystem theories: A Pattern. Kluwer Academic Publishers, p 383

Jørgensen SE (1993) State of the Art of Ecological Modeling. In: McAleer M (ed) Proceedings of the International congress on modelling and simulation, University of Western Australia, pp 455-481

Jørgensen SE, Mejer H (1979) A holistic approach to ecological modelling. Ecol Model 7:169-189

Jørgensen SE, Mejer H (1981) Exergy as a key function in ecological models. In: Mitsch W, Bosserman RW, Klopatek JM (eds) Energy and ecological modelling. Developments in environmental modelling, vol 1. Elsevier, Amsterdam, pp 587-590

Jørgensen SE, Nielsen SN, Mejer H (1995) Emergy, environ, exergy and ecological modelling. Ecol. Model., 77:99-109

Legendre L, Legendre P (1984) Ecologie Numérique I – Le Traitement Multiple des Données Écologiques. Masson, Paris, pp 1-197

Magurran AE (1988) Ecological Diversity and its Measurement. Croom Helm, London

Nielsen SN (1990) Application of exergy in structural dynamic modelling. Verh Int Verein Limnol 24: 641-645

Nielsen SN (1992) Strategies for structural dynamic modelling. Ecol Model 63:91-101

Nielsen SN (1994) Modelling structural dynamical changes in a danish shallow lake. Ecol Model 73:13-30

Nielsen SN (1995) Optimization of exergy in a structural dynamic model. Ecol Model 77:111-122

2.6 Thermodynamic Orientors: How to Use Thermodynamic Concepts in Ecology

Yuri Svirezhev

"Thermodynamics is full of highly scientific and charming terms and concepts, giving an impression of philosophical and scientific profundity. *Entropy, thermal death of the Universe, ergodicity, statistical ensemble*—all these words sound very impressive posed in any order. But, placed in the appropriate order, they can help us to find the solution of urgent practical problems. The problem is how to find this order..."

(from table-talks in Moscow)

"... nobody knows what entropy is in reality, that is why in the debate you will always have an advantage"

J. von Neumann

"Teleology is a lady no biologist can live without, but whose company seems shameful in society"

G. von Brükke

Abstract

In this chapter I will try to answer the following questions: How should we apply thermodynamic methods and concepts to ecology; how can we describe the ecosystem's behavior in the terms of physics (and particularly, thermodynamics); and what kind of physical criteria can be used for the estimation of anthropogenic impacts on ecosystems? From the viewpoint of thermodynamics, any ecosystem is an open system far from thermodynamic equilibrium, in which the entropy production is balanced by the outflow of entropy to the environment. I suggest the "entropy pump" hypothesis: that climatic, hydrological, soil and other environmental conditions are organized in such a way that only a natural ecosystem which is specific for these conditions can be in the dynamic equilibrium (steady state). In the framework of this hypothesis I can calculate the entropy production for an ecosystem under anthropogenic stress.

F. Müller M. Leupelt (Eds.)
Eco Targets, Goal Functions, and Orientors

By considering systems far from thermodynamic equilibrium, we can prove that the so-called exergy is a functional of a dissipative function, which is undertaken along the trajectory going from a thermodynamic equilibrium to a dynamic one. It was shown there is a close connection between the measure of an additional information (Kullback measure) and the exergy.

And finally I try to show the deep internal connection between Lyapunov functions which are the main instruments of the stability theory and basic thermodynamic concepts, especially in applications to ecology.

2.6.1 Introduction

Many studies are known which attempt to apply (directly or indirectly) thermodynamic concepts and methods in theoretical and mathematical ecology for the macroscopic description of biological communities and ecosystems. Such attempts can be divided into two classes.

The *first* class includes the direct transfer of such fundamental concepts as entropy, the First and Second laws of Thermodynamics, Prigogine's theorem, etc. into ecology. The literature on this subject is enormous, the most recent publications are Ulanowicz and Hannon (1987), Weber et al. (1988), Jørgensen (1992), Schneider and Kay (1994).

The *second* class includes some attempts to use the *methods* of thermodynamics, such as the Gibb's statistical method. In the 1940s a very elegant method for the construction of the formal statistical mechanics was proposed by Khinchin. This method can be applied to a wide class of dynamic systems, in particular, to Volterra's "prey-predator" system (Kerner 1957, 1959; Alexeev 1976). Unfortunately, none of these results can be interpreted satisfactorily from the ecological point of view (Svirezhev 1976).

Strictly speaking, there are no prohibitions in principle to applying thermodynamic concepts to such physical-chemical systems as ecological systems. The problem is the following: there is not a direct homeomorphism between the models (in a broad sense) in the thermodynamics and the models in ecology. For example, the model of ideal gas (the basic model of thermodynamics) cannot be applied directly to a population or, moreover, to a biological community. We try to explain why.

Among ecologists it have been taken almost as an axiom, that communities which are more complex in structure, more rich in comprising species, are necessarily more stable. This seems to be explained in the following way: Different species are differently adapted to environmental variations. Therefore, a variety of species may respond with more success to different environmental variations than a community composed of a small number of species, and hence the former will be more stable.

It must be intuitively clear that both an ecosystem and a biological community which exists sufficiently long time in a more or less invariant state (this property is often called "persistence"), should possess intrinsic abilities to resist perturbations coming from the environment. This ability is usually termed "stability". Though the notion seems obvious, it is quite a problem to provide it with a precise and unambiguous definition. So far, this heavily overloaded term found no establish ("stable") definition. For instance, the theory of stability, which can be considered as a branch of theoretical mechanics, is using about thirty different definitions of stability. So we can consider stability (and "entropy", too) as a "fuzzy" definition. Paraphrasing von Neumann we can say "... nobody knows what stability means in reality, that is why in the debate you will always have an advantage".

Among these definitions we can select the two large classes differing in respect of the requirements coming under the head of "stability". The first group of requirements concerns preservation of the number of species in a community. A community is stable if the number of member-species remains constant over sufficiently long time. This definition is the closest to various mathematical definitions of stability.

The second group refers rather to populations than to community which is considered to be stable when numbers of component populations do not undergo sharp fluctuations. This definition is closer to the thermodynamic (or more correctly, to the statistical physics) notion of system stability. In thermodynamics (statistical physics) a system is believed to be stable, when large fluctuations which can take the system far from the equilibrium or even destroy it, are unlikely (cf. Landau and Lifshitz 1964). Evidently the general thermodynamic concepts (for instance, the stability principle associated in the case of closed systems with the Second law and, in the case of open systems, with the Prigogine theorem) should be applicable to biological (and, in particular, ecological) systems.

Perhaps this was the motive that a specific diversity indices (in particular, the Shannon information entropy called also the *information diversity index*

$$D = -\sum_{i=1}^{n} p_i \ln p_i, \quad p_i = N_i / \sum_{i=1}^{n} N_i, \tag{2.6.1}$$

where n is the number of species (or some other groups) in a community, and N_i is the population size of ith species can be used as a measure of stability (Margalef 1951; MacArthur 1955).

In accordance with this "logic", the community is most stable if D is maximal. But, as may readily be shown, in this case the community structure is such that specimen of any species occur with the same frequency, since $\max_{p_i} D$ is attained at $p_i^* = 1/n$.

In other words, the diversity of a community is maximal when the distribution of species is uniform, i.e. when there are no abundant or rare species, and no structures. However, observations in real communities show that this is never the case, and that there is always a hierarchical structure with dominating species.

What is the cause of this ecological paradox? Probably, it is the formal application of models and concepts taken from physics and information theory to systems that do not suit. Both the Boltzmann entropy in statistical physics and the Shannon entropy in information theory make sense only for populations of weakly interacting particles. A typical example of such system is the ideal gas, the macroscopic state of which is an additive function of the microscopic states of its molecules.

Let us remember the original formulation of the Boltzmann entropy: $S_B \sim \ln W$ where W is a probability of state of the system. In this general formulation the Boltzmann formula is applied to any system, not only to systems with weak interaction. However, as soon as we use the standard formula $S = -k\sum_{i=1}^{n} p_i \ln p_i$ we use implicitly the classic thermodynamic model of an ideal gas.

The introduction of an entropy measure in to such sets is well founded. And furthermore, the stability of the equilibrium, when the entropy is maximal, is associated with the second law. On the other hand, the stable structure of a biological community is the consequence of interactions between populations rather than the function of characteristics of individual species, etc., i.e. a biological community is a typical system of strongly interacting elements. However, as soon as we become concerned with such systems the entropy measure is no longer appropriate. There is one more argument against the use of the diversity as a goal function relating to stability. The entropy increases (tending to a maximum) only for closed systems, but every biological system is an open system in a thermodynamic sense, so that its total entropy is changed in an arbitrary way. When in equilibrium (we speak of a dynamic equilibrium) the rate of the entropy production inside a system is positive and minimal. This is the Prigogine theorem. In this case the goal function relating to stability is the rate of the entropy production, not the entropy.

Notice, however, that in numerous competitive communities in the initial stages of their successions, far from climax, an increase in diversity may be observed. Diversity seems to be a "good" goal function for stability. Since in the initial stages, far from equilibrium, the competition is still weak and the community may well be regarded as a system with weak interaction.

2.6.2 The Physical Approach: Direct Calculation of the Entropy and the "Entropy Pump" Hypothesis

From the viewpoint of thermodynamics, any ecosystem is an open thermodynamic system. The climax of the ecosystem corresponds to the dynamic equilibrium (steady-state), when the entropy production in a system is balanced by the entropy flow from the system to the environment. This work is being done by the "entropy pump" (Svirezhev 1997).

Let us consider one unit of the Earth's surface which is occupied by a natural ecosystem which is maintained in the climax state. There is a natural periodicity in such a system (1 year); during this period the internal energy of the ecosystem is increased by a value of gross primary production. One part of this production is used for respiration with further transformation into heat, while another part (the net primary production), on the one hand, turns into litter and other forms of soil organic matter, and, on the other hand, is taken by consumers and for compensation of the respiratory losses.

Note that the part of net production which is taken by consumers, transfers sufficiently fast into heat since consumers often have very high R/B-coefficient. Therefore this part can be included into some general respiration losses for plants increasing their real respiration coefficient.

But, since the system is at dynamic equilibrium, an appropriate part of dead organic matter in litter and soil has to be decomposed (releasing a place for "new" dead organic matter). The "old" dead organic matter has "to be burned", so that its chemical energy is transformed into heat. The temperatures and pressures in the ecosystem and its environment are assumed to be equal, i.e. we consider an isotherm and isobaric process. Note that all the considerations are correct for terrestrial ecosystems and are not directly applicable to aquatic systems.

In the theory of open systems the total variation of entropy

$$dS(t) = d_i S(t) + d_e S(t), \tag{2.6.2}$$

where

$$d_i S(t) = dQ(t)/T(t)\,,\; dQ(t) \tag{2.6.3}$$

is the heat production caused by irreversible processes within the system and $T(t)$ is the current temperature (in K) at a given point of the Earth's surface. The value $d_e S(t)$ corresponds to the entropy of exchange processes between the system and its environment.

In fact, total heat production means heat emission of the plant metabolism (heat emitted during the process of respiration, R_v(t)) and heat emission of the consumer's metabolism ($R_c(t)$) and heat emission from the decomposition of "dead" organic matter ($D(t)$). Really $R_v(t) >> R_c(t)$ so that the total metabolism of the ecosystem is equal to the metabolism of its vegetation, in practice.

Integrating (2.6.2) with respect to a natural period (one year) we get

$$S(t+1) - S(t) = \int_t^{t+1} \frac{R_V(\tau) + D(\tau)}{T(\tau)} d\tau - \delta_e S, \tag{2.6.4}$$

where

$$\delta_e S = \int_t^{t+1} \frac{d_e S}{d\tau} d\tau. \tag{2.6.5}$$

Using the mean value theorem the integral in formula (2.6.4) can be rewritten in the form

$$\int_t^{t+1} \frac{R_v(\tau)}{T(\tau)} d\tau = \frac{1}{T(\theta_1^t)} \int_t^{t+1} R_v(\tau) d\tau = \frac{1}{T(\theta_1^t)} [P_0(t) - P_0^n(t)],\ \theta_1^t, \theta_2^t \in [t, t+1],$$
$$\int_t^{t+1} \frac{D(\tau)}{T(\tau)} d\tau = \frac{1}{T(\theta_2^t)} \int_t^{t+1} D(\tau) d\tau = \frac{1}{T(\theta_2^t)} \tilde{D}_0(t), \qquad (2.6.6)$$

The values $P_0(t)$ and $P_0^n(t)$ are annual gross (total) and net primary productions, the temperatures $T(\theta_1^t), T(\theta_2^t)$ are some "mean" annual temperatures which may be distinguished. In all these values the notation "t" means a number of the current year.

Although the ecosystem's state changes within a one-year interval, we can consider the climax natural ecosystem as a steady state if the time step is equal to one year. Since the system is in dynamic equilibrium (steady state) the "burned" part of the dead organic matter is $\tilde{D}_0(t) = P_0^n(t)$ and, moreover, in formula (2.6.3) $S(t+1) - S(t) = 0$.

Then

$$\delta_e S(t) = \frac{1}{T(\theta_1^t)} P_0(t) + [\frac{1}{T(\theta_1^t)} - \frac{1}{T(\theta_2^t)}] P_0^n(t). \qquad (2.6.7)$$

According to our assumption, the entropy $\delta_e S(t)$ must be "sucked out" by the solar "entropy pump" in accordance with the steady-state condition. Consequently, the power of this pump at some point of the Earth is equal to $\delta_e S(t)$. In the general case $T(\theta_1^t) \neq T(\theta_2^t)$.

However, if we take into account that not only photosynthesis and the respiration but also the decomposition of dead organic matter depend on the current temperature in a similar way, we can assume $T(\theta_1^t) \approx T(\theta_2^t) = T(t)$.

Apparently the best approximation for this "mean" temperature will be the *mean active temperature*, i.e. the arithmetic mean of all temperatures above 5°C, that provides a good indicator of biological activity. The next approximation may be the mean temperature of a vegetation period. Finally the annual mean temperature is perhaps only a rough approximation.

If we accept this assumption formula (2.6.7) can be written in the form

$$\delta_e S = \frac{P_0}{T} = \frac{r}{1-r} \frac{P_0^n}{T}, \qquad (2.6.8)$$

where r is a mean respiration coefficient (see, for instance, Larcher 1978).

Let us assume that the considered area is influenced by anthropogenic pressure, i.e. a flow of artificial energy (W) into the system takes place. We include in this notion ("the flow of artificial energy") both the direct energy flow (fossil fuels, electricity, etc.) and the inflow of chemical elements (pollution, fertilisers, etc.). We suppose that this inflow is dissipated inside the system, transformed into heat and, moreover, that it modifies the plant productivity.

Let the gross production of the ecosystem under anthropogenic pressure be P_1. We assume that despite anthropogenic perturbation the ecosystem is again in a steady state. When repeating the previous arguments we get

$$\delta_e^1 S = \frac{1}{T}(W + P_1). \tag{2.6.9}$$

We make a very important assumption and suppose that a part of the entropy released at this point by the "entropy pump" is still equal to

$$\delta_e S = P_0 / T. \tag{2.6.10}$$

Really we assume that the power of the local entropy pump corresponds to a supposed natural ecosystem situated in that location. In other words, *the climatic, hydrological, soil and other environmental local conditions are organized in such a way that only a natural ecosystem which is specific to this particular combination can be at the steady state without an environmental degradation.* This is the "entropy pump" hypothesis.

If we accept this hypothesis we must also assume that the transition from a natural to an anthropogenic ecosystem is performed sufficiently fast, so that the "adjustment" (i.e. parameters) of the entropy pump could not be changed. Therefore, in order that the other part of the entropy

$$\delta_e^2 S = \sigma = \delta_e^1 S - P_0 / T \tag{2.6.11}$$

should be compensated by the outflow of entropy to the environment, only one unique way exists. This compensation can occur only at the expense of environmental degradation ($\sigma > 0$), resulting, for instance, from heat and chemical pollution, and mechanical impact on the system. We assume also that the "natural" and "anthropogenic" ecosystems are connected by *the relation of succession.*

Let me include a few words about the relation of succession. Let us assume that the anthropogenic pressure has been removed. The succession from the anthropogenic ecosystem towards a natural one has started. The next stage of this succession would be a "natural" ecosystem which is typical in this locality. Thus, a successionally closed ecosystem as the first stage in the succession of an "anthropogenic" ecosystem after the anthropogenic stress has been removed.

What is a "dynamic" sense of the "successional closeness", why do we need this concept? The point is that in this approach we can compare only close steady states, their vicinities must intersect significantly, and the time-scale of a quasi-stationary transition (i.e. without serious perturbations of current equilibrium) from a natural to an anthropogenic ecosystem and *vice versa* must be small (in comparison to the time-scale of succession).

Let us consider the non-local dynamics of the system. We stop the flux of artificial energy into the ecosystem. As a result, if the ecosystem is not degraded, a succession will take place at the site which tends towards the natural ecosystem type specific for the territory (grassland, steppe, etc.). This is a typical reversible situation. Under severe degradation a succession would also take place, but towards another type of ecosystem. This is quite natural, since the environmental conditions have been strongly perturbed (for instance, as a result of soil degradation). This is an irreversible situation. So, the "successional closeness" concept means that we remain in a framework of "reversible" thermodynamics. And if there is no input of artificial energy, the steady state for a given site (locality) will be presented by the natural ecosystem, as the local characteristics of the "entropy pump" correspond exactly to the natural type of ecosystem.

Returning to (2.6.9) and (2.6.11) we get the formula for σ.

$$\sigma T = W + P_I - P_0 . \tag{2.6.12}$$

The values in (2.6.12) are not independent. For instance, P_I depends on W. Since we are not able to estimate these correlations within a framework of theory of thermodynamics, we have to use empirical correlations.

If we remember the sense of σ then it is obvious that the σ value can be used as the criterion for environmental degradation or as the "entropy fee" which has to be paid by society (really, suffering from the degradation of environment) for modern industrial technologies.

Of course, there is another way to balance the entropy production within the system. For instance, we can introduce artificial energy and soil reclamation, pollution control (or, generally, ecological technologies). Using the entropy calculation we can estimate the necessary investments (in energy units).

This approach has been applied to the analysis of several agricultural systems, in particular, to the analysis of crop production (maize) in Hungary in the 1980s (in details see Svirezhev 1997).

2.6.3 Systems far from Thermodynamic Equilibrium

Before introducing some special concepts like exergy, etc. we must remember that all of them consider the ecosystem as a system far from thermodynamic equilibrium. The "basic" variable for this theory is a rate of the entropy production, or a rate of the energy dissipation, the so-called dissipative function (β). Immediately a series of questions arises, dealing with the behaviour of the dissipative function β :

1. How can we calculate β for a system far from thermodynamic equilibrium, if we do not know the appropriate kinetic equations?

2. What can we calculate in this case?
3. What kinds of "thermodynamic" statements can be formulated in this case?

We will now attempt to answer these question.

Let us assume that we have a system of ordinary differential equations, which describe the dynamics (kinematics) of the considered system:

$$\frac{dC_i}{dt} = f_i(C_1,\dots,C_n),\ i = 1,\dots,n; \quad C_i \in P^n \tag{2.6.13}$$

where $\{C_i(t)\}$ is a vector of state variables and P^n is a positive orthant of phase space, i.e. all variables are non-negative. We assume also that the system (2.6.10) has only one stable equilibrium $\{C_{i*}\}$ and $C_i \Rightarrow C_{i*}$ for $t \Rightarrow \infty$ and for any $C_i, C_{i*} \in \Omega \subseteq P^n$.

As usually there is not a good description for the system of differential equations (2.6.12). In the best case we have a time-series of observations recorded in the course of transition from an initial state $\{C_{io}\}$ towards the stable equilibrium point $\{C_{i*}\}$. It corresponds to some solution $C = C(Co,t)$ of an unknown basic system of differential equations (2.6.13). Nevertheless, we can calculate the *total amount of dissipated energy*

$$L = \int_{t_0}^{\infty} \beta(t)dt \tag{2.6.14}$$

for the transition $C(t_0) \Rightarrow C^*(\infty)$ in one special case.

Let the system dynamics be a movement in a potential field with the chemical potentials

$$\mu_i = \mu_{io} + RTlnC_i,\ i=1,\dots,n, \tag{2.6.15}$$

where C_i are molar concentrations of corresponding components and R is the gas constant. We assume

$$\mu_{10} = \mu_{20} = \dots = \mu_{n0}, \tag{2.6.16}$$

i.e. the components C_i are substances of *identical* (or close) origin. Affinity for reaction (transition) $C_i \Rightarrow C_i^*$ is equal to

$$A_{i,i*} = ln(C/C_{i*}). \tag{2.6.17}$$

The initial values are arbitrary so that we can consider any point $C(t)$ (except the singular point C^*) as an initial point. Since the affinity for transition $C_i \Rightarrow C_i{}^*$ *is equal to* $A_{ii*} = ln(C_i/C_i^*)$ *then*

$$L = \int_{t0}^{\infty} \beta(t)dt = \int_{t0}^{\infty} \sum_{i=1}^{n} \beta_i(t)dt = \sum_{i=1}^{n} \int_{t0}^{\infty} ln\frac{C_i(t)}{C_i *}\frac{dC_i}{dt}dt =$$

$$= \sum_{i=1}^{n} \int_{C_i(t)}^{C_i(\infty)} (lnC_i - lnC_i *)dC_i = \sum_{i=1}^{n} \{-C_i ln\frac{C_i}{C_i^*} + (C_i - C_i^*)\}. \qquad (2.6.18)$$

We omit the factor RT in the expression for β, that is inessential when we consider an isotherm process. And finally

$$L = \sum_{i=1}^{n} L_i = -\sum_{i=1}^{n} \{C_i \, ln\frac{C_i}{C_i^*} - (C_i - C_i^*)\}. \qquad (2.6.19)$$

You can see that $L < 0$ for any $C_i > 0$, except $C_i = 0$, when $L = 0$. It means that the *value of total change for entropy of an open system far from thermodynamic equilibrium*, when it passes from some non-steady state to stable dynamic equilibrium, *is a negative value*. Also, this value does not depend on characteristics of this transition.

Certainly, *spontaneous* processes, when a system tends toward a stable dynamic equilibrium after small fluctuations, which appear inside a system, *are accompanied by an increase of entropy*. In our case the transition $C \Rightarrow C^*$ is not spontaneous *but forced*, it depends on interactions between the system and its environment.

It is obvious that the decrease of entropy (in similar transition processes) is a result of free energy consumption (by the system) from the environment. It is, in turn, a result of exchange processes, giving a negative contribution into the entropy production.

2.6.4. Exergy and Entropy: Exergy Maximum Principle

Let us suppose that the right-hand sides of equations 2.6.13 depend on some parameters $\alpha_1, \ldots, \alpha_m$, so that

$$\frac{dC_i}{dt} = f_i(C_1, \ldots, C_n; \alpha_1, \ldots, \alpha_m), \quad i = 1, \ldots, n. \qquad (2.6.20)$$

The vector of parameters α describes the state of the environment. It is obvious that the steady - state C^* depends on α .We consider the following *Gedanken-experiment:*

1. Let the current state of environment be described by the vector α^1, then
2. $C^*=C^*(\alpha^1)$.

3. We change the environment from the state α^1 to the state α^2 very quickly in comparison with the natural frequencies of the system.
4. We spend the energy (work) E^{12} to realise this change.

   After this change the state $C^*(\alpha^1)$ ceases to be a stationary state, and the system starts to evolve towards a new stationary (steady) state $C^*(\alpha^2)$. If we calculate the dissipative energy for this transition, we get

$$L^{12} = -\sum_{i=1}^{n}\{C_i^*(\alpha^1)\ln\frac{C_i^*(\alpha^1)}{C_i^*(\alpha^2)} - (C_i^*(\alpha^1) - C_i^*(\alpha^2))\}. \quad (2.6.21)$$

Since we cannot cancel the action of the second law then $E^{12} \geq -L^{12}$; and $\min E^{12} = -L^{12}$. We shall consider an extreme case and assume that $E^{12} = -L^{12}$.
If we assume also that the vector α^1 corresponds to the current state of the environment (the biosphere); the vector α^2 corresponds to some *pre-biological* environment, and $C^*(\alpha^2) = C^0$ are equal to concentrations of biogenic elements in some pre-biological structures, we see immediately that E^{12} is nothing more nor less than Jørgensen's exergy (Jørgensen 1992). Therefore,

$$Exergy(Ex) = \sum_{i=1}^{n}\{C_i \ln\frac{C_i}{C_i^s} - (C_i - C_i^0)\}. \quad (2.6.22)$$

Note that we omitted the factor *RT*.

Let us remember that the exergy is equal to the work which is necessary for a transformation of the environment to a pre-biological state. The latter can be considered as a thermodynamic equilibrium, i.e. it corresponds to the death of the system. In other words, the exergy is a necessary energy in order to kill the system, to destroy it.

Note (it is very important) that the work cannot be done on the system directly, it must be done on its environment, i.e. we cannot kill the system directly, in order to do this, we must change the environment in a hostile way (for the system). Therefore Jørgensen's "*exergy maximum principle*" postulates that this work must be *maximal.*

And finally I would like to call your attention to the following. There is a principal difference between the types of "Gedankenexperiment" ("experiment of thoughts") in classic thermodynamics and here. If in classic thermodynamics in order to change the state of the system we perform work on the system, then in non-equilibrium thermodynamics in order to obtain the same result we must perform the work on the system's environment.

2.6.5 Exergy and Information

Introducing the new variables

$$p_i = C_i / \sum_{i=1}^{n} C_i, \qquad \sum_{i=1}^{n} C_i = A \tag{2.6.23}$$

where A is the total amount of matter in the system, we can rewrite formula 2.6.22 in the form

$$Ex = A \sum_{i=1}^{n} p_i ln \frac{p_i}{p_i^0} + \{A ln \frac{A}{A_0} - (A - A_0)\}. \tag{2.6.24}$$

The vector $p=\{p_1,..., p_n\}$ describes the *structure* of the system, i.e. p_i are *intensive* variables. The value A is an *extensive* variable. The value

$$K = \sum_{i=1}^{n} p_i ln \frac{p_i}{p_i^0} \tag{2.6.25}$$

is the so-called *Kullback measure* which is very popular in information theory; $K \geq 0$ ($K=0$ iff $p=p^0$).

Let us bear in mind what the exact meaning of the Kullback measure is (see, for instance, Kullback 1959). Suppose that the initial distribution p^0 is known. Then we have got some additional information, and, in consequence, the distribution is changed from p^0 to p. So, $K(p, p^0)$ is the measure of this additional information. Note that K is a specific measure (per unit of matter). Then the product $A*K$ can be considered as a measure of the total amount of information for the whole system, which has been accumulated in the transition from some reference state corresponding to a thermodynamic equilibrium, i.e. some "pre-vital" state, as a reference to the current state of living matter. We can present the expression for exergy in the form:

$$\begin{aligned} &Ex = Ex_{inf} + Ex_{mat}, \quad where \\ &Ex_{inf} = A * K(p, p^0) \geq 0, \\ &Ex_{mat} = A * ln \frac{A}{A_0} - (A - A_0) \geq 0, \end{aligned} \tag{2.6.26}$$

i.e. as the sum of two terms: the first is a result of structural changes inside the system, and the second is caused by a change of total mass of the system.

If we accept Jørgensen's "*exergy maximum principle*" we must postulate that the exergy must increase during the system evolution, i.e. dEx/dt along the system trajectory. From (2.6.26) we have

$$dEx/dt = dEx_{inf}/dt + dEx_{mat}/dt = A\{\frac{dK}{dt} + (K + ln \frac{A}{A_0}) \frac{1}{A} \frac{dA}{dt}\} \geq 0. \tag{2.6.27}$$

Denoting $ln \frac{A}{A_0} = \xi$ and taking into account $A>0$ we get the evolutionary criterion in the form

$$\frac{dK}{dt} + (K+\xi)\frac{d\xi}{dt} \geq 0 . \qquad (2.6.28)$$

If the positiveness of $d\xi/dt$ means an increase of the total biomass in the course of evolution then the positiveness of dK/dt can be interpreted as an increase of the excess information contained in a biomass unit. It is obvious that if both the volume (mass) and its excess information increase the exergy also increases, it is trivial. If the total biomass is not changed ($d\xi/dt$=0) the system can evolve only if the excess information of the biomass increases. On the other hand, the excess information can be decreasing (dK/dt <0), but if the total biomass is growing sufficiently fast ($d\xi/dt$>>1) then the exergy grows and the system evolves. The evolution also occurs if the biomass is decreasing but the excess information of the biomass is growing (sufficiently fast). Finally there is one paradoxical situation in which the exergy increases while the total biomass and its excess information are decreasing. If $A<A_0$ then $\xi<0$, and $\xi(d\xi/dt)>0$. From (2.6.28) we have

$$\xi\frac{d\xi}{dt} \geq \left|\frac{dK}{dt}\right| + K\left|\frac{d\xi}{dt}\right| . \qquad (2.6.29)$$

It is obvious that this inequality can be realized if $|\xi| >> K$ and $|dK/dt| << 1$, i.e. the excess information increases very slowly remaining sufficiently low (K <<1). Since the condition $A<<A_0$ must be fulfilled in order that $|\xi| >> 1$ then we can say that the system has "paid" for its own evolution with its own biomass.

In the vicinity of thermodynamic equilibrium, at the initial stage of evolution $K \cong \xi \sim 0$. Then from (2.6.28) we get $dK/dt>0$, i.e. at the initial stage of evolution the system must increase the excess information of the own biomass in order to evolve.

It is interesting that the exergy maximum principle possesses certain selective properties. To clarify this we consider the system with constant total biomass ($\xi=const$, $d\xi/dt$=0). Then the state with exergy maximum corresponds to the maximum of K. It is easy to see that

$$\max_p K = \max_i (ln\, 1/p_i^0) \qquad (2.6.30)$$

It means that if the system is increasing the own exergy (without change of its biomass) in the course of its evolution then the system will eliminate all the components except one, namely, the element with minimal initial concentration. Usually in a pre-«biological» state (Jørgensen's "reference state") this is some "living" element or substance. In other words, if the system increases the exergy then it selects among its components those which have been presented in a minimal quantities at the beginning of evolution.

However, among these elements there are such ones which are necessary for life maintaining and the system must remain them. How to solve this contradiction? We can do this to introduce some constraints. For instance, it may be the requirement to maintain a certain level of the system's diversity. Formally it means (in this partial case) that we consider

$$max \sum_{i=1}^{n} p_i \, ln(p_i / p_i^0) \tag{2.6.31}$$

under the constraint

$$H = -\sum_{i=1}^{n} p_i \, ln \, p_i = const \tag{2.6.32}$$

In other words, the exergy must be increasing but not arbitrary, maintaining some certain level of diversity. I think there is here a very deep analogy with the Fisher's fundamental theorem of natural selection (cf., Svirezhev and Passekov 1990).

2.6.6 Lyapunov Functions as Goal Functions

One of the most important concepts in the theory of stability is that of Lyapunov functions. These are positive functions defined in a phase space of a dynamical system and possessing by the property of either monotonously increasing or monotonously decreasing along trajectories, and they can be considered as a special class of goal functions.

Let us return to the system (2.6.13) and check for it the following class of functions as a candidate for Lyapunov functions:

$$L = \sum_{i=1}^{n} C_i^* \varphi(C_i / C_i^*), \tag{2.6.33}$$

where

$$(\xi_i = C_i / C_i^*) \tag{2.6.34}$$

$$\varphi(1) = \frac{d\varphi}{d\xi}(1) = 0; \quad \frac{d^2\varphi}{d\xi^2} > 0 \ for \ any \ \xi \geq 0. \tag{2.6.35}$$

In other words, the function $\varphi(\xi)$ must be convex for positive ξ. It is obvious that $L(C^*) = 0$. Because the first variation of L in a vicinity of C^* is equal to

$$\delta L = \sum_{i=1}^{n} \frac{\partial L}{\partial C_i} \delta C_i = \sum_{i=1}^{n} \frac{\partial \varphi}{\partial \xi_i} \delta C_i, \tag{2.6.36}$$

then $\delta L(C^*) = 0$ for any variations δC^*. Calculating the second variation we get

$$\delta^2 L = \frac{1}{2}\sum_{i=1}^{n}\sum_{j=1}^{n}\frac{\partial^2 L}{\partial C_i \partial C_j}\delta C_i \delta C_j = \frac{1}{2}\sum_{i=1}^{n}\frac{d^2\varphi}{d\xi_i^2}(\xi_i)\frac{(\delta C_i)^2}{C_i^*} > 0 \quad (2.6.37)$$

for any non-zero variations δC. Thus, L is a convex function of C having an isolated minimum at the point C^* and increasing monotonously for any $C \in P^n$.

The equilibrium C^* is stable if the derivative dL/dt taken along the trajectory of (2.6.13)

$$\frac{dL}{dt} = \sum_{i=1}^{n}\frac{d\varphi}{d\xi}(\xi_i) f_i(C_1, \ldots, C_n) \leq 0. \quad (2.6.38)$$

for all trajectories (the Lyapunov stability theorem, see Malkin 1967).

On the other hand, the equilibrium C^* is unstable if the derivative $dL/dt \geq 0$ even for only one trajectory of (2.6.13) (the Chetaev instability theorem). Let

$$\varphi(\xi) = a(1-\xi)^2,\ a > 0,\ \textit{then the function}\ L = \sum_{i=1}^{n} a_i C_i^*(1-\xi_i)^2 \quad (2.6.39)$$

is a Lyapunov function. Setting $a_i = C_i^*/n$ we get

$$L = \frac{1}{n}\sum_{i=1}^{n}(C_i - C_i^*)^2, \quad (2.6.40)$$

i.e. the value L may be considered as a mean square measure of distance between a current state of the system and its equilibrium. If this distance decreases in time (i.e. the derivative $dL/dt < 0$ along the system trajectory) then in accordance with the Lyapunov stability theorem we can hope that the system moves to a stable equilibrium. Note that I speak of "hope", since we must check all the trajectories (or a statistically "sufficient" large number of them) in order to be able to speak correctly about a stability of equilibrium. If, on the other hand, this measure increases along some trajectory then (in accordance with the Chetaev instability theorem) the equilibrium is unstable and this trajectory resides from it.

All these results can be interpreted easily from a thermodynamic viewpoint. Indeed, in thermodynamics the value L is a mean square of fluctuations around an equilibrium or a *power of fluctuations*. Therefore, we can state that if a power of fluctuations decreases in time then the system moves to a stable equilibrium. Note that if the movement to a stable equilibrium can be called an *evolution of the system* then the foregoing statement can be reformulated in the following way: the power of fluctuations decreases during the process of the system's evolution. Since the state C^* is a *goal of evolution* in this sense it is natural to consider the function L as a *goal function*.

2.6.7 Lyapunov Function for Volterra Equations

Let the function φ *be* $\varphi(\xi) = \xi - ln\xi - 1$, then the corresponding Lyapunov function will be

$$L = \sum_{i=1}^{n} (C_i - C_i^*) - C_i^* \ln(C_i / C_i^*). \quad (2.6.41)$$

If our dynamical system is the system of Volterra equations

$$\frac{dC_i}{dt} = C_i(\varepsilon_i - \sum_{j=1}^{n} \gamma_{ij} C_j), \ i, j = 1, \ldots, n \quad (2.6.42)$$

with a community matrix $\Gamma = \|\gamma_{ij}\|$, we get immediately

$$\frac{dL}{dt} = -\sum_{i=1}^{n} \sum_{i=1}^{n} \gamma_{ij} \delta C_i \delta C_j, \quad (2.6.43)$$

and the non-trivial equilibrium C^* is always stable if the matrix Γ is positive definite. This is a well-known result in the general theory of Volterra systems (see Svirezhev and Logofet 1978).

2.6.8 Extreme Properties of Volterra Equations for Competing Species: One more Lyapunov Function

We consider again the Volterra system (2.6.42) assuming a symmetry for Γ, i.e. $\gamma_{ij} = \gamma_{ji}$, *and* $\varepsilon_i > 0$. Such a system is a very popular model for a community of competing species. The transformation $\eta_i = \pm\sqrt{C_i}, i = 1, \ldots, n$ transfers the positive orthant P^n into the complete co-ordinate space R_η^n in which the trajectories of the system (2.6.39) are trajectories of the steepest ascent for the function

$$W = \frac{1}{4} \sum_{i=1}^{n} \varepsilon_i \eta_i^2 - \frac{1}{32} \sum_{i=1}^{n} \sum_{j=1}^{n} \gamma_{ij} \eta_i^2 \eta_j^2 = \sum_{i=1}^{n} \varepsilon_i C_i - \frac{1}{2} \sum_{i=1}^{n} \sum_{j=1}^{n} \gamma_{ij} C_i C_j. \quad (2.6.44)$$

Then the system (2.6.41) can be rewritten in the gradient form

$$\frac{d\eta_i}{dt} = \frac{\partial W}{\partial \eta_i}, \ i = 1, \ldots, n; \quad \frac{dW}{dt} = \sum_{i=1}^{n} \frac{\partial W}{\partial \eta_i} \frac{d\eta_i}{dt} = \sum_{i=1}^{n} (\frac{\partial W}{\partial \eta_i})^2 \geq 0. \quad (2.6.45)$$

From (2.6.45) follows that the value W decreases in the process of the system's evolution attaining a maximum at the equilibrium, i.e. W may be considered as a goal function for the competitive community (in detail see Svirezhev and Logofet 1978).

Then the function

$$L = W(C) - W(C^*) \leq 0 \tag{2.6.46}$$

is a Lyapunov function for (2.6.41). Note that this Lyapunov function does not belong to the introduced class and, as opposed to those functions, the equilibrium C^* must not necessarily be non-trivial, it may be situated on appropriate co-ordinate hyperplanes. This implies that in the process of the community evolution one or several species are to be eliminated.

All these results also permit of a sensible interpretation. The value

$$R(C) = \sum_{i=1}^{n} \varepsilon_i C_i \tag{2.6.47}$$

in essence, accounts for the rate of biomass gain in the case that competition and any kind of limitation by resources are absent, and the growth is only determined by the physiological fertility and natural mortality of the organisms. Therefore it is natural to define the value R as the *reproductive potential* of the community.

The value

$$D = \frac{1}{2} \sum_{i=1}^{n} \sum_{j=1}^{n} \gamma_{ij} C_i C_j \tag{2.6.48}$$

may be used to measure the *rate of energy dissipation* resulting from inter- and intraspecific competition. Therefore we shall refer to D as the *total expenses of competition.* Hence the increase in D in the process of evolution may be interpreted as the goal of the community to maximise the difference between its reproductive potential and the total expenses of competition. This goal can be achieved in several ways: either the reproductive potential is maximised at fixed expenses of competition (r-strategy) or the competition expenses are minimised for a limited reproductive potential (K-strategy) as well. There may also be some intermediate cases.

Let us consider the problem from a thermodynamic viewpoint. In the vicinity of an equilibrium the entropy of an open system is presented in the form

$$S = S^{eq} + (\delta S) + \frac{1}{2}(\delta^2 S). \tag{2.6.49}$$

Differentiating (2.6.49) in respect to time we get ($\Delta S = S - S^{eq}$)

$$\frac{d\Delta S}{dt} = \frac{dS}{dt} = \frac{d(\delta S)}{dt} + \frac{d(\delta^2 S)}{dt}. \tag{2.6.50}$$

As usual we present the difference dS in the form $dS = d_e S + d_i S$ where $d_i S$ is the entropy change which takes place by means of internal processes and $d_e S$ is the entropy change caused by independent changes in the interaction between the system and its environment. If to assume that the difference d_iS has a higher order of smallness than d_eS then

$$\frac{d_eS}{dt}=\frac{d(\delta S)}{dt} \quad and \quad \frac{d_iS}{dt}=\frac{1}{2}\frac{d(\delta^2 S)}{dt}. \tag{2.6.51}$$

In accordance with the second law

$$\frac{d_iS}{dt}=\frac{1}{2}\frac{d(\delta^2 S)}{dt}\geq 0. \tag{2.6.52}$$

In thermodynamics the second variation $\delta^2 S$ is presented as a quadratic form of $\delta C_i = C_i - C_i^*$ and if it is negative definite then the corresponding equilibrium is stable (in a thermodynamic sense). Returning to our model we assume that the equilibrium C^* is non-trivial. Then, remembering that

$$\sum_{i=1}^{n}\gamma_{ij}C_j^* = \varepsilon_i;\ i=1,\dots,n, \tag{2.6.53}$$

we can rewrite the expression for $L = W(C) - W(C^*)$ in the form

$$L=-\frac{1}{2}\sum_{i=1}^{n}\sum_{j=1}^{n}\gamma_{ij}(C_i - C_i^*)(C_j - C_j^*) = -\frac{1}{2}\sum_{i=1}^{n}\sum_{j=1}^{n}\gamma_{ij}\delta C_i\delta C_j. \tag{2.6.54}$$

Setting $\delta^2 S = 2L$ we can consider $\delta^2 S$ as a Lyapunov function. In accordance with the Lyapunov stability theorem, if $\delta^2 S < 0$ (this inequality is a condition for thermodynamic stability) and $d(\delta^2 S)/dt \geq 0$ (this is the second law) then the equilibrium is stable (in a Lyapunov sense). It is obvious if the matrix Γ is positive definite then $d(\delta^2 S)/dt \geq 0$ (one of the sufficient conditions for the stability of C^*, see also Section 6). An interesting analogy between the theory of competing communities and the thermodynamics of chemical systems follows from the expression for $\delta^2 S$ in the previous formula:

$$\delta^2 S = -\frac{1}{T}\sum_{i=1}^{n}\sum_{i=1}^{n}\frac{\partial\mu_i}{\partial C_j}\delta C_i\delta C_j, \tag{2.6.55}$$

where

$$\mu_i = \mu_i(C_1,\dots,C_n) \tag{2.6.56}$$

is the chemical potential of ith component depending on all the concentrations in the general case (see, for instance, Rubin 1976). Comparing the expressions for $\delta^2 S$ we can see that the coefficients of competition are analogous to the partial derivatives of chemical potentials so that

$$\gamma_{ij} \sim \frac{1}{T}\frac{\partial \mu_i}{\partial C_j}. \tag{2.6.57}$$

I think that these and other analogies will be helpful if we intend to construct some form of phenomenological thermodynamics of biological communities.

Since L (and also correspondingly, $\delta^2 S$) is to increase when going further from the stable equilibrium then this may be regarded as a peculiar form of the *Le Chatelier principle*: any displacement from a stable equilibrium increases the competition expenses within the community.

2.6.9 Exergy as a Lyapunov Function

It is obvious that our class of Lyapunov functions is very large and so I would like to reduce it using some additional concepts, for instance, thermodynamic ones.

Before starting these thermodynamic speculations we consider the following function $\varphi(\xi) = \xi \ln \xi - \xi + 1$ (see 2.6.33). It is obvious that

$$\varphi(1) = d\varphi / d\xi(1) = 0; \; d^2\varphi / d\xi^2 > 0 \tag{2.6.58}$$

and the corresponding Lyapunov function will be

$$L = \sum_{i=1}^{n} \{C_i \ln(C_i / C_i^*) - (C_i - C_i^*)\}. \tag{2.6.59}$$

Comparing (2.6.59) and (2.6.22) we see that Jørgensen's *exergy* is formally a Lyapunov function for the system (2.6.13), i.e. $L = Ex$. In Jørgensen's interpretation the equilibrium C^* (a reference state) is identified with a thermodynamic equilibrium when the life is absent. Continuing the chain of logic we may say that the origin of life can be considered as the loss of stability for thermodynamic equilibrium and the movement of the system away from it along one trajectory. In this case (in accordance with the Chetaev instability theorem) if the exergy increases along this trajectory, i.e. the inequality

$$dL / dt = dEx / dt > 0 \tag{2.6.60}$$

takes place, then the thermodynamic equilibrium is unstable. It is easy to see that this is the other formulation for Jørgensen's maximal principle.

A few words about specific thermodynamic constraints. If we want to consider the exergy as a thermodynamic value the following relation must take place (see, for instance, Landau and Lifshitz 1964)

$$Ex(C_1,\dots,C_n) = A\,Ex(p_1,\dots,p_n), \quad \text{where } A = \sum_{i=1}^{n} C_i\,, \tag{2.6.61}$$

i.e. a value like a *specific exergy* (exergy per biomass unit) must make sense. If we remember the formulas (2.6.24), (2.6.26) we can see that

$$Ex(C) = AEx(p) + Ex(A), \tag{2.6.62}$$

i.e. the *exergy cannot be considered as a thermodynamic value* because the additivity (2.6.61) does not take place. We may save the situation in two ways:

1. If we assume that the total biomass is not changed, i.e. $A=const$ then

$$\begin{aligned} Ex(A) &= A \ln\frac{A}{A^*} - (A - A^*) = 0 \ \ and \ \ Ex(C) = AEx(p) \\ &= A \sum_{i=1}^{n} \{p_i \ln\frac{p_i}{p_i^*} - (p_i - p_i^*)\} = AK, \ \ since \ \sum_{i=1}^{n} p_i = \sum_{i=1}^{n} p_i^* = 1. \end{aligned} \tag{2.6.63}$$

   In other words we can consider the Kullback measure K as the specific exergy of one biomass unit.
2. As in thermodynamics of open systems we divide the change of total entropy (dS) into two parts: the first connected to internal processes (d_iS) and the second caused by exchange processes between the system and its environment (d_eS). In a similar way we may divide the total system exergy (Ex) into two parts (see Section 2.6.5):

$$\begin{aligned} &Ex = Ex_i + Ex_e, \ where \ Ex_i = Ex_{inf} = AK \ \ and \\ &Ex_e = Ex_{mat} = A\ln(A/A^*) - (A - A^*). \end{aligned} \tag{2.6.64}$$

   We identify the value Ex_i with internal processes changing only the system structure and the value Ex_e with external processes of exchange between the system and its environment leading only to the change in total biomass. In this case the value Ex_i can be considered as a thermodynamic value.

2.6.10 Conclusions

I have tried to demonstrate how to apply the concepts and methods of classical (and non-classical) thermodynamics to ecological problems. Ecosystems are systems far from thermodynamic equilibrium and when we try to calculate their entropy in a direct way we immediately run into such difficulties that the solution of the problem becomes unrealistic. I suggest the hypothesis of the "entropy pump" and the calculation of entropy production for ecosystems under anthropogenic stress in a circuitous way. Calculated in such a way, the entropy can be used as a measure of environmental degradation under anthropogenic impacts (for instance, intensive agriculture).

Recently Jørgensen (1992) suggested the exergy concept and based on this the method of optimal parametrisation for ecosystem models. From the "non-equilibrium thermodynamics point of view" it has been proved that the exergy concept is one possible corollary of thermodynamic extreme principles.

And finally I have tried to show the deep internal connection between Lyapunov functions which are the main instruments of the stability theory and basic thermodynamic concepts, especially in applications to ecology.

Acknowledgements. I am grateful to S.E. Jørgensen for his helpful comments and criticism. I am indebted also to Mrs. Alison Schlums for her careful linguistic editing of my manuscript.

References

Alexeev VV (1976) Biophysics of living communities. Uspekhi fisich eskikh nauk 120(4):647-676

Jørgensen SE (1992) Integration of ecosystem theories: a pattern. Kluwer Academic Publishers, Dordrecht Boston London

Kerner EH (1957) A Statistical mechanics of interacting biological species. Bull Math Biophys 19:121-146

Kerner EH (1959) Futher considerations on the statistical mechanics of biological associations. Bull Math Biophys 21:217-255

Khinchin AJ (1943) Mathematical foundations of statistical mechanics. Gostekhizdat, Moscow

Kullback S (1959) Information theory and statistics. Wiley, New York

Landau L, Lifshitz E (1964) Statistical physics. Nauka, Moscow

Larcher W (1976) Ökologie der Planzen. 2nd ed, Ulmer, Stutgart

MacArthur RH (1955) Fluctuations of animal population and a measure of community stability. Ecology 36:533-536

Malkin IG (1967) Theory of motion stability. Nauka, Moscow

Margalef RA (1951) A practical proposal to stability. Publ de Inst de Biol Apl Univ de Barselona 6:5-19

Rubin AB (1967) Thermodynamics of Biological Processes. Moscow Univ Press, Moscow

Schneider ED, Kay JJ (1994) Complexity and thermodynamics. Towards a new ecology. Futures 26(6):626-647

Svirezhev YM (1976) Vito Volterra and the modern mathematical ecology. In: Volterra V Mathematical theory of struggle for existence. Nauka, Moscow (the postscript to the Russian translation of this book)

Svirezhev YM (1990) Entropy as a measure of environmental degradation. Proc Int Conf Contaminated Soils, add. volume, Karlsruhe:26-27

Svirezhev YM (1997) Thermodynamics and ecology. Ecological Modelling (submitted)

Svirezhev YM, Logofet DO (1978) Stability of biological communities. Nauka, Moscow (English version: 1983, Mir, Moscow)

Svirezhev YM, Passekov VP (1990) Fundamentals of mathematical evolutionary genetics. Kluwer Academic Publishers, Dordrecht Boston London

Ulanowicz RE, Hannon BM (1987) Life and the production of entropy. Proc Royal Soc London 232:181-192

Weber BH, Depew DJ, Smith JD (eds) (1988) Entropy, information and evolution: new perspectives on physical and biological evolution. MIT Press, Cambridge, Massachusetts

2.7 Thermodynamic Orientors: A Review of Goal Functions and Ecosystem Indicators

Sven E. Jørgensen and Søren N. Nielsen

Abstract

This paper reviews a number of goal functions or ecological indicators: emergy, exergy, ascendency, overhead, ratio indirect to direct effect and specific (structural) exergy. The paper illustrates that the various indicators or goal functions to a high extent are related. The paper discusses the difference between the use of maximum exergy storage, maximum exergy destruction and minimum entropy as goal functions. It discusses how maximum exergy destruction for ecosystem under development and minimum entropy for mature ecosystems both are consistent with the principle of maximum exergy storage for ecosystems in all phases of development. It would therefore be advantageous to apply the principle of maximum exergy storage as ecological goal function and indicator and use it in a formulation of a tentative fourth law of thermodynamics (see Jørgensen and Nielsen, Sect. 2.4). As a conclusion of the paper is proposed a number of useful ecological indicators for an intensive and extensive description of an ecosystem under development or a mature ecological system. Most probably it is necessary to apply several ecological indicators simultaneously to get a proper description of the ecosystem.

2.7.1 Introduction

Goal functions are understood as mathematical functions, that can describe the direction of the ecosystem development. This should *not* be interpreted as ecosystems having predetermined goals, but rather that the self-organization ability of ecosystems makes it possible to meet perturbations by directive reactions which can be described by goal functions.

This volume has already mentioned several potential goal functions, proposed during the last 1-2 decades by several systems ecologists. It has furthermore been proposed (Jørgensen, 1994 and 1997), that the different goals functions at least to a certain extent are just different view points of the same matter. It is not surprising, that the very complex ecosystems need several different view points

F. Müller M. Leupelt (Eds.)
Eco Targets, Goal Functions, and Orientors

to be described properly, when a relatively simple physical phenomenon as light needs two descriptions: waves and particles!

At the workshop 'Unifying goal functions' it was agreed that *goal functions* should be reserved to optimization computation in models, first of all applied by development of structural dynamic models. When the concept was used in relation to the general development of ecosystems, *ecological indicators or orientors* should be used to stress that ecosystems don't have goals, but rather a *propensity* to develop in a given direction. In modelling, we attempt to apply a quantification of the most probable direction as a useful goal function to describe for instance the structural changes.

The following goal functions will be mentioned in this chapter: emergy, exergy, structural exergy, ascendency, both the total ascendency and the contributions from the through-flow, the ratio indirect to direct effect, maximum power and buffer capacities, defined as the relative change of a forcing function to the corresponding change of a state variable.

2.7.2 Review of some Important Goal Functions

The two goal functions emergy and exergy are based upon thermodynamics. Biomass or the free energy of combustion would only correspond to the energy of the ecosystem as "fuel", while emergy and exergy include an energy quality factor. *Emergy* considers how much solar radiation it costs to build a considered organism (biomass) (Odum 1983). The biomass or rather the free energy of the biomass is multiplied with a factor, found from the amount of solar energy it costs to construct one unit of energy in the considered organism. Patten (1992) shows how the cycling of energy in the ecological network can be used for calculation of emergy. The factors used in the calculations below (emergy 1) are based upon Brown and Clanahan (1992).

Exergy considers besides the free energy of the biomass the information embodied in the biomass structure (see Jørgensen and Nielsen Sect. 2.4) The difference between emergy and exergy is that while the prior considers how much solar energy it costs to construct the biomass at different trophic levels, the latter attempts to account for how far the system has been able to move away from thermodynamic equilibrium. That is the ability of the biomass of doing work including the work capacity stored in the information, i.e., the total result of the solar radiation. Different organisms can have different strategies to obtain a certain level of exergy. For instance a plant growing in the shade will have larger leaves and/or a higher chlorophyll concentration to be able to gain the same exergy as plants growing in full sunshine. The two types of plants will have the same exergy, approximately, but very different emergies.

Exergy seems therefore theoretically more correct, to be proposed as an ecological indicator or goal function, but emergy on the other side is relatively easy to interpret, when the energy network is known. In addition, when emergy is calculated for entire ecosystems, the differences in "strategies" between organisms at the same trophic level will to a certain extent neutralize each other. The difference is, that exergy gets the quality factor from the level of information stored in the genes, while emergy gets the quality factors from the costs of solar energy. The exergy storage is a measure of the result of the flow of exergy (low-entropy energy) through the system, while emergy is a measure of how much solar radiation the result has cost to obtain. It should therefore be expected that emergy and exergy are well correlated.

Bastianoni and Marchettini (1997) have calculated emergy and exergy for three lagoons. Two of them are artificial, built by man to treat more or less pre-treated sewage. One is fed with estuarine water mixed with more polluted effluent (named waste pond below). The second receives estuarine water and well treated sewage water after a treatment plant (denoted control pond). The last lagoon (Caprolace) is a natural system, placed in a national park. The authors used the ratio of emergy (unit seJ = solar equivalent Joule) to exergy to get further information on the state of the system. They showed that the density of solar equivalent required to maintain or create one unit of organization measured by the exergy content, decreases in the sequence waste pond > control pond > Caprolace.

The results are summarized in Table 2.7.1. The authors also show that the exergy increased and the emergy/exergy ratio decreased with time for the waste pond and control pond as a measure of an increased organization. These results are not surprising, as emergy accounts for the environmental cost, measured in solar equivalents, while exergy is a measure of the organization expressed as units of first class energy (potential for work).

Table 2.7.1 Emergy and exergy density and emergy/exergy ratios for three lagoons (after Bastianoni and Marchettini, 1997)

Lagoon	Energy density (100 MseJ /l)*)	Exergy density (TJ /l)	Emergy / exergy ratio 0.0001 seJ/J
Waste pond	31.65	0.38-0.85	37-84
Control pond	20.10	0.54-2.60	7.7-37
Caprolace	0.92	3.90-4.25	0.22-0.24

Exergy and emergy will increase with increasing concentrations of nutrients. If we, however, want to get a measure of the ability to use the *available resources*, we can apply specific or structural exergy, which is exergy divided by the sum of concentrations of all the components in the ecosystem that is ð c_i, where c_i symbolizes the concentration of the ith component. The specific exergy is with other words the exergy relative to the biomass concentration or nutrient level. It is an intensive attribute opposite exergy, which is an extensive variable. It accounts for the work capacity of the structure, including the contribution from the information, but it is independent of the nutrient level, i.e., the available resources.

Ulanowicz (1986) has introduced the concept of ascendency, A, (see Ulanowicz, Sect. 2.10) to account for the through-flow of energy in an ecosystem, T, and the network. Ascendency accounts for the size of the network, T, and for the information stored in the network, I. A is an extensive variable (dependent on the size of the system). The size term, T, is dominant in most calculations. (A-T) / T = I - 1 is an intensive (independent on the size of the system) attribute and accounts to a certain extent only for the structure or information embodied in the network, i.e., the complexity of the network, independent of the energy through-flow. It might therefore be expected that the specific exergy is well correlated to A-T/ T, while the exergy would be better correlated to the size term, T, although probably also well to A, as the size term is most dominant.

Patten (1991) has shown that *the indirect effects, I, dominate the direct ones, D;* see also Patten Sect. 2.8. The details of the calculations of the direct and the indirect effects are given in Patten (1991, 1993). In a eutrophication model, the direct effect on the influence of nutrients on the growth of phytoplankton is the growth rate of phytoplankton, regulated by the nutrient concentrations. The indirect effect is the role of the cycling of nutrients through the food chain and mineralization in an infinite number of times. The indirect effect is a result of the entire network and originates from all other links than the direct ones. The ratio indirect to direct effect will therefore give a measure of the network complexity. The more pathways the network has, the bigger role will the indirect effect play, compared with the direct one. The ratio I/D is therefore expected to be well correlated to specific exergy.

2.7.3 Comparison of Goal Functions by Model Application

Three different eutrophication models were used for a comparative examination; see Jørgensen (1994b). Model number one has only three state variables, namely phytoplankton, detritus and nutrients, while zooplankton has been added to model number two, and model number three contains zooplankton and fish in addition to the state variables of model number one. The equations for the three models can be found in Jørgensen (1994b). Different editions of the three models were developed by changing one of the key parameters and by changing the input of nutrients. Totally 16 cases have been examined. They are shown in Table 2.7.2. The key parameter for model 1 was considered the maximum growth rate of algae, μ-max. The key parameter for model number two is considered to be the maximum growth rate of zooplankton μZ-max. For the last model the growth rate of fish, PRED, was used as the key parameter. The last case study included in Table 2.7.2, case study 16, has not been included in the final analysis of the correlations, as this edition of model number two showed a chaotic behavior with many, violent and irregular fluctuations. Case studies 9 and 12 showed fluctuations, too, although without chaotic behavior, as the fluctuations were repeated regularly.

Table 2.7.2. Parameters and buffer capacities of the 15 examined models

Model	# of state variable	Variable parameter	Parameter value	Input conc. of nutrients	ß-Ex*)	ß-phyt¤)
1	3	μ-max	0.2	12	0.86	0.83
2	3	μ-max	0.25	12	0.90	0.90
3	3	μ-max	0.15	12	0.80	0.80
4	3	μ-max	0.4	12	0.93	0.93
5	3	μ-max	0.6	12	0.95	0.96
6	3	μ-max	1.0	12	0.96	0.97
7	3	μ-max	1.0	10	0.95	0.97
8	4	μZ-max	0.35	12	0.59	1.18
9	4	μZ-max	0.42	12	0.59	0.23
10	4	μZ-max	0.42	10	1.22	-0.86
11	5	PRED	0.14	12	0.78	-3,80
12	5	PRED	0.20	12	0.64	-0.96
13	5	PRED	0.115	12	0.77	-16.6
14	5	PRED	0.14	10	0.70	-3,22
15	5	PRED	0.14	15	0.88	-4.60
16	4	μZ-max	0.56	12	violent oscillations	

*) The change in exergy relative to the change in phosphorus input
¤) The change in phytoplankton concentration relatively to the change in phosphorus input

All model cases except case 16, were run until steady state or until the fluctuations were repeated several times (this was the case for study 9 and 12). Exergy, structural exergy, emergy (1), T, ascendency, A, the ratio indirect to direct effects, A-T/T and emergy (2) were calculated. The results are shown in Table 2.7.3.

The results of a correlation analysis of the computed goal functions for the 15 model runs are shown in Table 2.7.4.

Notice, that the correlation between structural exergy and I/D is 0.99. The correlation between exergy and ascendency is 0.971 and the corresponding relation is shown Fig. 2.7.1.

Table 2.7.3. The goal functions calculated for the 15 cases

state var.	exergy	struc. ex.	emergy 1	T	A	I/D	A-T/T	
3	36095	859	41682	10.29	17.95	1.68	0.7444	
3	37126	866	42881	10.59	18.77	1.69	0.7724	
3	34123	845	39397	9.74	16.55	1.65	0.6992	
3	38500	875	44462	10.98	19.81	1.68	0.8042	
3	39180	880	45265	11.19	20.31	1.67	0.8150	
3	39700	882	45387	11.32	20.71	1.67	0.8295	
3	32910	881	38008	9.39	15.65	1.67	0.6666	38008
4	704655	14867	142000	16.1	29.5	2.92	0.8323	82123
4	1728500	34960	264000	24.48	53.32	7.7	1.1781	84120
4	1470000	36000	229000	19.88	39.61	7.7	0.9925	69970
5	1995530	38989	602007	22.7	47.07	7.5	1.0736	93527
5	159669	3463	137870	11.44	21.2	1.8	0.8531	55130
5	1857800	37460	373544	23.2	48.8	8.1	1.1034	88594
5	1507661	35618	502379	17.56	32.4	7.5	0.8451	78542
5	2676843	41694	730720	30.0	64.6	7.5	1.1533	124090

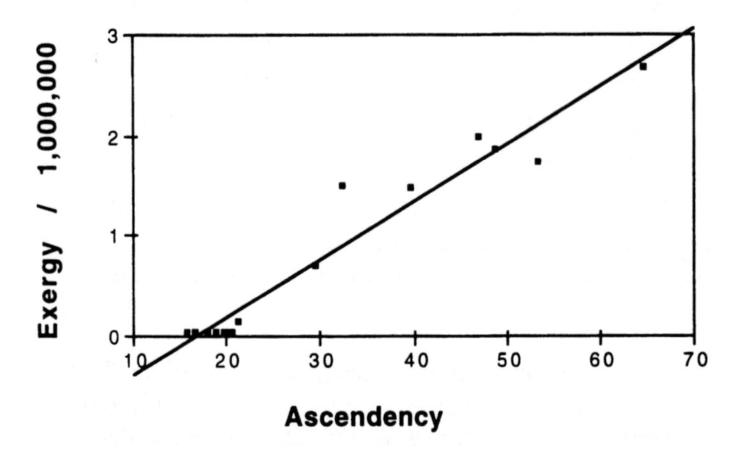

Fig. 2.7.1. The correlation between exergy and ascendency for the 15 case studies

A correlation analysis of the first 7 cases of model number one was also carried out. The results are summarized in Table 2.7.5. Particularly, the correlations between any pair of exergy, ascendency, T, emergy and A-T / T are very strong as indicated in Table 2.7.5. The relation between exergy and ascendency is shown in Fig. 2.7.2.

Table 2.7.4. Matrix of correlation coefficients for the 15 different case studies, representing 3 different models

ex	1,000	0,978	0,933	0,982	0,971	0,952	0,896	0,954
struc. ex	0,978	1,000	0,888	0,941	0,924	0,990	0,866	0,901
emergy (1)	0,933	0,888	1,000	0,880	0,862	0,841	0,757	0,917
T	0,982	0,941	0,880	1,000	0,998	0,912	0,944	0,960
A	0,971	0,924	0,862	0,998	1,000	0,899	0,959	0,946
I/D	0,952	0,990	0,841	0,912	0,899	1,000	0,859	0,841
A - T / T	0,896	0,866	0,757	0,944	0,959	0,859	1,000	0,860
emergy (2)	0,954	0,901	0,917	0,960	0,946	0,841	0,860	1,000

The results show that for all 15 cases there are good correlations between exergy, structural exergy, T, ascendency and I/D. Exergy also is well correlated with both emergies, although slightly better with emergy (2) than with emergy (1). Emergy is less correlated to the other concepts – lowest to I/D and A-T/T. Christensen, (1992) also has shown a relatively good correlation between ascendency and exergy for 42 different steady state models for a number of different types of ecosystems. Jørgensen (1993) showed that the correlation could be even improved (a correlation coefficient of 0.97 was obtained), when only models of the same type of ecosystem (15 different lakes) were considered. When we consider the same model but different cases, the correlations become even stronger. Exergy is calculated on the basis of the biomass and information stored in the biomass, while ascendency is calculated from the through-flow and the flows in the network. More biomass (and information) generates more flows and the more biomass a system must maintain, the more through-flow is needed for maintenance. The correlations between exergy and ascendency are therefore explainable.

Table 2.7.5. Matrix of correlation coefficients for the seven case studies of model 1

exergy	1,000	0,456	0,999	1,000	1,000	0,361	0,999
struc	0,456	1,000	0,447	0,456	0,461	0,351	0,436
emergy	0,999	0,447	1,000	0,999	0,998	0,378	0,998
T	1,000	0,456	0,999	1,000	1,000	0,356	0,999
A	1,000	0,461	0,998	1,000	1,000	0,355	0,999
I / D	0,361	0,351	0,378	0,356	0,355	1,000	0,372
A - T / T	0,999	0,436	0,998	0,999	0,999	0,372	1,000

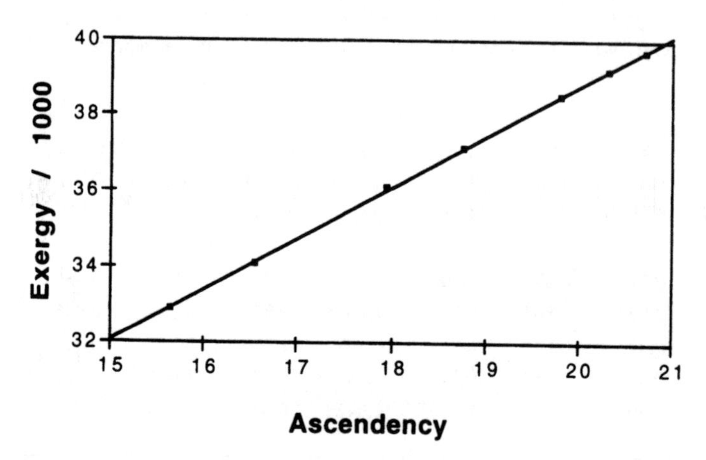

Fig. 2.7.2. The correlation between exergy and ascendency for model 1 (seven cases) is shown

The correlation of emergy to the other concepts became more acceptable, when different cases of the same model were examined. This is not surprising because the differences between the goal functions of the three models are wider than the differences between the goal functions of the same model, but with different parameters and/or inputs. It is, however, noticeable, that the correlation between I/D on the one side and exergy, specific exergy, T and ascendency on the other side is much weaker for the cases of the same model (see Table 2.7.5) than when all the results are considered: (compare Tables 2.7.4 and 2.7.5). This is according to the results obtained by Jørgensen (1992) and (1997): the same network gives independently of the input and to a certain extent of the parameters the same ratio I/D, while the exergy and to a less extent also the structural exergy will reflect the differences in inputs. The ratio I/D is dependent on the number of linkages in the network, but is very little influenced by any changes within the framework of the same network.

The general results seem to support the idea presented in the introduction to this chapter, that the different goal functions express different view points of the same matter, namely the *self-organization ability* of the ecosystem.

It was found that the goal functions were all relatively small for the case number 16, see Table 2.7.2. This is the case, where the system behaves chaotic, i.e., with violent and irregular fluctuations. Regular fluctuations opposite steady state situations seem to coincide with relatively low buffer capacities; see Table 2.7.2. It may be anticipated, that series of different buffer capacities should be included, if a correlation to the goal functions should be found. The present results do not give any clear picture of such relationships.

2.7.4 Other Comparisons of Goal Functions and Ecological Indicators

Various ecological indicators are generally used to assess the ecosystem health (Costanza 1992). These results may also be used to find correlations between two or more goal functions. The observed results may be summarized in the following points:

1. Exergy may be correlated to the sum of buffer capacities; see Jørgensen and Meyer, 1977.
2. Specific exergy accounts for the utilization of the available resources.
3. Specific exergy and overhead are following the same trends in many examinations.
4. Specific exergy, to a certain extent overhead, Finn's cycling index, Shannon's index, biomass relative to production and harvest relative to the through-put are all following the same trends for agricultural systems (Dalsgaard 1996) and seem all to be able to measure the sustainability of these systems, although the number of agricultural systems examined in this context is limited.

The ratio indirect to direct effects is not well correlated to the exergy (Jørgensen 1997), when the reactions of the same network to different inputs are compared, see also the results in Table 2.7.4. If on the other side the results of different networks are compared exergy and the ratio indirect to direct effects are well correlated which is also in accordance with the results shown in Table 2.7.3.

Nielsen and Ulanowicz (1997) have used the sensitivity of ascendency to examine the influence of increased input, increased output (energy dissipation, as the outputs cover the respiration). They found that the first and last factors will increase the ascendency, while increase of the energy dissipation will decrease the ascendency on the one hand. An increase of the energy dissipation above the average for one component (microorganisms which enhance the cycling of matter) on the cost of the dissipation of other components yields a higher ascendency. The same relationships are found between the size of inputs and outputs and exergy, when a specific model is applied. Increased input means more resources which will give higher exergy, while an increase of respiration will imply that more biomass is converted into heat and less exergy is stored. A faster mineralization will imply a higher concentration of the biological components on the cost of detritus, which will inevitably lead to higher exergy. These examinations confirm therefore the strong correlation between ascendency and exergy as found several times.

2.7.5 Exergy Storage and Exergy Destruction

When the first life emerged on earth about 3,5 billion years ago, the major challenge to the primeval life was to maintain, what was already achieved. This problem was eventually solved in the present form by introduction of DNA or DNA-like molecules which were able to store the already obtained information on how to convert inorganic matter to life-bearing organic matter. It requires, however, both matter and energy to continue the development, i.e., to continuously build more ordered structure of life-bearing organic compounds on the shoulders of the previous development. The question covered in this section is how the development proceeds when matter or / and energy become limiting constraints on the further development.

The amounts of elements on earth are constant, i.e., the earth is today a non-isolated system (energy is exchanged with space, but (almost) no exchange of matter takes place, when we leave out to account for the minor input from meteorites and the minor output of hydrogen), while energy is currently supplied by the solar radiation. The concept of the limiting element, i.e., that the element present in the smallest amount for the construction of organic life-bearing organic compounds relatively to its need determines the amount of the organic compounds formed, is embodied in the conservation laws; see Patten et al. (1997). Further development (evolution), when one or more elements are limiting is consequently not possible by formation of more organic life-bearing matter through uptake of more inorganic matter, but only by a better use of the elements, i.e., by a reallocation of the elements through a better organization of the structure which requires that more information is embodied in the structure. More ecological niches are thereby utilized, and more life forms will be able to cope better with the variability of the forcing functions (input environments). Exergy is not limiting in this phase of development, because each new unit of time brings more energy to combat the catabolic processes and to continue the energy consuming and exergy building anabolic processes. The exergy captured by the ecosystem, Ex_{cap}, is steadily increased under these circumstances (Kay and Schneider 1991 and 1992), because the structure able to capture exergy is increased. This is consistent with the results reported by Akbari (1995). He finds a positive relationship between caloric content (biomass, exergy) and canopy temperature differential of a plant community.

It is, however, not possible to capture more exergy by the ecosystem than the amount of exergy received by the solar radiation. Exergy captured will therefore also at a later stage of development, when a sufficiently large structure has been formed, become limiting. Under these circumstances, the additional constraint on the processes of the ecosystem is a better utilization of the incoming exergy, meaning that less of the exergy captured is used for maintenance and more is therefore available for further growth of the ordered structure, i.e., to increase the exergy stored in the system. Mauersberger (1982, 1985 and 1995), postulates that locally, and within a finite time interval, the deviation of the bioprocesses from a stable stationary state tends to a minimum. This is similar to, and a more formal version of, Lionel Johnson's least dissipation principle (Johnson 1995) which goes back to Lord Rayleigh (see also Patten et al. 1997). Mauersberger has used this optimization principle on several process rates as functions of state variables. As a result, he finds well known and well accepted relations for uptake of nutrients of phytoplankton, primary production, temperature dependence of primary production and respiration. It supports the use of the optimization principle locally and within a finite time interval.

Globally, the openness of ecosystems seems therefore to lead to a biological structure that is able to capture as much as possible of the available solar radiation, i.e., to optimize Ex_{cap} to cover the exergy degradation that is the exergy used for maintenance. Increased stored exergy, Ex_{bio}, implies increased Ex_{cap} (more structure to capture more of the incoming solar radiation which is the source of exergy), but it requires also that more exergy is degraded, as it costs more energy to maintain more structure. The excess entropy, which can be found from:

$$T_{ecosystem} * \text{excess entropy} = \text{exergy degraded to cover the maintenance} \quad (2.7.1)$$

is simultaneously locally minimized, which makes it possible to further increase Ex_{bio}, i.e. $\Delta Ex_{bio} > 0$, as the exergy captured after the maintenance has been covered can be used as ΔEx_{bio}). This leads inevitably to the question: how can we describe the results of the combat between anabolism and catabolism or between the above mentioned global and local optimization principles?

The two theories based on maximization of exergy storage and of exergy captured respectively are not consistent, when we have to describe the further development of a mature system, in which domain Mauersberger's minimum principle is valid. These considerations were leading to the following proposition, which is characteristic for the mature ecosystem:

Ecosystems locally decrease entropy (gain exergy) by transporting energy and matter from more probable to less probable spatial locations.

The generally observed tendency is that all gradients are steadily broken down. The heat is flowing from the highest to the lowest temperature to reduce the temperature difference. Positive electricity will neutralize negative electricity and the electrical charge will disappear. A gas is moving from a high pressure to a low pressure whereby the pressure difference will be reduced and finally disappear. Kay (1984) reformulated even the second law of thermodynamics to express, that "gradients inevitably will break down". The always observed positive production of entropy is a natural consequence of the steady reductions of gradients. Kay and Schneider (1990, 1991 and 1992) consider that the emergence of living organisms as yet another (more sophisticated) method to enhance this tendency of reducing (all) gradients. They use therefore the name "extended version of the second law of thermodynamics" for their hypothesis to underline this interpretation. The open question is however, whether life by Darwin's Theory gives survival, measured for the entire ecosystem by the exergy content, higher priority than to reduce the imposed gradients. Is life a manifestation of combating the reduction of the gradients represented by life? Or is life rather a manifestation of yet another sophisticated method to reduce the imposed gradients? Has the earth more exergy with life than it would have without life? as formulated by Kay (personal communication 1996). Interesting discussions on how to carry out this proposed comparison will probably follow, as it in best theoretical physical style will presume the application of "Gedanken-Experiment". This doesn't change the parallel development of the amount of exergy stored and the amount of exergy captured and mainly used for respiration, as ecosystems develop. As previously mentioned the conflict is limited to the further development of mature ecosystems towards even more stored information, but it can not be excluded that this conflict can be solved by calculations of the exergy of the earth with and without life as proposed by Kay.

Ulanowicz (1997) attempts to unify the two theories by hierarchy theory: A developing system occupies the focal level of our attention. It extracts exergy from the next higher scale at a rapid rate. Some of the available energy is used by the system to create ordered energetic structure (exergy) at the focal level. The bulk, however, is dissipated to the next lower level (the microscopic). Overall, exergy disappears. The idea to unify the two theories by application of hierarchy theory seems sound, but the exergy is gained and utilized in the first hand on the molecular level and should therefore also be extracted from that level by the higher levels. The exergy is "delivered" by photons which react on the molecular level. The biochemical maintenance processes require also exergy on the lowest (local) level. If the exergy demand for maintenance can be reduced, it will give more exergy for construction of structural order on the focal or even higher levels. When the energy demand corresponding to maintenance has been covered, it is possible to store surplus exergy in form of ATP-molecules, which can be used to construct organic molecules which are characteristic for the organic tissue. It is completely consistent with the tentative fourth law of thermodynamics and its interpretation of the theories presented by Kay and Schneider and Mauersberger as valid for ecosystems under development respectively for mature well-balanced systems.

2.7.6 Conclusions

Different goal functions, ecological indicators and orientors have been proposed, but as they all aim toward an expression for the information embodied in the ecological network, it is not surprising that they are strongly correlated. Exergy measures the information of the system, based upon the genes of the organisms, that are forming the ecological network. Structural exergy is exergy relatively to the total biomass, and should therefore be related to the information in the network, independent of the size of the network = biomass represented by the network. Emergy expresses the amount of solar energy, that it has cost to construct the network. Ascendency looks into the complexity of the networks itself, included the amount of mass or energy, that is flowing through the network. The ratio of indirect to direct effect measures the importance of cycling relatively to the direct cause and this ratio is therefore also a measure of the information of the network. The different concepts give slightly different information and are deducted from different viewpoints, but they all contribute to a pattern of ecosystem theory, as also proposed in the introduction to this chapter.

Table 2.7.6. Overview of applicable ecological indicators for an extensive and intensive description of ecosystems under development and of mature ecosystems

Ecosystem Stage	Extensive Description	Intensive Description
Under Development	Exergy captured Exergy stored Ascendency Biomass Indirect effect Emergy	Structural exergy Overhead Production/ biomass Cycling index Indirect effect/ direct effect A-T / T Emergy/ exergy
Mature	Exergy stored Ascendency Minimum excess entropy Indirect effect Emergy	Structural exergy Overhead Entropy production/ biomass Indirect effect / direct effect A - T / T Emergy / exergy

Particularly, the specific exergy and the indirect effect/direct effect ratio, when ecosystems (or models of ecosystems) with different complexity are compared, and exergy and ascendency, T and (A-T) / T for different flow situations in the same ecosystems (models), are well correlated. Structural exergy and I/D for different ecosystems (models) measure the complexity of the structure of the network, independent of the size (the total biomass or the total through-flow of energy). This strong correlation is therefore not surprising. Exergy and ascendency measure both the complexity, the size and the level of information of the system, although the latter is based upon the flows and the former on the storages. The strong correlation between the two concepts, confirmed a few times throughout the chapter, is therefore explainable.

As emergy measures the cost in solar equivalents and exergy the result stored in the ecosystem as biomass, structure and information, it is not surprising that the ratio emergy/exergy may be a strong indicator for the ecosystem development, as it was found by Bastioanoni and Marchettini (1997). The mature ecosystem will offer a high efficiency in the use of solar radiation to maintain and store the ecological structure; see Table 2.7.1.

Table 2.7.6 attempts to summarize the result of this chapter. For ecosystems at an early stage (under development) and at a mature stage (developed) applicable goal functions are proposed for a description of the ecosystem development. Both an extensive description considering the amount of the available resources (includes size terms) and an intensive description accounting for how well the ecosystem is able to utilize the available resources are taken into account in Table 2.7.6 (Jørgensen 1997).

References

Bastianoni S, Marchettini N (1997) Emergy/exergy ratio as a measure of the level of organization of systems. Ecological Modelling 99:33-40

Brown JH (1995) Macroecology. The University of Chicago Press, Chicago and London

Jørgensen SE (1994) Review and comparison of goal functions in system ecology. Vie Milieu 44 (1):11-20

Jørgensen SE (1994a) Fundamentals of Ecological Modelling (2nd Edition). Developments in Environmental Modelling 19. Elsevier, Amsterdam, New York, Tokyo

Jørgensen SE (1997) Integration of Ecosystem Theories: A Pattern (2nd Edition). Kluwer, Dordrecht

Kay J, Schneider ED (1992) Thermodynamics and Measures of Ecological Integrity. In: Proc. "Ecological Indicators". Elsevier, Amsterdam

Mauersberger P (1979) On the role of entropy in water quality modelling. Ecolog Modell 7:191

Mauersberger P (1982) Logistic growth laws for phyto – and zooplankton. Ecolog Modell 17:57-63

Mauersberger P (1983) General principles in deterministic water quality modelling. In: Orlob GT (ed) Mathematical Modelling of Water Quality: Streams, Lakes and reservoirs, International Series on Applied System Analysis 12. Wiley, New York, pp 42-115

Mauersberger P (1985) Optimal control of biological processes in aquatic ecosystems. Gerlands Beitr Geiophys 94:141-147

Mauersberger P, Straskraba M (1987) Two approaches to generalized ecosystem modelling: thermodynamic and cybernetic. Ecological Modelling 39:161-176

Odum HT (1983) Systems Ecology. Wiley Interscience, New York

Patten BC (1991) Network ecology: indirect determination of the life-environment relationship in ecosystems. In: Higashi M, Burns TP (eds) Theoretical Studies of Ecosystems: The Network Perspective. Cambridge University Press, pp 288-351

Ulanowicz RE (1986) Growth and Development, Ecosystems Phenomenology. Springer-Verlag, New York

2.8 Network Orientors: Steps Toward a Cosmography of Ecosystems: Orientors for Directional Development, Self-Organization, and Autoevolution

Bernard C. Patten

Abstract

A preliminary cosmography of ecosystems is developed consisting of 20 properties. Ecosystems are networks (P1), hierarchically organized (P2), of interacting agents; their abiotic agents respond to physical stimuli, whereas their biota make models and respond to these. Ecosystems are collections of their agents' environments (P3) superimposed on a large-number web (P4) of complex and intricate (P5) pathways. These pathways partition the constituent environments (P6), and involve both direct and indirect linkages (P7). The latter cause equalization of flows (network homogenization, P8) and amplification of inputs (network amplification, P9). Ecosystems, as dissipative structures, tend to increase their distance from thermodynamic ground (P10). Agents in ecosystems are quantitatively dominated by indirect interactions (P11). Ecosystem networks transform direct interactions into qualitatively different indirect interaction types (P12). Because of dominant indirect effects (P11, P12), determination in ecosystems is predominantly holistic (P13). This holism enables ecosystems to provide positive utility to their constituent biota beyond that obtained in direct interactions, which may be positive, neutral, or negative (network synergism, P14). Ecosystems are model-making complex adaptive systems (P15) which perpetually create and proliferate new niches for life (P16). They are also inheritance systems, containing both (internal) genotypes and (external) envirotypes associated with their constituent phenotypes (P17). Ecosystems are cybernetic systems with nondiscrete, distributed control (P18), under which they coevolve (P19). Finally, ecosystems are fundamental units for expansion of the "inner space" of reality (P20). These properties are presented as directional tendencies, that is, as developmental, organizational, and evolutionary "orientors."

F. Müller M. Leupelt (Eds.)
Eco Targets, Goal Functions, and Orientors

2.8.1 Introduction

For over ten years Sven Jørgensen (University of Copenhagen), Milan Straškraba (Czech Academy of Sciences) and I have been interacting around a set of ideas on the topic of "Ecosystems Emerging". We set out to explore how much of ecosystem properties could be derived from eight basic principles (conservation, dissipation, openness, growth, constraint, differentiation, adaptation, and coherence; Jørgensen et al. 1992). The exercise has proved challenging. Over the years it has produced lively interactions, and put us in touch with broader perspectives on ecosystems than we might otherwise have experienced. In "complexity science" (Waldrop 1992; Gell-Mann 1994; Kauffman 1993, 1995; Bak 1996), ecosystems are regarded as "far-from-equilibrium" complex adaptive systems expanding into the "adjacent possible" under "self-organized criticality" near the "edge of order and chaos." The book Complex Ecology (Patten and Jørgensen 1995) explores a variety of ecological perspectives on this subject. It is becoming clear in the investigation of ecological complexity that the popular goal of keeping ecology a strictly empirical science (Peters 1991), unaided by theory, is untenable in the long run. The realities of large scales, higher-order interactions, and indirect effects are incompatible with research styles that needlessly restrict the kinds of questions that can be asked and answered about complex natural phenomena (Brown 1995).

What shapes the development, organization, and evolution of ecosystems, where do they fit in the scheme of world organizations, and how can theory serve to clarify this? Can a cosmography of ecosystems be constructed? How might this be shaped by the "eco-targets", "goal functions", and "orientors" of this book? Broadly, these are directional criteria accounting for systems being pushed (by "forcing functions") or pulled (by "attractors") toward certain configurations, either with or without goals. In this chapter I will use Bossel's (1992) term "orientor" as a somewhat neutral concept with respect to "pushing" and "pulling." The chapter explores some properties of ecosystems that seem relevant to a cosmography, and at the end of each section an orientor statement is given.

2.8.2 Twenty Properties

Ecosystems, as complex adaptive systems, share in their complexity the following organizational properties:

Property 1: Ecosystems are Networks of Interacting Agents

Each ecosystem is a network of living and nonliving agents whose activities are determined by their interactions, involving the transfer and transformation of energy, matter, or information. Each agent is open to energy and matter, and changes by mapping inputs and states into new states and outputs. Agents are objects if nonliving and subjects otherwise. Objects respond to physical (ontic) inputs; subjects respond to both physical and modeled (epistemic, phenomenal) inputs. There are two kinds of causal interactions: transactions, involving direct exchange of energy or matter between two agents, and relations, which are indirect consequences of transactions. Transactions are conservative interactions whereas relations are informational and therefore nonconservative. Collections of interacting agents comprise networks, which may be explicit and physical, or implicit and aphysical (e. g., epistemic). Interactive networks delimited in time, space, or both are subnetworks, or systems. The boundaries of systems can be either real or imaginary. Causality can be direct (local, proximate, etc.), indirect (global, ultimate, holistic, etc.), or a combination of both. Every agent is itself a system, a bounded network of other lower-scale agents interacting within the confines of a defined system.

Orientor statement. Ecosystems are networks of interacting agents, biotic and abiotic, whose maximization of types and numbers gives the natural direction for change.

Property 2: Ecosystems are Hierarchically Organized

Reality and all its agents are hierarchically organized. Müller (1992) has reviewed the application of hierarchy perspectives to ecosystems. Hierarchical organization derives from the span of energy strengths between the four forces, and from the opposing tendencies of matter to spontaneously degrade and spontaneously self-organize. The bond energies associated with agent transactions are inverse to scale. Stronger transactions correspond to shorter time scales, smaller space scales, and higher behavioral frequencies. Weaker transactions entail longer times, larger spaces, and longer behavioral frequencies (Simon 1973). Thermodynamically, agents are dissipative (entropy-producing) systems whose degradation orients reality toward a global attractor of disordered randomness—called the state of thermodynamic equilibrium, maximum entropy, or heat death. Agents are also self-organizing (order-producing) systems whose nonrandomness represents departure from thermodynamic equilibrium. Departure sets up concentration and force gradients, producing states of disequilibrium referred to as exergetic, negentropic, antientropic, etc. These ordered states are maintained by work, involving energy input and expenditure, and entropy production and output. The two tendencies—to order (self-organization) and disorder (dissipation)—are diametrically opposed. Agents and their interactions tend to aggregate and lock-in from the "bottom–up" at discrete regions (scales) of spacetime. This produces scale-hierarchical organization in which parts become building blocks of wholes, within which as opposed to between which interactions are more concentrated. The scale-hierarchies are bi-directional; from any position they extend downward-and-inwardly to lower levels of organization, and upward-and-outwardly to higher levels. Every agent is a hierarchical system. All complex adaptive systems, spanning some 63 orders of energy magnitudes (the range of the four physical forces), contain many (a large number) levels of organization; this is true of ecosystems. Buss (1987) observed that in biological evolution natural selection operates on progressively higher units; lower units become fixed and resistant to further change, and it is such lock-in that makes the development of higher units possible.

This is a prescription for ecological organization also, involving autoevolution (without selection, Lima-de-Faria 1988) in addition to conventional organic evolution.

Orientor statement. Ecosystems are hierarchically organized systems in which the maximization of levels, involving fixation of lower levels leading to creative diversification of higher levels, provides a natural direction for change.

Property 3: Ecosystems are Collections of Environments

Each agent in a system is the locus of two environments produced by its external interactions with other agents. An ecosystem is a collection of such agents and their environments.

An agent's input environment is an afferent, convergent network from the past (history), and its output environment is an efferent, divergent network to the future. Each such environmental pair is defined by, and also defines, its agent.

Since causality cannot be propagated between agents more rapidly than the speed of light, as the remaining existence time of any agent diminishes with time, and as propagations of effects expand into the future beyond the existence time of their originating agents, input and output environments are bounded in spacetime by, respectively, converging and diverging light cones. Narrower restrictions of agents' environments to within systems containing the agents as members, are termed input and output environs. These are referenced to those particular agents in those particular systems.

The transactive input and output environs of all the agents in a system form past- and future-oriented partitions of the system (Patten 1978), meaning that a unique portion of the energy and matter of reality attaches to each agent in existence, and the transactive input and output environments of all agents taken together sum to the totality of physical reality.

Orientor statement. Ecosystems are sets of environments whose diversification, consistent with that of agents (Property 1) and levels (Property 2), sets natural directions for change.

Property 4: Ecosystems Contain a Large Number of Interactive Pathways

Agents in networks are interrelated by pathways with a number of special properties.

In the terminology of directed graphs (digraphs), a path in a network is a node-arc sequence; nodes (points) denote agents and arcs (arrows, or links) their direct interactions. Direct determination is manifested between agent pairs by transactions and implicit relations associated with direct paths (individual arcs); indirect determination is manifested by transactive sequences and implicit relations associated with indirect paths. The number of nodes in a path from an originating agent (as, in ecosystems, a primary producer) to a terminating agent (e. g., top carnivore) determines path length; direct paths or links (arcs) have length 1, and indirect paths lengths 2, 3, 4, ...

The number of paths between any two agents in a network increases exponentially as a function of path length, to the point where ecosystems contain enormous path numbers interconnecting their constituents. Generally speaking, long paths take commensurately longer times to travel from start to finish.

Transactive path strength is measured by the amount of substance carried. Strength lessens exponentially as path length increases because second-law dissipation accompanies every energy-matter transfer or transformation. The corollary is that direct links are the strongest in any path sequence they initiate, and subsequent lengthening of paths involves successive weakening of links.

Path-mediated transactions and relations between agents in ecosystems are complexly determined.

Orientor statement. Ecosystems contain large numbers of interactive pathways whose indefinite proliferation within and across scales is a perpetual tendency.

Property 5: Ecosystem Pathways are Complex and Intricate

Nature's "web of life" is a complex reticulum that can be decomposed into definable units based on conservative transactions, as shown in Figure 2.8.1a.

A path between any two agents in a network is simple or acyclic if it contains no repeated nodes (agents) along its length. A simple path from node j to node i can be notated α_{ij}; the simplest of these is a direct link, a_{ij}. The set of simple paths is $A_{ij} = \{\alpha_{ij}\}$.

A cycle is a compound path, one with the same originating and terminating node. A cycle of length 1 is a loop or self-loop, and in transactive networks this is used to denote an impedance of flow, that is, storage delay. Cycles (and loops) greatly complicate the pattern and temporality of substance transfers over pathways. A cycle at any node k may be labeled ω_{kk}, and the set of such cycles is Ω_{kk}.

Three primary cycle types can be recognized leading from agent j to agent i in a system: originating cycles that start at initiating nodes, j, but never touch the terminal or destination node, i ($\Omega_{jj|i}$, the set of cycles at j that do not touch i);

medial cycles associated with intermediary nodes, k, in paths from j to i but that do not touch either j or i (Ω_{kk}, $k \neq i,j$); and terminal cycles that start and end at destination nodes, i, but never touch the originating node, j (these are denoted Ω_{iiij}, the set of cycles at i that do not touch j). Nested cycles may occur at shared nodes within these primary cycle types, and nesting can be quite complex. Cycles are orienting in networks because they serve as attractors that tend to keep energy and matter flowing in regular circuits.

There are nine primary path categories to consider in understanding pathways between any selected pair of originating, j, and terminating, i, nodes in a network. These are: 1-direct links (arcs) from j to i (a_{ij}); 2-simple forward paths, including a_{ij} (A_{ij}; "\" in Figure 2.8.1a means a_{ij} is excluded since its is explicitly depicted); 3-compound forward paths (Ω_{ij}, containing medial cycles, Ω_{kk}, $k \neq i,j$); 4-originating node cycles (Ω_{jjji}); 5-direct feedback links from i to j (a_{ji}); 6-simple feedback return paths from terminal node i to originating node j (A_{ji}, where a_{ji} is included); 7-compound feedback return paths (Ω_{ji}, with medial cycles); 8-nonfeedback terminal node cycles (Ω_{iiij}); and 9–dissipative paths, ($\Delta_{[\bullet]i}$), that lead permanently away from the terminal node for other destinations [•] outside the path nexus interconnecting j and i and never return. Note that there are also dissipative paths that lead away from j and never reach i, but these are irrelevant to the considered interaction directed from j to i. Figure 2.8.1a shows all but the last of these path sets.

In addition, three secondary path sets can be constructed from the primary sets to account for substance flow from any originating to any terminating node in a network: first-passage paths carry material to the terminal node for the first time, recycle paths account for subsequent visits to the terminal node, and dissipative paths carry substance permanently away from the terminal node. First-passage paths are the set $\Phi = \{2\bullet(3\vee4)\}$, the concatenation (•) of category 2 with the disjunction ($\vee$) of 3 and 4; recycle paths are the set $P = \{6\bullet((7\vee8)\bullet\Phi)\}$.

Energy-matter transactions along first-passage paths carry mode 1 determination, recycle paths are responsible for mode 2 determination, and dissipative paths carry mode 3 determination away from the terminal node nexus and eventually to the system exterior. Numerically, at steady state, mode 1 = mode 3 transactions.

Orientor statement. Ecosystem pathways are complex and intricate, and the tendency for them to become moreso, consistent with proliferation (Property 4), is a natural direction in development, organization, and evolution.

Property 6: Ecosystem Pathways Partition Constituent Environs

Each environ encompasses a nonoverlapping share of physical reality.

The three modes of network determination described above form a partition of the j'th contribution to the energy and matter in both the input and output environs of each i, as illustrated in Figure 2.8.1b. The input environ of each terminal node contains two partition elements: afferent mode 1 paths from all agents whose efferent paths reach the node in question, and afferent mode 2 paths. The output environ of each terminal node also consists of two partition elements: efferent mode 2 paths, and mode 3 paths.

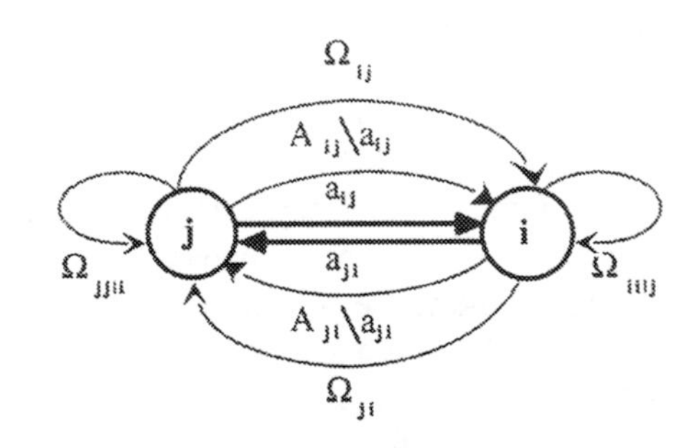

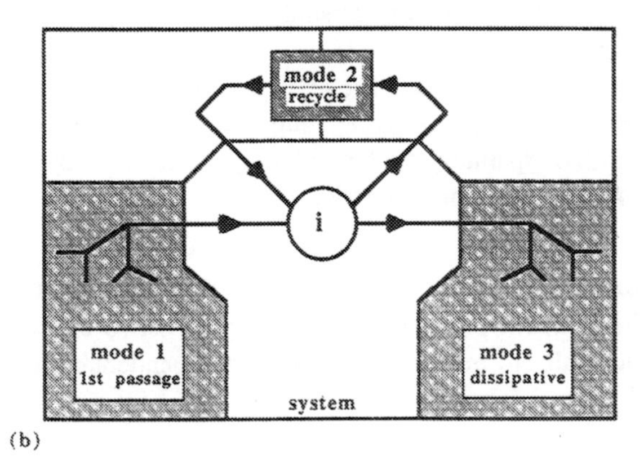

Fig. 2.8.1. **(a)** Categories of paths leading from agent j to agent i in an arbitrary network (subscripts are best read from right to left): a_{ij} and a_{ji} = direct links (path length 1); $A_{ij}\backslash a_{ij}$ and $A_{ji}\backslash a_{ji}$ = acyclic paths of lengths > 1, i. e., excluding (\) direct links; Ω_{ij} and Ω_{ji} = paths with medial cycles (not touching i or j); and Ω_{jjii} and Ω_{iiij} = cycles at j and i, respectively, not reaching the other (these sets differ at j and i for different selections of i and j, respectively). **(b)** Path partitioning of the input and output environs associated with agent i in a system: Mode 1 = first passage paths, Φ_{ij}, $\forall j$; Mode 2 = recycling paths, $(\Omega_{iiij}\backslash\Phi_{ij}\bullet\Omega_{ji})\bullet((A_{ij}\cup\Omega_{ij})\bullet\Omega_{jjii})$; Mode 3 = dissipation paths, $\Delta_{[\bullet]ji}$, which leave i for other destinations [•] and never return.

The completeness of networks implied by accounting for all paths of all kinds and lengths from all nodes (seen as originating) to all other nodes (seen as terminating) in a system is a property referred to as transitive closure. Coherent behavior, and many of the orientor tendencies of systems discussed in this book, would be impossible without this property.

Orientor statement. Maximization of cycling, and thus of the mode 2/mode 1 ratio in input environs, and the mode2/mode 3 ratio in output environs, is an ecosystem orientor.

Property 7: Ecosystem Agents Interact by Direct and Indirect Pathways

A venerable theme in philosophy and ecology is that all things in nature are connected together—the "web of life" idea.

If direct interactions (arcs) in an ecosystem network are represented by ones in an adjacency matrix, $A = (a_{ij})$, and noninteractions by zeros, then typically only a small percentage of the entries in such matrices is nonzero. That is, most agents do not interact directly with one another. Adjacency matrices can be multiplied m times, A^m, to enumerate indirect paths of lengths m = 2, 3, 4, ..., etc. When this is done, depending on the structure of A, most or all of the cells in the power matrices tend to fill up, that is, become nonzero. To this extent, all the agents in the system are mutually dependent, over indirect pathways even though not directly linked.

A necessary condition for all agents to be mutually dependent—universal interconnection—is a spanning cycle, one that includes all agents as touching nodes. Transactive spanning cycles naturally exist in ecosystems to the extent that matter undergoes biogeochemical cycling in the effectively closed ecosphere of the planet.

Orientor statement. Agents, particularly biotic units, operate atomistically to maximize their direct dependencies in ecological networks, but ecosystems self-organize holistically to maximize the indirect dependencies among their agents. Cycling (Property 6) is a major factor in this.

Property 8: Ecosystems Tend to Equalize Energy-Matter Flows between Constituents

This property is *network homogenization* (Patten et al. 1990).

It represents a major departure from traditional trophic-dynamics (Lindeman 1942), which holds that inefficient energy transactions along food chains cause progressive energy shortages at succeeding trophic levels. Ecosystems are widely understood to run out of energy after only a small number of transfer steps beyond photosynthetic fixation. That models tend to be made in accordance with the accepted paradigm of strongly acyclic food-chaining (mode 1, simple paths) and weak food-cycling (mode 2) serves to reinforce the presumption. However, when cycling is realistically introduced, indirect effects become dominant (Property 11, below), and energy flows become more equally distributed over all the agents at all trophic positions in ecosystems. The basis for this is that energy and matter are not uniformly introduced into ecosystems as inputs to all agents. In autotrophic ecosystems, primary producers only receive solar inputs which usually are massive compared to other sources. In heterotrophic systems, allochthonous organic matter inputs are initially usable by only a few trophic guilds. Therefore, in ecosystem networks the initial distribution of newly introduced bioenergy tends to be highly heterogeneous. However (Patten 1985; Higashi et al. 1993; and elsewhere), the process of achieving transitive closure in networks with high cycling takes substance everywhere and tends to homogenize the embodied distribution at dissipation.

Transitive-closure flow matrices representing these relationships tend to have equal values, rowwise and columnwise, denoting approximately equivalent flows from all agents as sources to all others as destinations in the system. Thus, ecosystems tend by their indirect path structure to distribute initially heterogeneous incoming energy and matter more or less uniformly over the interconnection network. This is network homogenization. To the extent that it is true, it denies (at least for the idealized steady-state case, which was also the Lindeman case) the main pillar of traditional trophic-dynamic ecology – that ecosystems run out of energy after only a few transactional steps due to low transfer efficiencies along food chains. The short, discrete food chains and small acyclic food webs on which the classical theory rests are really subpaths and subwebs embedded in much more complex reticular structures (Property 5), with combinatorially great numbers of transfer pathways (Property 4). The conclusion from network homogenization is that most organism groups in ecosystems tend to have about equal access to initially unequally distributed resources.

Orientor statement. Ecosystems tend to develop network configurations that equalize the flows of metabolizable energy-matter between their constituent agents. Orientation to cycling (Property 6) is a critical factor in this.

Property 9: Ecosystems Derive More than Face Value from their Inputs

This property is *network amplification* (Patten et al. 1990).

In it, each unit of resource input to an agent in a network is "amplified" to produce, at steady state, more than one unit of throughflow derived from the unit input. The basis of this property is mode 2 cycling, as explained in the cited reference. At the terminus of mode 1 paths, less energy always remains than at path origins due to first-passage dissipation (Patten and Higashi 1995). Consequently, the first passage flow to i from j must always be less than the flow introduced at j; the first-passage mode is strictly nonamplifying (even though in complex systems this mode would typically include cycling at both originating and medial nodes). This is one of the many paradoxes of network organization. In mode 2, the free energy associated with the chemical bonds of cycling organic matter is carried around as internal energy within the matter itself, and is progressively dissipated as this is metabolized and remetabolized through the ecosystem to dissipation (mode 3).

Network amplification is a truly remarkable property, a subtle accomplishment of network organization that does not, though on the surface it seems to, violate any of the laws of thermodynamics. Based on it, ecosystems can be seen as tending to derive more than face value from their inputs.

Orientor statement. Ecosystems tend to develop network organizations that, through cycling (Property 6), enhance network amplification.

Property 10: Ecosystems Increase Distance from Thermodynamic Ground

As stated in the Introduction, complex adaptive systems are said to be poised in their dynamics on the edge between order and chaos (Kauffman 1993, 1995), in a state of self-organized criticality (Bak 1996). As members of this class, ecosystems also self-organize and evolve toward order at the balance point of a thermodynamic dialectic that is inherent in their network organization (see Figure 2.8.2.).

In network dynamics, order and organization are produced first by first-passage transactions of energy and matter. This is because as substance is passed along trophic and nontrophic chains implicit relational information develops also as more degrees of freedom become established between interacting agents. As the transactions develop arithmetically, interconnecting some but not all of the agents directly, corresponding relations develop geometrically to interconnect all or most of them indirectly. This development of relational connectance faster than transactional connectance represents antientropic (or negentropic, exergetic, ascendant, emergetic, etc.) climb to network order. This is the direction of ecosystem development – increasing distance from thermodynamic ground. Exergy comes from thermodynamics, negentropy from thermodynamics and information theory, and ascendancy and emergy from ecological network analyses. A common basis in network properties for exergy, emergy, and ascendancy as extremal principles or "goal functions" for ecosystems has been suggested by Patten (1995). A quantifiable degree of order is attained in the first-passage mode, shown as the ascending staircase in Figure 2.8.2. This implies ordering movement against the second-law dissipative gradient. We can therefore recognize a first phase of network organization, in which far-from-equilibrium distance is gained away from thermodynamic ground. Exergy, emergy, ascendancy, and negentropy (as well as many of the other orientors in this book) are all acquired in this phase.

The acquisition of initial order in mode 1 has a maintenance requirement against second-law degradation. This is fulfilled by mode 2 cycling, in which the order and organization gained in mode 1 are maintained by recycling transactions. This maintenance of achieved states far from thermodynamic equilibrium represents a gradient-maintenance phase of network organization (refer to Fig. 2.8.2.).

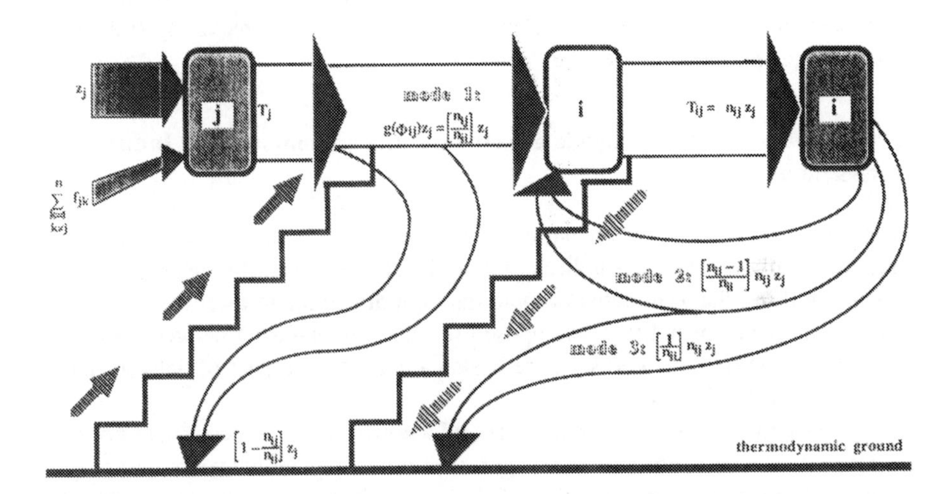

Fig. 2.8.2. Depiction of conservative energy–matter transactional relationships in the three modes of network organization interconnecting any two system nodes, j and i. At the left is input from the environment. The throughflow-stream carries this input to the right toward i, with losses out of the system (to "thermodynamic ground") along the way. The flow that reaches i (shown in the center as the "ghost" of actual i further to the right) for the first time is mode 1, or first-passage, flow. This flow has traveled for different durations over the set (Φ_{ij}) of first-passage paths identified in Figure 2.8.1a. The stair steps upward signify first passage "hill climbing" which increments the thermodynamic state of i upward away from equilibrium. In this sense mode 1 transfers are antientropic and gradient acquiring in the network relationship of j as energy–matter source to i. The flow depicted further to the right is the throughflow at i derived ultimately from z_j. This is made up of both mode 1 first-passage and mode 2 cycling flows. As cycling serves to maintain i at the distance from thermodynamic equilibrium attained at the terminus of mode 1 (at the top of the ascending stairsteps to the left), mode 2 can be thought of as the gradient-maintaining mode of network transactive organization. Mode 3 is then the entropic or dissipative mode of gradient degradation, taking the flow at i derived from j back down the stairstep to thermodynamic ground. The fact that mode 1 = mode 3 satisfies the conservation principles for energy and matter. The relations depicted in this diagram are canonical for transactional networks of arbitrary complexity interconnecting any two agents, source j and terminal i, within a system.

Finally, the order and organization reached and maintained in modes 1 and 2 are destroyed by mode 3 dissipative flows; this amounts to an entropic or gradient-destruction phase of network organization, and is shown as the descending staircase in Fig. 2.8.2. Quantitatively, mode 1 = mode 3, in accordance with the conservation principles for mass and energy, and the second law of thermodynamics.

Ecosystems tend to develop internal organizations that maximize both mode 1 and mode 2 transactions. This is the direction of their normal development. In moving away from equilibrium in mode 1, and maintaining the departure in mode 2, they increase their distance from thermodynamic ground.

Orientor statement. Ecosystems tend to develop networks that maximize both first-passage (mode 1) and (consistent with Properties 6–9) recycling (mode 2) interchanges between their agents.

Property 11: Ecosystem Agents are Quantitatively Dominated by Indirect Interactions

There is no satisfactory short name for this property, as there was for network homogenization, network amplification, and (later, Property 14) network synergism. I will use the term network quantitative indirectness to refer to the quantitative dominance of indirect, as opposed to direct, interactions between coupled agents in ecosystem networks. There is also a qualitative aspect of network indirectness, and this follows as Property 12.

Network indirectness is both a cause and a consequence of ecosystem holism. Direct binary transactions between agents set up ecological networks, but play a minor role in actual determination. This is quantitatively dominated, at steady state, by indirect interactions (Patten 1982a,b, 1984, 1991; Higashi and Patten 1986, 1989). The rationale behind this conclusion is described in these references.

The same holds true for relational interactions because, as the transactions that underlie these increase, relational indirectness increases even faster. Thus, agents in ecosystems are dominated by indirect interactions more than by the interactions that directly link them.

Orientor statement. Ecosystems self-organize to maximize the indirect dependencies, first transactive and then relational, among their agents.

Property 12: Ecosystems Qualitatively Transform their Agent Interactions

This property is *network qualitative indirectness*. It means that direct interactions often will differ qualitatively from their indirect counterparts just described as quantitatively dominant. This is demonstrated by employing a value measure, utility, derived from a cost–benefit version of input–output analysis (Patten 1991, 1992). Both direct and ultimate (direct plus indirect) utilities can be calculated.

Their signs (+, –, and 0) can be used to denote nine ecological interaction types (the prominent ones are named): (0,0) = neutralism, (0,+), (0,–), (+,0), (–,0), (+,–) = nihilism (e. g., predation), (–,+) = altruism, (+,+) = mutualism, (–,–) = competition. Model analyses show that ultimate interaction types often differ from the direct types that establish the network. Thus, networks can transform their agents' effective interactions in the shift from local (direct) to global (direct plus indirect) scales. Ecosystems qualitatively transform the direct vs. indirect interactions between their agents.

Orientor statement (see "network synergism", Property 14). Ecosystems tend in self-organization to develop networks that maximize positive ultimate interactions between their constituent agents.

Property 13: Ecosystem Determination is Predominantly Holistic

Whole → part (holistic) determination follows as a direct corollary to the previous two properties.

A number of network characteristics tend to increase the dominance of indirect vs. direct transactions. M. Higashi (see Patten 1991, pp 302-305) described several of these in terms of indirect (I)/direct (D) ratios. These are system *size* (number of agents), *connectivity* (density of direct transactions), *storage* (flow impedance), feedback and nonfeedback *cycling* (the two forms described in connection with Figure 2.8.1a), and *strength of direct interactions.* The last is paradoxical. Higashi (see Patten 1991:344-346) showed that strength of indirect transactions increases as a parabolic function of that of direct transactions. This above list of properties contributing to the tendency for I/D > 1 is probably not exhaustive.

Altogether, network indirectness (Properties 11 and 12) and its structural causes (above) diminish the importance of direct causal determinism in nature. Binary direct transactions establish the connective tissue of reality, but once this structure is in place and operating, control shifts away from local events and to the whole–part relation. Wholeness dominates partness to an extraordinary degree in this because networks are pervasive.

Therefore, holistic determination is the dominant mode of being and becoming in connected nature. This is probably true both within and across scales—from the interior of the cell to throughout the ecosphere. At ecological scales, ecosystems are plausible operational units of whole → part determination.

Orientor statement. From the maximization of indirect effects (Properties 11 and 12), the maximization of holistic determination follows as an orientor for ecosystems.

Property 14: Ecosystems Provide Positive Utility to their Biota

This is called network synergism. It derives from utility input–output analysis of resource benefit–cost relationships. Resource transactions are those for which a gain is positive and a loss negative (Patten 1991, 1992); in transaction of a poisonous substance, for example, the opposite is true.

Living agents in ecosystem networks receive two kinds of utility advantages from resource transactions:

1. |Benefit (B)/cost (C)| ratios computed from sums of positive (B) and negative (C) utilities conveyed between agents in a system always increase in comparisons of direct vs. ultimate transactions. In the former, such ratios are either 1.0 in absolute value or indeterminate (0/0), corresponding to the three kinds of proximate interactions definable by conservative resource transfers—(0,0), (+,–), and (–,+). This is because a quantity of substance gained by one agent is exactly given up by another as the source; conservative transactions are zero-sum, and thus for direct interactions $|B| = |C|$. For ultimate transactions, however, $|B/C|$ ratios are >1.0, since $|B| > |C|$ always holds. Therefore, utility is gained from the indirect transactions propagated over the network.
2. Interaction types tend to shift from less positive proximate transactions (predation (+,–) being the commonest) to more positive ultimate ones (especially mutualism, (+,+)). Fath and Patten (1997) have explored the algebraic basis for these tendencies and demonstrated that they are not fortuitous.

The first property always holds, and the second has some exceptions, but only for depauperate model networks. These results lead to the conclusion that under holistic determination (Property 13), conferred by both quantitative and qualitative dominance of indirect transactions (Properties 11 and 12), organisms benefit from collective life in ecosystems vs. if they lived more autonomously with less interdependence. Through the indirectness inherent in their network organization, ecosystems convey positive utility to their living constituents.

Orientor statement. The direction of ecosystem development is toward maximum network synergism.

Property 15: Ecosystems are Model-Making Complex Adaptive Systems

Ecosystems are coherent agglomerations of elements of the lithosphere, atmosphere, hydrosphere, biosphere, and noosphere (see later, Property 20) such that, in nature, there is always a place for everything (accommodation property) and everything is in its place (consistency property). This could only be described as a marvel of natural design.

Ecosystems achieve accommodation and consistency by adaptive self-organization. The capacity to adapt – adaptability – involves the revision, rearrangement, and reordering of constituents and their interrelations according to past experience. These rearrangements are not only adjustments to current circumstances, but are also in essence "predictive models" of what the future of reality will be like. In accordance with Ashby's (1952) law of requisite variety, and Conrad's (1983) adaptability theory of biological organization, these "models" must encompass sufficient variety (information) to meet the environmental variety that will arrive in the unknown future. The models are, in effect, implicit and explicit assumptions about how things work, and will work, encoded (see below, Property 17) at any given time in the extant agents of ecosystems and in the network configuration of these agents' interactions.

Living subjects (Property 1), as previously stated, make models of reality and respond to the resultant phenomenal (i. e., epistemological, as opposed to purely physical) inputs with intentionality (and often proactivity) to stay alive, derive benefits from the physical resource base, and propagate their derived fitness to future generations. It is ecosystems that match these models to projected future reality, in part through the mechanism of natural selection. Selection meshes the living and nonliving components of ecosystems, and hones them to mutual accommodation and consistency. It revises the living agents within ecological networks in accordance with uncontrollable changes in nonliving ones, retrofitting life to nonlife in the ongoing process of autoevolutionary self-design. This process engages the whole ecosystem, and is not passive (Property 16, next). Ecosystems, therefore, are model-making complex adaptive systems.

Orientor property. Obviously, the better and more predictive the "models" the more fitted an ecosystem will be to its physical circumstances, and the better able it will be to realize its goal functions.

Property 16: Ecosystems Create and Proliferate New Niches

In environ theory (Patten 1978), ecological niches are proximate restrictions of input and output environs (Patten 1981, Patten and Auble 1981). Niches form the leading edges, in spacetime, of these environs. Niche and environ creation are progressive in ecosystem development and evolution, and their proliferation leads to complexification.

Living and nonliving agents, interacting together, create new possibilities for new agents and new network configurations in ecosystems. Proximally, these new possibilities are fundamental niches (Hutchinson 1957). When these become occupied by agents, each such object or subject is immersed in the reality flowstream and automatically acquires input and output environs (environments bounded by a system-of-definition). Environs are extended niches because they encompass both the direct (niche) and indirect (extended) determinants (afferent) and consequences (efferent) of behavior. The proximate portions of environs are therefore realized niches (Hutchinson 1957).

In ecology the niche concept is limited to living agents. For example, the affinity of hydrogen and oxygen to form water is conditioned by factors like temperature, pressure, concentrations, other chemical bonding, etc., but these conditions are not referred to as the multidimensional fundamental niche of one element (say oxygen) that can be filled by the other (hydrogen). The restriction of niches to living things in ecological applications can be followed here without loss of relevant generality.

Patten and Auble (1981) showed that the proximate portion of input environs which delivers direct inputs to organisms corresponds to the place or habitat niche of Grinnell (1917). The proximate portion of output environs which receives direct outputs from organisms is the role or function niche of Elton (1927). Organism impacts cause niche changes in a process called niche construction (Odling-Smee 1988, 1996). As agents and their environs change and evolve, the niche configuration of ecosystems also changes and evolves. The changing "habits" of structure and function, and pattern and process, which confine and canalize dynamical (state-space) behavior, are encompassed in the progressive niche concept of Vandermeer (1972). Realized habitat niches are proximate restrictions of input environs, realized role niches are proximate restrictions of output environs, and progressive niches created by the dynamics of these input-output niches are shaped by the dynamics of corresponding whole extended environs. Referring to Property 14, niches convey direct utility to organisms, while environs convey ultimate utility. Thus, the classical niche concepts of ecology are all congruent within the formalism of state-space environmental system theory. Ecosystems in their ongoing processes create and proliferate new niches – new conditions for life.

Orientor statement. Ecosystems maximize niche creation.

Property 17: Ecosystems are Inheritance Systems

From a non-reductionistic point of view, organisms can be said to "inherit" not only their genes but also their environments from their ancestors. Genes code from inside, in accordance with conventional biology. Environments, correspondingly, code from outside. This is unconventional. Both the environmental and genetic codes converge from up- and downscale in phenotypic expression, and are inherited.

For living agents, reality is semiotic, a universal network of encoded signs and signals that interpenetrates all levels of biological organization. The significance of these signs and signals is given by species-specific epistemologies that are inherent in the notion that living things make models (Property 1) and, within physical (niche) constraints, respond to a mixture of phenomenal (modeled) and physical inputs, achieving some degree of liberation from the latter in the process.

Downward-inwardly the reality code originates with the rule structures (natural laws) and regularities associated with elementary particles and fundamental forces. It continues upscale with the periodic table, and the known laws for atomic and molecular assembly as given in physics and chemistry. It reaches its most widely recognized expression as such in biology – in macromolecular patterns and pathways in cells, in ontogenetic development, and in organismic traits and fitnesses to which the DNA-based genetic code gives rise. All these are progressive examples of lower-level hierarchical lock-in (Property 2).

Upward-outwardly the "coding" of reality is less evident, and less thought about in this way because of the bias to reductionism. It is nevertheless expressed just as really in the regularities and rule structures of macroscopic levels of organization – in the earth's orbit around the sun, and its axial rotation; in the rhythms of solar radiation exposure that these phenomena set up; in the energetic and biogeochemical transactions of the ecosphere; and in the ordered habitats and niches, habituated patterns and processes, and aperiodic chance disturbances impressed upon established structure and function, that select and organize everyday life in ecosystems. Aquatic and terrestrial ecosystems of many different kinds all "code" for their respective kinds of organisms, but we do not interpret reality in this way. This coding is to us "selection", but in essence it is little different from the relationship between DNA and the set of proteins "selected" (or "coded for") by this particular molecule. A noteworthy form of rule-structuring in humans and other social organisms is "culture"—expected or conditioned behavioral responses to particular semiotic (phenomenal) stimuli. The qualities of culture as inheritance systems have been well-noted in behavioral biology literature.

In the reproduction, growth, and development of living subjects, dual inheritance paralleling rigorous state-space determinism is the rule. The input (Z) × state (X) dyad from system theory (Property 1) has its biological counterpart in "nurture" (Z) × "nature" (X), and in this latter expression of the deeper principle both exogenous (environmental, or input-related) and endogenous (genetic, or state-related) coding contribute to the synthesis of new organisms. The genotype–phenotype pair of classical genetics is therefore an incomplete specification of determinate reproduction; an external *envirotype* is needed to complete the mechanism.

The genotype and such an envirotype are potentially immortal, but the phenotype formed at the locus of their intersection, coded both from within and without, is mortal. Discontinuous input and output of discrete phenotypes, by birth and death, and immigration and emigration, is the mechanism by which the biota and abiota of ecosystems are adaptively fitted together as described under Property 15. This articulation is closer to Darwin's original concept of "fitness" within "the entangled bank" than neo-Darwinian usage; we have resorted to the more awkward "fittedness" to make the distinction (Patten 1991).

The encoding envirotypes of phenotypes are their environs – the extended realized niches (Property 16) that are the mode 2 (cycling) portion of the partition units (Property 3) of the ecosystems which the phenotypes inhabit. The proximate niches of environs have their internal counterpart in the genomes of the genetic mechanism. In their coding function, envirotypes are the niche-constructing "selective forces" that in conventional genetics comprise the instrument of natural selection.

Niche-constructing phenotypes (Property 16) materially alter and shape their output environs, which wrap-around to contribute parts of the phenotypes' input environs at later points in spacetime. To the extent of this causal closure – output contributing to input – organisms can direct their own evolution by managing the system of selection. The Weissmann barrier (Bus 1987) may prevent the "inheritance of acquired characteristics" via the cell, but the Lamarckian imperative can still be met through the environmental route of niche-constructing the envirotype, the environmental code. This must be accomplished by mechanisms of indirect control, however. In general terms, it can be concluded that ecosystems are true systems of inheritance.

Orientor statement. Consistent with the cycling characteristics of networks registered in many of the foregoing properties (esp. 5–9), ecosystems self-organize to maximize the wrap-around of their biotic agents' output environs to input environs, making environment, particularly its portions niche-constructed by organisms (Property 16) to form envirotypes, heritable.

Property 18: Ecosystems are Cybernetic Systems with Distributed Control

Regulation and stasis in ecosystems, damping of variations, and general "balance-of-nature" phenomena are achieved by network-mediated decentralized control, rather than by discrete controllers (Patten and Odum 1981).

In control theory, the behavior of dynamical systems is regulated by a two-stage process – error generation followed by error correction. It is a principle of cybernetics that there is no capability for control without error.

The essence of control is negative feedback in which deviations of state or output dynamics from reference behavioral trajectories are damped. Output is measured against the reference set point (corresponding to the "eco-targets", "orientors", and "goal functions" of this book) by the controller, generating a deviation measure. When this exceeds the error limits set into the controller, the device activates a corrective mechanism which damps the deviation. The term "negative" feedback derives from the fact that deviations are damped; if amplified, this is positive feedback as seen, for example, in geometric growth processes.

There are no known discrete controllers in ecosystems as in organisms and machines. "Keystone species", possible examples of such controllers, are questionably so, given the plasticity of ecosystems to reorganize after such species are removed. In general, control functions in ecosystems are served more by a looser kind of "swarm intelligence" (Varela and Bourgine 1992; Hoffmeyer 1995) that is implicit in the network organization. Rather than being based on direct feedback, control in ecosystems is manifested implicitly through indirect feedback inherent in network complexity. Control functions are diffusely spread out over the entire web, exploiting cycling (Properties 5–9) and dominance of indirectness (Properties 11–13). Even if keystone species did exist, their mechanism of control would involve large webs that connect them to the rest of the ecosystem. Controllerless control can be referred to by a variety of terms: indirect, diffuse, distributed, decentralized, implicit, or remote control.

Its central feature is that it is network based, with or without (but generally without) explicit controllers. The network basis for distributed control lies in the feedback component of mode 2 cycling (Property 5). In general, if agent i is more important in the transactive output environ of j than in j's input environ (or alternatively, j is more important in the input environ of i than in i's output environ), then j controls i. If the other way around, then i controls j. In both cases the control is complexly mediated by the pathways encompassed in the path sets of Fig. 2.8.1.a. The control is distributed, or indirect, both in time and over network space because of the many pathways of different kinds and lengths involved in reaching the prerequisite condition for steady state, transitive closure.

Orientor statement. Achievement of maximum distributed control through network self-design is an orientor of ecosystem development.

Property 19: Ecosystems are Units of Coevolution

Organisms and their environs must coevolve or the species in question will go extinct. See Patten (1982c, pp 194-196) for the rationale behind this proposition.

Organism–environment (or, by Property 17, phenotype–envirotype) coevolution applies even to cases where species or life-cycle stages are distributed temporally or spatially over a variety of different ecosystems. As all the environs within a particular ecosystem at a given time form a system partition, they and their contained and defining agents, however transitory or ephemeral, and the interactions between them, must coevolve together as a collective unit in operative intervals of spacetime. Ecosystems, taken as wholes, then, are coevolutionary units.

Orientor statement. Ecosystem organization enhances the coevolution of organisms and their environs within them.

Property 20: Ecosystems Expand the Inner Space of Reality

Ecosystems, as complex adaptive systems, are steady-state seeking (e.g., climax communities), but in general they never get or remain there. They change and complexify perpetually, and in the process generate perpetual novelty and expansion of "inner space" in the universe.

From physics, the hierarchical universe (Property 2) is spatially expanding. The "upward–outward" expansion of outer space has been occurring for 8–25 billion years since the Big Bang at a speed given by the still uncertain Hubble constant. When that becomes fixed, so too will the age of the universe be finally established. There is also a "downward–inward" expansion at the other end of the hierarchy that receives little attention as such. This is evident from the experience of life on earth, and it developed in three phases:

1. *Ontic, prebiotic.* Before life began, the ecosphere of the planetary surface consisted of an array of open systems involving the phases of matter deployed as an interdigitating lithosphere, hydrosphere, and atmosphere. Evolution of the planet over geologic time as a complex adaptive system created many new configurations of matter and energy, as abiotic processes are also known to do elsewhere in the cosmos. Novelty and physicochemical diversity increased with time. The new configurations involved consistent relationships between agents, interactions, niches, and environs – imposed and encoded by the laws of physics and chemistry. Rigorous input–state–output determinacy of agent–environ transactions and relations (Property 1) was established under these laws, and the "coding" for consistency that determinacy implies made the feedback portions of environs heritable envirotypes (Property 17). Abiotic ecosystems encompassing multiple agent–envirotype interactions became established well before the advent of life and passed on the environments they created to the next collections of interactions – the next ecosystems – that followed them. The earth's surface evolved toward anisotropy, accruing ever more antientropic structure, heterogeneity, and gradients (Property 10) at all organizational scales. Gradient destruction in accordance with the second law of thermodynamics drove the assembly and maintenance of new physicochemical structure, and new environs associated with new evolved objects. In the evolution from simple inorganic components of the "prebiotic soup" to more complex organic substances, physical spacetime became more and more filled with an increasingly diverse array of objects and their concomitants: structures, forms, and surfaces; spaces, interstices, and microniches; forces and gradients, and activity and change – inner space was expanding.
2. *Ontic, biotic.* With the arrival of modeling-based life this process exploded. Mineralogical life may have formed first, in accordance with Cairns-Smith's (1982) clay theory. Then, carbon-based proteinoids, coacervates, and other organic macromolecules may have followed (Fox and Dose 1977; Casti 1989). Self-sustaining autocatalytic sets of chemical compounds became established, membranes formed and demarcated them as systems. With semipermeability, inclusion and exclusion were realized; the cell habit developed, stabilized interior environs, and evolved first in big steps and then small. Organelles differentiated and gave rise to specialized cell function; the eucaryotic habit emerged from procaryotic. Early biochemical diversification and adaptive radiation gave way to lower-level lock-in (Property 2) to a few fixed patterns reflected in chemo- and photosynthetic production, autotrophic and heterotrophic metabolism, and genetic encoding of experience as we know these in modern phyla today. Free-living cells formed symbiotic alliances; in the evolution of individuality (Buss 1987) multicellular organisms emerged and, based on the requirement to satisfy physical nonisolation and openness, their populations grew into the niches of the planetary surface, and further diversified these as the higher-level associations of communities and

ecosystems. Natural selection (or equivalently, environmental coding; Property 17) of individuals became the way of evolutionary descent. The *biosphere* was formed, and it progressively complexified through the interactions of carbon-based life with itself and the rocks, air, and water of the abiotic planet. Biogenic "niche construction" (Property 16), on both sides of the semipermeable membranes that circumscribed the building blocks of life, produced niche diversification at all scales – macroscopic and external, and microscopic and internal. The inner space of the universe became further expanded.

3. *Epistemic.* The origin of life as model-making objects ("subjects", Property 1) carried with it the simultaneous origin of implicit *epistemologies*. However rudimentary at first, as primitive representations that could free life from purely physical determination, models introduced subjectivity into the dynamics of existence. The earliest models, as those of today, were objective and material in their physical basis. But operationally, they were informational and contextual in their expression of and intersection with ontic reality.

All models have in common that they encode experience, and always involve signs, signals, syntaxes, semantics, and an ability to decode and derive meaning from what is encoded. Semiochemicals must have been among the earliest forms of chemistry-based epistemologies. They still exist widespread among the phyla, and highly refined by their evolutionary descent (Eisner and Meinwald 1995). Nucleic acids, enzymes, hormones, pheromones, and allelopaths are all examples in this category. Cellular irritability, tropisms, nervous nets and brain function in sensory-motor coupling, cognition, awareness, and ultimately the mental states of consciousness and rationality – all these have the original chemical, and in many cases electrochemical, basis expressed now in highly refined forms. The most advanced expression of epistemic phenomena we see, not surprisingly, as occurring in our own species – manifested as culture in mythology, religion, science, music, art, literature, even drug-altered mental states, and lately, "virtual reality."

Of course we cannot be sure of this, locked as we are into our own cognition. Each species may sense its own epistomological development, and its worldview derived through this, as the most refined by experiential criteria that completely escape others. There is, it seems, no way for any species to know this or be able to resolve it. What does seem clear is that all the epistemologies on earth, of all the species in existence at any given time, comprise another compartment of the planetary surface, a biologically extended version of Teilhard de Chardin's (1975) anthropocentric noosphere. In its expression far beyond its physical basis and bounds, and in its ability to interact with and exert great influence upon ontic reality, this collective noosphere introduces subjectivity of most diverse kinds into reality, and through the physical impacts of this acquires a legitimate ontic status. The noosphere so conceived represents a profound, unique, almost aphysical, but at the very least qualitatively different, expansion of the inner space of the universe – an expansion that is uniquely biologically given.

In summary, the ecosystems on earth form an *ecosphere* consisting of the five interdigitating realms: Lithosphere, atmosphere, hydrosphere, biosphere, and noosphere. In its informational attributes, this ecosphere has been called the *semiosphere* (Hoffmeyer 1996). All the earthspheres, but most proximately the biosphere and noosphere, contribute by their evolutionary complexification to expansion of the inner space of reality. Because the biosphere and noosphere are manifestations of life, and because individualized life is still nonisolated and open and has prior dependency on ecosystems, it is ecosystems that in the first instance can be accorded responsibility for expanding the inner space of reality.

Orientor statement. Ecosystems develop, self-organize, and evolve in directions that tend to maximize expansion of the inner space of reality.

2.8.3 Conclusion

This book is devoted to an exploration of the proposition that teleology in different degrees operates broadly in the dynamics of natural phenomena. This chapter has presented a set of properties which appear to give directional tendencies to the development, self-organization, and autoevolution of ecosystems.

Many of these properties strain credulity in relation to where ecology stands right now in its development as a science. They are network-given, empirically intractable, and in their holistic dimensions not well understood at the present time. They can, however, perhaps serve to guide future development toward a more balanced mix of empiricism and theory, and reductionism and holism.

Concerning empiricism and theory, science must realize, and then come to terms with the realization, that not everything relevant to know about ecology can be learned through empirical observations alone. Across the orders of magnitude from 10^{-42} to 10^{21} eV, the energy levels of the weakest (gravity) and strongest (strong nuclear) physical forces, there is just too much hierarchical complexity that is for all practical purposes empirically intractable. Concerning reductionism and holism, many of the above-discussed network properties derive from constituted wholes and are lost when these are decomposed into parts so the networks dissolve. Reductionism, put quite simply, destroys emergence, and thus much of what is critical to be known about reality and the human role in it is beyond the reach of this dominant approach of science.

In the Introduction a series of questions was posed: What shapes the development, organization, and evolution of ecosystems? Where do they fit in the spectrum of world systems? Can theory clarify this? Can a cosmography of ecosystems be constructed? How might the "eco-targets, goal functions, and orientors" of this book fit into such a scheme?

That there is direction in the cosmos seems clear from how events are perceived to unfold through time. To what extent these directions are pushed by the forces external to systems vs. pulled by the kinds of internal attractors explored in this book remains for the future to decide.

References

Ashby WR (1952) An Introduction to Cybernetics. Wiley, New York

Bak P (1996) How Nature Works, the Science of Self-Organized Criticality. Copernicus/ Springer–Verlag, New York

Bossel H (1992) Real-Structure Process Description as the Basis of Understanding Ecosystems and Their Development. Ecol Mod 63:261–276

Buss LW (1987) The Evolution of Individuality. Princeton University Press, Princeton NJ

Brown JH (1995) Macroecology. University of Chicago Press, Chicago

Cairns-Smith AG (1982) Genetic Takeover and The Mineral Origins of Life. Cambridge University Press, Cambridge

Casti JL (1989) Paradigms Lost, Images of Man in the Mirror of Science. William Morrow, New York

Conrad M (1983) Adaptability, the Significance of Variability from Molecule to Ecosystem. Plenum, New York

Eisner T, Meinwald J (eds) (1995) Chemical Ecology. The Chemistry of Biotic Interaction. National Academy Press, Washington DC

Elton CS (1927) Animal Ecology. Sidgwick and Jackson, London

Fath BD, Patten BC (1997) Network Synergism: Emergence of Positive Relations in Ecological Models. Ecol Mod, in press

Fox SW, Dose K (1977) Molecular Evolution and the Origin of Life. Marcel Dekker, New York Basel

Gell-Mann M (1994) The Quark and the Jaguar, Adventures in the Simple and the Complex. Freeman, New York

Grinnell J (1917) The Niche-Relationships of the California Thrasher. Auk 34:427–433

Higashi M, Patten BC (1986) Further Aspects of the Analysis of Indirect Effects in Ecosystems. Ecol Mod 31:69–77

Higashi M, Patten BC (1989) Dominance of Indirect Causality in Ecosystems. Amer Nat 133: 288–302

Higashi M, Patten BC, Burns TP (1993) Network Trophic Dynamics: the Modes of Energy Utilization in Ecosystems. Ecol Mod 66:1–42

Hoffmeyer J (1995) The Swarming Cyberspace of the Body. Cybernetics and Human Knowledge 3:16–25

Hoffmeyer J. (1996) The Global Semiosphere. In: Rauch I, Carr GF (eds) Semiotics Around the World: Synthesis in Diversity. Proc Fifth Congress Int Assoc Semiotic Studies, Berkeley, California, June 13–18, 1994. Mouton de Gruyter, Berlin

Hutchinson GE (1957) Concluding Remarks. Cold Spring Harbor Sympos Quant Biol 22:415–427

Kauffman SA (1993) The Origins of Order. Self-organization and Selection in Evolution. Oxford University Press, New York

Kauffman SA (1995) At Home in the Universe. The Search for Laws of Self-Organization and Complexity. Oxford University Press, New York

Lima-de-Faria A (1988) Evolution Without Selection. Form and Function by Autoevolution. Elsevier, Amsterdam

Lindeman RL (1942) The Trophic-Dynamic Aspect of Ecology. Ecology 28:399–418

Jørgensen SE, Patten BC, Straskraba M (1992) Ecosystems Emerging: Toward an Ecology of Complex Systems in a Complex Future. Ecol Mod 62:1–27
Müller F (1992) Hierarchical Approaches to Ecosystem Theory. Ecol Mod 63: 215–242
Odling-Smee FJ (1988) Niche-Constructing Phenotypes. In: Plotkin, HC (ed) The Role of Behavior in Evolution. MIT Press, Cambridge, Massachusetts, pp 73–132
Odling-Smee FJ (1996) Niche Construction. Amer Nat 147:641–648
Patten BC (1978) Systems Approach to the Concept of Environment. Ohio J Sci 78:206–222
Patten BC (1981) Environs: the Superniches of Ecosystems. Amer Zool 21:845–852
Patten BC (1982a) Indirect Causality in Ecosystems: its Significance for Environmental Protection. In: Mason WT, Iker S (eds) Research on Fish and Wildlife Habitat. Commemorative Monograph in Honor of the First Decade of the US Environmental Protection Agency. Office of Research and Development, US Environmental Protection Agency, EPA-600/8-82-022, Washington, DC, pp 92–107
Patten BC (1982b) On the Quantitative Dominance of Indirect Effects in Ecosystems. In: Lauenroth WK, Skogerboe GV, Flug M (eds) Analysis of Ecological Systems: State-of-the-Art in Ecological Modelling. Elsevier, Asterdam, Pp 27–37
Patten BC (1982c) Environs: Relativistic Elementary Particles for Ecology. Amer Nat 119:179–219
Patten BC (1984) Toward a Theory of the Quantitative Dominance of Indirect Effects in Ecosystems. Verh Gesellschaft für Ökologie 13:271–284
Patten BC (1985) Energy Cyling in the Ecosystem. Ecol Mod 28:1–71
Patten BC (1991) Network Ecology: Indirect Determination of the Life–Environment Relationship in Ecosystems. In: Higashi M, Burns TP (eds) Theoretical Ecosystem Ecology: The Network Perspective. Cambridge University Press, London, pp 288–351
Patten BC (1992) Energy, Emergy and Environs. Ecol Mod 62:29–69
Patten BC (1995) Network Integration of Ecological Extremal Principles: Exergy, Emergy, Power, Ascendency, and Indirect Effects. Ecol Mod 79:75–84
Patten BC, Auble GT (1981) System Theory of the Ecological Niche. Amer Nat 117: 893–992
Patten BC, Higashi M (1995) First Passage Flows in Ecological Networks: Measurement by Input–Output Flow Analysis. Ecol Mod 79: 67–74
Patten BC, Higashi M, Burns TP (1990) Trophic Dynamics in Ecosystem Networks: Significance of Cycles and Storage. Ecol Mod 51:1–28
Patten BC, Jørgensen SE (eds) (1995) Complex Ecology: The Part–Whole Relation in Ecosystems. Prentice Hall, Englewood Cliffs, New Jersey
Patten BC, Odum EP (1981) The Cybernetic Nature of Ecosystems. Amer Nat 118:886–895
Peters RH (1991) A Critique for Ecology. Cambridge University Press, Cambridge
Simon HA (1973) The Organization of Complex Systems. In: Pattee HH (ed) Hierarchy theory, the Challenge of Complex Systems. Braziller, New York, pp 1–27
Teilhard de Chardin P (1975) The Phenomenon of Man. Harper And Row, New York
Vandermeer JH (1972) Niche Theory. Ann Rev Ecol Syst 3:107–132
Varela FJ, Bourgine P (1992) Toward a Practice of Autonomous Systems. MIT Press, Paris
Waldrop MM (1992) Complexity. The Emerging Science at the Edge of Order and Chaos. Simon and Schuster, New York

2.9 A Utility Goal Function Based On Network Synergism

Brian D. Fath and Bernard C. Patten

Abstract

Network synergism is the property inherent in all complex adaptive systems that the direct resource transactions between organisms and their environments, when integrated across a whole-system organization, translate into integral (direct plus indirect) relationships that are more positive than the local ones (Fath and Patten, in press). Ecosystems on the whole provide hospitable conditions for life. These positive relationships are observed in the integral utility of the system. An input-output analysis based methodology that measures the total integral utility has been developed to model these synergistic relationships (Patten 1991, 1992). In this paper, we demonstrate how network synergism arises in simple systems, and compare its behavior with several other proposed ecological goal functions namely trophic transfer efficiency (Odum 1969), cycling (Finn 1976), maximum power (Lotka 1922), maximum indirect effects (Patten 1995), and connectivity.

2.9.1 Introduction

2.9.1.1 Goal Functions

The purpose of the "Unifying Goal Functions" workshop was to discuss the existence and usefulness of ecological extremal principles. As participants in the workshop, we presented research regarding the property of network synergism. *Network synergism* is the property inherent in all complex adaptive systems that the direct resource transactions between organisms and their environments, when integrated across a whole-system organization, translate into integral (direct plus indirect) relationships that are more positive than the local ones (Fath and Patten in press). These positive relationships are observed in the integral utility (relative net flow intensity) of the system. We did not promote network synergism as a goal function during the conference. However, if ecosystems organize to maximize ecological utility then we would expect its behavior to be similar to that of

F. Müller M. Leupelt (Eds.)
Eco Targets, Goal Functions, and Orientors

other candidate goal functions. Here, we compare the behavior of ecological utility with several other proposed ecological goal functions.

During the workshop it was agreed that the label "goal function" should be applied only to models and not to natural systems because of the teleological overtones associated with implicating nature as a goal seeking entity. However, the property network synergism can be applied as a modeling goal function or as a phenomenological extremal principle (Patten 1995). Conference participants were asked to consider an extensive list of candidate goal functions. A parameter was classified as a possible goal function if it was perceived to have an inherent direction during ecosystem development. The goal functions were separated hierarchically according to the extent of their application from species to populations to communities to ecosystems. Integral utility is a measure of the overall system organization and function, and therefore, is an ecosystem level property. A benefit-cost indexbenefit-cost is described to quantify the total system level synergism. Some other ecosystem level parameters include trophic transfer efficiency, cycling, power, indirect effects, and connectivity. Here the behavior of network synergism is compared with these other parameters for various conditions using simple ecological models.

2.9.1.2 Transactions and Relations

We define transactions and relations to provide a framework for thinking about the direct and indirect interactions between organisms in an ecosystem. A *transaction* is the direct, observable transfer of conservative resources between two organisms, and a *relation* is the direct and indirect consequence of these transfers. For example, predation is the relation associated with the transaction of feeding. Many relations arise only indirectly from the associated transactions. Two organisms without a direct transaction between them compete if they use the same resource. The common prey resource mediates the competition. Ecosystems are vast collections of transactions and relations expressed across many scales of organization.

Resource transactions couple biota and abiota together in interactive networks that constitute an ecosystem. All ecosystems have a primary input transfer of energy, usually the photosynthetic capture of solar radiation, and a final output transfer of energy often in the form of unusable heat. Therefore, as open systems, they are driven away from thermodynamic equilibrium. In response to these transactions, quantitative and qualitative relations are established in the system. The qualitative relations can be used to determine the value-oriented direct and indirect interaction types between system components (Patten 1991, 1992; Fath and Patten in press). The integral quantitative relations give a weighted value of the total positive and negative utility in the system. These values are used to determine the relative amount of network synergism in each system and to compare trends in network synergism for various systems. The integral qualitative and quantitative relations can be investigated by " environ analysis environ ."

2.9.1.3 Environ Utility Analysis

Input-output analysis was developed by Leontief (1966) to analyze the interdependence of industries in an economy (Miller and Blair 1985). Input-output analysis can also be used to analyze the interdependence of organisms in an ecosystem (Hannon 1973). An environmental extension of input-output analysis, environ theory (Patten 1978), uses each component's input and output *environs* to describe the distribution of resources throughout the interior of a system. An input environ is the network of all direct and indirect within-system transactions into a component. An output environ is the network of direct and indirect transactions leading away from a component. In environ analysis, organisms couple together to form ecosystems through the transactions expressed within their input and output environs. Output and input environs capture both bottom-up and top-down processes in the sense of being source and terminal oriented, respectively. That is, an output environ issues from every component (source), and an input environ ends at every component (terminus).

For a value-oriented environ analysis, we borrow a second concept from economics, utility. *Utility* is defined as the value of a specific product or currency relative to the value of an entire asset or throughflow. It is the relative value of the net flow between two components. For example, $1 has more utility to a laborer in a day $10 were earned compared to the utility of the same dollar during a day $100 were earned. In environ utility analysis, the currency, flow of energy, nutrients, or other matter between two components in a model, is normalized by the total flow through the component.

In model systems previously studied, the total utility assessed over the entire web of interactions is greater than that associated with individual components (Patten 1991, 1992). That is, integral (direct plus indirect) relations are more positive than direct transactions. This result provided the origin of the term *network synergism.*

2.9.2 Examples

2.9.2.1 Three-Component Food Chain

To demonstrate how network synergism arises in simple systems, we apply environ utility methodology to a three-component food chain. In this model, component x_1, which receives a constant input, is grazed by component x_2, which is preyed upon by component x_3 (Figure 2.9.1). A dimensional direct flow matrix, $F = (f_{ij})$, is constructed based

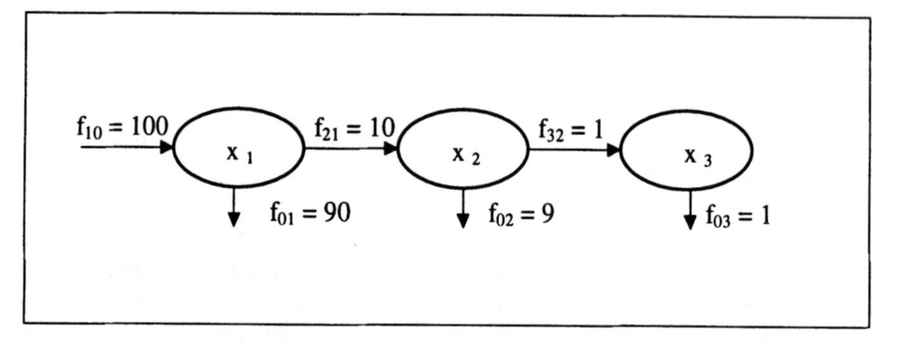

Fig. 2.9.1. Three-component food chain

on the transactions within the system. The precise units of flow are irrelevant for this example as long as we assume they are consistent. By our convention, flows are from columns j = 1,...,n to rows i = 1,...,n. In Fig. 2.9.1, $f_{21} = 10$ and $f_{32} = 1$. All other within system flows are zero. The flow matrix for this network is:

$$\mathbf{F} = \begin{bmatrix} 0 & 0 & 0 \\ f_{21} & 0 & 0 \\ 0 & f_{32} & 0 \end{bmatrix} = \begin{bmatrix} 0 & 0 & 0 \\ 10 & 0 & 0 \\ 0 & 1 & 0 \end{bmatrix}. \tag{2.9.1}$$

The net direct transaction between each pair of components is given by a net flow matrix $\Delta = (\delta_{ij}) = (f_{ij} - f_{ji})$. This skew symmetric matrix $(\delta_{ij} = -\delta_{ji})$ is normalized by the steady-state throughflow at each component, T_j, to represent the direct ecological utility. At steady-state, the inflows and outflows are balanced,

$$T_i = \sum_{j=0}^{n} f_{ij} = \sum_{j=0}^{n} f_{ji} \tag{2.9.2}$$

where f_{i0} is the inflow to i from the environment and f_{0i} is the flow from i to the environment. For Fig. 2.9.1, $\mathbf{T} = [100,10,1]$. A similar normalization (Ulanowicz and Puccia 1990) originally motivated this approach. Normalization gives a direct utility matrix, **D**, where $d_{ij} = (f_{ij} - f_{ji})/T_i$.

The element d_{ij} is interpreted as direct utility because the net flow between i and j is expressed relative to the total flow at i. For Fig. 2.9.1,

$$\mathbf{D} = \begin{bmatrix} 0 & \frac{-f_{21}}{T_1} & 0 \\ \frac{f_{21}}{T_2} & 0 & \frac{-f_{32}}{T_2} \\ 0 & \frac{f_{32}}{T_3} & 0 \end{bmatrix} = \begin{bmatrix} 0 & -0.1 & 0 \\ 1 & 0 & -0.1 \\ 0 & 1 & 0 \end{bmatrix}. \tag{2.9.3}$$

D gives only the direct utilities in the system. Powers of **D** identify indirect utilities corresponding to resource flows of the same power. For example, $\mathbf{D}^2$ gives utilities associated with transaction sequences of length 2, and $\mathbf{D}^3$ gives utilities over paths of length 3, etc.. An integral utility matrix, **U**, which accounts for the contributions of all direct and indirect interactions, is found by summing all powers of **D** (Eq. 2.9.4). **U** is an integral utility matrix because its elements represent the total nondimensional utility expressed between the components by powers of **D** (Patten 1991, 1992).

$$\mathbf{U} = \mathbf{D}^0 + \mathbf{D}^1 + \mathbf{D}^2 + \mathbf{D}^3 + \mathbf{D}^4 + \ldots \tag{2.9.4}$$

Each component in the system is actively open to exchange with its external environment. Some components explicitly receive external inflows, and all components transmit outflows to the environment. This potential for openness, which is a fundamental property of system structure and function, is represented in the first term of the power series, $\mathbf{D}^0=\mathbf{I}$, where **I** is the n×n identity matrix. An analytic representation of **U** is given by Equation 2.9.5.

$$\mathbf{U} = \sum_{k=0}^{\infty} \mathbf{D}^k = (\mathbf{I}-\mathbf{D})^{-1} = [\det(\mathbf{I}-\mathbf{D})]^{-1} \operatorname{adj}(\mathbf{I}-\mathbf{D}) \tag{2.9.5}$$

The determinant (det(•)) and adjoint (adj(•)) can be calculated algebraically, but they become computationally very complicated as n increases. The infinite power series in Equation 2.9.4 converges when the eigenvalues, λ, of **D** are less than one in magnitude. For the above example, λ = 0, 0.4472i, and -0.4472i so the power series converges (The nonzero eigenvalues of D are always imaginary, but we are only concerned with the magnitude.). The following algebraic and numeric integral utility matrices are calculated:

$$\mathbf{U} = \frac{1}{\left(1+\frac{f_{21}}{T_1}+\frac{f_{32}}{T_2}\right)} \begin{bmatrix} 1+\frac{f_{32}}{T_2} & \frac{-f_{21}}{T_1} & \frac{f_{21}f_{32}}{T_1T_2} \\ 1 & 1 & \frac{-f_{32}}{T_2} \\ 1 & 1 & 1+\frac{f_{21}}{T_1} \end{bmatrix} = \begin{bmatrix} \frac{11}{12} & \frac{-1}{12} & \frac{1}{120} \\ \frac{5}{6} & \frac{5}{6} & \frac{-1}{12} \\ \frac{5}{6} & \frac{5}{6} & \frac{11}{12} \end{bmatrix}. \quad (2.9.6)$$

In this food chain model, only the first component receives input from the environment (Figure 2.9.1), so $f_{21}/T_2 \equiv 1$ and $f_{32}/T_3 \equiv 1$. These simplifications are used in Equation 2.9.6. **U** can be used to identify the qualitative integral relations in the system. The quantitative integral relations expressed in the dimensionalized integral utility matrix, **Y** = (υ_{ij}), are defined in Equation 2.9.7.

$$\mathbf{Y} = \mathrm{diag}(\mathbf{T}) * \mathbf{U} \quad (2.9.7)$$

Applying Equation 2.9.6 to this example gives:

$$\mathbf{Y} = \frac{1}{\left(1+\frac{f_{21}}{T_1}+\frac{f_{32}}{T_2}\right)} \begin{bmatrix} \left(1+\frac{f_{32}}{T_2}\right)T_1 & -f_{21} & \frac{f_{21}f_{32}}{T_2} \\ T_2 & T_2 & -f_{32} \\ T_3 & T_3 & \left(1+\frac{f_{21}}{T_1}\right)T_3 \end{bmatrix} = \begin{bmatrix} \frac{275}{3} & \frac{-25}{3} & \frac{5}{6} \\ \frac{25}{3} & \frac{25}{3} & \frac{-5}{6} \\ \frac{5}{6} & \frac{5}{6} & \frac{11}{12} \end{bmatrix}. \quad (2.9.8)$$

We use these two integral utility matrices to determine the qualitative and quantitative relations in the system. Component x_2 receives positive utility from component x_1 (υ_{21} is +) and component x_3 receives positive utility from component x_2 (υ_{32} is +). Opposite signs occur in the reciprocal relations (υ_{12} and υ_{23} are -). This represents the difference of being predator or being prey. The integral relation between components x_1 and x_3 is positive (υ_{13} and υ_{31} are positive although there is no direct link between them (d_{13}=d_{31}=0). In general, the nonzero elements in **D** (or **Δ**) are greater in magnitude than the corresponding elements in U (or Y) because the utility is more evenly distributed in **U** (or **Y**). However, if we want to look at the organization of the entire system, then we are concerned mostly with comparing the overall positive or negative quality of the direct and integral utility matrices. The index used to measure the total amount of system-wide synergism is a ratio of total positive to negative utility in the system (Patten 1991, 1992). This ratio is expressed in the following benefit-cost ratio (*b/c*):

$$\frac{b}{c} = \frac{\Sigma + \text{utility}}{\Sigma - \text{utility}} \quad (2.9.9)$$

If $b/c > 1$, then network synergism occurs. When comparing two systems, a greater b/c value corresponds to a higher level of synergism. Since Δ is skew-symmetric, transactions in the direct utility matrix are zero-sum and the magnitudes of positive and negative utility are equal. In Δ, $b/c = 11/|-11| = 1$. However, the integral utility matrix is not zero-sum because it incorporates cycling and indirect effects in the open network. In **Y**, $b/c = 111.75/|-9.17| = 12.19$. Since $b/c > 1$ here, network synergism is demonstrated.

2.9.2.2 Three-Component Complete Digraph

The three-component food chain is a subsystem of the complete 3x3 graph (Figure 2.9.2). The structure in Figure 2.9.2 also contains as subsystems all one-, two-, and three-component systems.

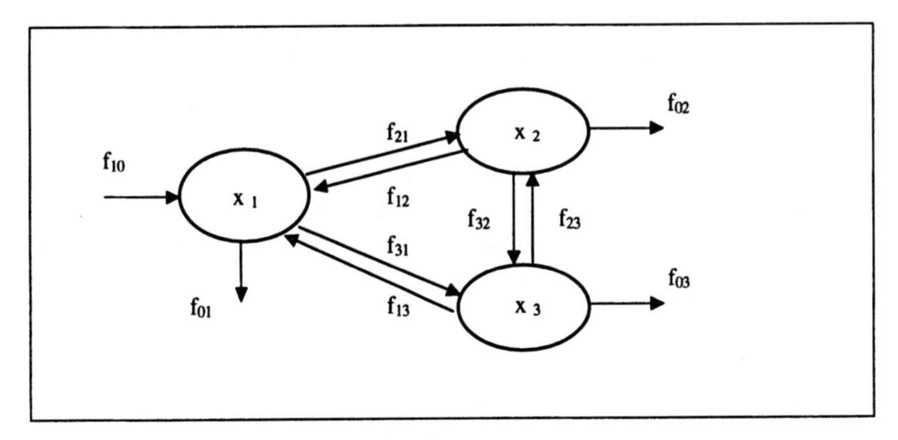

Fig. 2.9.2. Three-component complete digraph

The general **Y** matrix for Figure 2.9.2 is given in Equation 2.9.10.

$$
\mathbf{Y} = \frac{1}{\det(\mathbf{I}-\mathbf{D})}
\begin{bmatrix}
T_1 + \dfrac{T_1(f_{23}-f_{32})^2}{T_2T_3} & f_{12}-f_{21}+\dfrac{(f_{13}-f_{31})(f_{32}-f_{23})}{T_3} & f_{13}-f_{31}+\dfrac{(f_{12}-f_{21})(f_{23}-f_{32})}{T_2} \\
f_{21}-f_{12}+\dfrac{(f_{13}-f_{31})(f_{32}-f_{23})}{T_3} & T_2+\dfrac{T_2(f_{13}-f_{31})^2}{T_1T_3} & f_{23}-f_{32}+\dfrac{(f_{12}-f_{21})(f_{31}-f_{13})}{T_1} \\
f_{31}-f_{13}+\dfrac{(f_{12}-f_{21})(f_{23}-f_{32})}{T_2} & f_{32}-f_{23}+\dfrac{(f_{12}-f_{21})(f_{31}-f_{13})}{T_1} & T_3+\dfrac{T_3(f_{12}-f_{21})^2}{T_1T_2}
\end{bmatrix}
\tag{2.9.10}
$$

where

$$\det(\mathbf{I}-\mathbf{D}) = \frac{T_1T_2T_3 + T_1(f_{23}-f_{32})^2 + T_2(f_{31}-f_{13})^2 + T_3(f_{21}-f_{12})^2}{T_1T_2T_3}. \tag{2.9.11}$$

It can be shown that the diagonal elements of **Y** are always positive ($\upsilon_{ii} > 0$) and det(**I–D**) > 0 for all static open networks of any size or complexity (Fath and Patten, in press). By summing all elements in **Y**, we see that the direct flow terms cancel and integral utility is determined by the indirect interactions. Also, if a direct flow between two components is absent, then the indirect relations determine the integral relation in the system. Summing the elements of **Y** in Equation 2.9.10 gives:

$$\sum_{i=1}^{3}\sum_{j=1}^{3} v_{ij} = \frac{1}{\det(I-D)}\left(\begin{array}{l} T_1 + T_2 + T_3 + \dfrac{T_1(f_{23}-f_{32})^2}{T_2T_3} + \dfrac{T_1(f_{23}-f_{32})^2}{T_1T_3} + \dfrac{T_1(f_{23}-f_{32})^2}{T_1T_2} \\ + \dfrac{2(f_{13}-f_{31})(f_{32}-f_{23})}{T_3} + \dfrac{2(f_{12}-f_{21})(f_{23}-f_{32})}{T_2} + \dfrac{2(f_{12}-f_{21})(f_{31}-f_{13})}{T_1} \end{array}\right) \tag{2.9.12}$$

It can be shown that the right-hand side of Equation 2.9.12 is always positive, indicating that network synergism always occurs for the complete 3×3 system and all of its subsystems. Analytic extensions to higher order systems may be possible, but are beyond the scope of this paper.

2.9.3 Comparison Of Utility With Other Goal Functions

There have been previous attempts to compare some of the various ecosystem extremal principles (Jørgensen 1992, 1994, Patten 1995). Here, we compare network synergism to a few of these ecosystem parameters to see how the parameter behavior changes under various conditions. The benefit-cost ratio in Equation 2.9.8 is used in the comparison to indicate the magnitude of network synergism. In particular, we look at other ecosystem level properties dealing with flows of energy through the system, namely total system throughflow (power), total transfer efficiency, the ratio of indirect to direct flow, cycling, and connectivity. The last property, connectivity, is not related to the flow through system, but rather to the structure of the system. It has been proposed that all these parameters will increase during the course of ecosystem development. Each parameter is defined in terms of the compartmental flow-storage models used above.

Power, which is work per unit time, can be measured as the total energy flow in an ecosystem. Total system throughflow (TST) is the sum of the throughflows at each compartment (Equation 2.9.13).

$$\mathrm{TST} = \sum_{i=1}^{n} T_i \tag{2.9.13}$$

The direct transfer efficiency, $\mathbf{G} = (g_{ij})$, is the amount of flow along any one arc divided by the throughflow of the source component: $g_{ij} = f_{ij}/T_j$. This is identical to the standard Lindeman efficiency used in the ecological literature. The total system transfer efficiency (TSTE) is calculated by summing all the elements of **G** (Equation 2.9.13).

$$\mathrm{TSTE} = \sum_{i=1}^{n} \sum_{j=1}^{n} g_{ij} \tag{2.9.14}$$

TSTE is not itself an efficiency parameter, but rather is the sum of all the efficiencies in the system, and therefore, can be greater than one. For example, the complete 3×3 system has 12 connections. If the efficiency along each link is 0.10, then TSTE=1.2.

Parallel to the argument we made above using **D** to find the integral utility matrix **U**, **G** can be used to find the integral flow matrix, **N**:

$$\mathbf{N} = \sum_{k=0}^{\infty} \mathbf{G}^k \tag{2.9.15}$$

The direct flows in the system are represented in **G** and the integral flows in **N**. Indirect effects concern relationships at a distance in flow networks and are represented in Equation 2.9.15 when $k > 1$. Therefore, the total indirect flow in the system is equal to the integral flow minus the direct flow and initial condition (indirect = **N–G–I**). An indirect to direct (*i/d*) flow parameter is found by taking a ratio of the sum all the elements of **N–G–I** and all the elements of **G**:

$$\frac{i}{d} = \frac{\sum_{i=1}^{n} \sum_{j=1}^{n} (n_{ij} - g_{ij} - i_{ij})}{\sum_{i=1}^{n} \sum_{j=1}^{n} g_{ij}}. \tag{2.9.16}$$

The integral flow matrix, **N**, can also be used to calculate cycling in a system. If a diagonal element of **N** exceeds 1, then cycling has occurred at that component. Cycled total system throughflow ($\mathrm{TST_C}$) is:

$$\mathrm{TST_C} = \sum_{i=1}^{n} \left(\frac{n_{ii} - 1}{n_{ii}} \right) T_i, \tag{2.9.17}$$

and its ratio to TST is the cycling index (Finn 1976):

$$CI = \frac{TST_c}{TST}. \tag{2.9.18}$$

Finally, connectivity refers to the structural aspect of the system and is measured by the number of links in the system digraph.

It is our hypotheses that the benefit-cost ratio (*b/c*) will increase with increased efficiencies, cycling, throughflow, indirect effects, and connectivity. In order to test these hypotheses, the complete three-component digraph (Figure 2.9.2) was used for all tests except connectivity as the standard structure to compare the various parameters. This structure was held unchanged and the amount of flow between the components was changed by randomly choosing values for the direct transfer efficiencies, g_{ij}, in the system. For a given set of randomly chosen direct transfer efficiency values, all the ecological parameters were calculated. In this manner, different parameter values were gathered over a wide range of transfer values for the same system structure. The different parameters were plotted against each other to observe any emergent patterns. The input into the system was also held constant for all the runs except one in which the parameters were compared for changing input. To test the connectivity, the above mentioned parameters were calculated for all nontrivial subsets of the complete four-component system and compared to the number of links in each structure.

2.9.4 Results

We used the cycling index (CI) as the baseline to compare the behavior of the other ecological parameters, so we tested first to see how cycling responds to the total system transfer efficiency (TSTE). As expected, CI increased as the TSTE increased (Figure 2.9.3) because a greater value for TSTE indicates more flow stays in the system along each path before it leaves the system as output and the CI is therefore greater (no distinction is made as to the form of the output). It has been stated that the indirect-direct flow ratio will increase as the cycling in the system increases (Higashi and Patten 1989). Also, the maximum power principle states that the total system throughflow (TST) of energy in a system will be maximized. We find that both *i/d and* TST increased as CI increased (Figure 2.9.4).

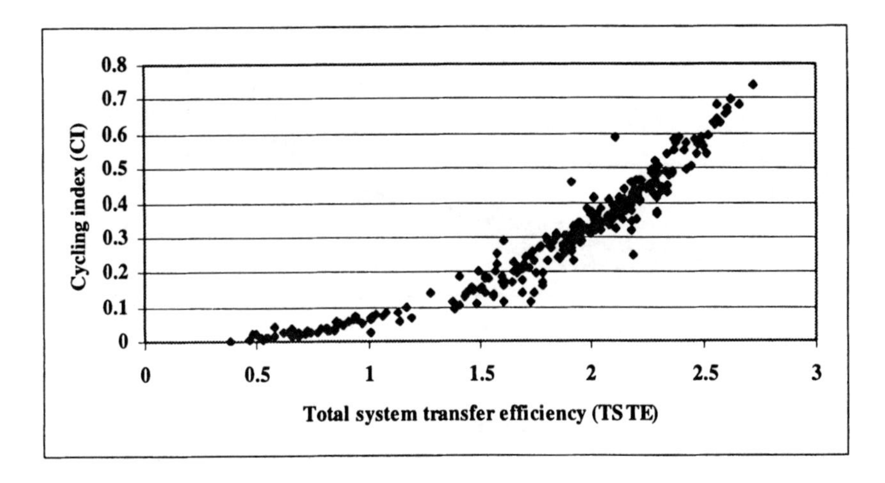

Fig. 2.9.3. Cycling index versus total system transfer efficiency for a Monte Carlo simulation of randomly generated transfer efficiencies using the complete three-component system for the model structure

Again, this is because in a system with a greater CI the flow stays in the system longer resulting in both greater indirect flow and TST. One might expect the utility benefit-cost ratio (*b/c*) to increase with increased CI, however, the correlation between *b/c* and CI is less obvious (Figure 2.9.5). This is due to the fact that, for the complete three-component system, *b/c* is dependent on two conflicting processes, total system throughflow and relative net flow (Equation 2.9.12). As one of these increases the other generally decreases and visa versa. For systems with a low CI (<10%), *b/c* seems to decrease. Systems with very little cycling behave similarly to a food chain and can have a very high value of *b/c* because the utility measure is dominated by the relative net flow between any two components (Equation 2.9.12). Therefore, the linear chain model can have a high *b/c* and low cycling. Relative net flow and cycling are inversely related. However, for higher cycling values (>10%), the minimum value of *b/c* slopes slightly upward as CI increases because the throughflow terms (1st three terms in Equation 2.9.12) begin to dominate the system. For a given system, total system throughflow increases as CI increases. The tradeoff between these two aspects is evident in Fig. 2.9.5, and therefore, *b/c* is not as strongly correlated to CI as *i/d* and TST.

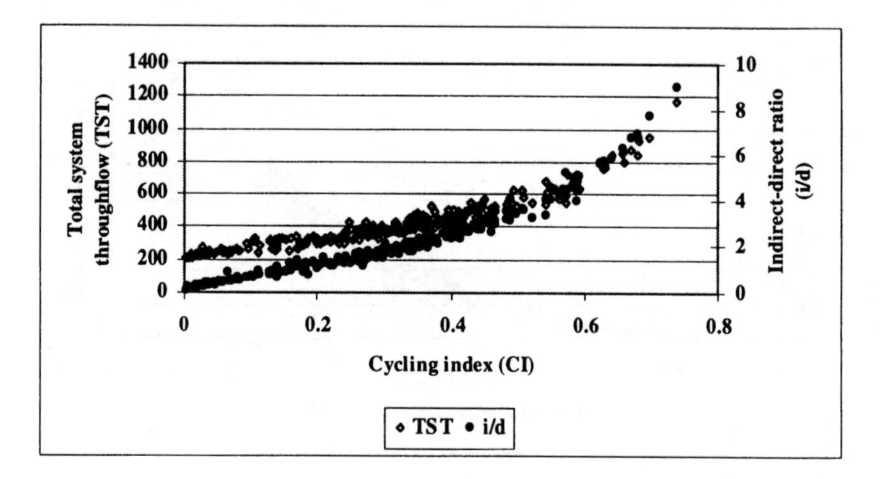

Fig. 2.9.4. Total system throughflow and indirect-direct flow ratio versus cycling index for a Monte Carlo simulation of randomly generated transfer efficiencies using the complete three-component system for the model structure

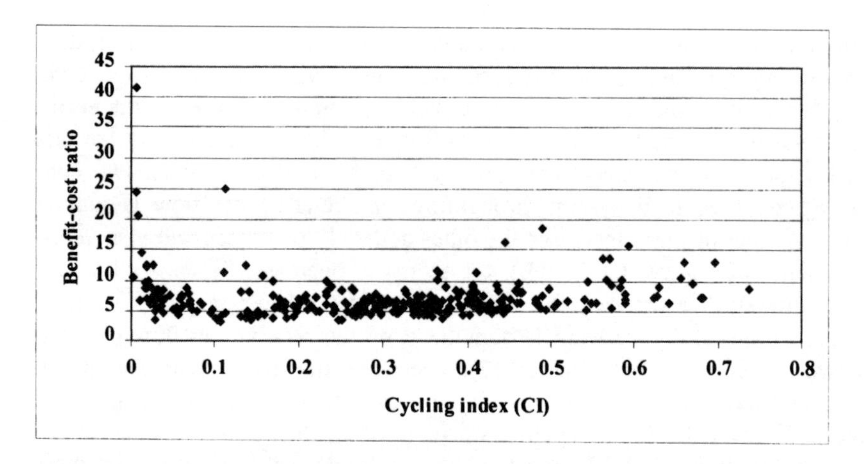

Fig. 2.9.5. Benefit-cost ratio versus cycling index for a Monte Carlo simulation of randomly generated transfer efficiencies using the complete three-component system for the model structure

All of the above comparisons were done using data that were generated by randomly choosing values for the transfer efficiencies. Another set of tests were done while holding the transfer efficiencies along each link constant, $g_{ij} = 0.10$. Under these conditions, we wanted to see the effect of changing the input and structure of the system. Increasing the input does not change the values for CI, *i/d*, and *b/c* because these parameters are all relative to the total flow so their values were not affected by a change in the input.

The only parameter affected by changing the input was TST because a system with greater initial input has greater throughflow, all other things being equal.

To investigate how TST, CI, *i/d*, and *b/c* change with respect to various structures, we looked at all non-trivial structures of the four-component digraph. Again, we assume external inflow only into the first component. A non-trivial system is one in which all four components have some flow through them. Therefore, they must be connected by direct or indirect links from the first component (i.e. if the components were buttons and the links threads tied between the buttons, picking up the first button would result in picking up all four). For a four-component system, a minimally connected structure has three links of which there are 16. There is only one maximally connected system that has 12 links. In all, there are 2432 possible non-trivial structures. As the connectivity of the system increased, there is a general increase in TST, CI, and *i/d* (Fig.s 2.9.6 - 2.9.8).

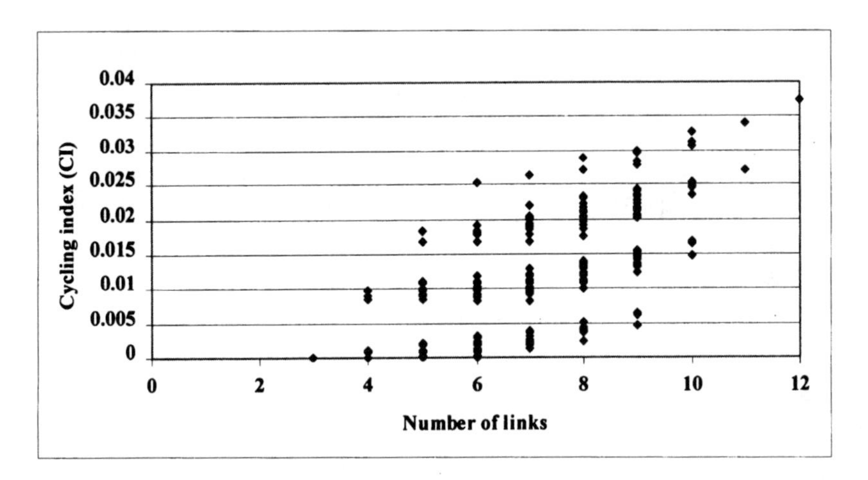

Fig. 2.9.6. Cycling index for all nontrivial subsystems of the four-component system in which all the links between component transfer efficiency, $g_{ij} = 0.10$

The transfer efficiency along each link was constant, $g_{ij} = 0.10$, and therefore, in general a system with more links (higher connectivity) has a greater value in its flow parameters. In most cases, the parameter values for these structures come in discrete groupings. The different groupings are dependent on the order of connections in the system. For example, the two extreme cases in which the four components are minimally connected are (1-2-3-4), and (1-2, 1-3, 1-4). In the first case the flow to component is $(f_{10})*(0.1)^3$, wheras, in the second case the flow to component four is $(f_{10})*(0.1)$. Clearly, TST is greater in the second case.

Adding another connection to both structures will increase the TST in both systems, but it would still be greater in the second case. In fact, we see that the flow in the latter case with 4 links would be greater than that in the former case with 5 links. So the relation between links and flow properties is not monotonic. Again, the relation between connectivity and *b/c* is less obvious (Figure 2.9.9). There is no apparent pattern between them.

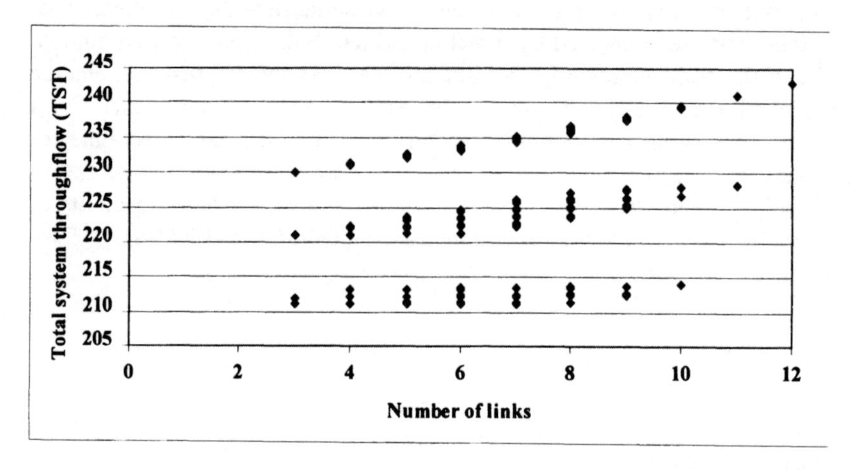

Fig. 2.9.7. Total system throughflow for all nontrivial subsystems of the four-component system in which all the links have the same between component transfer efficiency, $g_{ij} = 0.10$

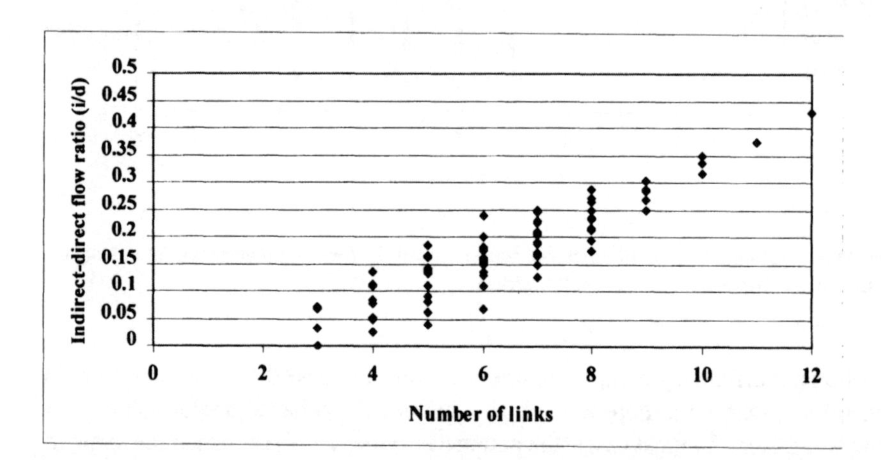

Fig. 2.9.8. Indirect-direct flow ratio for all nontrivial subsystems of the four-component system in which all the links have the same between component transfer efficiency, $g_{ij} = 0.10$

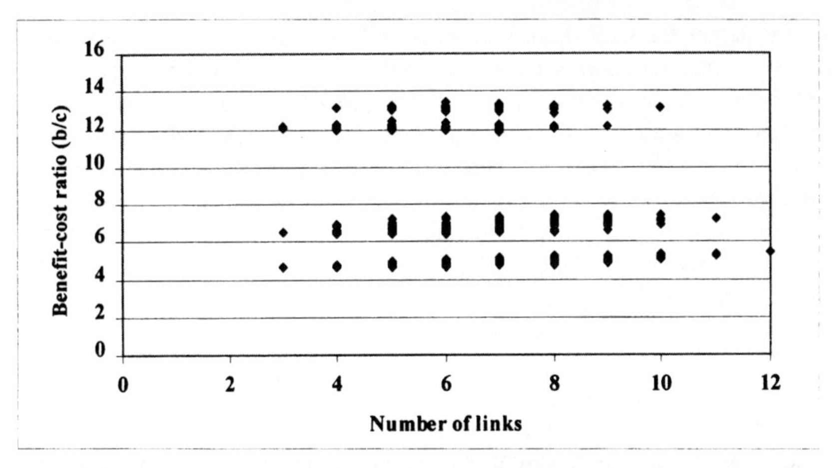

Fig. 2.9.9. Benefit-costs ratio for all nontrivial subsystems of the four-component system in which all the links have the same between component transfer efficiency, $g_{ij} = 0.10$

Again, the discrete levels are due to different classes of systems, but now corresponding to the number of reciprocal flows (f_{ij}–f_{ji}) between two components. Since utility measures the relative net flow between components, systems without reciprocal flows tend to have higher *b/c*. Therefore, as the number of connections increases there is a greater probability of reciprocal flows. Because each link has the same transfer efficiency along it, the net flow terms cancel when reciprocal flows exist. This has the tendency to keep *b/c* low. If the connections had variable transfer values (here they are assumed to be the same), then it may be possible to have a system with both high net flow and cycling.

2.9.5 Discussion

In conclusion, environ utility theory provides a method for determining qualitative and quantitative relations derived from conservative transactions between all components of integrated systems. Network synergism is an inherent property of networks arising from direct transactions of energy or matter between the component parts of systems. We believe this holds over many levels of organization, from cells to ecosystems, and that it is an important cause and consequence of the self-organization widely ascribed to complex adaptive systems. Organisms operating in the zero-sum, "eat-and-be-eaten", world of ecosystems tend to enjoy dominantly positive integral relationships.

The ratio of positive to negative integral utility offers another perspective to evaluate ecosystem development. The parameters cycling, throughflow, and indirect effects are correspondingly related to the within system transfer efficiencies and the connectivity of the system structure. However, the relation of the utility parameter with these properties is less apparent. It is only weakly related to cycling and seems to not be related to system connectivity. These differences arise because the other parameters are all dependent on maximizing the system throughflow. Utility is not dependent on maximizing total flow through the system but rather on optimizing throughflow and net flow between components. Therefore, a system in which one of these properties is very high, such as a food chain, is likely to have a high value of positive utility because the net flow is at a maximum. This subtle difference makes it a useful parameter in determining the relationship between two components in a system and may give it a role to play in the overall suite of goal functions.

Acknowledgements: The comments of two anonymous reviewers were very helpful and appreciated. We also like to thank F. Müller and M. Leupelt along with the Ökologie-Zentrum at Christian-Albrechts-Universität for sponsoring and organizing the "Unifying Goal Functions" workshop.

References

Fath BD and Patten BC (in press) Network synergism: emergence of positive relations in ecological systems.

Finn JT (1976) Measures of ecosystem structure and function derived from analysis of flows. J Theor Bio 56:363-380

Hannon B (1973) The structure of ecosystems. J Theor Bio 41:535-546

Higashi M and Patten BC (1989) Dominance of indirect causality in ecosystems. Am Nat 133:288-302

Jørgensen SE (1992) Integration of ecosystem theories: a pattern. Kluwer, Dordrecht, Netherlands

Jørgensen SE (1994) Review and comparison of goal functions in systems ecology. Vie et Milieu 44:11-20

Leontief WW (1966) Input-output economics. Oxford University Press, New York

Lotka AJ (1922) Contribution to the energetics of evolution. Proc Natl Acad Sci USA 8:148-154

Miller RE and Blair PD (1985) Input-output analysis: foundations and extensions. Prentice Hall, Inc, Englewood Cliffs, New Jersey

Odum EP (1969) The strategy of ecosystem development. Sci 164:262-270

Patten BC (1978) Systems approach to the concept of environment. Ohio J of Sci 78:206-22

Patten BC (1991) Network ecology: indirect determination of the life-environment relationship in ecosystems. In: Higashi M, Burns T (eds) Theoretical studies of Ecosystems: the network perspective. Cambridge University Press, New York

Patten BC (1992) Energy, emergy and environs. Ecol Modell 62:29-69

Patten BC (1995) Network integration of ecological extemal principles: exergy, emergy, power, ascendency, and indirect effects. Ecol Modell 79:75-84

Ulanowicz RE and Puccia CJ (1990) Mixed trophic impacts in ecosystems. Coenosis 5:7-16

2.10 Network Orientors: Theoretical and Philosophical Considerations why Ecosystems may Exhibit a Propensity to Increase in Ascendency

Robert E. Ulanowicz

Abstract

Three disparate metaphors have dominated the discourse on ecosystem dynamics: The ecosystem as (1) machine, (2) organism, or (3) stochastic assemblage. Motivated, in part, by ambiguities of this nature, Karl Popper suggested the notion of "propensity" to generalize the Newtonian concept of force. Using propensities one can articulate a theory of ecosystem development that encompasses all three analogies. Probabilistic indices borrowed from information theory can be used to quantify the degree of trophic constraint operating in an ecosystem, the amount of flexibility available for it to adapt to new circumstances, and, ultimately, the propensity for each transfer to occur. Consequently, the ascendency of an ecosystem may be defined as the flow-averaged system level propensity for activity. Under this rubric, the observed propensity for ascendency to increase over time becomes the probabilistic counterpart for living systems to Newton's second law in mechanics.

2.10.1 Introduction

In his recent book, "An Entangled Bank", Joel Hagen (1992) traces the development of the concept of ecosystem. Throughout his narrative he lays great emphasis on the role that metaphors have played to express various concepts. From among the various metaphors for ecosystems, it is possible to identify three enduring themes: The ecosystem as (1) machine, (2) organism, or (3) chance assembly.

Of the three analogies, only that of the machine has deep roots in the modern synthesis. Thus do Depew and Weber (1994) argue that Darwin himself inherited his strictly mechanical vision of evolution from Newton via Malthus and Smith. George Clarke (1954), in his textbook on ecology, went so far as to depict ecosystem populations and processes as the gears and wheels of a machine. Elsewhere, the mechanical aspects of ecosystem behavior have been emphasized in

F. Müller M. Leupelt (Eds.)
Eco Targets, Goal Functions, and Orientors

recent years by Connell and Slatyer (1977) and most notably in the technocratic visions of Howard Odum (1960).

The machine notwithstanding, the analogy that motivated most of American ecology early this century was the simile of ecosystem as organism. It should be noted that it was precisely the Aristotelian notions of organicism that had been so vehemently eschewed by the early leaders of the modern movement, such as Francis Bacon and Thomas Hobbes (see below). Hagen traces the prominence of organismal thought in ecology to Frederic Clements (and Shelford 1939), who identified the romanticism of Jan Smuts (1926) as a leading influence in his thinking. Clements' notions have been defended in the face of withering criticism by such eminent personalities as G. Evelyn Hutchinson and Eugene Odum (1977).

Ironically, the strongest opposition to organicism in ecology has come not from the defenders of newtonianism, but rather from latter-day nominalists. In the eyes of these critics the structure and function of ecosystems have been greatly exaggerated (at best). What Clements regarded as intricate coordination among the biotic elements of a forest, Gleason (1917) viewed as only a random assembly of plant species. During the 1950's there was a significant defection of American ecologists from the Clementsian viewpoint towards the nominalist persuasion. Some see changing social fashions behind the shifts in metaphors for ecosystems (Barbour 1996; Schwarz, this volume).

Metaphors are intended to be loose analogies. That is, there must be some correspondence between the entities being compared, but it is also understood that significant differences remain. A fortiori, commonalties and discrepancies will also exist among the three metaphors. To discuss better the relationships among the three metaphors applied to ecosystems, it helps to focus upon one of them as a point of reference, describe briefly its fundamental tenets, and then discuss how the remaining two differ or agree as regards those items. (Table 2.10.1)

Ulanowicz (1997), following the lead of Depew and Weber (1994), has characterized the mechanical, or Newtonian perspective as based upon five fundamental postulates about the universe. Most importantly, the Newtonian world is assumed to be closed to all causes, save for those that are either material or mechanical (efficient). All other types are strictly enjoined. While the nominalists recognize material and mechanical causes as legitimate, they regard the world as being less lawful than Newtonians would maintain (see "determinism" below). The organicists have been heavily influenced by Aristotle's ideas on causality and allow, in addition to the material and mechanical, the workings of formal and final causes in nature.

Perhaps the four categories of causation are best illustrated in the (unfortunately, unsavory) example of a military battle. The material causes are taken to be the swords, guns, tanks or other ordnance used in the fray. The efficient agents are the individual soldiers who wield the swords, pull the triggers or drive the tanks in their efforts to inflict terrible harm upon their adversaries.

Formal agencies can be either static or dynamical. The former is revealed in the influence that the landscape and topography has upon the conflict, whereas the latter exists in the ever-changing spatial juxtaposition of the armies with respect to each other. The final cause of the battle usually extends beyond the battlefield and is seen in the social, economic and political factors that brought the armies against each other.

Table 2.10.1. Comparisons of Outlooks

Mechanism (Newtonianism)	Organicism (Holism)	Stochasticism (Nominalism)
Material, Mechanical	Material, Mechanical, Formal, Final	Material, Mechanical
Atomistic	Integral	Atomistic
Reversible	Irreversible	Irreversible
Deterministic	Plastic	Indeterminate
Universal	Hierarchical	Local

The Newtonian world is considered to be atomistic. Newtonian systems can be divided for the sake of study, and the working of the ensemble is regarded to be naught but that of the individual parts in concert. Of course, atomism is the keystone of nominalism. At the other end of the spectrum is the organicist outlook, whereby living systems are considered to be integral and indivisible. Organicists always regard information about system parts acting in isolation to be insufficient for a full explanation of the dynamics of the whole.

Newtonianism alone assumes that the world is reversible. Newton's laws are all reversible with respect to time. A motion picture taken of a Newtonian event, such as the collision of two billiard balls, looks qualitatively the same, whether it is run forward or in reverse. Neither organicism nor stochasticism assumes reversibility.

To a newtonian the world is deterministic. Given the specifications of a system and its environment at any one time, the newtonian holds that it is possible in principle to predict the state of the system at any later time, insofar as the behavior of the environment during the interim is known. Any discrepancies between newtonian predictions and subsequent observations are ascribed to ignorance. At the other extreme is the stochastic, who regards order as passing illusion. Organicists fall somewhere between these two poles and usually regard systems as being what might best be termed "plastic". Organicists' opinions range the gamut from Clements with his rigid, almost mechanical version of holism, to Lovelock (1979) and his more liberal vision of the biotic control of the physical environment, where little in the way of a fixed endpoint is evident (other than life itself).

Finally, newtonian laws are considered to be universal across all scales of time and space. Nominalists patently eschew universals in favor of concentrating solely upon the individual, local event. Again, organicism falls between these extremes, taking what has been described as a "hierarchical" view of the world. The hierarchical worldview involves more than simply the (epistemic) nesting of biological order into sequences, such as biome, ecosystem, population, organism, etc. More importantly, hierarchicalists assume that there exist laws and regularities that pertain to each scale, but that the influence of any particular law diminishes in proportion to how remote is the scale at which the particular law was formulated from that of the event to which it is to be applied (Allen and Starr 1982). For example, to most biologists (but not to all, e.g. Wilson 1975) it would appear ill-advised to try to relate higher-level social behavior to some genetic antecedents. Physicists, on the other hand, mix gravitational and quantum-level phenomena with apparent abandon.

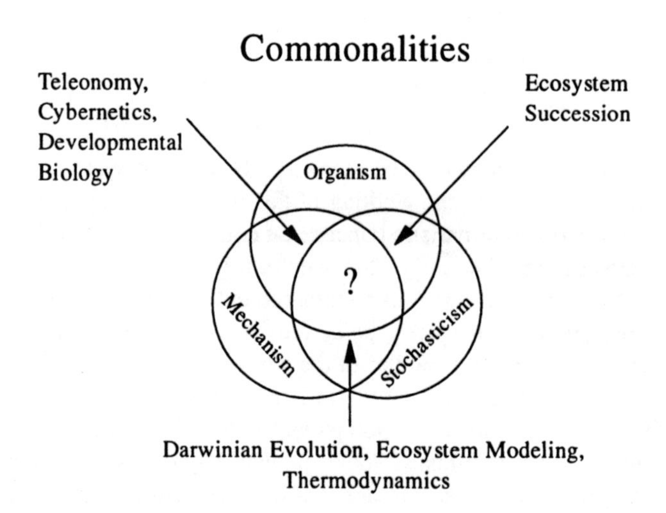

Fig. 2.10.1. Venn Diagram showing the relationships among the three perspectives prevalent in ecosystems science

By now it should be apparent that the three perspectives on ecosystems have certain features and assumptions in common (Fig. 2.10.1). Both the newtonian and stochastic outlooks, for example, are causally closed and regard systems as atomistic; the organismal and stochastic viewpoints both assume events are irreversible, etc. It should not be too surprising, therefore, to encounter narratives that are built upon these overlaps. For example, classical cybernetics and developmental biology combine some aspects of organisms with simple mechanisms. Darwinian evolution, conventional ecosystem modeling and even thermodynamics all include elements both of mechanism and chance. Ecosystem succession seems to involve both organic and stochastic dynamics.

The question remains whether any discipline can build upon the tripod of all three perspectives. Judging from how fragments particular to each of the three viewpoints are found throughout ecology, it would seem that only an amalgam of all three perspectives will suffice as a foundation for ecology. It will be argued here that such a trilateral common ground does exist. Furthermore, it will be suggested that the best approach to deriving an amalgamated ecosystem theory might be to concentrate first on the marriage of mechanism and chance at the macroscopic level into what Karl Popper (1990) has called "propensities". To this very general and abiotic concept can be subsequently appended the organic notions of cybernetics and autocatalysis. Finally, to implement these ideas it will be necessary to derive an explicit mathematical form by which propensities may be quantified and estimated.

2.10.2 A World of Propensities

Perhaps the central tenet in the modern synthesis is that all events are the results of material particles acting according to newtonian-like laws. Nothing happens that is not lawfully ordained. It is the portrait of a closed universe. As LaPlace (1814) speculated, if some superior intelligence could have complete knowledge of the positions and momenta of all particles in the universe, it would be able to hindcast all of history and forecast all of the future. We now realize that LaPlace was far too optimistic. Both the formulation of the second law of thermodynamics by Carnot (1824) and the more recent development of quantum physics have raised doubts about the plausibility of a closed universe. But the newtonian vision of a closed world has been altered only marginally as a result of these challenges. Conventional wisdom now allows as how causality may deteriorate at the edges of reality (i.e., at very small and very large scales), but the universe at the scale of direct human observation is still assumed to operate in newtonian fashion. For example, biology's "Grand Synthesis" permits chance to operate at the molecular scale of the genome, but restricts the remainder of the living world to lawful behavior (Ulanowicz 1997).

In the neo-Darwinian scenario for evolution the observer is always being forced to switch back and forth abruptly from the stochastic netherworld of Boltzman (the genome) to the deterministic theater of Newton (the phonome and its environment). It is a rather schizoid depiction of reality, prompting some to seek a vision that is more inclusive of both true accident and limiting constraint. For example, Charles Sanders Pierce (1877) has suggested that the world is causally open, and Karl R. Popper (1990) has concluded that a full grasp of the nature of evolution is impossible under the assumption of a closed universe. To

Popper the key to his wider vision of nature lies in expanding the narrow newtonian concept of force. Forces are but special, limiting cases of a broader notion that he has called "propensities".

A propensity is the tendency for a certain event to occur under given circumstances. There are two noteworthy features of this definition. First is the notion of chance or probability; second is the requirement that surroundings always be taken into account. In the absence of both of these conditions, propensity degenerates into the conventional notion of a force. I.e., a force is a propensity that acts in isolation.

As an example, if situation B *always* follows upon condition A, then one may conclude that B is lawfully related to A in a way that allows one to search (or define) a causal "force" behind the transition. In terms of probabilities, this situation is trivial: Given A, B always follows-without exception. The conditional probability of B happening, given that A has been observed becomes unity [$p(B|A)=1$].

When phenomena are truly in isolation, as is nearly the case with gravitational attraction between heavenly bodies, or when ensembles of events are virtually independent of each other (so that the atomistic postulate holds), then it suffices to describe events in terms of newtonian forces. Ecologists are keenly aware that such conditions rarely occur in the field, and true isolation is usually difficult, if not impossible, to achieve in the laboratory. A more realistic scenario is that when A happens, then B occurs *most of the time.* But not always! On occasion, A is followed by C, or by D, or by E, etc. As a consequence, $p(B|A)<1$, and $p(C|A)$, $p(D|A)$, etc., are all >0. Popper attributes these latter, non-zero conditional probabilities to unavoidable "interferences" among processes.

Popper is always relating propensities to conditional probabilities, but he never claims that conditional probabilities by themselves quantify propensity. He regards the quantification of propensities as an unsolved problem and counsels only that "We need to develop a calculus of conditional probabilities." Hence, the task remains to derive a formula involving conditional probabilities that fully characterizes the notion of propensity and merges smoothly with existing phenomenology.

At this point it should be noted that Popper has advocated a subtle, but very significant shift regarding the interpretation of probabilities. Since their inception, probabilities have been invoked to treat an observer's ignorance about the details of a situation. For example, one says that the probability that any given number will turn up after the throw of a die is 1/6. Presumably, if one knew the exact translational and rotational momenta, as well as the exact position and attitude of the die at any given instant during its trajectory, then that information, along with the values of certain parameters such as the elasticity of the die and the surface, the precise shape of the die, etc., one would be able to predict which face would show.

Popper, however, is advocating the use of conditional probabilities, not simply to cover the ignorance of the observer, but also to pertain to a degree of indeterminacy *inherent in the situation itself.* This shift is from the epistemic toward the ontological. Some readers probably will be reluctant to accept such interpretation and cling instead to the faith expressed by LaPlace. It is worthwhile noting, however, that the argument from knowledge of detail is predicated upon the assumption of atomism, which is not an element of the organic perspective.

Finally, Popper emphasizes that propensities, unlike forces, cannot exist divorced from their surroundings. An essential element of propensities is their context, which invariably includes other propensities. Thus, one sees how circularity is actually built into the definition of propensities, and one may exploit that circularity to help articulate Popper's "calculus of conditional probabilities".

2.10.3 Propensities in Propinquity

What, indeed, does happen when many propensities interact with each other? One may begin by focusing upon bilateral interactions. If unilateral effects can be characterized as being either positive (+), negative (-), or neutral (0), then the nature of bilateral interactions can be denoted by couples of unilateral interactions, i.e., predation (+,-), competition (-,-), neutralism (0,0), etc. Of the nine possible couples, mutualism (+,+) exhibits singular characteristics that impart to its participants the advantage to persist, on the average, beyond the duration of components engaged in other types of interactions. Furthermore, it may be argued that ensembles of mutualistic interactions, or what in chemistry is called "autocatalytic configurations", exhibit behaviors that are decidedly nonmechanical (i.e., organic) in nature (Ulanowicz 1989, 1997).

To illustrate and study the nature of autocatalysis it helps to consider a triad of processes, A, B, and C. It is assumed that the activity of A has a *propensity* to increase the activity of B. B, in turn, exerts a similar propensity upon C, which has the same effect upon A. Thus, the indirect effect of A upon itself is positive, giving rise to autocatalysis. It should be noted, however, that, unlike in chemistry, A, B, and C do not have to be mechanically linked. That is, the activity of A does not have to abet that of B in every instance–just in most.

To elaborate upon the special attributes of autocatalysis, one may cite at least eight significant properties (Ulanowicz 1989, 1997):

1. That autocatalysis is *growth enhancing* is virtually tautological in that activity anywhere in the loop tends to increase activity in all the other members.
2. It is *selective,* because perturbations that enhance catalysis are rewarded, whereas those that impair activity are decremented. For example, suppose a chance perturbation occurs to element B that happens to increase either its sensitivity to A or the catalytic effect it has upon C. Then the effects of that change will be propagated around the loop in such a way that the activity levels of all elements will be augmented. In particular, the perturbation in B will be rewarded. The opposite occurs when a random change to B either de-

creases the catalytic effect of A or diminishes B's effect upon C. The same feedback results in B's receiving less catalysis from A.

3. Autocatalysis is *symmetry-breaking* in that the selection it engenders defines a preferred direction for change. The direction of catalysis (A --> B --> C) prevails strongly over anything occurring in the counter sense.
4. Selection favors those changes that bring more material and energy into the autocatalytic cycle, thereby inducing what may be called *centripetality*. As a particular example of the selectivity, suppose that the change to B is such that it brings in more of the material and energy necessary to support B's activity. This change will be rewarded. By induction one may conclude that the reward structure for *each* component works to increase the input of elements necessary for the functioning of that compartment. The net result is that the autocatalytic loop resembles the focus of a radial pattern of centripetal flows.
5. Selection and centripetality thus work to *induce competition* between autocatalytic cycles and favor the replacement of components by taxa that are more efficient at sustaining autocatalytic activity. Particular compartments may come and go like actors in a play, whilst the overall feedback structure (the play itself) persists. In other words, the overall cycle is likely to have a characteristic lifetime that exceeds that of any particular constituent.
6. The combined attributes of selection, centripetality and longer lifetime all point toward of a degree of *autonomy* that the autocatalytic configuration as a whole possesses from the properties and histories of its parts. (No one is suggesting the complete autonomy of higher level processes - they remain dependent upon events at the lower levels. The former, however, can no longer be reduced entirely to the latter. I.e., the autocatalytic system is not atomistic.)
7. The foregoing considerations should make it clear that autocatalysis can be characterized as *emergent,* in the sense that some or all of the foregoing properties might be missed if one were to observe only a part of the cycle. By considering only a fragment of an autocatalytic cycle one might mistakenly be led to identify an input as an autonomous initial cause and an output as a determined terminal effect. As soon as one increases the scope of observation so as to encompass all members of the loop, however, the interdependence of such causes and effects becomes apparent, and the foregoing attributes begin to emerge.
8. Finally, autocatalysis is *formal* in the sense that the cycle is a relational form of individual processes.

As to the overall effects these combined properties have upon the development of a flow network, it may be said that they change both its extensive and intensive natures. Extensive properties are those that depend upon the size of the system, and the growth enhancing nature of autocatalysis acts like a ratchet to push the activity level of the cycle ever higher. Meanwhile, selection and associated properties change the qualitative (intensive) character of the network by "pruning" away (or at least diminishing) those elements of the network that are less engaged in autocatalytic activities. The net effect of indirect mutualism is depicted schematically in Fig. 2.10.2. Fig. 2.10.2a represents an inchoate network with ill-defined transfers among the components. After autocatalysis has increased the activity level of the system (indicated by the thicker arrows) and pruned away the autocatalytically less efficient links, the network comes to resemble more the configuration in Fig. 2.10.2b.

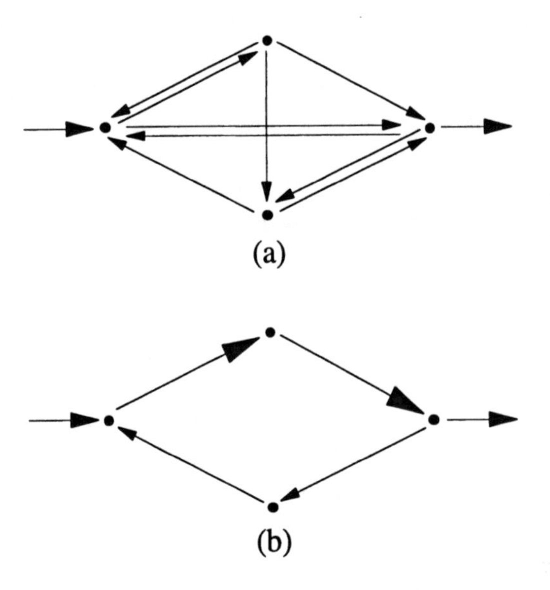

Fig. 2.10.2. Schematic representation of the effects of autocatalysis upon flow networks: (a) A typical inchoate starting network. (b) Same system after autocatalysis has reinforced certain pathways and winnowed others.

2.10.4 Quantifying Constraint and Freedom

Regarding the transition from configuration 2.10.2a to 2.10.2b, it is obvious that flow is more constrained in the latter. That is, for any compartment in 2.10.2a there are more possibilities for transfer (or, more generally, influence) than are evident in 2.10.2b. All other things being constant (an important assumption that will be discussed below), the effects of autocatalysis and competition are progressively to constrain influence to operate along those pathways that are most efficient at autocatalysis. Hence, to quantify the transition from 2.10.2a to 2.10.2b, one must derive expressions for the relative amounts of constraint and freedom in each case.

Quantifying constraint is a task accomplished most effectively by using information theory. Most unfortunately, there is a widespread opinion that information theory, because it was first formulated in communication theory, has only metaphorical application outside of that narrow domain. This is a very mistaken and counterproductive mind set. Fundamentally, information theory is about quantifying changes in probability assignments, and is legitimately applicable anywhere one can define a proper probability (Tribus and McIrvine 1971; Ulanowicz 1986). To help emphasize this generality, the following derivation will be accomplished using a lexicon of constraint and freedom, rather than the conventional terms, information and uncertainty. This alternative nomenclature also has the advantage of emphasizing that freedom may stem, in part, from the inherent nature (ontology) of a situation, and not entirely from inadequacies on the part of the observer (epistemology).

Following the lead of Boltzman, the *indeterminacy* of any particular outcome is defined to be proportional to the negative logarithm of the probability of that outcome, i.e.,

$$f_i = -k \log p(A_i) \tag{2.10.1}$$

where $p(A_i)$ is the probability that event A_i will occur, f_i is the indeterminacy of A_i and k is a scalar constant (which will be discussed presently). Now $p(A_i)$ is the unconditional probability that A_i will occur under all possible conditions. If, in some instances, B_j were to occur just prior to A_i, then one could speak of the *conditional* probability, $p(A_i|B_j)$, which , in general, would be different from $p(A_i)$. As in Equation 2.10.1, the indeterminacy of A_i consequent to B_j (call it f_{ij}) would be

$$f_{ij} = -k \log p(A_i|B_j) \tag{2.10.2}$$

If B_j somehow constrains A_i, the situation becomes less indeterminate, and one would expect f_{ij} to be smaller than f_i.

Hence, one may speak of the constraint, C_{ij}, that B_j exerts upon A_i as

$$C_{ij} = f_i - f_{ij} \tag{2.10.3a}$$

$$= -k \log p(A_i) - [-k \log p(A_i|B_j)] \tag{2.10.3b}$$

$$= k \log[p(A_i|B_j)/p(A_i)]. \tag{2.10.3c}$$

It is worthwhile noting that $C_{ij} = C_{ji}$. This is a result of Bayes' Theorem, which may be invoked to show that

$$\frac{p(A_i|B_j)}{p(A_i)} = \frac{p(A_i|B_j)}{p(B_j)} = \frac{p(A_i, B_j)}{p(A_i)p(B_j)}, \tag{2.10.4}$$

where $p(A_i, B_j)$ is the (symmetrical) joint probability that A_i and B_j occur together. In words, the constraint that B_j the exerts upon A_i is equal to the constraint that A_i exerts upon B_j. One may regard this symmetry as the probabilistic analog to Newton's Third Law, which states that for every action there is an equal and opposite reaction.

To estimate the aggregate constraint that all A_i and B_j are exerting upon each other, one weights each C_{ij} by the probability $p(A_i, B_j)$ that A_i and B_j co-occur, and then sums these products over all combinations of i and j:

$$C = k \sum_i \sum_j p(A_i, B_j) \log[p(A_i, B_j)/p(A_i)p(B_j)], \tag{2.10.5}$$

where C represents the average mutual constraint at work in the system as a whole.

To estimate C for a given system, one first must attach specific physical events to A_i and B_j. One convenient identification is with the transfers of some particular form of material or energy. Thus, A_i might represent a quantum of medium entering compartment i. B_j is associated with a quantum of medium leaving compartment j. In order to account for all medium passing through an n-compartment system, one must include exchanges with the external world. This can be done by letting j=0 be the origin of all external inputs to the system, and i=n+1 be the sink for all exports (Hirata and Ulanowicz 1984). If the amount transferred from i to j is denoted by T_{ij} , then one possibility for estimators for the probabilities in Equation 2.10.5 might be

$$p(A_i, B_j) \sim T_{ij}/T \tag{2.10.6}$$

$$p(A_i) \sim \left(\sum_{q=1}^{n+1} T_{iq}\right)/T \tag{2.10.7}$$

$$p(B_j) \sim \left(\sum_{p=0}^{n} T_{pj}\right)/T, \tag{2.10.8}$$

where

$$T = \sum_{p=0}^{n} \sum_{q=1}^{n+1} T_{pq} \tag{2.10.9}$$

is a measure of aggregate activity, called the "total system throughput". (These probabilities could have been defined in a number of ways. Ulanowicz and Abarca-Arenas [1997], for example, use compartmental stocks to estimate apriori rates of exchange, but the original definitions, cast wholly in terms of flows, are retained here for the sake of simplicity.)

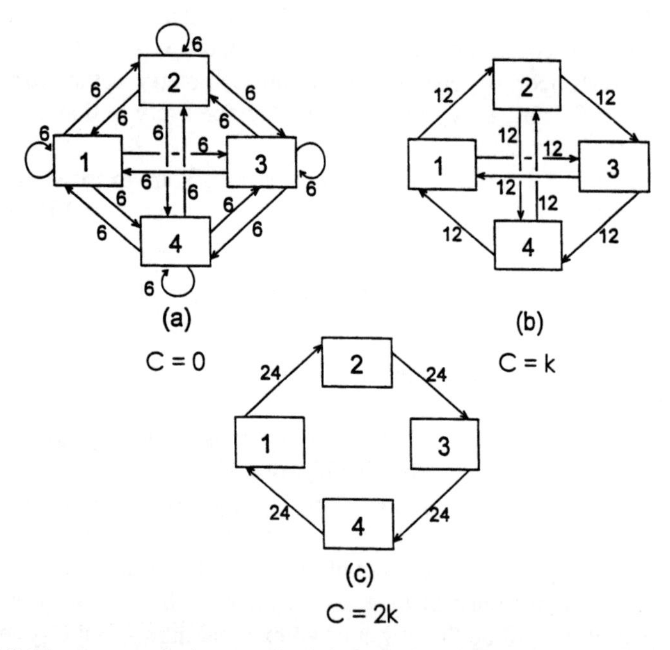

Fig. 2.10.3. Three progressively constrained configurations of flow among four compartments: (a) The wholly equivocal scenario. Medium is equally likely to flow to any component in the system. (b) Flow exiting any compartment is constrained to flow to only two other compartments. (c) Fully constrained configuration. Each compartment can contribute to only one other component. The value of "C" or "constraint" is calculated from formula (2.10.10)

Substitution of (2.10.6) - (2.10.9) into (2.10.5) yields

$$C = k \sum_{i=0}^{n} \sum_{j=1}^{n+1} (T_{ij}/T) \log[T_{ij} T / (\sum_{p=0}^{n} T_{pj})(\sum_{q=1}^{n+1} T_{iq})] \qquad (2.10.10)$$

as the estimate of constraint inherent in any network of quantified flows. Fig. 2.10.3 illustrates that C behaves in the desired manner. In Fig. 2.10.3a, there is maximal indeterminacy about where a quantum of medium will flow next. In Fig. 2.10.3b medium is more constrained in where it may flow. Finally, flow in Fig. 2.10.3c is maximally constrained: from any given compartment flow may proceed to only a single prescribed recipient.

Although C characterizes the intensive consequences of increasing autocatalysis, it does not address the extensive change, namely, the increase in system activity. Also, the scalar constant, k, remains undefined. Both of these deficiencies can be corrected by using the scalar constant to impart physical dimensions to the constraint index (Tribus and McIrvine 1971).

By setting k=T one scales the constraint index by the total system activity and simultaneously eliminates the denominator from the multiplier of the logarithm. This scaled measure of constraint is renamed the system "ascendency". It quantifies both the "size" of the system and the degree to which the system activity is organized by its internal constraints.

$$A = \sum_{i=0}^{n} \sum_{j=1}^{n+1} T_{ij} \log[T_{ij} T / (\sum_{p=0}^{n} T_{pj})(\sum_{q=1}^{n+1} T_{iq})] \qquad (2.10.11)$$

Ascendency was formulated initially as a phenomenological index that encapsulates several of the criteria that Odum (1969) had identified as characteristics of systems in the later phases of ecosystem succession (Ulanowicz 1980). His list of 24 criteria can be summarized in terms of four tendencies: Natural ecosystems tend to increase in species richness, predator/prey specificity, internalization, and cycling. *Ceteris paribus*, increases in each of these attributes contribute to a higher ascendency. A larger number of species means that all summations in formula (2.10.11) will be extended. Narrower predator/prey specificity was seen in Fig. 2.10.3 to be an explicit contribution to C, the measure of constraint. Finally, both internalization and cycling contribute toward greater system activity (total system throughput.) Whence, observation suggests that "in the absence of major perturbations, ecosystems naturally tend towards configurations of ever-greater ascendency".

2.10.5 Caveat

Ascendency seems to provide an appropriate goal function with which to describe ecosystem development. While this is true to an extent, it is imperative to emphasize immediately two important disclaimers.

This first caveat is that ascendency tells only a part of the story of ecosystem development. If ecosystems were fully constrained, then the metaphor of the machine would have been sufficient. But organic behavior requires certain freedom from total constraint. Of course, without constraint there would be no system worth studying (the nominalist extreme). Without sufficient freedom, however, a system becomes brittle (Holling 1986) and unable to adapt to a changing environment. It dies or collapses to some inchoate configuration.

Fortunately, it is straightforward to quantify the freedom still possessed by A_i and B_j beyond their mutual constraint. In fact, the residual freedom that A_i possesses in the presence of B_j already has been defined as, f_{ij}, the indeterminacy of

A_i consequent to B_j. (See equation 2.10.2). Calculated in similar fashion, f_{ji} expresses the indeterminacy (freedom) of B_j in the presence of A_i. If one adds f_{ij} to f_{ji}, substitutes the probability estimators (2.10.6) - (2.10.8) into the sum, averages the outcome using the joint probabilities, and scales the result by the total system throughput, one arrives at an expression for the residual freedom (Φ) of the system,

$$\Phi = - \sum_{i=0}^{n} \sum_{j=0}^{n+1} T_{ij} \log[T_{ij}^2 / (\sum_{p=0}^{n} T_{pj})(\sum_{q=1}^{n+1} T_{iq})]. \tag{2.10.12}$$

The quantity Φ is complementary to the ascendency and is termed the system overhead (Ulanowicz and Norden 1990). The ascendency and the overhead together quantify the structured complexity of the system, X, which includes both organized and inchoate attributes,

$$X = A + \Phi. \tag{2.10.13}$$

The complementarity expressed in this definition of complexity (2.10.13) signifies that an increase in ascendency could adversely affect the system overhead. Thus, if more ascendency signifies a tighter degree of organization, one immediately realizes that there can be "too much of a good thing". It cannot be emphasized enough that too high a proportion of ascendency might impair system integrity by crowding out system overhead, which functions as a "strength in reserve" that is essential to a system for adaptation and survival.

The second warning is that ascendency, in spite of being a surrogate for constraint and efficiency, is itself a non-mechanical attribute. Ascendency, after all, is based upon a probabilistic rather than a mechanistic depiction of reality. Any directed change, such as the tendency for living systems to increase in ascendency, should not be likened to a mechanical goal function, such as the Hamiltonian operator. Probabilistic goal functions do not "drive" the system toward a fixed, pre-determined endpoint, as is the common, deterministic notion of a goal function. Rather, probability operators behave more like Bossel's (1987) "orientator" functions, which merely guide the system along a vague direction. It appears ascendency should be regarded as such an "orientator" function. As such, it is immune to most of the criticisms leveled against goal functions for introducing teleology into biology.

2.10.6 Quantifying Propensities

The reader should recall that the above definition of autocatalysis incorporated the notion of propensities in a cyclical juxtaposition. It should not be too surprising, therefore, to find that the quantification of propensity lies buried somewhere within the ascendency calculus.

To identify the explicit formula for propensity, it helps to recall some early definitions underlying irreversible thermodynamics that were laid down by Onsager (1931). Onsager generalized the work of earlier phenomenologists, such as Fourier, Fick and Ohm. Each of these investigators had attached a putative force to an observed flow. Thus did Fourier show how the rate of thermal conduction varies in proportion to the negative of the imposed gradient in temperature. Similarly, Fick traced the origins for mass diffusion to a gradient in species concentration, and Ohm connected electrical current to a concomitant gradient in voltage. Onsager was able to show how, by careful choice of dimensions, one could identify a thermodynamic "force" conjugate to each simple physical flow in such a way that the product of each force-flux pair has the dimensions of production of entropy $[ML^2T^{-3}\theta^{-1}]$.

Following Onsager's lead, the aggregate entropy production for any system of processes became the sum of all force-flux pairs that comprise the system. Any calculation with the form

$$\sum_i Flow_i \; x \; Force_i \tag{2.10.14}$$

became known as a "power function". One immediately notes that the equation for ascendency (2.10.11) has this form. Each flow, T_{ij}, is multiplied by a corresponding logarithmic term, and the results are summed. Ascendency is what is known as a "quasi-power function" (James Kay, personal communication). In drawing this analogy, each logarithmic term becomes the homolog of a thermodynamic force, and the temptation is to equate the counterparts. But the probabilistic nature of the ascendency, as just mentioned, precludes calling any of its components a force in the mechanical sense of the word. Rather one should regard the term more as a tendency, or a *propensity* for the transfer T_{ij} to happen.

That is, one identifies

$$\log[T_{ij} T / (\sum_{p=0}^{n} T_{pj})(\sum_{q=1}^{n+1} T_{iq})] \tag{2.10.15}$$

as the propensity for medium to flow from i to j.

Under this rubric, ascendency appears as the system-averaged propensity for activity to occur. Furthermore, the propensity for ascendency to increase over time becomes the propensity of a propensity-in loose analogy with Newton's second law, which defines force as something proportional to acceleration – the rate of change of a rate of change. Whence the statement, "In the absence of additional external influences, the ascendency of a living system has a propensity to increase" comes to bear loose analogy to both Newton's first and second laws. It resembles the first law (A body in motion will continue along a straight line, unless acted upon by an external force.) in that it describes what happens in the absence of new external influence. Like Newton's second law, it is a second-order statement, describing the propensity of a propensity. It differs markedly, however, from Newtonian dogma because it is highly non-conservative. Just as entropy may appear *ex nihilio*, ascendency may do likewise. It is the law of a new order-the living order-that ecosystems exhibit the propensity to increase in ascendency.

References

Allen TFH, Starr TB (1982) Hierarchy. University of Chicago Press, Chicago. p 310

Barbour MG (1996) American ecology and American culture in the 1950s: Who led whom? Bull Ecol Soc Am 77(1):44-51

Bossel H (1987) Viability and sustainability: Matching development goals to resource constraints. Futures 19:114-128

Carnot S (1824) Reflections on the Motive Power of Heat (translated 1943). ASME, NY. 107 p

Clarke GL (1954) Elements of Ecology. John Wiley, NY

Clements FE (1916) Plant Succession: An Analysis of the Development of Vegetation. Carnegie Institution of Washington, Washington, D.C.

Clements FE, Shelford VE (1939) Bio-ecology. John Wiley and Sons, New York, p 425

Connell JH, Slatyer RO (1977) Mechanisms of succession in natural communities and their role in community stability and organization. Am Nat 111:1119-1144

Depew DJ, Weber BH (1994) Darwinism Evolving: System Dynamics and the Genealogy of Natural Selection. MIT Press, Cambridge, MA, p 588

Gleason HA (1917) The structure and development of the plant association. Bull Torrey Botanical Club 44:463-481

Hagen JB (1992) An Entangled Bank: The Origins of Ecosystem Ecology. Rutgers University Press, New Bunswick, NJ, p 245

Hirata H, Ulanowicz RE (1984) Information theoretical analysis of ecological networks. Int J Systems Sci 15:261-270

Holling CS (1986) The resilience of terrestrial ecosystems: local surprise and global change. pp 292-317. In: Clark WC, Munn RE (eds) Sustainable Development of the Biosphere. Cambridge University Press, Cambridge, UK

LaPlace PS (1814) A Philosophical Essay on Probabilities. Dover Publications, Inc., NY, p 196

Lovelock JE (1979) Gaia: A New Look at Life on Earth. Oxford University Press, NY, p 157

Odum EP (1969) The strategy of ecosystem development. Science 164:262-270

Odum EP (1977) The emergence of ecology as a new integrative discipline. Science 195:1289-1293

Odum HT (1960) Ecological potential and analogue circuits for the ecosystem. Am Sci 48:1-8

Onsager L (1931) Reciprocal relations in irreversible processes. Physical Review A 37:405-426

Pierce CS (1877) The fixation of belief. Popular Science Monthly 12:1-15

Popper KR (1990) A World of Propensities. Thoemmes, Bristol, p 51

Smuts JC (1926) Holism and Evolution. MacMillan, NY, p 362

Tribus M, McIrvine EC (1971) Energy and information. Sci Am 225(3):179-188

Ulanowicz RE (1980) An hypothesis on the development of natural communities. J theor Biol 85:223-245

Ulanowicz RE (1986) Growth and Development: Ecosystems Phenomenology. Springer, NY

Ulanowicz RE (1989) A phenomenology of evolving networks. Systems Research 6:209-217

Ulanowicz RE (1997) Ecology, the Ascendent Perspective. Columbia University Press, NY

Ulanowicz RE, Abarca-Arenas LG (1997) An informational synthesis of ecosystem structure and function. Ecological Modelling 95:1-10

Ulanowicz RE, Norden J (1990) Symmetrical overhead in flow networks. Int J Systems Sci 21 (2):429-437

Wilson EO (1975) Sociobiology. Harvard University Press, Cambridge, MA, p 697

2.11 Applying Thermodynamic Orientors: Goal Functions in the Holling Figure-Eight Model

Brad Bass

Abstract

The Holling figure-eight model is proposed as a description of ecosystem dynamics that can incorporate the complex behavior that is germane to realistic biological systems. The figure-eight includes the two phases of exploitation and conservation as well as destructive events, such as fire, and system reorganization following such an event. The Holling figure-eight is underlain by two goal functions: the maximization of exergy consumption, the maximization of exergy storage and the emergence of a self-organized critical state at the conservation state.

Exergy is defined as the quality of energy or the amount of work that can be done with a particular input of energy. Self-organized critical systems are sensitive to small changes that can cause rapid changes in the system. They are characterized by a log-linear relationship between the magnitude and the frequency of a particular event. The conservation phase of the Holling figure-eight is characterized as a balance between the attractor of exergy maximization and thermodynamic equilibrium. It is argued that this be explained with reference to the self-organized critical state between species abundance and the community mass respiration rates.

Recent analysis of the eigenvalues of ecological models suggests that even though a system may appear to be resistant to a particular perturbation, if it is in the basin of attraction of a semi-stable attractor, it will eventually respond but at a much later point in time. This provides further insight into management decisions that interfere with the Holling figure-eight at the conservation phase. The explanations of each goal function and semi-stability are followed by some discussion on the relevant management implications.

F. Müller M. Leupelt (Eds.)
Eco Targets, Goal Functions, and Orientors

2.11.1 Introduction

Holling (1986) proposed a four-phase conceptual model of ecosystem dynamics that includes exploitation and conservation as well as destructive and renewal components to explain the failure of many natural resource management schemes (Fig. 2.11.1). The model is drawn as a sideways figure-eight i.e. ∞. There are two dimensions in this model: connectivity (abscissa), and nutrient availability (ordinate).

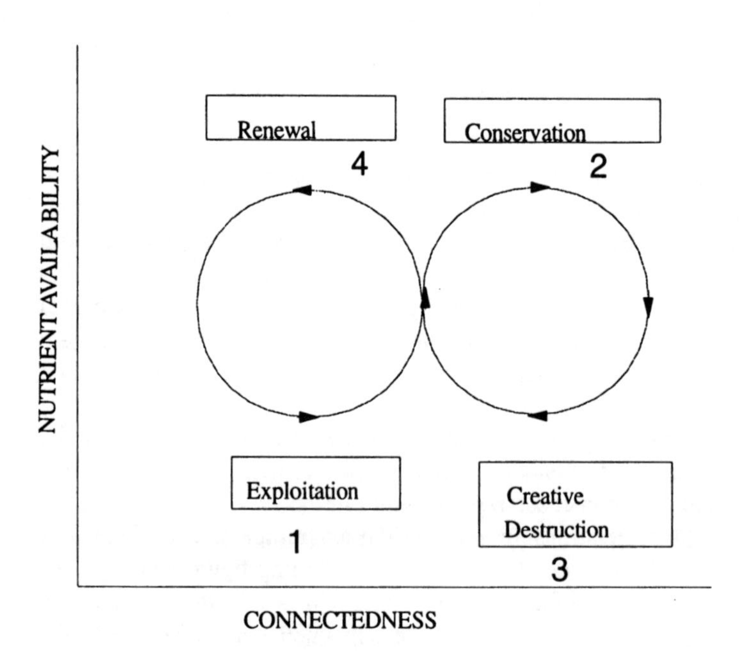

Fig. 2.11.1. Holling figure-eight model

Kay (1994) and an Ontario Ministry of Natural Resources (OMNR) report on State of the Landscape Reporting (OMNR 1996) suggest that the Holling figure-eight is an appropriate conceptual model for resource and ecosystem management because Holling's figure-eight embeds the role of destruction into ecosystem dynamics. Holling (1986) and the OMNR report (OMNR 1996) suggest that this dynamic is repeated at different spatial and temporal scales. The OMNR report (OMNR 1996) proposes that the Holling figure-eight is the conceptual linkage between structure and ecological process both in ecology and the social realm. This extension of the Holling figure-eight has been used to devise a framework for indicator development to assess recent changes to environmental regulations in the Province of Ontario, Canada (OMNR 1996).

This paper developed out of discussions of Working Group V, at *The Work-shop on Atmospheric Change and Biodiversity: Formulating a Canadian Science Agenda* (Bass 1996). Working Group V, addressed the impacts of atmospheric change on biodiversity from the perspective of self-organizing systems. One of the challenges put to the workshop was to develop a single indicator to monitor the impacts of multiple atmospheric stresses on biodiversity. Working Group V, addressed issues of self-organization and began with a discussion of the Holling figure-eight as a conceptual model for understanding how climate change and other atmospheric stresses interact with ecosystems and biodiversity. The linkage between Holling's figure-eight and preserving biodiversity under atmospheric change, and the development of the indicator, required an understanding of the two goal functions, underlying Holling's figure-eight: the maximization of exergy storage and consumption and the emergence of a self-organized critical relationship between species abundance and community mass respiration rates.

Exergy is defined as the amount of work that can done with an available source of energy. Schneider and Kay (1994) demonstrate that gradients of exergy drive the evolution of ecosystems. As ecosystems evolve, exergy is used to build new and more complex structures that store larger amounts of exergy. As the ecosystem matures, it stores more exergy than can be used to create new structures. This storage becomes the fuel for another self-organized dissipative structure such as fire or an insect pest. After the destructive event, decomposers release the exergy that was stored in the biomass which is made available for the next exploitation phase.

Self-organized criticality or a self-organized critical relationship exists when two characteristics are related logarithmically such that the graphical relationship is a straight line with a slope of -1 (Fig. 2.11.3). In self-organized critical systems small instabilities may result in very large disturbances that cascade throughout the whole system. Over time, as the system recovers it will return to the self-organized critical state defined by the above-mentioned graphical relationship. As more exergy is consumed and stored, the system is expected to be characterized by fewer animal species of larger body size. This trend is counter balanced by small environmental perturbations which favor more animal species of smaller body size. Choi (1994) demonstrated that the relationship between body size and species diversity is self-organized critical. Bass et al. (1998) used these two goal functions to develop a single indicator of change in biodiversity based on the measurement of outgoing longwave radiation.

Holling used the figure-eight to illustrate how resource systems, such as forests, could be over-managed to the extent that the shift from conservation to creative destruction is delayed. In unmanaged systems, the creative destruction is relatively small in space and time, and it is a necessary component of ecosystems adaptation. In reviewing several case studies, Holling pointed out that the longer this shift is delayed, the greater the destruction is, and the more surprising it is, when it does occur. Similarly, in self-organized critical systems, the longer the system remains at the self-organized critical state, despite the occurrence of minor instabilities, the larger and more surprising the eventual disturbance. The discovery of semi-stability in ecological simulation models provides additional insight into this aspect of the Holling figure-eight.

This paper discusses the two goal functions that underlie the Holling figure-eight and the additional insights of the discovery of semi-stability. It is difficult to define the two goal functions and semi-stability without some definitions from the language of complex systems. The definitions are provided in the following section, followed by a description of the Holling figure-eight, the two goal functions, and semi-stability. A discussion of the management implications follows a presentation of each goal function and semi-stability.

2.11.2 The Language of Complex Systems

Ecosystems are complex systems (Kay 1994; Hansell et al. 1997). *Complexity* is defined as a wide range of emergent, system-level, behaviors that can include self-organization, rapid changes of state and historical dependence (Judd 1990). Complex systems are also open systems, and complexity arises from the interaction between different components, not solely from system structure (Hansell et al. 1997). Describing change in complex systems requires a different description of change as well as a new vocabulary (Hansell et al 1997).

The first stage is to identify those variables that describe the state of the system in order to define the *state space* or the boundary within which the system must operate. The second stage is to define areas of the state space that *attract* the system. *Attractors* are defined as points in state space where the system is drawn to following a perturbation. They are points of stable equilibrium in the state space.

An attractor can also be thought of as a graphical tool that provides a different way of viewing the evolution of system through time. It shows how the different data relate to each other, and the location that appears to "attract" the data on the graph or in the state space (Kiel 1994). The *basin or domain of attraction* is a region of the state space within which a perturbation will cause a movement towards an attractor. Points on the rim of the basin of attraction may form an unstable region, or an area from which the data can escape the attractor, in state space. This unstable region can be called a *repellor* (Byers and Hansell 1996).

A "model" for describing system dynamics, based on Prigogine and Nicolis (1971), accounts for the emergence of new attractors in complex systems. Prigogine and Nicolis (1971) demonstrated that as complex chemical compounds are bombarded by internal and external events, the organization may break down and lead to chaos. Eventually, new compounds will form into completely new and more complex structures. The important point in understanding complexity is that as new structures emerge from within the system, the change can be discontinuous and it can follow a period of chaos. Consider the Bénard cell, a structure that occurs as a gradient is imposed across the boundaries of a system containing a liquid or a gas, such as a pot of soup or the atmosphere.

As the bottom of the pot is heated, a gradient is imposed from the bottom (hot) to the top (cold). As the gradient increases, the system resists movement away from equilibrium. With a strong enough gradient, structures emerge in the liquid which are required to dissipate the heat from the bottom to the top of the pot. New structures and even more complex structures will emerge as the gradient increases further, but the change is discontinuous. Consequently, the system becomes increasingly organized as the steepness (difference between the hot and cold boundary layers) of the gradient becomes larger, in order to continue to dissipate the influx of energy, hence the name *dissipative structure*. The structures *self-organize* in response to the need to dissipate the heat.

Dissipative structures emerge or self-organize in a narrow window where the energy input is high enough, but not too high (Kay 1994). If the energy input falls below a minimum threshold, the structures cannot emerge, or existing structures cannot be supported. If the energy input crosses a critical threshold, it overwhelms the ability of the organized structures to dissipate the energy and remove the gradient. Thus, at this point the behavior of the system appears to be highly unpredictable or chaotic. *Self-organization*, in a dissipative structure is the process by which the internal structure and in some cases, the function of a system is modified in order to handle increasing gradients of exergy.[1]

In a living system, the ability of an incoming gradient to establish disorder is decreased by increasing system throughput and degrading or dissipating the exergy input. Schneider and Kay (1994) view ecosystems as dissipative structures organized to effectively degrade exergy. As a corollary, material flow cycles in the ecosystem will tend to be closed to maintain the supply of material that is necessary for the exergy degrading process. Any evolutionary or adaptive strategy can enhance the potential for a system and its components to survive, if its net effect increases the ability to degrade incoming exergy. This can be observed through the emergence of more complex structures with increased diversity, and a greater number of hierarchical levels, as ecosystems develop.

Essentially, a living system develops more structure for dissipation by increasing the utilization or consumption rate of *exergy*. While exergy is often defined as the quality of energy, or the amount of work that can be extracted from an energy source; it can be more precisely defined as the sum of the available free energies. Exergy also allows for a restatement of the second law of thermodynamics.

When referring to isolated systems, the second law of thermodynamics has been generally expressed as follows (Sussman, 1972):

- Heat flows spontaneously from a hot object to a cooler object, and not in the reverse direction.

---

[1] A more complete discussion is found in Kay (1984).

- All possible spontaneous changes increase the disorder or entropy of the universe.

In other words, all closed systems move from order to disorder. In considering open systems, Schneider and Kay (1994) restated the second law of thermodynamics and its implications:

As the amount of available exergy increases, the more likely it is that a self-organizing dissipative system will emerge to capture it. (In biology, an organism is an example of a self-organizing dissipative system.) The further a system is moved from thermodynamic equilibrium (as measured by the exergy content in the system), the stronger is the tendency to return to thermodynamic equilibrium. Thus, structures emerge to dissipate the increasing gradient; when this gradient is removed, the structures disappear and the system returns to equilibrium.

Schneider and Kay (1994) applied a thermodynamic analysis to ecosystem, based on this restatement of the second law of thermodynamics. However, for biological systems, the notion of dissipative structures and the analogy of the Bénard cell are a starting point, but not sufficient for dealing with questions of ecosystems. The Holling figure-eight model has been shown to be a useful framework for considering some of these issues.

2.11.3 The Holling Cycle

Holling (1986) reviewed a number of resource management case studies involving situations where an ecosystem collapsed or was seriously damaged. He formulated a conceptual model of ecosystem dynamics that stressed discontinuous change as an internal property of the systems and provided insight as to when such an event would occur. The model includes four phases. The first two phases determine ecosystem succession: an exploitation phase related to r-strategists and a conservation phase related to K-strategists. The third phase is one of discontinuous change caused by events such as a fire or a pest outbreak, while the fourth phase is one of reorganization or renewal as resources that are released through the previous change are now made available for the exploitation function. These four phases are linked in a figure-eight (Fig. 2.11.1).

The third and fourth phases suggest an ecosystem dynamic beyond exploitation and succession. The third phase occurs with increasing connectedness and exergy storage in the maturing ecosystem. The connectedness becomes the "backcloth" (Atkin 1974, 1978) that allows the rapid spread of a disturbance resulting in an abrupt change in the system. Holling terms this "creative destruction" after Schumpeter (1950). The massive disturbances associated with phase 3 may cause significant losses both in terms of area and species, the previously accumulated yet unavailable exergy, or stored capital, is now increasing in its availability. This stored capital is made available through decomposition, phase 4, and retained through mechanisms, some of which are the colloidal behavior of soil, rapid nutrient uptake by remaining vegetation and reduced rates of nitrification (Marks and Bormann 1972).

2.11.4 The First Goal Function: the Maximization of Exergy Storage and Consumption

Kay reinterprets the Holling figure-eight in terms of exergy, the first goal function in order to explain why the Holling cycle occurs and to link it to a broader system perspective (Fig. 2.11.2). Much of this discussion is based on Kay's thermodynamic interpretation. The y-axis represents the rate at which a system uses exergy while the x-axis represents the exergy stored in the biomass of that system.

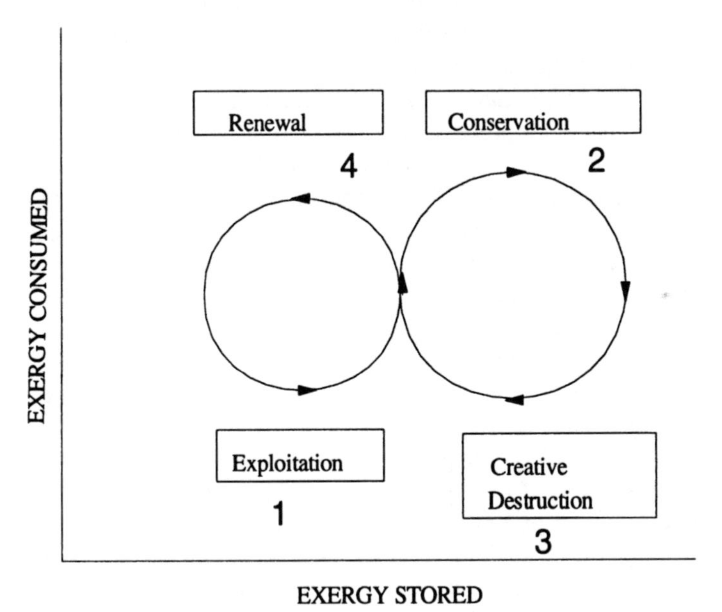

Fig. 2.11.2. Holling's figure-eight model recast in thermodynamic terms (Kay 1994)

In the exploitation phase, given sufficient material and biological information, dissipative processes emerge which utilize the exergy from the incoming solar energy. The rate of exergy consumption increases followed by additional biomass and stored exergy. A positive feedback loop emerges; as exergy storage becomes larger, the ecosystem develops more structure which allows the system to better utilize incoming exergy, leading to the further development of more structure. This is the *First Thermodynamic Branch.* The source of exergy from this branch is derived from the input of solar energy.

With increasing exergy storage (i.e. more biomass), the probability that some self-organizing process, such as fire or a pest outbreak, will emerge to consume this exergy also increases. In the conservation phase, the exergy storage reaches a maximum which amplifies the risk of the emergence of a new dissipative process since the system is as far out of equilibrium as is possible. An event such as a fire moves the system to a new state that is closer to thermodynamic equilibrium. Thus maximum exergy storage implies maximum thermodynamic risk.

Maximum thermodynamic risk and thermodynamic equilibrium are both attractors at phase 2 of the Holling figure-eight model, but eventually the equilibrium attractor dominates. It will be argued later that in fact an ecosystem has many smaller attractors at smaller spatial scales. The tension between thermodynamic risk and thermodynamic equilibrium would naturally occur on these smaller scales. The system itself is on a larger attractor that remains stable, unless an attempt is made to maintain the thermodynamic risk by attempting to preserve the system in the conservation phase.

In the release phase, the exergy that was stored in the biomass becomes available for use. The self-organizing process, (i.e. fire or pest outbreak), does not store biomass and exergy, rather exergy is made available from the dead biomass and stored nutrients. This is the *Second Thermodynamic Branch*, and the exergy is derived completely from stored exergy. In the reorganization phase, while the stored exergy is consumed, decomposition processes provide the raw materials to start the First Thermodynamic Branch.

Ecosystems self-organize through the capture of resources (exergy and material) which are used to build more structure and degrade exergy. The whole process is not driven by exergy, but rather by gradients of exergy. This latter point is illustrated in Schneider and Kay (1994) using the data of Luvall and Holbo (1989). As previously noted, dissipative structures emerge or self-organize in a narrow energetic window. If one type of structure predominates, then the system becomes overextended or brittle and cannot fully utilize the available exergy or material (Kay 1994). If the structure cannot fully utilize the inputs, then some other structure that is better adapted will emerge to displace it.

In resource management, one objective that is often pursued, is to preserve one specific resource, be it a tree or a fish species. At some point, this structure is overwhelmed by another emergent structure such as a fire or an insect pest. Management objectives that focus on maintaining some fixed state of an ecosystem, maximizing some structure such as biomass or productivity, or minimizing pest outbreaks of fire will lead to a larger loss than would naturally occur because this type of management encourages the storage of excess exergy that cannot be consumed or degraded by the structures that are preserved (Kay 1994). Management must focus on facilitating the Holling figure-eight, not maintaining the conservation stage.

The creative destruction highlights the importance in preserving biodiversity. In the movement from the decomposition and retention of nutrients to exploitation, changes in the physical context such as the atmosphere which impacts upon the availability of nutrients, environmental conditions or information could cause a shift in the ecosystem such that a new system emerges. This new ecosystem represents a different first thermodynamic branch, which has adapted to a changed physical context or occurs because of a loss of much of the genetic information that was contained in the previous system. At this stage biodiversity is crucial, because it acts as genetic library as well a record of the history of previous adaptations.

2.11.5 The Second Goal Function: Self-Organized Critical Relationships

Patterns of size and species abundance, at different spatial scales can be expressed as a log-linear relationship with scaling exponents near -1 (Fig. 2.11.3). This pattern may emerge because ecosystems evolve to a self-organized critical state (Choi 1994).

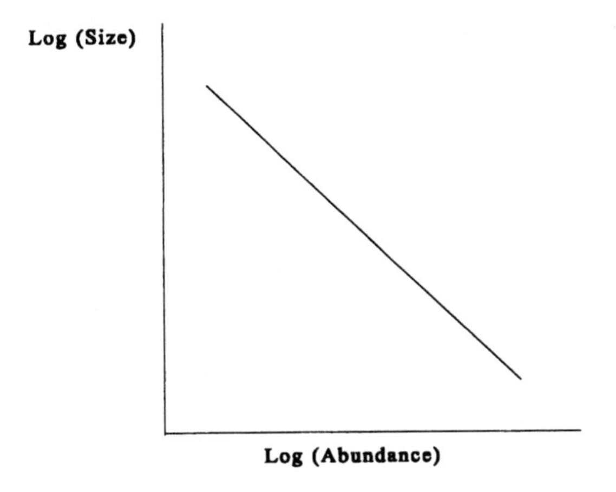

Fig. 2.11.3. Log-linear relationship between species size and abundance

A similar relationship exists between body size and temperature, i.e. body size generally decreases as temperature increases, and this pattern is detectable across many different taxonomic and geographic scales (Choi 1994). Most of the data were collected on poikilotherms. Amongst homeotherms, this pattern is known as Bergmann's rule and is clearly an important and perhaps analogous pattern. There are many potential factors that are at play in regulating this empirical pattern. The most proximate (physiological) explanation is due to the elevated rates of metabolic costs due to a higher temperature environment.

Choi's explanation of these relationships links community energetics and the mechanism of succession (Choi 1994). Body size is a proxy for mass specific respiration rate, which in turn is an estimate of the rate of exergy dissipation (Zotin and Zotin 1982). Estimates based upon allometric physiological relationships indicate that systems that are dominated by rapidly growing organisms are energetically *leaky*, meaning higher rates of total and mass specific exergy dissipation. This quantity decreases with growth or aging in an individual, which is in accordance with Prigogine's least dissipation principle (Prigogine and Nicolis, 1971).[2]

Community mass specific respiration rates approach a minimum with a scaling exponent of -1, the slope in Fig. 2.11.3, suggesting that exponents near -1 are favored from a thermodynamic perspective. Thus as community mass specific respiration rates decrease with maturity, the ecosystem is less *leaky* which corresponds to the increase in stored exergy at the conservation phase of the Holling figure-eight.

Based on data from several ecosystems, Choi (1994) hypothesized that the log-linear relationship between body size and abundance or temperature is indicative of a spatio-temporal fractal pattern which Bak et al. (1989) labeled self-organized criticality. Fluctuations across many different scales will affect the form of this log-linear relationship. When perturbations are extremely strong, the system behavior becomes highly non-linear, even chaotic. Self-organized criticality reasserts itself after a perturbation (Bak at al. 1989). Choi (1994) and Bass et al. (1998) suggest that perturbed ecosystems return to the self-organized critical state as which corresponds to the Holling figure-eight after the destructive event.

Bass et al. (1998) used both goal functions to develop an indicator of change in biodiversity based on measuring the leakiness of an ecosystem. Such an indicator could complement any government policy aimed at preserving biodiversity. However, there are other management implications. Self-organized critical states will eventually collapse, although predicting when this will occur is difficult. The longer the interval between collapses, the larger the collapse.

Bak et al. (1989) demonstrate this with a sandpile, the prototypical self-organized critical system. They construct a "sandpile" in which one grain of "sand" at a time is added to the top of the pile.[3] There are always small avalanches, but eventually a large avalanche will occur, destroying a good part of the pile. The larger the pile is, the higher level of the destruction. An analysis of the second goal function reaches a similar conclusion to that of the previous section a management philosophy that aims to facilitate the Holling figure-eight.

[2]Prigogine's least dissipation principle applies to non-equilibrium dissipative systems near a local steady state. Prigogine demonstrated that these systems will change in such a manner that the rate of energy dissipation is reduced which corresponds to increasing exergy storage.

[3]Bak et al. (1989) constructed a "sandpile" out of plastic balls.

Maintaining the system in the conservation phase will result in larger losses than would otherwise be expected.

2.11.6 Semi-Stable at the Conservation Phase

Nonlinear models may appear to be stable for long periods of time, even with significant changes in the values of parameters, and then suddenly shift to a new attractor. In these cases, with a shift in a model parameter, the previous attractor is still attractive but not stable. Byers et al. (1992) demonstrate that age structured biological population models may remain close to the old attractor, or equilibrium values in the time domain, for hundreds or thousands of generations after a shift in parameters that should have imposed a new set of equilibrium values. Another example is the persistence glacial landscapes, such as boreal spruce bogs, in southern regions characterized by much warmer climates (Hansell et al. 1997). In these cases, the old attractor is semi-stable.

Semi-stability can also be defined with respect to the attractor of a mathematical model. If an attractor is stable, the basin of attraction is shaped such that the system is continuously drawn to that point just as a marble is always drawn to the bottom of a bowl. If the shape of the basin is "flattened", the system can leave the attractor and shift to a new attractor with a "deeper" basin of attraction. Mathematically, for non-linear models, the stable attractor is characterized by the eigenvalues of the Jacobian being less than zero.[4]

For the sake of illustration, assume a simple linear model with a fixed-point attractor where the system is drawn to one point. Then, the geometric interpretation for each eigenvalue (Seydel 1994) is a straight line, for each state variable, that passes through the point. The point cuts the eigenvalue in half, and each half-ray is a trajectory of the system. If the absolute value is large (small), the model reaches the attractor quickly (slowly). An eigenvalue less than zero, for a continuous time system, indicates that the trajectory of the system is headed towards the attractor, and an eigenvalue that is greater zero indicates that the trajectory is headed away from the system. For a nonlinear version of this system the trajectories are not straight lines, but would be tangent to those straight lines at the point.

In a nonlinear continuous time system, if the state space is characterized by an attractor, then all trajectories are headed towards the attractor. In other words, volumes contract around an attractor, and all eigenvalues are less than zero. The exact opposite occurs with a repellor, and all eigenvalues are greater than zero. If a semi-stable attractor appears, at least one trajectory behaves as if it is on a

[4]The Jacobian, **J** is a matrix where the entries are the first-order partial derivatives of a series or ordinary differential equations or an iterated function system ($x_{t+1} = f[x_t]$). Geometrically, the Jacobian describes the shift and deformation of a volume with a rotation in coordinates. The entries are $\delta x_i/\delta x_j$ or the change in the ith element of the vector field with a change in the jth variable. In modelling, then the Jacobian represents a state-transition matrix (Morrison 1991).

repellor, or at least one eigenvalue is greater than zero. In the basin of attraction of a semi-stable attractor there exists a small number of directions where the volume is expanding, not contracting around the attractor. In a discrete time, nonlinear system if one the absolute value of an eigenvalue is greater than one, than that trajectory is moving away from the attractor.

A semi-stable attractor is no longer stable, but depending on the magnitude, the behavior of the system may appear stable for many iterations of the model. Byers and Hansell (1996) provide an example of an fixed-point attractor for a discrete time model in three-dimensional space. This Jacobian has the following eigenvalues: (0.1, 0.1, 0.99999). This is asymptotically stable. If the eigenvalues change to (0.1, 0.1, 1.00001) the system is unstable although the transient model behavior will remain indistinguishable from the transient behavior the previous model represented by the previous eigenvector (0.1, 0.1, 0.99999) for quite some time.

The length of time that a system can spend on a semi-stable attractor can be depicted by examining the residence times around the old, or now semi-stable, attractor (Hansell et al. 1997). The residence or dwell time that the system spends in the neighbourhood of the semi-stable attractor is the amount of time spent in the basin of attraction. The relationship between the residence times T, and the relevant parameter p, is characterized by the following power law distribution or log-linear relationship, and approximates a negative exponential distribution (Byers and Hansell 1996).

$$T \sim (p - p_c)^{-1} \tag{2.11.1}$$

The variable p_c is the threshold value of a critical variable which corresponds to the rim of the basin of attraction of the semi-stable attractor, and p is the current position, in the state space. As the distance between p and p_c increases, the dwell times decrease until the system leaves the semi-stable attractor. Plotting the logarithm of the dwell times versus the logarithm of $(p - p_c)^{-1}$ provides a graph that resembles Fig. 2.11.3, a straight line with a slope of -1.[5]

There are several implications of semi-stability. When a system arrives at a stable attractor, it will remain in that basin of attraction until that attractor is destabilized by large perturbation. If the attractor is semi-stable, the system will leave that basin of attraction, despite the absence of a large perturbation. The difference between stability and instability has implications for the conservation phase of the Holling figure-eight. What if the conservation phase is a semi-stable attractor, or how could it be transformed from a stable to a semi-stable attractor? For example, Hansell et al. (1997) note the remarkable stability of certain glacial landscapes, in non-glacial climates, long after the last glacial period. Climatic changes engendered by the current interglacial should have been a large enough perturbation to shift this landscape to another basin of attraction. If this system is

[5]This may also be a self-organized critical relationship, but this has not yet been demonstrated.

on a semi-stable attractor, can other ecosystems that are deemed sensitive to climate change arrive in the basin of attraction of a semi-stable attractor?

Holling (1986) and Kay (1994) argue that attempts to maintain a system in the conservation phase interfere with the role of creative destruction, and that the figure-eight proceeds on much smaller spatial scales in unmanaged systems. Left to its own devices, the system will go through the Holling figure-eight more often, but usually over a smaller spatial extent. In a sense, the perturbations are not large enough to alter the whole landscape which suggests that different parts of the system are on different yet smaller, perhaps semi-stable, attractors. These smaller attractors are subsumed by a larger attractor, i.e. the whole system, which maintains its stability. Management attempts to preserve a conservation phase, where they are successful, transforms the larger system attractor into a large semi-stable attractor.

The possibility of semi-stability opens up a new, albeit limited role for ecosystem simulation models. If a model of a particular system can be written as series of ordinary differential equations, for continuous time simulation, or as series of iterated functions, for discrete time simulation, then it can be analyzed for semi-stability. The state variables of this model, such as species population, are used to define the state space of the basin of attraction, and thus the Jacobian matrix which is necessary to identify and analyze the eigenvalues. Even if an analysis of the eigenvalues suggests that the attractor is stable, the parameters can be perturbed in order to try to destabilize the attractor or create a semi-stable attractor.

If the size of the perturbation is small, and a semi-stable attractor emerges, then even though the current observations do not suggest that the system is in the basin of attraction of a semi-stable attractor, this could be in error due to background noise, measurement error or measurement uncertainty. In this case, it would be prudent to act as if the system had arrived in the basin of attraction of a semi-stable attractor. It also implies that small changes in management could create a semi-stable attractor. If a semi-stable attractor emerges with a large perturbation of a parameter, it implies that significant changes to factors such as harvesting, waste management, and stocking may result in deleterious environmental and economic effects at a much later point in time. Whether or not it is possible to construct a realistic ecosystem model is a valid concern but is beyond the scope of this paper.

2.11.7 Conclusions

An analysis of ecosystem dynamics must account for how changing gradients of exergy can induce self-organizing processes in complex systems, but this is not sufficient. Holling (1986) suggests that ecosystems proceed through four phases described in the figure-eight diagram: exploitation, conservation, release and reorganization. The figure-eight includes destructive events as a constituent part of ecosystem dynamics. The figure-eight is underlain by two goal functions: the maximization of exergy storage and exergy consumption and self-organized criticality. The first goal function implies that as new, more complex dissipative structures emerge, so that an ecosystem can utilize and dissipate increasing amounts of exergy. The second goal function implies that as new dissipative structures emerge, a self-organized critical relationship emerges between species abundance and body size.

Reinterpreted in terms of exergy storage and consumption, the system is in balance between the two attractors of exergy maximization and thermodynamic equilibrium, in the conservation phase. Choi (1994) and Bass et al. (1998) suggest that this balance is a state of self-organized criticality which characterizes mature ecosystems as the tendency towards larger organisms is balanced by the impact of random perturbations which favors smaller organisms. Both goal functions lead to the same conclusion that regarding management, the goal should be to facilitate the Holling figure-eight otherwise the damage to events such as fire will be much larger than expected.

The discovery of semi-stable attractors in ecological models suggests that an ecosystem is characterized by many small attractors. This discovery offers further insight into the Holling figure-eight and why a management philosophy aimed at preserving the conservation phase cannot succeed. Holling (1986) maintains that the figure-eight will happen on much smaller spatial scales if no attempt is made to block the release phase. This suggests that an ecosystem is comprised of several smaller attractors, perhaps semi-stable attractors, being destabilized at different times. Although a smaller attractor may be destabilized, it is subsumed by a larger attractor which remains stable. Maintaining the system at the conservation stage changes this larger attractor into a semi-stable attractor, but the system cannot remain at this attractor indefinitely, thus the damage occurs at a much larger spatial scale. Semi-stability also creates an important management role for simulation models. Models can now be used to indicate the existence of a semi-stable attractor or whether management actions could shift the system onto a semi-stable attractor. Models can now be further linked to the quality of measurement as well, since some analyses of semi-stability may suggest a sensitivity to measurement uncertainty. Semi-stability further explains the existence of surprise, since the departure from a semi-stable attractor cannot be predicted.

This type of discussion is largely theoretical, but it does have some practical import. Bass et al.(1998) have used the two goal functions to develop a single indicator of changes in biodiversity. The Holling figure-eight is an important component in a framework to develop indicators of for assessing environmental impacts of government policies (OMNR 1996). Moore (1996) has adapted some principles of ecology into a very similar four-phase model as the basis of a successful business strategy, thus returning to the roots that the Holling figure-eight shares with evolutionary economics (Schumpeter 1950).

Acknowledgments: The original paper was given in James Kay's absence due to illness and hence contains much material from his work and presentations which I have attended, as well as some material of my own. I would like to thank James Kay for sharing his thoughts on the Holling figure-eight and for several helpful comments on an earlier draft of this manuscript. I would like to thank Ted Byers for clarifying several aspects of semi-stability, and I would also like to thank Jae Choi, Ian Craine and Roger Hansell for several lively discussions around these ideas. The synthesis of material is solely my own. Finally, I would like to thank Alannah Naber and Wolfgang Kramer for their helpful comments on earlier drafts of this manuscript, and Felix Müller and Maren Leupelt for their patience.

References

Atkin R (1974) Mathematical Structure in Human Affairs. Heinemann Educational Books, London

Atkin R (1978) Time as a Pattern on a Multidimensional Backcloth. Journal of Self-organized criticalityial and Biological Structures 1:281-295

Bak P, Chen J and Creutz M (1989) Self-organised Criticality in the Game of Life. Nature 342:780-781

Bass B (1996) Working Group Five Report. In: Munn RE (ed) Atmospheric Change and Biodiversity: Formulating a Canadian Science Agenda. Institute for Environmental Studies, University of Toronto, pp 46-56

Bass B, Hansel RIC and Choi J (1998) Towards a Simple Indicator of Biodiversity. Journal of Environmental Monitoring and Assessment 49 (in press)

Byers RE, Hansell RIC and Madras N (1992) Stability-like Properties of Population Models. Theoretical Population Biology 42:10-33

Byers RE and Hansell RIC (1996) Implications of Semi-stable Attractors for Ecological Modelling. Ecological Modelling 89:59-65

Choi J (1994) Patterns of Energy Flow and its Role in Structuring Communities: Studied with Fish Communities in Lakes of the Bruce Penninsula and Manitoulin Island. PhD Dissertation, University of Toronto, Toronto

Hansell RIC, Craine IT and Byers RE (1997) Predicting Change in Non-linear Systems, Journal of Environmental Monitoring and Assessment 46:175-190

Holling CS (1986) The Resilience of Terrestrial Ecosystems: Local Surprise and Global Change. In: Clark WC and Munn RE (eds) Sustainable Development of the Biosphere. IIASA, Cambridge University Press, pp 292-317

Kay JJ (1984) Self-Organization in Living Systems. Doctoral thesis, Dept of Systems Design Engineering. University of Waterloo, Waterloo, Canada

Kay JJ (1994) Some Notes on: The Ecosystem Approach, Ecosystems as Complex Systems and State of the Environment Reporting. Dept of Environment and Resource Studies, University of Waterloo, Waterloo

Kiel LD (1994) Managing Chaos and Complexity in Government. Jossey-Bass, San Francisco

Judd K (1990) Chaos in Complex Systems. In: Dynamics of Interconnected Biological Systems, Birkhäuser, Berlin

Marks PL and Bormann FH (1972) Revegetation Following Forest Cutting: Mechanisms For Return to Stead-state Nutrient Cycling. Science 176:914-915

Morrison F (1991) The Art of Modeling Dynamic Systems. John Wiley & Sons, New York, NY

Moore JF (1996) The Death of Competition. Harper Collins Publishers, New York, NY

Ontario Ministry of Natural Resources (1996) State of the Landscape Reporting: The Development of Indicators for the Provincial Policy Statement under the Planning Act. Prepared by Boyle M, Kay JJ and Pond B

Prigogine I and Nicolis GQ (1971) Biological Order, Structure and Instabilities. Quarterly Review of Biophysics 4:107-148

Schneider ED and Kay JJ (1994) Complexity and Thermodynamics: Towards a New Ecology, Futures 24(6):626-647

Schumpeter JA (1950) Capitalism, Socialism and Democracy. Harper, New York, NY

Seydel R (1994) Practical Bifurcation and Stability Analysis. Springer-Verlag, New York

Sussman M (1972) Elementary Thermodynamics. Addison-Wesley, Reading, MA

Zotin RS and Zotin AI (1982) Kinetics of Constitutive Processes during Development and Growth of Organisms. In: Lamprecht I and Zotin AI (eds) Thermodynamics and Kinetics of Biological Process. Walter de Gruyter, New York, NY, pp 423-435

2.12 Quantifying Ecosystem Maturity – a Case Study

Werner L. Kutsch, Oliver Dilly, Wolf Steinborn and Felix Müller

Abstract

In this paper Odum's maturity concept, which is one of the basic sources of the orientor and goal function approach, is analyzed on the basis of empirical data from the terrestrial sites of the project "Ecosystem Research in the Bornhöved Lakes District". The maturity concept is linked with the thermodynamic non-equilibrium principle. Both theories are tested on three levels: ecosystems, plant community succession, and microbial processes. Specific indicators are used to test the hypotheses. Instead of a continuous long-term observation of one ecosystem, different neighboring developmental stages are compared. In all cases it is obvious that the empirical data support the hypotheses: There are ecosystem properties which are regularly optimized throughout a normal ecosystem development. These potential indicators can be taken to proof the ecological significance of the non-equilibrium principle of thermodynamics.

2.12.1 Introduction

In his "maturity concept" Odum (1969) postulated a general "strategy of ecosystem development", which is a well ordered process of community development resulting from changes in the physical environment caused by the ecosystem itself. This development culminates in a (meta)stable ecosystem in which biomass, symbiotic function and resource utilization are maximized. Younger pioneer or developing stages can be distinguished from mature phases by various properties, which are termed orientors, attractors or goal functions (Fig. 2.12.1). Odum's concept that has been often and controversially discussed has become a central part of theoretical ecology (e.g. Ulanowicz 1986; Schneider 1988; Herendeen 1989; Baird et al. 1991; Christensen 1992; Jörgensen 1992; Breckling 1992; Schneider and Kay 1994; Müller 1996), but unfortunately, there is a lack in concrete long-term research, which could quantify the maturing ecosystem properties. As a preliminary attempt to extend that small empirical background, six different ecosystems from the Bornhöved Lake District are compared in the following case study: a beech forest, a wet alder forest, a wet and a dry grass-

F. Müller M. Leupelt (Eds.)
Eco Targets, Goal Functions, and Orientors

land, and two crop fields under different management regimes. Although the maturity concept was originally formulated to describe the temporal development of single ecosystems, we will investigate it in a comparative manner, replacing time by space to rank ecosystems with reference to maturity. This case study is a continuation of former ecosystem comparisons which were conducted to test ecological theories with data from the Ecosystem Research Center of the University of Kiel (e.g. Müller 1996; Schrautzer et al. 1996; Kappen et al. in print; Reiche et al. in print).

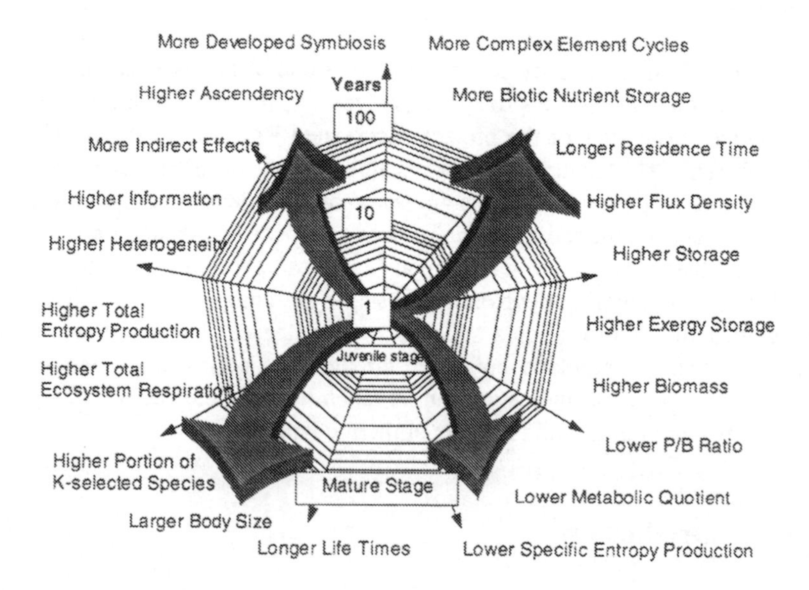

Fig. 2.12.1. A compilation of some ecological orientors

Following the assumption that, due to a hierarchical self-similarity and a functional generality, constituent parts often reflect general properties of ecosystems (Heal and Dighton 1986; Elliott 1997), whole ecosystems will be compared as well as subsystems such as primary producers and destruents. In addition, the empirical data will be used to test the potential of combining the apparently contrary concepts of 'succession', which is based on community ecology, and 'maturity', which is based on an ecophysiological ecosystem concept (see also Bröring and Wiegleb this volume). As a connecting tool, ecophysiological features of single species or functional groups will be related to tendencies in ecosystem development to explain community succession.

2.12.2 Materials and Sites

The research sites are parts of the main research area of the Kiel Ecosystem Research Center. They are located in the Bornhöved Lake District, 30 km south of Kiel (54°06'N, 10°14'O). This landscape was formed glacially during the Pleistocene and consists of morainic hills and lakes. The ecosystems are located along two west-east running catenas, following the slope of a kames hill to Lake Belau (Fig. 2.12.2, for more details see Hörmann et al. 1992).

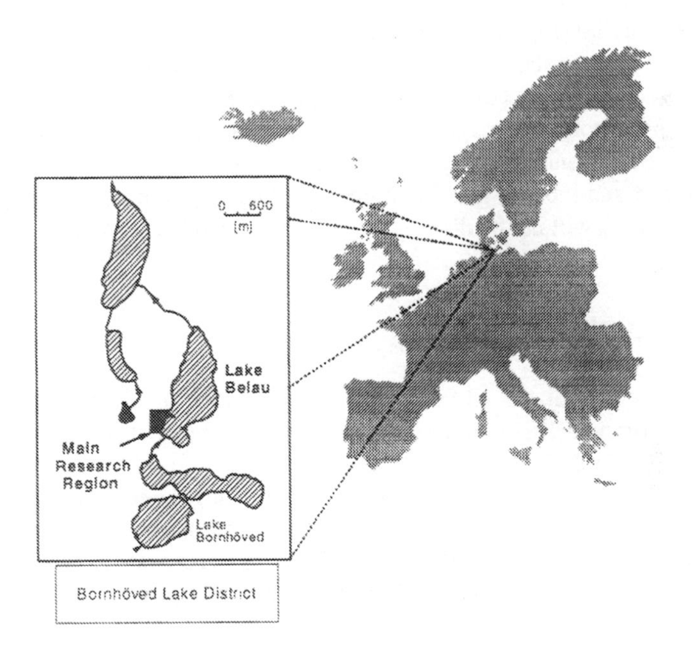

Fig. 2.12.2. The main research area of the project 'Ecosystem Research in the Bornhöved Lakes District'

The beech forest located at the top of the hill is a typical *Asperulo-Fagetum* with a mean tree age of about 100 years. The alder forest, growing on the west shore of Lake Belau is an *Alnetum glutinosae* with an average tree age of about 60 years. The beech forest is the zonal climax plant community, whereas the alder forest is an azonal stable successional state. Both forests are typical of the North German landscape, the first representing well-drained conditions, while the latter is highly influenced by surface-near groundwater dynamics. The landuse regimes of the crop fields are (i) maize monoculture (*Zea mays* L.) with addition of cattle slurry plus mineral fertilizers and (ii) crop rotation (1990: *Avena sativa* L. and *Sinapis alba* L. as intercropping, 1991: *Beta vulgaris* L., 1991/92: *Secale cereale* L. and *Lolium multiflorum* L. as intercropping, 1993: *Zea mays* L.), fertilized by organic manure as well as mineral nitrogen. The dry grassland is a *Lolio-Cynosuretum*, regularly fertilized with mineral N, whereas the non-fertilized, wet grassland has been classified as *Rannunculo-Alopecuretum*, partly dominated by *Alopecurus pratensis* L. According to FAO (1988) the soils were classified as dystric-cambic Arenosol (beech forest, field maize monoculture), eutri-cambic Arenosol (field crop rotation), cumuli-aric Anthrosol (dry grassland), and as fibric Histosol (alder forest) and eutric Histosol with a highly decomposed, dark grey to black topsoil in the wet grassland (Schleuß 1992).

As a basic assumption for the following site comparisons the crop fields are defined as pioneer stages (Grime 1979; Kappen et al. in print), the dry grassland is defined as an intermediate, and the beech forest as a mature stage of ecosystem development. In terms of the community concept they represent two ruderal, an intermediate, and the climax community of a secondary zonal succession. The alder forest and the wet grassland represent two stages in an azonal succession. Therefore they will not be directly compared to the other systems.

2.12.3 Basic Methods

Most of the data discussed in the following sections are derived from aggregations of comprehensive measurements that have been surveyed by numerous colleges (Table 2.12.1).

Therefore, only those methods which are of focal interest to the specific results have been summarized in the following. Transpiration and total evapotranspiration were modelled by Herbst (1997) for three ecosystems with an extended SVAT-model (Shuttleworth and Wallace 1985). Net primary production (NPP) was measured by harvesting the agricultural fields (Kutsch 1996; Wachendorf 1996) and the grasslands (Sach 1997; Weisheit 1995). In the beech forest NPP and standing wood biomass (WB) were determined by analyses of litter fall (Lenfers 1994) and stem diameter increment (preliminary data by Schrautzer and Wellbrock). The measurements were integrated in a forest growth model (Hoffmann 1995). Belowground production was not measured in all cases but estimated as 16.6 % of aboveground NPP (Hoffmann 1995). In the alder forest NPP and WB were determined as parts of an integrated study of leaf and fine root production and tree ring analysis (Eschenbach et al. 1997).

The microbial biomass content was estimated using both substrate-induced respiration (SIR) and fumigation-extraction (CFE) (Anderson and Domsch 1978; Vance et al. 1987), while the microbial basal respiration was measured at 22° C and a water content higher than 40 % water capacity. Basal respiration was related to the CFE-biomass content to estimate the qCO_2.

Table 2.12.1. Applied methods and their authors

Parameter	Method/Sensor	Authors	References
Energy Budget and Microclimate			
Net Radiation	TNR (Delta-T)		
Global radiation	Li 200 SA (LiCor)		
Ppfd	Li 190 SB (LiCor), GaAsP-Photodiodes (Hamamatsu)	Becker, Beinhauer, Eschenbach, Herbst, Hinrichs, Hollwurtel,	Schaefer 1991; Hörmann et al. 1992; Herbst 1997
Air/soil temperature	Pt 100 (Degussa)	Kraus, Kutsch,	
Air Humidity Profiles	Hmp 35 (Vaisala)	Schaefer, Weisheit,	
Soil heat flux	Heat Flux Plate (Thies, Göttingen)	Vanselow	
Storages			
Phytomass	Yield and dry weight measurement, litter sampler	Eschenbach, Kutsch, Lenfers, Lalubie, Mette, Sach, Weisheit	Hörmann et al. 1992
Wood biomass	Stem diameter (beech)	Schrautzer, Wellbrock	Hoffmann 1995
	Tree ring analysis (black alder)	Eschenbach, Werner	Eschenbach et al. 1997
Root biomass	Harvesting, Root length density	Middelhoff	Eschenbach et al. 1997
Biomass Fauna	Kempson extraction	Hanssen, Hingst, Irmler	Irmler 1995
Microbial Biomass	SIR, CFE	Dilly, Kutsch	Dilly 1994; Kutsch 1996
Soil organic carbon	Dry incineration	Kutsch, Schleuß, Wachendorf,	Schleuß 1992; Kutsch 1996; Wachendorf 1996
Total soil nitrogen	Micro-Kjieldahl-procedure (Tecator)	Wachendorf	Wachendorf 1996
Phytomass carbon and nitrogen content	C-H-N-Analyser	Eschenbach, Lenfers, Mette, Weisheit, Wiebe	Mette 1994
Microbial nitrogen	C-H-N-Analyser	Dilly	Dilly 1994
Fluxes			
Leaf transpiration	Steady State Porometer (Li 1600. LiCor)	Vanselow, Herbst	Herbst 1995; Herbst & Vanselow 1997
Leaf CO_2-exchange	Open system with infrared gasanalyser, CO_2/H_2O-Porometer (both WALZ)	Eschenbach, Kutsch, Ludwigshausen, Steinborn, Weisheit,	Eschenbach 1996; Eschenbach et al. 1997; Weisheit 1995
Branch respiration	Open system with infrared gasanalyser	Kutsch, Steinborn	Eschenbach et al. 1997
Root respiration	Open system with infrared gasanalyser	Kutsch, Staack	Kutsch 1996
Soil respiration	Open system with infrared gasanalyser	Kutsch, Wötzel	Kutsch 1996; Kutsch & Kappen 1997
Production soil fauna	Kempson extraction	Hanssen, Hingst, Irmler	Irmler 1995
Element concentration in the soil solution	Ceramic cup samp-lers with constant under-pressure, rapid flow autanalyser	Aue, Rambow, Schimming, Wetzel	Reiche et al. 1996; Wetzel 1997; Schimming et al. i.p.

Total ecosystem respiration was estimated with an integrated model of microbial respiration, root respiration and aboveground plant respiration (Kutsch and Kappen 1997).

The model was calibrated with field measurements of soil and branch respiration (Kutsch 1996; Eschenbach et al. 1997) or by applying laboratory measurements of basal respiration. If no measurements were available values were taken from the literature (Gansert 1995).

The soil properties were investigated by Schleuß (1992) and Wachendorf (1996). Soil organic carbon was determined by dry incineration at 1200° C and calorimetric measurement of the CO_2 with a Ströhlein-aparatus (Coulomat 702). Total soil nitrogen was measured by a Micro-Kjieldahl-procedure using a Flow-Injection-Analyzer (Tecator). The nutrient dynamics were estimated by sampling wet deposition, throughfall, stemflow and by modelling the dry deposition (Spranger 1992, Branding 1997), by measuring the element concentrations in the soil water at different depths and groundwater (Lilienfein 1994; Rambow 1995; Schimming et al. 1995; Wetzel 1997), and by regularly taking vegetation samples (Schimming et al. 1995; Schimming et al. in print). The chemical analyses were conducted with a Rapid Flow Autoanalyser (Tecator).

2.12.4 Results

2.12.4.1 Ecosystemic Tendencies

As has been discussed in detail in the preceding papers, the trends of ecosystem maturation can be studied successfully, from the thermodynamic perspective (e.g. Joergensen and Nielsen this volume; Svirezhev this volume), investigating ecosystems as open systems far from thermodynamic equilibrium. Schneider and Kay (1994) stated that complex structures will develop spontaneously within energetic or materialistic gradients which are caused by the environment of the systems. Self-organization leads to structures that dissipate the exergy more and more efficiently due to a developing system of complex, small, internally interacting gradients. Thus, increasing complexity will reduce the probability of collapse, which is a potential consequence of the steep external gradients the system is exposed to (Kappen et al. in print; Müller 1996). Mature systems are able to utilize the incoming exergy in an optimized manner, to capture a higher exergy amount and to transform it into structure more efficiently in an increasing number of degrading steps. Thus the incorporated exergy may be kept longer inside the system and involved in more metabolic processes than in young or stressed ecosystems. As a result of their non-equilibrium principle Schneider and Kay (1994) defined 7 attributes that typically are optimized in ecosystems within successional processes. These features are listed and characterized by respective indicators that have been derived from the Bornhöved data sets (Table 2.12.2).

Table 2.12.2. Ecosystemic indicanda and indicators based on the thermodynamic non-equilibrium principle of Schneider and Kay (1994). In some cases direct measurements of an orientor are not possible. In these cases only certain aspects of an orientor can be reflected and some indications must be assigned to a qualitative level. E.g. network analyses have been carried out for the aquatic project sites only. Therefore, the trophic features and the cycling characteristics can only be deduced from structural features: If biodiversity is high, there is a high probability for an intensive realization of the cycling principle because all organisms of an ecosystem have to be integrated in the energy and nutrient flow schemes. The ecosystem features written in italic letters, will be quantified but are not yet available.

Orientor / Indicandum / Maturity characteristics	Potential Indicators	Data Source
1. Energy Capture		
	Energy Balance	Hollwurtel, pers. comm., Hörmann et al. 1992; Herbst 1997
- *Surface Temperature*	*Remote Sensing*	*Barkmann i.p.*
2. Energy Flow Activity		
- *Number of Cycles*	*Network Analysis*	
- *Length of Cycles*	*Network Analysis*	
- Material Cycling		
- Loss Reduction	Matter Balances	Schimming et al.1995; Schimming et al. i.p.; Wetzel 1997
- Storage	Matter Storage	Schleuß 1992, Wachendorf 1996, Wetzel 1997, this paper
- *Residence Times*	*Network Analysis*	
3. Average Trophic Structure		
- *Chain Length*	*Network Analysis*	
- *Average Trophic Level*	*Network Analysis*	
- Trophic Efficiencies	Production/Storage Ratio	this paper
	Respiration/Storage Ratio	this paper
4. Respiration and 5. Transpiration		
	Respiration	Kappen et al i.p., this paper
	(Evapo)transpiration	Herbst 1997
6. Biomass		
	Phytomass	Kappen et al. i.p.
	Zoomass	Irmler 1995
	Microbial Biomass	Dilly 1994
7. Organism Types		
	Numbers of Species	Hörmann et al. (1992)

The respective data for these indicators are shown in Table 2.12.3. Transpiration as a central part of the energy balance is not only a preferred process of entropy export but also linked with other orientors. It is connected with the opportunity to regulate the water balance of the ecosystem, which includes the ability to control nutrient fluxes and minimize losses (Rapport et al. 1997), and with cooling effects that are important variables of the long-term viability of the whole system. Transpiration made up 35% of total evapotranspiration in the maize monoculture field, 63 % in the beech forest and 70 % in the alder forest (Herbst 1997). This indicates an increase during maturation, as hypothised above.

Community complexity can be taken as a long-term indicator for the maturity of the systems. It is represented by the numbers of species in Table 2.12.3. This is not a measurement of the whole community but only takes plants and some focal groups of the soil fauna into account. The obvious differences between the (young and disturbed) agroecosystems (crop rotation field and grasslands) and the (more mature) forests result from the systems' ages as well as agricultural practice.

A consequence of the increasing bioceonotic complexification which regularly takes place during the maturation process, is the growing number of community members which have to be provided with energy, water and nutrients. Thus, the intra-organismic paths of energy and matter become more and more significant throughout the development. Consequently, more and more biotic storages are involved in the ecological processes. This ecosystem property can be indicated by biomass measurements. Total phytomass highly depends on the life forms of the dominant species and on the actual land use regimes. In our data sets, the total phytomass is lowest in the crop rotation field and highest in the beech forest, where a high amount of wooden biomass is stored. Due to its intensive management the maize monoculture field has the highest phytomass of the systems with non-wooden life forms. The biomasses of macro- and mesofauna (Irmler 1995; Hanssen and Hingst 1995; Hingst, unpublished data) in total as well as the Testaceae biomass which represents one part of the microfauna, increases from crop rotation field < dry grassland < beech forest in the zonal systems. The azonal systems do not show a clear tendency but it is obvious that the peat systems (alder forest and wet grassland) provide much more suitable habitats than the dry ecosystems. Lumbricidae and diptera especially have high masses in these systems. The microbial biomass (Dilly 1994) is lowest in the maize monoculture field, highest in the wet grassland and intermediate in the other systems without a clear tendency. Soil organic carbon (Schleuß 1992; Wachendorf 1996) and total stored organic carbon – defined as the sum of soil organic carbon and wooden biomass, both remaining stable in a medium time scale - increase from maize monoculture field < crop rotation field < dry grassland < beech forest < alder forest < wet grassland. The total ecosystem organic carbon as well as the portion of intraorganismic carbon also increase during maturation of the zonal ecosystems.

The extremely high organic matter storage in the soils of the wet alder forest and the wet grassland results from the high water table and its effects on the process of mineralization.

Table 2.12.3. Some ecosystem properties from the Bornhöved Lake District

Indicator	Unit	Field MM*	Field CR*	Dry GL*	Beech Forest	Wet GL*	Alder Forest
Energy Balance							
Total Evapotranspiration	mm y^{-1}	429			617		768
Transpiration	mm y^{-1}	150			388		538
Transpiration	% Total Evapotranspiration	35			63		70
Structures							
Number of Species (Plants)		9	10	15	44	26	30
Number of Species (Soil fauna)		n.d.	173	257	366	265	484
Storages							
Total Phytomass	t C ha^{-1}	11	6.5	9.9	131.2	6.6	81.5
Wood Biomass	t C ha^{-1}	0	0	0	128.5	0	75
Biomass Macrofauna	kg dM ha^{-1}	n.d.	1.38	6.28	24.1	46.6	41.4
Biomass Mesofauna	kg dM ha^{-1}	n.d.	0.73	2.69	6.67	4.99	9.31
Biomass Testacea	kg dM ha^{-1}	n.d.	0.4	0.6	2.6	2.36	1.9
Microbial Biomass	t C ha^{-1}	0.67	0.92	0.96	0.93	2.32	0.93
Soil Organic Carbon	t C ha^{-1}	56	60	72	80	370	176
Total Ecosystem Organic Carbon (TEC) (except Fauna)	t C ha^{-1}	67	66.5	81.9	211.2	376.6	257.5
Total Stored Organic Carbon (TSOC)	t C ha^{-1}	56	60	72	208.5	370	251
Biotic Carbon	% TEC	17.4	11.1	13.3	62.6	2.3	31.9
Total Phytomass	t N ha^{-1}	0.24	0.21	0.17	1.12	0.13	1.04
Wood Biomass	t N ha^{-1}	0	0	0	1.02	0	0.89
Microbial Biomass	t N ha^{-1}	0.15	0.2	0.2	0.16	0.41	0.16
Total Soil Nitrogen (TSN)	t N ha^{-1}	4.06	4.85	4.58	5.03	22.35	11.08
Total Ecosystem Nitrogen (TEN)	t N ha^{-1}	4.3	5.06	4.96	6.15	22.89	12.28
Total Stored Nitrogen (TStN)	t N ha^{-1}	4.06	4.85	4.58	6.05	22.35	11.97
Biotic Nitrogen	%TEN	9	8.1	7.4	20.8	2.4	9.8
Fluxes							
NPP/TSOC	(t C ha^{-1} y^{-1})/ (t C ha^{-1})	0.196	0.108	0.062	0.037	0.016	0.034
Total Ecosystem Respiration (TER)	t C ha^{-1} y^{-1}	9.9	6.7	n.d.	9.4	n.d.	19.7
Relative Ecosystem Respiration (TER/TSOC)	(t C $ha^{-11}$$y^{-1}$)/ (t C ha^{-1})	0.18	0.11	n.d.	0.04	n.d.	0.08
Production Soil Fauna	g C m^{-2} y^{-1}	n.d.	0,5	1	22	7	4

*Abbrevations: MM: maize monoculture, CR: crop rotation, GL: grassland; n.d. not determined

Three features of the system fluxes seem to be useful indicators to evaluate maturity (Table 2.12.3). Due to the increasing storage of living and dead biomass, the production/storage ratio and the specific respiration (Kutsch 1996; Eschenbach et al. 1997) decrease throughout maturation of the zonal ecosystems. The increasing productivity of the soil fauna (Irmler 1995) also indicates higher recycling activities within the system during maturation. However, anthropogenic activities as fertilization affect these trends. Landuse of course also plays an important role for nutrient losses.

Wetzel (1997) showed higher nutrient leaching rates from arable soils than from soils of the mature forest systems. E.g. the leaching rates of nutrients such as Ca, Mg, Na, S, P, Cl, and N are in the mean 2.6 times higher in the crop rotation field than they are in the beech forest. Complex structures, high storages, and mature food webs thus indeed result in a reduction of irreversible losses of nutrients. On the other hand, forest degradation and soil acidification are recently increasing processes that provoke high leaching rates in the forest ecosystems, as well. Thus, disturbance leads to a retrogression whereby the values of the maturity indicators can be reduced.

2.12.4.2 Successional Trends in the Vegetation during Ecosystem Maturation

The sequential modifications of the vegetation during ecosystem maturation have been particularly described as changes in species composition or life forms (Lüdi 1930, Grime 1979, Dierßen 1990, Dieschke 1994). Single species or plant communities exhibit ecophysiological properties which reflect the effects of the environmental constraints and the internal competition. The interactions between organisms, populations and species during succession cause processes of adaptation, which can be defined as the genetic ability of species to adjust to specific environmental situations. Thus, ecophysiological properties can be used as a tool to integrate vegetation processes on the community level with the respective trends in matter and energy fluxes. One concept of succession (Grime 1979) postulates competition, stress, and disturbance as selecting criteria for different plant strategies. The respective 'actors' are called 'competitors', 'stress-tolerators', and 'ruderals'. Plant species can be ranked in a triangle which shows the relative influence of the three criteria on their strategies. Besides the three primary strategies, which cover the angle areas, four secondary strategies have been described for the intermediate areas (Table 2.12.4; Fig. 2.12.3). In Grime's model of vegetation processes the relative importance of the criteria change during ecosystem maturation and succession of the respective plant community is induced. Secondary succession, which starts after a major disturbance, reveals a characteristic sequence of annuals, perennial herbs, shrubs, and trees, if further disturbance does not occur (Fig. 2.12.3).

Table 2.12.4. Grime's strategy types

Strategy type	Abbreviation	Examples
Ruderals	R	*Poa annua, Veronica persica*
Competitors	C	*Alopecurus pratensis, Urtica dioica, Sambucus nigra, Betula pubescens*
Stress-tolerators	S	*Milium effusum, Vaccinium myrtillus*, most Lichens
Competitive ruderals	C-R	*Galium aparine, Rannunculus repens, Hordeum vulgare, Secale cereale, Zea mays,*
Stress-tolerant ruderals	S-R	*Anemone nemorosa*, most Bryophytes
Stress-tolerant competitors	C-S	*Carex acutiformis, Mercurialis perennis, Fagus sylvatica*
'C-S-R' strategists	C-S-R	*Festuca ovina, Lotus corniculatus, Glechoma hederacea*

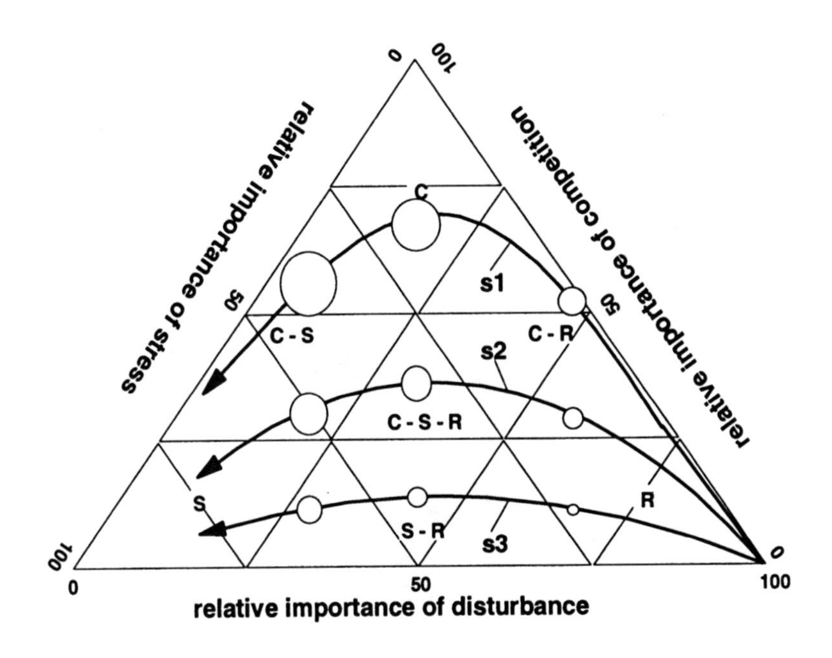

Fig. 2.12.3. Diagram representing the path of vegetation succession under conditions of high (s1), moderate (s2) and low (s3) potential productivity. The size of plant biomass at each stage of succession is indicated by the circles. From Grime (1979)

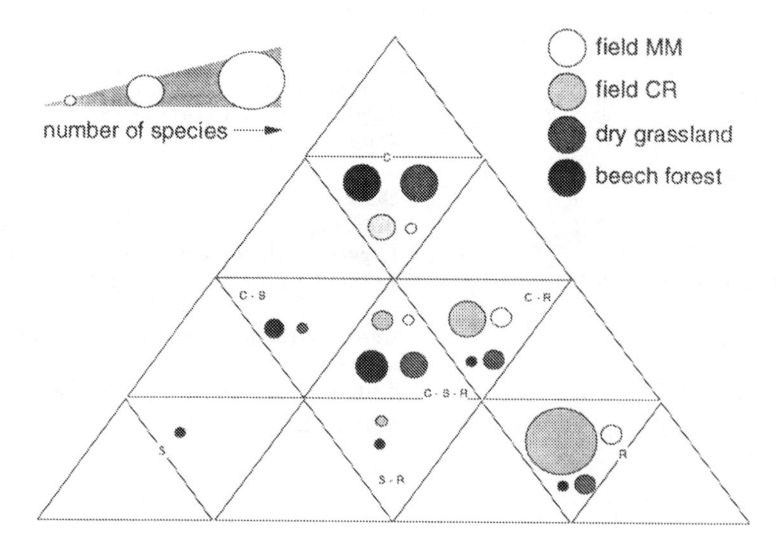

Fig. 2.12.4a. Frequency of species belonging to the different strategy types for the four zonal ecosystems

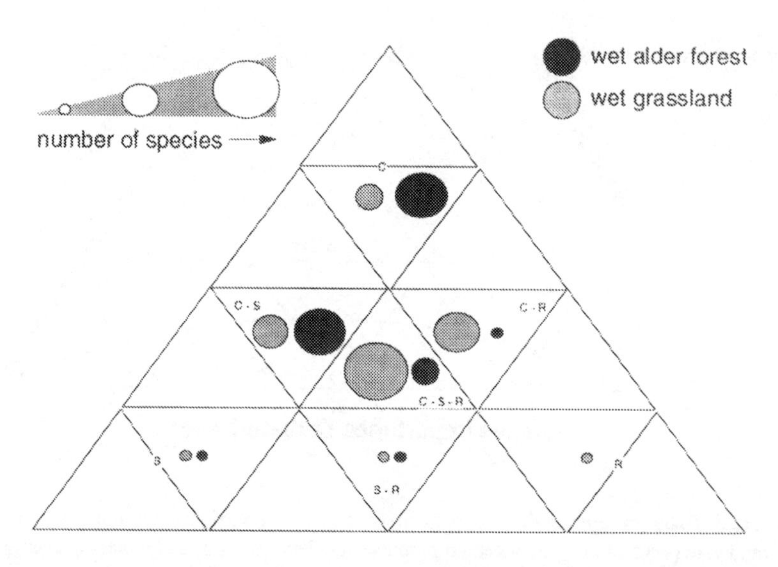

Fig. 2.12.4b. Frequency of species belonging to the different strategy types for the two azonal ecosystems

This reflects a change of the predominating criterium of selection from disturbance to competition, taking place in the first period of a succession. The relative importance of competition in this intermediate phase depends on the potential productivity of the habitat. During the later phases stress-tolerance becomes progressively more important due to shading by a closed canopy, a high respiration required to maintain biomass, a low proportion of available nutrients, and further trends of ecosystem development, as mentioned above.

In a vegetation analysis (Schrautzer and Breuer, unpublished data) the frequency of plant species of every strategy type were determined in each of the ecosystems under consideration (Fig. 2.12.4).

In both crop fields most of the species belonged to the R- and the C-R-type. Only a few C and C-R-S-strategist could be found. C-S and S-types were completely absent. The total number of species was higher in the crop rotation field than under monocultural landuse. Dry grassland and beech forest revealed a maximum of species within the C- and C-R-area with the dry grassland showing a tendency towards C-R and R, and the beech forest showing a tendency towards C-S and S. These results confirm Grime's (1979) model of a change in the relative importance of selecting criteria during succession.

In Table 2.12.5, life forms, properties of the life-cycles, and some ecophysiological features of the dominant plant species in the considered ecosystems are listed. *Zea mays* in the monoculture and *Secale cereale* as one crop of the rotation field are annuals, *Beta vulgaris* (crop rotation field) is biannual. *Dactylis glomerata, Agropyron repens, Holcus lanatus* and *Alopecurus pratensis* as dominant species of the grasslands are perennial herbal plants with a longevity of approximately 20 years. *Fagus sylvatica* and *Alnus glutinosa* are trees with a longevity of about 200 and 120 years, respectively. The herbal species *and Alnus glutinosa* flower yearly or biannually, *Fagus sylvatica* intermittently. Depending on the life form, the maximum plant biomass varied between 6.5 t C ha^{-1} for the crop rotation and 131 t C ha^{-1} in the beech forest. The maximum net photosynthesis rate at ambient CO_2 concentration was highest for *Zea mays* in the monoculture (Hörmann et al. 1992) and *Secale cereale* in the crop rotation field (Kutsch, unpublished data). The four grass-species of the grasslands revealed intermediate values (Weisheit 1995) as well as *Alnus glutinosa* in the wet alder forest (Eschenbach 1996), whereas *Fagus sylvatica* had the lowest values (Kutsch and Eschenbach, unpublished data). The relative growth rates (RGR) could only be estimated on an annual time scale by dividing NPP (Table 2.12.3) by maximum plant biomass. This RGR is not comparable with the RGR defined by Grime (1979) for seedlings, but it reveals the same tendency for early and late successional plants. This procedure resulted in a value of 1 for the annuals on the crop fields, 0.5 to 0.9 for the grassland species and below 0.1 for the trees. The C/N ratio of green leaves did not vary between the species of the different ecosystems, whereas the reallocation of nitrogen did. The C/N ratio of the litter in autumn was 35 for *Zea mays*, 46 for *Dactylis glomerata,* 59 for *Fagus sylvatica,* but only 27 for *Alnus glutinosa*. It seems that *Alnus glutinosa* did not reallocate much nitrogen due to N_2-fixation of the Alder-*Frankia*-symbiosis.

In spite of some restrictions due to the fact that the species in the agricultural ecosystems are products of an anthropogenic selection (e.g. *Zea mays* is not an indigenous species), tendencies in the features of the dominant species during ecosystem maturation could be found (Table 2.12.6).

Table 2.12 5. Ecophysiological features of the dominant plant species in the six ecosystems under consideration

Indicator	Unit	Field MM	Field CR	Dry Grassland	Beech Forest	Wet Grassland	Alder Forest
Dominant Species		*Zea mays*	*Secale cereale, Beta vulgaris*	*Dactylis glomerata, Agropyron repens*	*Fagus sylvatica*	*Holcus lanatus, Alopecurus pratensis*	*Alnus glutinosa*
Longevity	years	1	1-2	20 (?)	200	20 (?)	120
Flowering		yearly	yearly - biennial	yearly	intermittent	yearly	yearly
Max. Net Photosynthesis Rate (Dominant Species)	μmol CO_2 m^{-2} s^{-1}	40	26	15.5 - 21.5	5.5 - 9.4	15.5 - 21.5	16
Max. Plant Biomass	t C ha^{-1}	11.0	6.5	9.9	131.2	6.6	81.5
NPP	t C ha^{-1} y^{-1}	11	6.5	4.9	8.1	5.9	8.6
Relative Annual Growth Rate	(t C ha^{-1} y^{-1})/ (t C ha^{-1})	1	1	0.5	0.05	0.9	0.07
Min. Leaf C/N-Ratio (Dominant Species)	g C g^{-1} N	23	n.d.	17	23	18	19
Leaf C/N-Ratio at the end of the vegetation period (Dominant Species)	g C g^{-1} N	35	n.d.	46	59	n.d.	27

n.d.: not determined

They are in accordance with successional trends as described by Grime (1979), Heal and Dighton (1984), and Larcher (1994), and thus these factors also support the maturity concept of Odum (1969).

Table 2.12.6. Some tendencies throughout succession found in our observations compared with properties of ruderal, competitive and stress-tolerant plants (after Grime 1979; Larcher 1994) and some vegetation trends described by Heal and Drighton (1986)

Parameter	Ruderal	Competitive	Stress-tolerant	Tendency in our observations
Growth forms	herbs	herbs, shrubs, trees	lichens, herbs, shrubs, trees	herbs -> trees
Perennation	dormant seeds	dormant buts and seeds	stress-tolerant leaves and roots	dormant seed -> dormant buds
Longevity in the established phase	very short	relative short - long	long - very long	increase
Frequency of flowering	high	usually each year (established plants)	intermittent over a long life-history	increase
Max. net assimilation rate per unit leaf area (Larcher 1994)	high	high	low	decrease
Maximum potential relative growth rate	rapid	rapid	slow	decrease
Proportion of annual production devoted to seeds	large	small	small	decrease
litter C:N ratio		increase		increase
simple sugar content		decrease		n.d.
lignin content (Heal and Dighton 1986)		increase		n.d.

n.d.: not determined

2.12.4.3 Trends in the Soil Organic Matter Subsystem and Features of the Microbial Communities

Since late successional plants produce litter which is poor in nutrients and simple sugars but high in lignin (Heal and Dighton 1986) they change the physico-chemical environment of the top soils. These processes cause modifications in the composition as well as in the activity of the microbial communities favoring K-selected organisms (Heal and Dighton 1986; Gerson and Chet 1989). Table 2.12.7 shows features of microbiota according to the theories of r- and K-selection. They represent the more "classical" definitions originating from population growth equations (Witthaker 1975; May 1980).

These definitions, which focus on the number of individuals in populations, rarely consider ecophysiological features such as biomass content or effectiveness. In addition, microbial communities may be characterized by the following ecophysiological properties: (i) the metabolic quotient (qCO_2) according to Anderson and Domsch (1986) and (ii) the SIR/CFE ratio according to Dilly and Munch (1995a).

Table 2.12.7. Properties of r- and K-selected microorganisms and trends in litter quality during ecosystem succession

Parameter	r-selected microorganisms	K-selected microorganisms	Source
Generation time	short	long	Gerson & Chet 1981
Spore germination	rapid	slow	Gerson & Chet 1981
Mycelial growth rate	high	low	Gerson & Chet 1981
Presence	sporadic	continuous	Gerson & Chet 1981
Ressource exploitation	quick	slow	Gerson & Chet 1981
Substrate	sugars, cellulose	cellulose, lignin	Gerson & Chet 1981, Heal & Dighton 1986
substrate availability	ephemeral	abundant, ever-available	Gerson & Chet 1981

The qCO_2 as the biomass-related respiration quantifies the efficiency of the populations. Decreases of the qCO_2 throughout ecosystem successions have been reported by Anderson and Domsch (1990) and Insam and Haselwandter (1989). They indicate the same trend as the total ecosystem R/B-ratio, which decreases, maximizing the amount of biomass per available energy. Consequently, a low qCO_2 suggests microbial communities dominated by K-strategists. In the dry systems under consideration, qCO_2 in the topsoils decreased from monoculture field > crop rotation field > dry grassland > beech forest (Fig. 2.12.5b) while conversely microbial biomass increased in this order (Fig. 2.12.5a). In the wet grassland an extremely high microbial biomass with a low qCO_2 was found. The alder forest revealed a microbial biomass in the same order of magnitude as the beech forest but the highest qCO_2 of all sites.

Substrate-induced respiration (SIR) and chloroform fumigation-extraction (CFE) are two different methods of determining microbial biomass (Dilly and Munch 1995b). CFE reflects the whole biomass, whereas SIR preferentially measured those parts of the communities which quickly react to an added easily decomposable energy source. High SIR/CFE-ratios indicate a microbial community dominated by r-selected species (Dilly 1994). The highest SIR/CFE-ratio was found in the topsoil of the monoculture field, the lowest ratios were found in the beech forest and the wet grassland. The crop rotation field, the dry grassland and the alder forest took an intermediate position. We assume that the microbial community in the topsoils of the beech forest and the wet grassland are dominated by K-selected species, whereas r-selected species become more and more dominant in the alder forest, the crop rotation field, the dry grassland and the monoculture field.

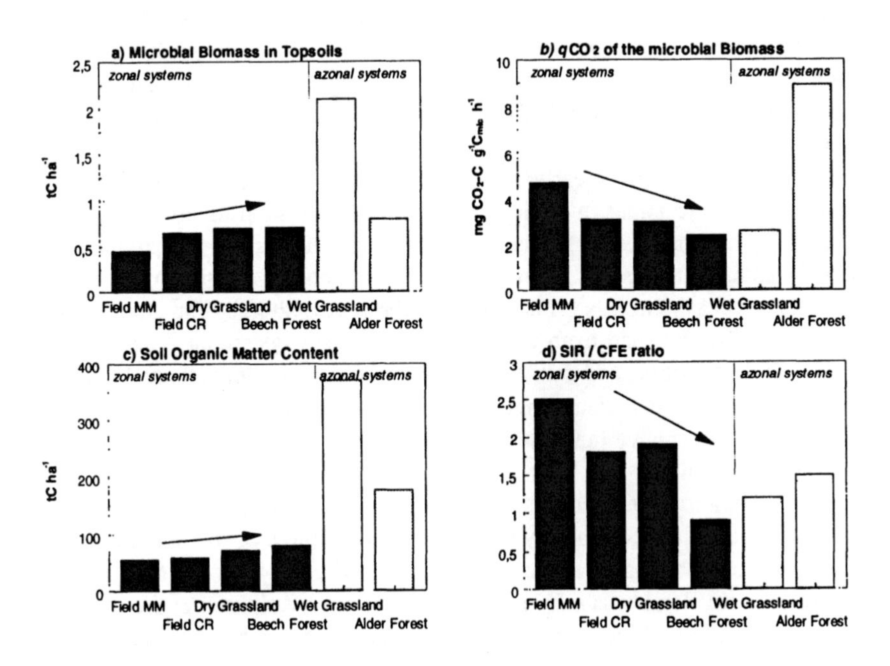

Fig. 2.12.5. Microbiological features in topsoils of different ecosystems of the Bornhöved Lakes District. Arrows indicate tendencies throughout succession for zonal systems.

2.12.5 Conclusions

On the basis of empirical data from the Kiel Ecosystem Research Center some hypotheses of Odum's maturity concept and its thermodynamic continuation in the nonequlibrium principle were tested by a comparison of neighboring ecosystems that represent different successional stages. This approach seems to be basically correct because agricultural systems represent perpetuated pioneer stages of a secondary succession, but it is limited by the fact that agricultural systems are driven to high production and yield by external factors. A 'pure' pioneer stage in a natural succession would have lower inputs of nutrients and as a consequence a lower primary production and lower leaching rates than the crop fields in this study. In spite of this methodological limitation, this study reveals many tendencies during ecosystem maturation leading to 7 functional orientors which were derived from the non-equilibrium principle (Schneider and Kay 1994) and the ecological law of thermodynamics after Jörgensen (1992). The higher species numbers indicate that the flow diversity within the trophic networks – as well as the structural diversity – increase. The ranking of additional orientors such as the storage capacity also supports the theory of ecosystem maturation. We can formulate the hypothesis that the longer a system's growth rythms and the temporal growth extents are (forest > grassland > crop fields), the more evidently the orientors will fit into the optimization scheme of the non-equilibrium principle.

Nevertheless, the study clearly showed that only those ecosystems can be compared, which are equal in environmental constrains such as climate, water table, or soil properties. Clear tendencies turned out only for the four zonal ecosystems. In addition the internal comparison of the azonal systems showed that a relative young ecosystem in a primary succession represented by the alder forest can hardly be distinguished from an older but disturbed system such as the wet grassland. Soil microbial and faunal biomass, qCO_2 in the topsoil, or total organic carbon in the soil indicate that the wet grassland is more mature than the wet alder forest. Extensive land use does not seem to alter the maturity state of an ecosystem very much.

These investigations were conducted on three nested scales, summarizing ecosystem properties, plant community properties, and features of the microbiological communities. The basic assumption that constituent parts reflect general properties of ecosystems was supported by several observations. I.e. the relative ecosystem respiration showed the same tendency as the metabolic quotient of microbiota in the soil. Both are highest in the crop fields and decrease during maturation.

Besides the comparison of the orientors, it was attempted to test the potential of a combination of structural and functional concepts of ecosystem development. The investigation of successional properties in the vegetation subsystems could be used to classify the compared units in accordance with the successional concept of Grime (1979). Ecophysiological features which were used to characterize the dominant species of different developmental stages (Table 2.12.5 and 2.12.6) also are in accordance with the maturity concept. They reflect the successional adjustment of the vegetation to the changing constraints during ecosystem maturation and can be interpreted as an 'acclimation on the ecosystem level.' Nevertheless, species composition cannot be exactly predicted during ecosystem maturation because succession is also driven by some stochastic processes.

Also the integration of microbial characteristics can be regarded as another successful example for the combination of structural and functional concepts, distinguishing microbial r- and K-strategies in the investigated ecosystems and coupling these features with the ecophysiological measures of qCO_2 and the SIR. Although it is often stated that functional and structural comprehensions can hardly be compared, the presented investigation shows that the combination of succession theory and thermodynamically based developmental concepts opens a wide range of future research and application in the field of systems ecology.

One possible application of the maturity concept may be the integration into the paradigm of 'ecosystem health' (Costanza et al. 1992). Rapport et al. (1997) defined 'ecosystem health' as 'an emerging integrative science dealing with the health of regional systems. Akin to other health sciences, e.g. human health, the concept does not conjure up a single model, or a single method. Rather it speaks to a cluster of concepts and methods contributing to the goal of promoting viable life systems.' However, ecosystem health requires measurable indicators. Describing the development of natural ecosystems, the trends during ecosystem maturation as described above can be used as indicators for the health of ecosystems or for the comparative assessment of different land use forms respectively. Agricultural practices cause deviations from those trends and prevent maturation. It locks the ecosystem in an early state of maturation. However, agriculture is essential to feed people. But by the degree of deviation it is possible to distinguish between those kinds of land use that greatly harm nature and those which can be accepted from an ecological point of view. In addition, a group of indicators derived from the maturity concept may serve as an early warning system with regard to health of rural areas. As an example the two arables of this study were compared by an assortment of indicators which were derived from the maturity concept. The beech forest, which was shown to be the most mature system of this study, was used as the reference. Seven indicators were selected and depicted as a "quality amoeba" diagram (ten Brink 1991; de Jong 1992) in Fig. 2.12.6. The value of the beech forest was set to 1. The tendency during ecosystem maturation of each indicator was figured from the inner part of the circle outwards. When the value of the orientor increases throughout ecosystem maturation, a point closer to the center represents a lower value. When the value of the indicator decreases throughout ecosystem maturation, it represents a higher value. So for both cases it can be stated: the nearer a point to the center of the circle – or the smaller the area of the amoeba – the more the ecosystem is deviated from its natural trends. Fig. 2.12.6 clearly shows the differences between both fields: with the exception of the biotic N content the values of the crop rotation field are closer to those of the beech forest and the area of the amoeba is larger. Consequently, crop rotation with organic manure and intercropping represents a more healthy land use system than maize monoculture. Further research is required to define distinct boundaries, which allow to declare the maize monoculture treatment as 'unhealthy' and call for rehabilitation and changes in the use of this field.

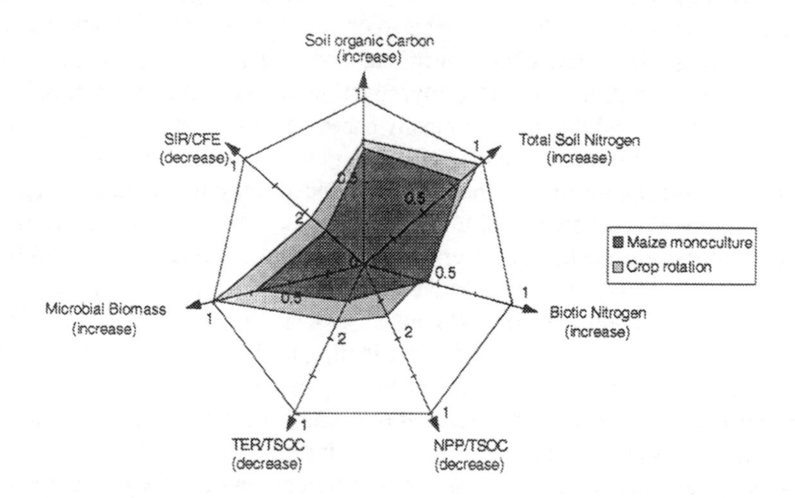

Fig. 2.12.6. Integrative comparison between two agricultural systems regarding to seven different properties, which change throughout ecosystem maturation. Tendencies of change are depicted from the center of the circle outwards and remarked in parentheses. The values are related to the beech forest, which represents a value of one for each property. NPP/TSOC: Net primary production per total stored organic carbon. TER/TSOC: Total ecosystem respiration per total stored organic carbon. SIR/CFE: Biomass estimated by substrate induced respiration per biomass estimated by chloroform fumigation extraction.

Finally, we can conclude, that there is a unique strategy of ecosystem development during maturity, if disturbances or destructive Holling-phases (see Bass, this volume) are excluded. This strategy can be characterized by certain ecosystem properties, which function as orientors of systems dynamics. Based on our data, we can state that the orientation principle is a valid theory, which will imply a high degree of potential applicability.

References

Anderson JPE, Domsch KH (1978) A physiological method for measurement of microbial biomass in soils. Soil Biol Biochem 10:215-221

Anderson TH, Domsch KH (1986) Carbon assimilation and microbial activity in soils. Zeitschrift für Pflanzenernährung und Bodenkunde 149:457-468

Anderson TH, Domsch KH (1990) Application of eco-physiological quotients (qCO_2 and qD) on microbial biomass from soils of different cropping histories. Soil Biol Biochem 22:251-255

Baird D, McGlade JM, Ulanowicz RE (1991) The comparative ecology of six marine ecosystems. Phil Trans R Soc London 333:15-29

Brink B ten (1991) The AMOEBA approach as a useful tool for establishing sustainable development? In: Kuik O, Verbruggen H (eds): In search of indicators of sustainable development. Kluwer, Dordrecht, pp 71-87

Branding A (1997) Die Bedeutung der atmosphärischen Deposition für die Forst- und Agrarökosysteme der Bornhöveder Seenkette. EcoSys Suppl 14:1-122

Breckling B (1992) Uniqueness of ecosystems versus generalizability and predictability in ecology. Ecological Modelling 63:13-27

Christensen V (1992) Network analysis of trophic interactions in aquatic ecosystems. PhD Thesis, Royal Danish School of Pharmacy, Copenhagen

Costanza R, Norton BG, Haskell BD (eds.) (1992): Ecosystem health. New goals for environmental management. Island Press, New York

De Jong F (1992) Ecological quality objectives for marine coastal waters: the Wadden sea experience. Internat J estuar Coast Law 7/4:255-276

Dierschke H (1994) Pflanzensoziologie: Grundlagen und Methoden. Ulmer, Stuttgart

Dierssen K (1990) Einführung in die Pflanzensoziologie (Vegetationskunde). Wissenschaftliche Buchgesellschaft. Darmstadt

Dilly O (1994) Mikrobielle Prozesse in Acker-, Grünland- und Waldböden. EcoSys Suppl 8:1-127

Dilly O, Munch JC (1995a) Ecological meanings of different methods for the estimation of microbial biomass and activity in soils. Abstracts. 7th International Symposium on Microbial Ecology, Santos (Brasil), p 84

Dilly O, Munch JC (1995b) Microbial biomass and activities in partly hydromorphic agricultural and forest soils in the Bornhöved Lake Region of Northern Germany. Biology and Fertility of Soils 19:343-347

Elliott ET (1997) Rationale for developing bioindicators of soil health. In: Pankhurst CE, Doube BM, Gupta VVSR (eds) Biological indicators of soil health, CAB International, Wallingford, pp 49-78

Eschenbach C (1996) Zur Ökophysiologie der Primärproduktion der Schwarzerle. In: Pfadenauer J et al. (eds) Verhandlungen der Gesellschaft für Ökologie, Band 26. Fischer, Stuttgart, pp 89-95

Eschenbach C, Middelhoff U, Steinborn W, Wötzel J, Kutsch W, Kappen L (1997) Von Einzelprozessen zur Kohlenstoffbilanz eines Erlenbruchs im Bereich der Bornhöveder Seenkette. EcoSys Suppl 20:121-132

FAO (1988) Soil map of the world. Revised legend. World soil resources report 60. Rome

Gansert D (1995) Die Wurzel- und Sproßrespiration junger Buchen (Fagus sylvatica L.) in einem montanen Moder-Buchenwald. Cuvillier, Göttingen

Gerson U, Chet I (1981) Are allochthonous and autochtonous soil microorganisms r- and K-selected? Rev Écol Biol Sol 18:285-289

Grime JP (1979) Plant strategies and vegetation processes. Wiley, London

Hanssen U, Hingst R (1995) Einfluß systementlastender Nutzungsformen auf die biozönotische Struktur im Feuchtgrünland. Mitt DGaaE 9:475-480

Heal, OW, Dighton, J (1986) Nutrient cycling and decomposition in natural terrestrial ecosystems. In: Mitchell, MJ, Nakas, JP (eds): Microfloral and faunal interactions in natural and agro ecosystems. Nijhoff & Junk Dordrecht, pp 14-73

Herbst M (1995) Stomatal behaviour in a beech canopy: an analysis of Bowen ratio measurements compared with porometer data. Plant, Cell and Environment 18:1010-1018

Herbst M (1997) Die Bedeutung der Vegetation für den Wasserhaushalt ausgewählter Ökosysteme. PhD-Thesis, University of Kiel

Herbst M, Vanselow R (1997) Transpiration, Bodenverdunstung und Gesamtverdunstung in einem Maisfeld - gleichzeitige Messungen auf verschiedenen Maßstabsebenen. EcoSys Suppl 20:71-77

Herendeen R (1989) Energy intensity, residence time, exergy, and ascendency in dynamic ecosystems. Ecol Model 48:19-44

Hörmann G, Irmler U, Müller F, Piotrowski J, Pöpperl R, Reiche EW, Schernewski G, Schimming CG, Schrautzer J, Windhorst W (1992) Ökosystemforschung im Bereich der Bornhöveder Seenkette. Arbeitsbericht 1988 -1991. EcoSys 1:1-338

Hoffmann F (1995) FAGUS, a model for growth and development of beech. Ecol Model 83:327-348

Insam H, Haselwandter K (1989) Metabolic quotient of the soil microflora in relation to plant succession. Oecologia 79:174-178

Irmler U (1995) Die Stellung der Bodenfauna im Stoffhaushalt schleswig-holsteinischer Wälder. Faun Ökol Mitt 18:1-200

Jörgensen SE (1992) Integration of Ecosystem Theories: a Pattern. Kluwer Academic Publ, Dordrecht

Kappen L, Kutsch WL, Müller F, Eschenbach C (in print). Hierarchical process interactions in the terrestrial carbon cycle. submitted to Ecological Modelling

Kutsch WL (1996) Untersuchungen zur Bodenatmung zweier Ackerstandorte im Bereich der Bornhöveder Seenkette. EcoSys Suppl 16:1-125

Kutsch WL, Kappen L (1997) Aspects of carbon and nitrogen cycling in soils of the Bornhöved lake district. II. Modelling the influence of temperature increase on soil respiration and organic carbon content in soils under different managements. Biogeochemistry 39:207-224

Larcher W (1994) Ökophysiologie der Pflanzen. Ulmer, Stuttgart

Lenfers UA (1994) Stoffeintrag durch Streufall in verschiedenen Waldökosystemen im Bereich der Bornhöveder Seenkette. Diploma-Thesis, University of Kiel

Lilienfein M (1991) Zum Stofftransport in der wasserungesättigten Zone und im Grundwasser im Bereich der Bornhöveder Seenkette. PhD-Thesis, University of Kiel

Lüdi W (1930) Die Methoden der Sukzessionsforschung in der Pflanzensoziologie. Handb biol Arbeitsmeth 11(5):527-728. Berlin, Wien

May RM (1980) Theoretische Ökologie. Verlag Chemie, Weinheim

Mette R (1994) Ertragsstruktur und Mineralstoffaufnahme von Mais und Hafer im Einflußbereich von Wallhecken. Verlag UE Grauner. Stuttgart

Müller F (1996) Emergent properties of ecosystems – consequences of self-organizing processes? Senckenbergiana maritima 27:151-168

Odum EP (1969) The strategy of ecosystem development. Science 164:262-270

Rambow K (1995) Untersuchungen zum Stoffverhalten in einer forstlich und einer landwirtschaftlich genutzten Braunerde im Bereich der Bornhöveder Seenkette – Messung und Simulation. PhD-Thesis, University of Kiel

Rapport DJ, McCullum J, Miller MH (1997) Soil health: ist relationship to ecosystem health. In: Pankhurst CE, Doube BM, Gupta, VVSR (eds) Biological indicators of soil health. CAB International, Wallingford, pp 29-47

Reiche EW, Schimming CG, Mette R, Schrautzer J (1996). Nitrogen balances of ecosystems, landscapes, and watersheds. In: Von Cleemput et al. (eds) Progress in Nitrogen Cycling Studies. Kluwer Academic Publ, Dordrecht, pp 365-369

Reiche EW, Müller F, Kerrinnes A, Dibbern I (in print) Components of heterogenity in forest soils and understorey communities. In: Tenhunen JD (ed) Ecosystem properties and landscape function in Central Europe, Ecological Studies, Springer, Heidelberg

Sach W (1997) Vegetation und Nährstoffdynamik unterschiedlich genutzten Grünlandes in Schleswig-Holstein. PhD-Thesis, University of Kiel

Schaefer W (1991) Automatische Erfassung von Umweltdaten in der "Ökosystemforschung im Bereich der Bornhöveder Seenkette". Proceedings 6. Symposium "Informatik für den Umweltschutz", Informatik-Fachberichte 296:211-220

Schimming CG, Mette R, Reiche EW, Schrautzer J, Wetzel H (1995) Stickstoffflüsse in einem typischen Agrarökosystem Schleswig-Holsteins. Meßergebnisse, Bilanzen, Modellvalidierung. Z Pflanzenernähr Bodenk 158:313-322

Schimming CG, Schrautzer J, Reiche EW, Munch JC (i.p.) Nitrogen retention and loss from ecosystems of the Bornhöved Lakes Region. In: Tenhunen JD (ed) Ecosystem properties and landscape function in Central Europe, Ecological Studies, Springer, Heidelberg

Schleuß U (1992) Böden und Bodenschaften einer norddeutschen Moränenlandschaft. EcoSys Suppl 2:1-185

Schneider ED (1988) Thermodynamics, information, and evolution: new perspectives on physical and biological evolution. In: Weber BH, Depew DJ, Smith JD (eds) Entropy, Information, and Evolution: New Perspectives on Physical and Biological Evolution. Cambridge Univ Press, Cambridge, pp 108-138

Schneider ED, Kay JJ (1994) Life as a manifestation of the second law of thermodynamics. Math Comput Model 19:25-48

Schrautzer J, Asshoff M, Müller F (1996) Restoration strategies for wet grasslands in Northern Germany. Ecol Engineering 7:255-278

Shuttleworth WJ, Wallace, JS (1985) Evaporation from sparse crops – an energy combination theory. Quaterly Journal of the Royal Meteorological Society 111:839-855

Spranger T (1992) Erfassung und ökosystemare Bewertung der atmosphärischen Deposition und weiterer oberirdischer Stoffflüsse im Bereich der Bornhöveder Seenkette. EcoSys Suppl 4:1-153

Ulanowicz RE (1986) Growth and development: Ecosystem phenomenology. Springer, Berlin Heidelberg New York

Vance ED, Brookes PC, Jenkinson DS (1987) An extraction method for measuring soil microbial biomass C. Soil Biology and Biochemistry 19:703-707.

Wachendorf C (1996) Eigenschaften und Dynamik der organischen Bodensubstanz ausgewählter Böden unterschiedlicher Nutzung einer norddeutschen Moränenlandschaft. EcoSys Suppl 13:1-130

Weisheit K (1995) Kohlenstoffdynamik am Grünlandstandort – untersucht an vier dominanten Grasarten. PhD-Thesis, University of Kiel

Wetzel H (1997) Prozeßorientierte Deutung der Kationendynamik von Braunerden als Glieder von Acker- und Waldcatenen einer norddeutschen Jungmoränenlandschaft – Bornhöveder Seenkette. PhD-Thesis, University of Kiel, Kiel

Witthaker RH (1975) Communities and Ecosystems. Macmillan Publ, New York

2.13 Case Studies: Orientors and Ecosystem Properties in Coastal Zones

Daniel Baird

Abstract

Carbon flow networks of 11 coastal ecosystems were constructed from comprehensive data bases collected over the past decade. These models include upwelling systems, shallows seas, large and small estuaries, a coastal lagoon, and a beach/surf zone. The trophic and cycling structure of each model was analysed by means of network analysis in an attempt to identify on or more system level properties for inter-system comparison. results from these analyses show that although a number of very useful system level indices, such as total system throughput, development capacity, ascendency, the magnitude of recycling, P/B ratio, etc., can be calculated, the identification of one universal orientor is not yet possible. The calculated values of the different system attributes for the different ecosystems vary greatly between them. However, dimensionless indices, such as the normalised A/C ratio, showeda declining trend from upwelling systems to small esturies, whilst the FCI ratio followed an inverse trend. This paper examines a number of system properties and their value towards a better understanding of ecosystem behaviour and structure.

2.13.1 Introduction

The marine coastal zone contains globally a great diversity of ecosystems. This zone includes the region from the shelf break to the landward margin of coastal dunes; it includes estuaries, coastal lagoons, wetlands, sandy beaches, the surf zone, large shallow seas and coastal upwelling systems. The variety of identifiable ecosystems along continental and island margins is large and driven by complex sets of physical and chemical processes, which is reflected in the great diversity of plan and animal life. Coastal ecosystems interact across common boundaries through the exchange of energy, material and biota, drastically in species composition and diversity, food-web structure, rates of production and decomposition, and energy flow pathways. Coastal ecosystems are complex entities, with individualistic and characteristic internal organization, and have

F. Müller M. Leupelt (Eds.)
Eco Targets, Goal Functions, and Orientors

developed unique ways and means to transfer and distribute energy and material within themselves. They function, however, by means of inter-related processes such as the cycling of material, energy transfer, predation, competition, etc., with the goal to optimize these in such a way as to maximize self-organization and functional behaviour.

Despite these differences in physico-chemical processes, biotic diversity and system topology, it is not inconceivable to suggest that coastal ecosystems function by means of similar processes, and that they exhibit properties common to all. The functioning of all natural ecosystems is, after all, subject to the laws of thermodynamics, the process of recycling, and in the transfer of energy by various degrees of efficiency. These ecosystems exist, however, at different levels of dynamic equilibria, and develop and change at different rates according to external and internal forcing functions. It is therefore the objective of this paper to examine a number of coastal ecosystems in an attempt to identify system level properties common to all, yet characteristic to each in terms of magnitude.

2.13.2 Study Sites, Methods and Assumptions

A brief description of each and reference to the relevant literature from which the information was extracted follows:

1. Southern Benguela upwelling system (eastern Atlantic Ocean, 31° S to 35°S). The Benguela ecosystem in the south-east Atlantic Ocean is one of the four eastern boundary current regions in the world. It is bounded by the Aghulas Current retroflection area to the south (35°S) and the warm Angola Current to the north (18°S). The Benguela system is divided into two ecological subsystems separated by a zone of semi-permanent upwelling between 23°S and 31°S off southern Namibia, which effectively prevents the movement of biota between the northern and southern subsystems (Shannon 1985; Shannon and Field 1985; Crawford et al. 1987). The physical, chemical and biological properties of the Benguela system are described in Shannon (1985); Chapman and Shannon (1985); Shannon and Pillar (1986); Crawford et al. (1987). The southern Benguela system supports significant purse-seine and demersal fisheries (David 1987). Data used in this paper to construct flow networks are from i.e. Berg et al. (1985); Field et al. (1989); Lucas et al. (1987); Duffy et al. (1987).
2. Peruvian upwelling system (between 4°S and 14°S). The Peruvian upwelling system supports one of the most productive fisheries in the world, and for some time the largest single fisheries (the anchoveta, *Engraulis ringens*) (Pauly and Tsukayama 1987; Jarra-Teichmann and Pauly 1993). This system is part of the larger Humbolt Current System in the south-western Pacific and extends for 2100km along the east coast of Peru. Data for this paper are from

Jarre et al. (1989); Pauly and Tsukayama (1987); Jarra-Teichmann and Pauly (1993).

3. Swartkops estuary (32°52'S, 25°39'E). The Swartkops is a small, temperate, turbid, well-mixed estuary which has been well studied during the past 15 years. The estuary is located on the south east coast of southern Africa near the city of Port Elizabeth and discharges into the Indian Ocean. The river and estuary flows through heavily populated urban areas, and is subject to agricultural, industrial and domestic pollution. It attracts large numbers of migrating Palearctic birds during the austral summer months. The salinity ranges from 35 at the mouth to about 10 at the head and the water temperature varies between 28°C to 13.5°C. Most of the results relevant to this paper are from Baird et al. (1988); Martin and Baird (1987); Baird et al (1986); Baird and Ulanowicz (1993).
4. Kromme estuary (34°08'S, 24°51'E). The estuary is shallow (average depth at low water spring of 2.75m) and well-mixed. The river and estuary flow through a sparsely populated region with no industries and little agricultural activity in the catchment of flood plain. The temperature fluctuates between 11.7°C in winter and 28°C in summer. Two large impoundments have been constructed in the catchment during the late seventies and early eighties, resulting in little freshwater inflow since 1983 and a homogeneous salinity regime of between 32 and 38 throughout the system. Data for this paper were obtained from Bickerton and Pierce (1988); Baird et al. (1991, 1992); Heymans and Baird (1995); Baird and Heymans (1996).
5. The Ythan estuary (20°00'W, 57°20'N) is located about 20km north of Aberdeen, Scotland. The estuary is tidal for about 8km from the mouth and receives about 2000m3 of primary treated sewage daily from upstream town. The salinity ranges from 5 at the head to 35 at the mouth, whilst the temperature fluctuates from 3°C in winter to 20°C in summer. Information on the tidal characteristics, standing stocks and flow network can be obtained from Baird and Milne (1981); Raffaelli and Hall (1996).
6. The Ems estuary (6°54'E, 53°26'N) is a shallow semi-diurnal tidal system draining into the Wadden Sea. The estuary contains large tidal flats of about 246km2, is rich in benthic and pelagic life and receives large amounts of nutrients from the catchment. The water temperature ranges from 4 - 20°C. Information on the food web characteristics of the system is mainly from Baird et al. (1993); Baretta and Ruardij (1998).
7. The Baltic Sea (54°50' - 60°04'N and 16° - 227°45'E) is a large, enclosed, non-tidal estuary; it is the largest brackish water area in the world (Janson 1972). This paper refers to the Baltic proper, one of the three basins that constitute the Baltic Sea, which has become slightly eutrophic and anoxic (Stigebrand and Wulff 1987). Information on the standing stocks and carbon flow network are from Wulff and Ulanowicz (1989); Baird et al. (1991).
8. Chesapeake Bay (36°50'W, 39°40'N). Chesapeake Bay is the largest drowned river valley estuary in the USA This paper refers to the mesohaline region

which encompasses about 48% of the total area of the bay and in which the salinity ranges from 6 to 18°C. The data for this paper was mainly obtained from Baird and Ulanowicz (1989).

9. The Sundays beach/surf-zone ecosystem is situated on the south-east coast of South Africa, just north of Port Elizabeth. The ecosystem is a high-energy environment with surf zone of an intermediate type characterised by bars, channels and rip currents (McLachlan and Bate 1984). The beach becomes dissipative during storms (McLachlan 1990). The sand is fine to medium, well sorted, and primarily quartz. The climate is temperate and the seawater temperature ranges between 10o and 26oC (monthly means: 16 - 21°C). Phytoplankton is the only primary producer in this system, with diatoms (60%) dominating. The fauna of the system includes three distinct trophic assemblages, macroscopic fauna, interstitial fauna and the microbial loop in the water. Food inputs to the intertidal macrofauna come from the sea, about 93% as particulate organics, both living and detritus, and at least 7% as carrion (McLachlan et al. 1981). Primary production is utilized by the interstitial fauna of the sediments, by the macroscopic fauna, by the microbial loop, and as export from the system. The macrofauna include *clams (Donax serra, D. sordidus)*, mysids (*Gastrosaccus psammodytes*), gastropods (*Bullia* sp.), swimming crabs (*Ovalipes punctatus*) and various isopods (*Eurydice longicornis*, *Excirolana natalensis* and *Pontogeloides latipes*). Ichthyoplankton include mullet (*Liza richardsonii*), sand shark (*Rhinobatos annulatus*) and other benthic feeding, zooplankton feeding, piscivorous and omnivorous fish species. The system is considered to be self-sustaining, exporting carbon to the nearshore zone (McLachlan and Bate 1984)
10. A coastal dune system of approximately 30km long and 3km wide runs parallel to the sea along the northern shore of Algoa Bay, from Sundays River Mouth in the West (33°43'S, 25°51'E) to Woody Cape in the East (33°46'S, 26°19'E). The dune system consists of dunes parallel to the coast with ephemeral slacks in between. The slacks comprise xerophytic plants and are inhabited by numerous insects species, while meiofaunal communities occur in the sandy substrate. Birds and small mammals feed on the plant and insects in the slacks. The input into the slacks is mainly wind-driven detritus. The biomass and trophic interactions were studied by Elliot (1996).
11. The Bot River coastal lagoon (34°21'S, 19°04'E) is closed, semi-permanently from the sea by a dune barrier which is artificially opened every 3-5 years in winter due to high water levels, mass mortality of fish, and excessive siltation. The physical, chemical and biological characteristics of the system are described by, for example, Bally (1985); Bally et al. (1985); Bally and McQuaid (1985); De Decker and Bally (1985); Koop et al. (1983); Bennet (1989); Bennet and Branch (1990).

Carbon flow models were constructed for each of the systems described above which illustrate the standing stocks of the various living and non-living components of each system, and the flows between them. Each system contained three non-living compartments namely dissolved organic carbon (DOC), suspended particulate organic carbon (susp POC), and sediment particulate organic carbon (sed POC). Comparison of ecosystems are complicated by having different degrees of aggregation (Mann et al. 1989), and to overcome this species having the same mode of feeding and which obtain their food from common prey sources have been grouped together in respective model compartments. A further criterion for system comparisons is that the topology of the networks should be consistent from one to the other. In this paper the systems contain each between 14 and 16 compartments (including the three non-living ones), who were also assumed to be in a steady state by balancing the inputs of each compartment with corresponding outputs. The same currency, carbon, was used for biomass and flows in each network. The trophic and cycling structure of each system was analysed and system properties calculated by algorithms described by Kay et al. (1989) and Ulanowicz and Kay (1991). The following indices were obtained from the analyses: Several global measurements of system organization such as: total system throughput (T), which measured as the sum of all flows through all compartments in the system and depicts the extent of total activity transpiring in the system. It is to some degree a measure of the total power generated within the system (Odum 1971); network ascendency (A) which incorporates both the size and organization of flows into a single measurement, and is expressed as the product of T and the mutual information inherent in the flow network (Ulanowicz 1986); development capacity (C) which measures the potential for a particular system to develop, and which would be the natural limit to the ascendency. The A/C ratio thus becomes the fraction of system development actually realised. It has also been argued that highly organized systems would tend to internalize most of its activity, so that it is relatively indifferent to outside supplies and demands.

The A_i and C_i indices thus represents internal exchanges alone (Ulanowicz and Norden 1990) and considered to be the quantity most representative of the system's development status (Field et al. 1989; Baird et al. 1991); system overheads (o), given numerically by the difference C - A, and which represents the amount of the development capacity that does not appear as an organized structure or constraints. The overhead represents the cost to the system to operate the way it does. The magnitudes and uncertainties of the imports, exports and respirations contribute to the overheads. On the other hand the overhead represents the degree of freedom that the system has at its disposal for the ascendency to increase; the Finn Cycling Index (FCI) which gives the proportion of total system activity (T) that is devoted to the recycling of energy and material in a system (Finn 1976). The cycling index measures the retentiveness of the system - the higher the FCI, the greater the proportion cycled; the Average Path Length (APL) which measures the average number of transfers a unit of flux will experience from its entry into the system until it leaves the system. The index is derived from APL = ((T-Z)/Z) where Z is the sum of all exogenous inputs: the

Trophic Efficiency of a system and derived from the Lindeman Spine which results from the mapping of the network of transfers into a linear food chain with discreet trophic levels. The log mean of the net transfer between these discreet levels yield the average efficiency by which material and energy is transferred within the system (Baird et al. 1991). The efficiency by which the net's primary production (the NPP Efficiency) products are directly utilized by herbivorous consumers in the system has been utilized by herbivorous consumers in the system has been determined from network output. The Production/Biomass (P/B) ratio, considered to be a system property and an index of system development has also been calculated. This ratio is expected to be high in immature, developing systems, but to diminish as the system grows and matures (Odum 1969; Christensen 1995).

2.13.3 Results and Discussion

From the biomass information and flow matrices a few general but characteristic features of the systems under discussion may be recognized. Firstly, all systems possess a functionally property, the P/B ratio, which not only differs between systems, but are also different for different domains of the ecosystem system.

We can recognize three general domains namely:

1. the pelagic domain, consisting of phytoplankton, free living bacteria, micro- and meso-zooplankton, where the collective mean annual P/B ratio for these communities is consistently >1;
2. the benthic domain comprizing of all macro-benthic and meiofaunal communities. This highly diverse community, particularly in estuaries, has a mean annual P/B ratio close to 1;
3. the nektonic domain containing fish, birds and marine mammals, and has a mean annual P/B ratio of <1.

The variability of the P/B ratio in the various sub-sections of the ecosystems underlines the problematic nature of a single ratio for the whole system. The P/B ratios for the eleven systems are given in Table 2.13.1, which illustrates the large variability in this property between the various systems. Upwelling systems appear to have high ratios, whereas that of small but productive estuaries (e.g. the Ythan, Krom, Swartkops), have lower P/B ratios. The Baltic Sea is intermediate to the upwelling and surf zone on the one hand, and typical estuaries on the other. The high P/B ratios for upwelling systems reflect characteristically high pelagic production, also a typical feature of the surf zone where most of the production is attributed to bacteria and phytoplankton (Heymans and McLachlan 1996). Both these systems are driven and maintained by physical processes, such as upwelling and waves respectively. Benthic production dominates in most estuarine systems where the main energy flow pathway is normally through detritus. Odum (1969) suggested that the P/B ratio should decline as a system develops and matures over time. The P/B ratio results presented here (see Table 2.13.1) would indicate upwelling systems as "immature", whereas estuaries are supposedly less so. Baird et al. (1991) argued that it may not be the case, and that the P/B ratio may not be a reflection of system maturity but rather an indication of adaptation to the physical environment. The P/B ratio may therefore not be an appropriate orientor of the *system as a whole,* but is nevertheless very useful in the context of individual communities and/or domains of individual ecosystems.

The second general observation one can make about coastal ecosystems is that the main pathways of energy flow are distinctly different for different systems. Estuaries generally show a strong coupling between benthic and pelagic domains. Phytoplankton production is usually much lower than that of macrophytes and benthic micro-algae, and much of this production is channelled via detritivores. The detritus food web is well developed in estuaries through which a large proportion of total energy flows, and where detritivory exceeds herbivory by far (Baird and Ulanowicz 1993). Pelagic production appears to be high in upwelling and the surf zone, with benthic production considerably lower.

Table 2.13.1 Some properties of selected coastal ecosystems

Property	System 1.	2.	3.	4.	5.	6.	7.	8.	9.	10.	11.
Area (km^2)	220,000	82,000	4	3.2	2.4	500	257,000	5,980	20	10 (m^2)	13,6
NPP Eff (%)	40	82	38	9.2	10.5	98	87	42	40	0.22	12
ø Path Lgth	2.54	2.24	3.95	2.38	2.86	3.42	3.27	3.61	2.3	1.06	3.22
No. of Cycles	1	15	15	19	15	26	22	20	187	9	5
FCI (%)	0.01	0.04	43.8	25.9	28	28	22.8	29.7	12.8	0.5	36
Troph. Eff (%)	12.1	3.7	4.1	6.1	6.6	12.5	16.2	8.9	2.6	5.8	5.6
TST*	2,977	48,430	17,541	16,879	9,350	1,283	2,577	11,224	6,900	439,550	152,340
A/C	50.6	47.6	28.1	33.7	34.4	38.3	55.6	49.3	45	59.4	48
A/C	45.1	45.2	30.4	29.4	33.1	37.1	38.5	35.1	40.1	39.4	57
P/B annual	78.7	78.4	3.67	4.3	5.4	14.6	29.3	51.5	146	7.2	8.7

System: **1.** Southern Benguela upwelling system, **2.** Peruvian upwelling system, **3.** Swartkops estuary, **4.** Kromme estuary, **5.** Ythan Estuary, **6.** Ems Estuary, **7.** Baltic Sea, **8.** Chesapeake Bay, **9.** Sundays Beach/Surf-zone, **10.** Coastal Dune system, **11.** Bot River coastal lagoon;
* ($MGcM^2d^1$)

The pelagic food web is well developed, and the benthic-pelagic coupling is not as strong as in estuarins and large shallow seas. The surf zone systems are highly dependent on adjacent systems for the import and export of energy and material (McLachlan and Illenberger 1986), and are, just as upwelling systems, driven by physical processes. Coastal lagoons show some similarity with estuaries in that macrophyte production is high, phytoplankton productivity is low, and that the detritus food web is well developed. Coastal lagoons, however, appear to be subject to eutrophication due to point and non-point sources of nutrients, and that exchanges with the adjacent marine environment is not possible.

The above review of superficial characteristics of coastal ecosystems does not provide us, however, with quantifiable system properties. I am of the opinion that information indices derived from network analysis provide a starting point for global system comparisons, and from which we may arrive at properties common to all ecosystems. Some of these properties have been used by various authors (e.g. Field et al. 1989; Warwick and Redford 1989; Baird and Ulanowicz 1993, 1995; Christensen 1995; Baird and Heymans 1996) in attempts to compare systems spatially or temporally. A number of these system properties are listed in Table 2.13.1 and illustrated in Fig. 2.13.1.

Although considered to be a system characteristic, total system throughput, T, is not considered to be useful as an orientor because its value depends on the sum of flows in a system, which can fluctuate dramatically between systems. There also appears to be great variability in values of most of the other properties listed in Table 2.13.1. For example, the efficiency of the utilization of plant production (the NPP Efficiency coefficient) vary between 0.2% in a coastal dune system, to 80% in the Peruvian upwelling system. There also seems to be large variation in the mean trophic efficiency of the ecosystems, with no clear trend emerging (see Fig. 2.13.1). The APL seems to be less complex in upwelling systems than in estuaries. The amount of material recycled, a universal property of all ecosystems and expressed here by the FCI, show however, an increasing trend from upwelling systems to estuaries. There also seems to be a positive correlation between the APL and FCI values, namely a corresponding increase in FCI with APL. This trend was also observed by Baird et al. (1991) and Christensen (1995).

The dimensionless A/C ratio, and particularly the normalized internal Ai/Ci ratio, are considered promising indices to compare the growth and development status of different ecosystems (Mann et al. 1989), and has been used as such by, for example, Baird et al. (1991), Baird and Ulanowicz (1993), Christensen (1995). Fig. 2.13.1 shows a descending trend in the A/C ratio from upwelling to small estuaries. The higher Ai/Ci values for coastal lagoon and dune systems are unexpected, as both of these systems appear rather simple. This trend, in the normalized A/C ratio, appear to be the inverse of the FCI ratio. Whether this rather tenuous relationship is a meaningful one, or will hold as a common indicator of system development needs further scrutiny as more information becomes available.

The question remains whether we can formulate common rules for coastal systems, or characterize them by means of a universal property, or properties. It is unlikely that a common orientor will easily be identified, but it is more likely that a number of properties, viewed in relation to each other, may provide the insight we seek to evaluate a system, or to use in inter-system comparative studies. The data analysed here for the different systems are based on many field observations and measurements. The variability in the global indices shown by means of network analysis indeed reflect this variability between the various coastal systems, and is not an artefact of the methods used.

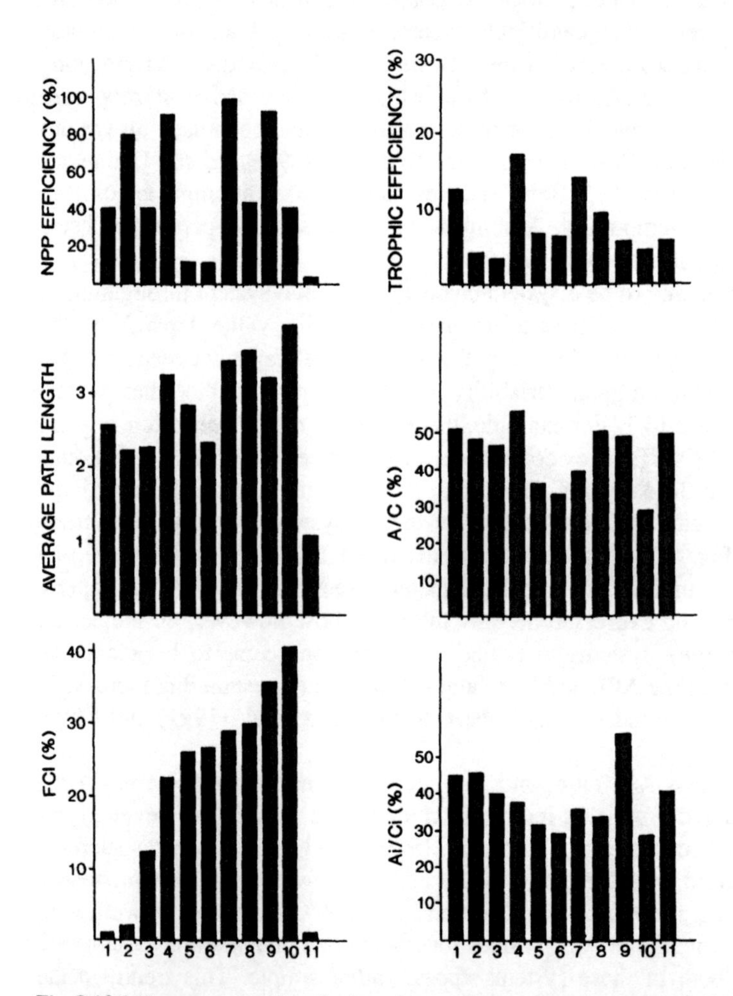

Fig. 2.13.1. System properties of selected coastal ecosystems. (1 Benguela Upwelling System; 2 Peruvian Upwelling System; 3 Surf Zone; 4 Baltic Sea; 5 Ythan Estuary; 6 Krom estuary; 7 Ems estuary; 8 Chesapeake Bay; 9 Coastal Lagoon; 10 Swartkops estuary; 11 Dune system)

A number of system attributes have been identified and which can be quantified, as discussed throughout this volume, and which can possibly be considered as "emergent properties". Coastal ecosystems provide a diverse array of systems to examine for goal functions. A growing body of information is necessary to understand the fundamental behaviour of ecosystems better.

References

Baird D, McGlade JM and Ulanowicz RE (1991) The Comparative Ecology of Six Marine Ecosystems. Phil Trans R Soc Lond B333:15-29

Baird D, Marais JFK and Bate GC (1992) An Environmental Analysis for the Kromme River Area to Assist in the Preparation of a Structure Plan. Institute for Coastal Research, University of Port Elizabeth, South Africa Report C16:1-56

Baird D and Ulanowicz RE (1993) Comparative Study on the Trophic Structure, Cycling and Ecosystem Properties of Four Tidal Estuaries. Mar. Ecol. Prog. Ser. 99:221-237

Baird D and Ulanowicz RE (1989) The Seasonal Dynamics of the Chesapeake Bay. Ecol Monogr 59:329-364

Baird D, Hanekom NM and Grindley JR (1986) Report no. 23, Swartkops. In: Heydorn AEF and Grindley JR (ed) Estuaries of the Cape, Part II. Synopses of Available Information on Individual Systems. CSIRO Res Rep 422:1-82

Baird D, Marais JFK and Martin AP (eds) (1988) The Swartkops Estuary. Proceedings of a symposium, South Africa Natn Scient Prog Rep 10:1-101

Baird D, Heymans JJ (1996) Assessment of Ecosystem Changes in Response to Freshwater Inflow of the Kromme River Estuary, St. Francis Bay, South Africa: A Network Analysis Approach. Water SA 22:307-317

Baird D, Milne H (1981) Energy Flow in the Ythan Estuary, Aberdeenshire, Scotland. Estuar coast Shelf Sci 13:455-472

Bally R (1985) Historical Records of the Bot River Estuarine System. Trans Roy Soc South Africa 45:323-345

Bally R and McQuaid CD (1985) Physical and Chemical Characteristics of the Waters of the Bot River Estuary, South Africa. Trans Roy Soc South Africa 45:317-322

Baretta J and Ruardij P (eds). (1988) Tidal Flat Estuaries: Simulation and Analysis of the Ems estuary. Ecol Stud 71:1-353

Bennet BA (1989) The Diets of Fish in Three South-Western Cape Estuarine Systems. South Africa, J Zool 24:163-177

Bennet BA and Branch GM (1990) Relationships Between Production and Consumption of Prey Species by Resident Fish in the Bot, A Cool Temperate South African Estuary. Estuar coast Shelf Sci 31:139-155

Bergh MO, Field JG and Shannon LV (1985) A Preliminary Carbon Budget of the Southern Benguela Pelagic Ecosystem. Int Symp Upw West Africa, Inst Inv Pesq, Barcelona, Vol. 1, pp 281-304

Bickerton IB and Pierce S (1988) Estuaries of the Cape Part II. Synopsis of Information on Individual Systems. CSIRO Report No. 33: Krom (CMS 45), Seekoei (CMS 46), and Kabeljous (CMS 47), pp 56

Christensen V (1995) Ecosystem Maturity – Towards Quantification. Ecolog Model 77:3-32

Crawford RLM, Shannon LV and Pollock DE (1987) The Benguela Ecosystem. Part IV. The Major Fish and Invertebrate Resources. Oceanogr Mar Biol 25:353-505

David JHM (1987) Diet of South African fur seal (1974-1985) and an Assessment of Competition with Fisheries in South Africa. S Afr mar Sci 5:693-713

Duffy DC, Siegfried WR and Jackson S (1987) Seabirds as Consumers in the Southern Benguela Region. S Afr Mar Sci 5:771-786

Elliot BL (1996) Dune Hummocks as Ecological Units in the Alexandria Coastal Dunefield: Patterns of development and succession. MSc Thesis, UPE, pp 6

Field JG, Moloney CL and Atwood CG (1989). Network Analysis of Simulated Succession after an Upwelling Event. In: Wulff F, Field JG and Mann KH (eds) Network Analysis in Marine Ecology: Methods and Applications. (Coast estuar Stud 32) Springer, Heidelberg, pp 131-159

Finn JT (1976) Measures of Ecosystem Structure and Functions derived from Analysis of Flows. J theor Biol 56:363-380

Heymans JJ and McLachlan A (1996) Carbon Budget and Network Analysis of a High-energy Beach/Surf-zone Ecosystem. Est Coast Shelf Sci 43:485-505

Jansson BO (1972) Ecosystem approach to the Baltic Problem. Bull ecol Res Comm 16:1-82

Jarre A, Muck P and Pauly D (1989) Interactions between Fish Stocks in the Peruvian Upwelling Ecosystem. ICES symposium: Multispecies Models Relevant to Management of Living Resources. Paper No. 27:1-18

Jarre-Teichmann A, Pauly D (1993) Seasonal Changes in the Peruvian Upwelling Ecosystem. In: Christensen V, Pauly D (eds) Trophic Models of Aquatic Ecosystems. ICLARM Conf Proc 26:307-314

Kay JJ, Graham LA and Ulanowicz RE (1989) A Detailed Guide to Network Analysis. In: Wulff F, Field JG and Mann KH (eds) Network Analysis in Marine Ecology: Methods and Applications. (Coast. est. Stud. 32), Springer-Verlag, Heidelberg, pp 15-61

Koop K, Bally R and McQuaid CD (1983) The Ecology of South African Estuaries Part XII: The Bot River, a Closed Estuary in the South Western Cape. South Africa. J Zool 18:1-10

Lucas MI, Probyn TA and Painting SJ (1987) An Experimental Study Of Microflagellate Bactovory: Further Evidence for the Importance and Complexity of Microplankton Interactions. S Afr Mar Sci 5:791-810

Martin AP and Baird D (1987) Seasonal Abundance and Distribution of Birds on the Swartkops Estuary, Port Elizabeth. Ostrich 58:122-134

McLachlan A (1990) The Physical Environment. Chapt. 2 In: Brown AC and McLachlan A (eds) Ecology of Sand Shores. Elsevier, Amsterdam New York, pp 5-39

McLachlan A and Bate G (1984) Carbon Budget for a High Energy Surf Zone. View Milieu 34(2/3):67-77

McLachlan A, Erasmus T and Dye AH et al. (1981) Sandy Beach Energetics: An Ecosystem Approach Towards a High Energy Interface. Est Coast Shelf Sci 13:11-25

Odum EP (1969) The Strategy of Ecosystem Development. Science 164:262-270

Pauly D and Tsukayama I (1987) The Peruvian Anchoveta and its Upwelling System: Three Decades of Change. ICLARM Stud Rev 15:1-351

Raffaelli DG, Hall SJ (1996) Assessing the Relative Importance of Trophic Links in Food Webs. In: Polis GA, Winemiller KO (eds) Food Webs:... . Chapman and Hall, pp 185-191

Shannon LV (1985) The Benguela Ecosystem. Part 1. Evolution of the Benguela; Physical Features and Processes. Oceanogr Mar Biol 23:105-182

Shannon LV and Field JG (1985) Are Fish Stocks Food Limited in the Benguela Pelagic System? Mar Ecol Prog Ser 22:7-19

Shannon LV and Pillar SC (1986) The Benguela Ecosystem. Oceanogr Mar Biol 24:65-170

Stigebrandt A and Wulff E (1987) A Model for the Dynamics of Nutrients and Oxygen in the Baltic Proper. J mar Sci 42:729-759

Ulanowicz RE (1986) Growth and Development: Ecosystem Phenomenology. Springer-Verlag, New York

Ulanowicz RE and Norden JS (1990) Symmetrical Overhead in Flow Network. Int J Systems Sci 21:429-437

Ulanowicz RE and Kay JJ (1991) A Package for the Analysis of Ecosystem Flow Networks. Environ. Software 6:131-142

Warwick RM and Radford PJ (1989) Analysis of the Flow Network in an Estuarine Benthic Community. In: Wulff F, Field JG, Mann KH (eds) Network Analysis in Marine Ecology: Methods and Applications. (Coast est Stud 32), Springer-Verlag, Heidelberg, pp 220-231

Wulff F and Ulanowicz RE (1989) The Comparative Anatomy of the Baltic Sea and the Chesapeake Bay Ecosystems. In: Wulff F, Field JG, Mann KH (eds) Network Analysis in Marine Ecology: Methods and Applications. (Coast est Stud 32) Springer Heidelberg, pp 232-258

2.14 Case Studies: Modeling Approaches for the Practical Application of Ecological Goal Functions

Søren N. Nielsen, Sven E. Jørgensen and Joao C. Marques

Abstract

Some of the goal functions suggested by various authors within the science of modern ecosystem theory have been applied to case studies. This paper presents a review of the current state of application. The applications can be divided in three major directions. The first direction deals with the introduction of goal functions in models and studies of their phenomenological behavior compared with actual observations of ecosystem development. In the second type of cases, the goal functions are introduced and an actual is optimization carried out. Third, studies have been performed where comparisons with other ecological concepts have been made. The various concepts, when applied to ecosystems, are in many cases in agreement with each other. Meanwhile, differences indicate that it may be necessary to apply several of the functions at the same time in order to gain a full understanding of the system. Also the role of goal functions to the future development and improvement of models is discussed and a framework to a future strategy is laid out.

2.14.1 Introduction

During the recent decades it has become clear that the traditional deterministic approach for the establishment of ecosystem models is not adequate. In an increasing number of situations, model predictions have been observed to fail. Perhaps, this is not surprising after all when taking into account the complexity of such systems as well as the relations and feed-back mechanisms which are involved.

The models have typically failed when large shifts in external or internal relations have occurred. By external relations we here refer to changes in the set of variables describing the governing forces acting on the system, usually referred to as either forcing or control functions. By internal relations we refer to changes in parameter sets caused by shifts in species composition or shifts in the descrip-

F. Müller M. Leupelt (Eds.)
Eco Targets, Goal Functions, and Orientors

tion of the kinetics as well as the internal relationships, i.e. the trophic structure of the ecosystem.

In short, the models fail because they lack one important feature of natural ecosystems which can be summarized under one word: adaptability. Adaptability in the sense used here covers not only the possibility of a single organism or process to adjust, answering back to changes with proper arrangements, but it also covers the higher-level perspective, that the resulting "sum" of all adaptations carried out, also must be of benefit to the ecosystem as a whole. Hence, this results in a demand for the future models to be more flexible. On the basis of this demand the approach of Structural Dynamic Modeling was developed.

Structural Dynamic Models are able to adjust the ecosystem's structure and dynamics in accordance with a trade off between internal capacities for change matched with demands induced by the prevailing conditions. The need for adjustment implicitly bears the requirement of some sort of function, in mathematics such type of function is usually referred to as goal function, that can serve to evaluate the state of the system, i.e. if one state is performing "better" than the other. Such statement results in the risk of being accused for teleology, idealism, vitalism, creationism or other that are not compatible with the traditional perception of science. Meanwhile, this should not be the case. The opinion of the authors is that no interference with the ecosystem from the outside is presumed, other than the forcing functions. What happens in the system is solely determined by the compartments of the system, their existing possibilities, as well as its new ones, and the network, the processes and feed-back interactions alone. The combination and structure is subjected to selection where one performance can be evaluated as better than the other, in accordance with a fully materialistic derived principle, e.g. principles derived form the thermodynamic laws.

This is the basic philosophy and justification of applications of goal functions in ecosystem models. That, by its self-organizing properties, the system governs itself (teleonomy) so that it happens that system performance is reflected in, and can be expressed by, a mathematical function that is continuously increasing during an undisturbed development of the system. Thus, it seems that the given function becomes optimized, and it can thus serve as a goal function to the system's development.

Several proposals of functions to be applied as goal functions for the development of biological systems at various levels can be found in literature (Aoki 1987, 1988; Mauersberger 1995; Odum 1983, 1996; Hall 1995; Ulanowicz 1986, 1997). The goal functions mainly stem from two directions of modern ecosystem theory. The one direction is derived from network and graph theory. The other takes its starting point in the thermodynamic laws and especially their possible extensions to situations far from-equilibrium.

The application of goal functions for optimization of models has mainly happened only in the area of thermodynamic understanding of ecosystems, and the major part is using the concept of the exergy (Jørgensen 1982; Jørgensen and Mejer 1981; Mejer and Jørgensen 1979; Nielsen 1992). Unfortunately, neither have any of the principles derived from network theory (Higashi and Burns 1991), nor any of the other thermodynamically derived expressions been applied. In principle, we can identify two directions in the application of the exergy concept, the exergy storage (Jørgensen 1992, 1994; Nielsen, 1992) and the exergy degradation perspective, respectively (Kay and Schneider 1992, 1994; Schneider and Kay 1994a,b,c, 1995). Likewise, the examples of application also seem to be dominated by aquatic models, whereas terrestrial applications are still lacking, although phenomenological studies are promising (see example *d* below).

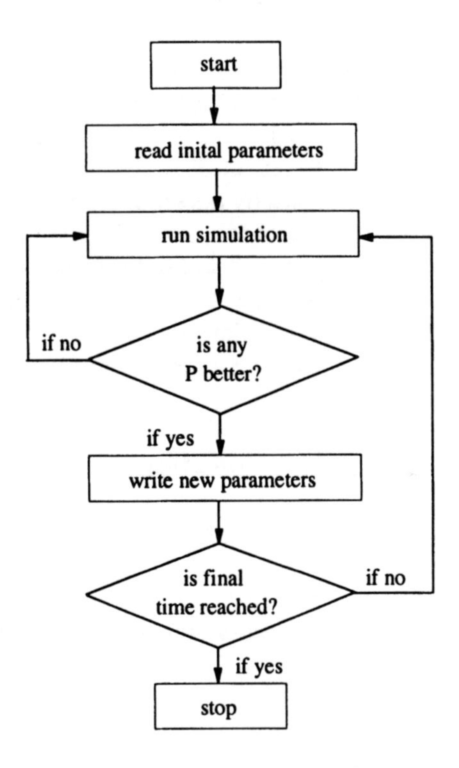

Fig. 2.14.1. Flow chart showing the procedural steps behind the implementation of optimization in a structural dynamic model. Intervals and allowed degree of manipulation of parameters may be varied during the run.

In the following, an attempt to review the existing applications of goal functions, according to the knowledge of the authors, has been made. The major experiences have been described and the possible conclusions for the applications in the future have been pointed out.

Basically, the known examples of application can in principle be divided in three types:

- Phenomenological applications

- Optimization applications
- Applications relating to other principles

In the phenomenological applications the goal functions have been implemented into models, and their possible role and importance has been evaluated based on studies of their behavior, especially during structural dynamic shifts in the model.

In the optimization examples the goal function has been coupled to the model through an optimization algorithm. This allows the parameters to be flexible during a model run. The principle behind the optimization procedure is shown in Fig. 2.14.1. The resulting change and adjustment of parameters is thought to mimic the adaptational properties of organisms and the ecosystem.

In the comparative applications, goal functions have been confronted, in order to analyze whether they are contradictory, or what different features of the ecosystem behavior they are able to reflect. It is already proposed that the various goal functions are not telling us exactly the same thing about a given ecosystem. Rather, each of the functions tells its own story. Together, they will form a pattern and in order to get the full picture it will be necessary to apply more of them at the same time to the same case. The above division of the examples has been implemented in the following description, although in some cases no clear distinction can be made, since all perspectives are included in one paper.

2.14.2 Phenomenological Applications

In the following studies goal functions have been applied and their behavior during ecosystem evolution and development has been observed. The examples presented here are some of the first attempts to evaluate goal functions, especially exergy, as indicators of ecosystem state. The examples given in this section are from Danish shallow lakes, Lake Annone, Roskilde Fjord and some terrestrial systems.

a. Danish shallow lakes. Attempts to develop structural dynamic models able to demonstrate the development of Danish shallow lakes during an eutrophication process have been carried out. In addition, attempts to predict and understand the development following a process of de-eutrophication (cleaning), e.g. by means of biomanipulation, have been carried out. The model development was started as a co-operation with the National Environmental Research Institute and was constructed using knowledge and data from a database under development. The observations, together with expert knowledge, indicated the possibility to develop a general structural dynamic model for shallow lakes, including the shifts between, first, 6, at a later stage, 9, types of phytoplankton, 2 types of zooplankton, and one zooplanktivorous fish compartment. Structural dynamic models or (SD-models) are models that are able to simulate shifts in species composition and trophic structure of the ecosystem. The stepwise development of this type of models can be found in Nielsen (1990a, b, 1992b, 1994, 1995), a synopsis is found in Nielsen (1992a). The major results showed, that it was possible to make SD-models on the basis of deterministic parameters, but in getting closer and closer to the situation of simulating a case-study, the generality of the model, initially observed, was lost (Nielsen 1994, 1995). Of importance to the application of goal functions, it was found that by excluding "winning" species or types of phytoplankton from the model, the model would always afterwards simulate lower values of exergy. Exergy was calculated according to Jørgensen and Mejer (1981). Remaining results are presented under example *e* below.

b. Lake Annone. In the eastern basin of Lake Annone, Italy, a mass extinction of the fish population was caused by gill infection disease at the end of the season 1975. Subsequently, an increase in the zooplankton population, both in size and biomass, as well as a decrease in phytoplankton biomass, was observed. By the use of a simple model, it was found that the parameter sets calibrated were not only giving the highest exergy but also giving results for state variables that were in correspondence with the observations from the western basin of the lake (Jørgensen and de Bernadi, in print).

c. Roskilde Fjord. The application of exergy in order to explain the observed geographical distribution pattern of macrophytes was performed in connection with a EU financed project having as target areas, three estuaries in Denmark, Portugal and Italy, respectively, see examples below. In Denmark, the target area was Roskilde Fjord, where the competition between the green algae *Ulva lactuca* and the phanerogame *Zostera marina* was to be analyzed. A SD-model was developed that basically confirmed the above results found for phytoplankton in Danish shallow lakes. The established macrophyte societies developed towards increasing exergy during their development, the species able to demonstrate the highest exergy also eventually became the dominant species of the system (Nielsen, 1997).

d. Terrestrial systems. The following appears to be the only known example of application and observation of goal functions to terrestrial ecosystems. The attempts have been made by Schneider and Kay (1994b,c), thus representing the exergy degradation perspective, and are based on devices and techniques described by Luvall and Holbo (1991). The basic assumption is that evapotranspiration on the one hand provides a fancy transport system to the plant population, and on the other, at the same time results in a cooling of the system proportional to the activities undertaken, and thus to the exergy degraded by the system. The more biomass, the more exergy degraded, the more cooling of the system. A measure of the cooling for a given terrestrial system is found by calculations made on different spectra, obtained by either remote sensing or devices that can be mounted on minor air-crafts. The coolest systems are expected to be found where exergy degradation is highest, i.e. the areas of tropical rain forests on the earth. This is confirmed by observations made by the methods described above. The above phenomenological studies serve as justifications of using exergy, storage or degradation, as indicator of ecosystem state and as goal function.

2.14.3 Optimization Application

In the following studies goal functions have been applied and their behavior during ecosystem development was observed in Lake Væng, Lake Søbygård in some population models, in the Lagoon of Venice and Lake Balaton.

e. Lake Væng. In this study the final model developed was used as a test example for a more dynamic application of the exergy principle. The approach is explained by Fig. 2.14.1. In this case, it was found that adjustment of parameters is carried out by manipulating the most sensitive parameters of the model aiming at a higher exergy. The importance of the choice of the optimization interval and allowed percentage of adjustment of parameters, as well as the importance of intrinsic parameters of the applied mathematical algorithm, was investigated. The optimization results showed to be sensitive to all of the factors mentioned (Nielsen 1995).

f. Lake Søbygård. Exergy was applied to elucidate the changes occurring in Lake Søbygård, Denmark. In this case, structural dynamic changes were induced by a long-term change in the fish population. The changes observed were due to a failing recruitment of planktivorous fishes. The lowering in the fertility of fish was a result of increasing pH caused from eutrophication. The lowering in fish density, allowed an increase in the zooplankton population. This in turn leads to lower concentration of phytoplankton but with a larger size. The larger size leads to reduced grazing (growth rate of zooplankton) but also higher loss rates due to sedimentation. The results of this exercise fall in two parts. First, applying exergy optimization, it was possible to simulate the observed change in parameters as a result of the changes in the fish population. Second, it was observed, that keeping the parameters of 1985 in the model, i.e. imitating that species composition was maintained, and thus using a strictly deterministic model, this would have lead to a highly unstable, maybe chaotic, situation of the ecosystem (Jørgensen 1992 a, b).

g. Population models. The hypothesis has also been tested on two minor population models. In a two species predator-prey model (Allen 1985) it was shown that exergy optimization would result in a correction of the model parameters in accordance with what was known to occur as a result of random mutations (Jørgensen 1994, p 504).

In another example, where exergy optimization was applied to a three-species competition model, it was shown that exergy optimization, in the initial phase of development, when the system is rich in resources, is carried out by adjustment of competition parameters. As system resources become scarce, carrying capacity parameters are adjusted (Jørgensen 1994, pp 504-506).

h. Lagoons of Venice. The application of exergy in order to explain the observed geographical distribution patterns of macrophytes was performed in connection with a EU financed project having three estuaries in Denmark, Portugal and Italy, respectively, as target areas. In the case of Lagoons of Venice an SD-model was developed to analyze the competition between the green algae, *Ulva rigida* and the rooted macrophyte, *Zostera marina.* Since optimization is generally time consuming, among other things, because the computer in this case has to optimize many parameters at many places in space, it would be nearly impossible to make a dynamic optimization on all the points of investigation in the lagoon. A very elegant solution to this was made, by coupling the essential parameters of the developed SD-model, to one property of the organisms, the surface to volume ratio, SA/V-ratio, by allometric principles. Applying this approach it was possible to predict the optimal SA/V-ratio and in turn translate it into species that should be present at the investigated localities. The predictions matched observations with 87-95% (Coffaro 1996).

i. Lake Balaton. A structural dynamic model including exergy maximization has been applied to explain observed data from Keszthely Bay in Lake Balaton. In this case it is hypothesized that intermediate disturbance, IDH (intermediate disturbance hypothesis, Connel 1978) is the explanation of the observed high bio-diversity in the area. According to IDH, either higher or lower disturbance frequencies should result in lower bio-diversity. The developed model was used for optimization of exergy over a period of 7 months using an allowed change of 10% of 6 parameters every 6th day. Several runs were made taking into account the effect of the frequency with which wind disturbance occurred in the lake. To summarize, the exergy levels found are higher when wind disturbance of the lake is occurring with a certain frequency, when compared to steady mixing, and optimization of parameters are allowed (Jørgensen and Padisàk 1996).

2.14.4 Applications Relating to Other Principles

Above, it has already been shown, in two examples, that exergy relates to at least two other concepts found in modern ecosystem theory, chaos theory and the intermediate disturbance hypothesis, IDH, respectively. In the following relations to other concepts from modern ecosystem theory are summarized.

j. Buffer capacity. At early states the relation between the choice of model complexity , buffer capacity and exergy was analyzed (Jørgensen 1982). The results indicate that exergy also contributes to higher buffer capacity of the system, and thus relates to widely discussed concepts such as stability, resistance and resilience.

k. Chaos. Lately it has been shown that exergy also relates to concepts of modern mathematics and physics like chaos theory (Jørgensen 1995). The analysis and conclusion made during the study (example f) above was extended and indicates that exergy governs the ecosystem and its parameters to perform "at the edge of chaos".

l. Ascendency. Attempts have been made to investigate the correlations to other measures of ecosystem performance like ascendency proposed by Ulanowicz (1986, 1997). Investigations may be found in Christensen (1992) and Salomonsen (1992). The most extensive treatment, based on several case studies and taking advantage of the software ECO-PATH, is Christensen (1992). The results show that exergy is related especially to the relative ascendency of the system. Recently it has been shown that mechanisms argued to be of benefit to the thermodynamical function of an ecosystem also will contribute positively to the ascendency of the system (Nielsen and Ulanowicz, in print). Investigations of ascendency and various related concepts can be found in a list of papers by Ulanowicz et al. (Ulanowicz 1995, 1996; Ulanowicz and Wolff 1991; Ulanowicz and Arbarca-Arenas, in print; Ulanowicz and Baird, in review).

m. Diversity. Logically, it should be reasonable that thermodynamic functions and diversity measures might also find some connection. The initiatives, known to the authors, taken in this direction are quite new and were carried out, based on data from the Mondego estuary at Figueira da Foz, Portugal, in connection with the above mentioned project of comparing European estuaries. Two diversity measures were compared with different forms of calculating exergy (Marques et al., in print; Marques et al., this volume). The results are not clear which calls upon for further investigations.

n. Integration. Only one example of an application of goal function to evaluate ecosystems highly connected with socio-economic activities of humans is known. This study was carried out in the Phillipines and based on the analyses of the nitrogen cycles of 4 different farms which represented, from a subjective point of view, various levels of integration, e.g. differences in numbers of goods produced and cycling of resources in the system. The various levels of integration were reflected in measures such as cycling index, ascendency and exergy (Dalsgaard 1996).

2.14.5 Discussion and Conclusion(s)

Before more overall conclusions for the future are made, it may be worth while dwelling a moment to draw the major conclusions on the basis of the exercises presented.

1. The indications that exergy, storage or degradation, or at least a thermodynamic interpretation of ecosystems, represent a promising approach, are so clear that this direction seems worth to pursue. Ecosystems may be disturbed and exergy decrease locally in time and space, but as soon as this disturbance stops, the system will try to re-establish and exergy will increase.
2. The optimizations have to be constrained. Mathematical algorithms, in general, tend to optimize parameters into unrealistic values. Mathematical algorithm have to be constrained either internally, i.e. included in the algorithm, or there has to be a constraint from the outside.
3. There is a demand for biological realism in the way the algorithms are implemented. This means that it might show necessary to optimize different (types of) parameters with different speed, magnitude, and maybe at different intervals. This indicates a need for more sophisticated and hairy routines than the ones implemented.
4. There still seems to be a long way to simulate abrupt shifts in ecosystem structure. The optimization might be working around a local minimum in a landscape full of local minima. How does one move, if possible, to another minimum? Or does the single minimum represent a dead-end, that is strongly history dependent, and from which only major disturbances, like a catastrophe, can change the situation?

To summarize, the above mentioned points leave us with some major problems to be solved, and indeed this would be the case for any goal function to be implemented. The problems seem to be divided into two major areas, but both relate to the algorithms used. The first problem is strictly related to mathematics. The challenge is to find a feasible algorithm, constrained, and at the same time able to deal with a "stiff" system. The other problem will be to imitate ecosystem features as closely as possible and simulate also the major structural shifts. This problem maybe solved by loosing the modeling even further from its thighs to the organismic way of thinking that still lies behind many modeling efforts, and enhancing the efforts in attempts taking a starting point in the function of the system. We may in this case allow the algorithms to make larger adjustments than at the organism level. This may also be reached by integrating the modeling procedure more with tools like *artificial intelligence* (Kompare 1995), neural networks, game theory, cellular automata, or other recent strategies of modeling.

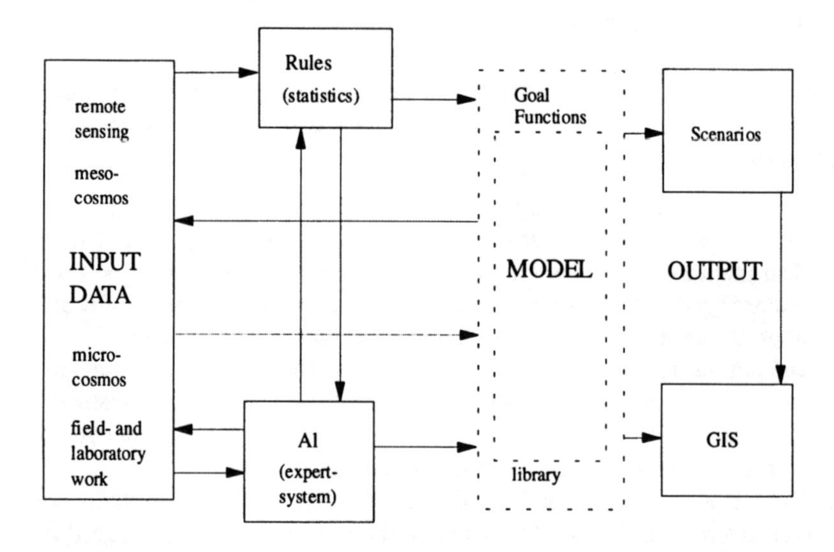

Fig. 2.14.2. Illustration of a future framework for environmental modeling. Increased use of tools like artificial intelligence and expert systems, as well as an increased importance of a feedback loop ensuring the production of an appropriate empirical data-background should be stressed.

This leaves us with the consideration how a future model complex will look like. A conceptualization of a suggested solution is shown in Fig. 2.14.2. The core is still a model, but the model is more flexible than the type of models used hitherto. Thus, the outer boundary symbolizes a gross model, consisting of a library of submodels, kinetic equations, parameters, etc. The actual model used represents a subset of the gross model. Production of input data, left side of figure, is taking place at many levels of hierarchy following a research strategy put forward by Kemp et al. (1980). The input data are put forward to the model in the traditional deterministic way. But the data are also subjected to further interpretation by use of artificial intelligence or other methods of derivation of rules. The outputs are delivered either 1) from the model or 2) via simulations of scenarios. After this data are passed on to a GIS system or an other pixel-based product for presentation.

References

Allen PM (1985) Ecology, Thermodynamics and Self-Organization: Towards a New Understanding of Complexity. In: Ulanowicz RE and Platt T (eds) Ecosystem Theory for Biological Oceanography. Can Bull Fish Aquat Sci 123:3-26

Aoki I (1987) Entropy Balance in Lake Biwa. Ecol Modelling 37:235-48

Aoki I (1988) Entropy balance laws in ecological networks at steady state. Ecol Modelling 42:289-303

Christensen V (1992) Network analysis of trophic interactions in aquatic ecosystems. ICLARM Contribution No. 835

Coffaro G (1996) Modelling Primary Producers Dynamics in the Lagoon of Venice. Ph.D. thesis. The Royal Danish School of Pharmacy, Copenhagen

Connel J (1978) Diversity in Tropical Rain Forests and Coral Reefs. Science 199:1304-1310

Dalsgaard JPT (1996) An ecological modelling approach towards the determination of sustainability in farming system. Ph.D. Thesis. The Royal Danish School Of Pharmacy, Copenhagen

Hall CAS (Ed) (1995) Maximum Power. The Ideas and Applications of H.T. Odum. University Press of Colorado, Colorado, p 393

Higashi M, Burns TP (eds) (1991) Theoretical studies of ecosystems: The network perspective. Cambridge University Press, Cambridge, p 364

Jørgensen SE, de Bernadi R (in print) The Application of a Model with Dynamic Structure to simulate the effect of Mass Fish Mortality on Zooplankton Structure. Ecol Modelling

Jørgensen SE (1982) Exergy and Buffering capacity in ecological systems. In: Mitsch WJ, Ragade RK, Bosserman RW, Dillon Jr JA (eds) Energetics and Systems. Ann Arbor Science place of publishers, pp 61-72

Jørgensen SE (1992a) Integration of Ecosystem Theories: A Pattern. Kluwer, p 383

Jørgensen SE (1992b) Development of Models Able to Account for Changes in Species Composition. Ecol Modelling 62:195-208

Jørgensen SE (1992c) The shifts in Species Composition and Ecological Modelling in Hydrobiology. Hydrobiologia 239:115-129

Jørgensen SE (1994) Review and Comparison of Goal Functions in Systems Ecology. VIE MILIEU 44(1):11-20

Jørgensen SE (1995) The Growth Rate of Zooplankton at the Edge of Chaos: Ecological Models. J Theor Biol 175:13-21

Jørgensen SE, Mejer HF (1981) Exergy as a Key Function in Ecological models. In: Mitsch WJ, Bosserman RW Klopatek JM (eds) Energy and Ecological Modelling. Developments in Environmental Modelling vol. 1. Elsevier Scientific Publishing Company p 839 pp 587-590

Jørgensen SE, Padisàk, J (1996) Does the Intermediate Disturbance Hypothesis Comply with Thermodynamics? Hydrobiologia 323:9-21

Kay JJ, Schneider ED (1992) Thermodynamics and Measure of Ecosystems Integrity. In: McKenzie DH, Hyatt DE, McDonalds VJ (eds) Ecological Indicators. Vol 1. Proceedings of the International Symposium on Ecological Indicators. Elsevier, Fort Lauderdale, pp 159-182

Kay JJ, and Schneider ED (1994) Embracing Complexity – The Challenge of the Ecosystem Approach. Alternatives 20(3):32-38

Kemp WM, Lewis MR, Cunningham JJ, Stevenson JC, Boynton WR (1980) Microcosmos, Macrophytes, and Hierarchies: Environmental Research in the Chesapeake Bay. In: Giesy jr. JP (ed) Microcosms in Ecological Research. US Technical Information Center, US Department of Energy, Symposium Series 52 (CONF-7811011)

Kompare B (1995) The Use of Artificial Intelligence in Ecological Modelling. Ph.D. thesis. The Royal Danish School of Pharmacy, Copenhagen

Luvall JC, Holbo HR (1991) Thermal Remote Sensing Methods in Landscape Ecology. In: Turner M, Gardner RH (eds) Quantitative Methods in Landscape Ecology, Ecological Studies 82. Springer Verlag, New York, pp 127-152

Marques JC, Pardal MA, Nielsen SN, Jørgensen SE (in print) Analysis of the Properties of Exergy and Biodiversity Along an Estuarine Gradient of Eutrophication. Ecol Modelling 102:155-167

Mauersberger P (1995) Entropy control of complex ecological processes. In: Patten BC, Jørgensen SE (eds) Complex Ecology. The Part-Whole Relation in Ecosystems. Prentice Hall, Englewood Cliffs, New Jersey, pp 130-165

Mejer H, Jørgensen SE (1979) Exergy and Ecological Buffer Capacity. In: Jørgensen SE (ed) State-of-the-art. Ecol Modelling 7:829-846

Nielsen SN (1990a) Recent Developments in Structural Dynamic Models. In: Fenhann J, Larsen H, Mackenzie, GA, Rasmussen B (eds) Environmental Models: Emissions and Consequences. Risp International Conference 22-25 May 1989 Developm in Environm model 15. Elsevier, Amsterdam

Nielsen SN (1990b) Application of Exergy in Structural-Dynamic Modelling. Verh Internat Verein Limnol 24:641-645

Nielsen SN (1992a) Application of Maxiumum Exergy in Structural Dynamic Models. Ph.D. thesis. National Environmental Research Institute, Copenhagen

Nielsen SN (1992b) Strategies in Structural Dynamic Modelling. Ecol Model 63:91-101

Nielsen SN (1994) Modelling Structural Dynamical Changes in a Danish Shallow Lake. Ecol Modelling 73:13-30

Nielsen SN (1995) Optimization of Exergy in a Structural Dynamic Model. Ecol Model 77:111-122

Nielsen SN (in print) Examination and optimization of different exergy forms in macrophyte societies. Ecol Modelling

Nielsen SN, Ulanowicz RE (in print) In the Consistency between thermodynamical and network approaches to ecosystems. Proceedings from ISEM conference, Beijing, 1995. Ecol Modelling

Odum HT (1983) Systems Ecology. Wiley Interscience, New York, p 644

Odum HT (1996) Envrionmental Accounting. Emergy and environmental decision making. John Wiley and Sons, New York, p 370

Schneider ED, Kay JJ (1994a) Exergy degradation. Thermodynamics and the development of ecosystems. In: Szargut J, Kolenda Z, Tsatsaronis G, Ziebik A (eds) Proceedings of the International Conference ENSEC '93 (Energy systems and Ecology Cracow, Poland July 5-9, 1993) 1:33-42

Schneider ED, Kay JJ (1994b) Complexity and thermodynamics. Towards a new ecology. Futures 26(6):626-647

Schneider ED, Kay JJ (1994c) Life as a Manifestation of the Second Law of Thermodynamics. Mathl Comput Modelling 19(6-8):25-48

Schneider ED, Kay JJ (1995) Order from Disorder: The Thermodynamics of Complexity in biology. In: Murphy MP, O'Neill LAJ (eds) What is Life: The Next Fifty Years. Reflections on the future of Biology. Cambridge University Press, Cambridge, pp 161-173

Salomonsen JS (1992) Examination of properties of exergy power and ascendency along a eutrophication gradient. Ecol Modelling 62:171-181

Ulanowicz RE (1986) Growth and Development. Ecosystems Phenomenology. Springer Verlag, New York, p 203

Ulanowicz RE (1995) Trophic Flow Networks as Indicators of Ecosystem Stress. In: Polis G, Winemiller K (eds) Food Webs: Integration of Patterns and Dynamics. Chapmann-Hall, New York, pp 358-368

Ulanowicz RE (1996) The Propensities of evolving systems. In: Khalil EL, Boulding KE (eds) Evolution Order and Complexity. Routledge, London, pp 217-233

Ulanowicz RE (1997). Ecology the Ascendent Perspective. Columbia University Press, p 201

Ulanowicz RE and Wolff WF (1991) Ecosystem flow networks - loaded dice. Mathematical Biosciences 103:45-68

Ulanowicz RE and Arbarca-Arenas LG (in print). An informational synthesis of ecosystem structure and function. Ecol Modelling, forthcoming

Ulanowicz RE and Baird D (in review) Nutrient controls in ecosystem dynamics. J Marine Systems, forthcoming

2.15 Case Studies: Soil as the Interface of the Ecosystem Goal Function and the Earth System Goal Function

Steven Cousins and Mark Rounsevell

Abstract

Ecosystem boundaries are chosen such that a domain is defined where a single ecosystem goal operates as the system attractor. Other goals may still operate on this area because the ecosystem contains semi-autonomous parts which are the organisms in a food web. It is possible for parts to have separate goals when the parts are separated from the whole by buffers or stores. The directed movement of water is defined here as the earth system goal. Ecosystems as goal directed objects are defined as the abundance of different sized organisms on the territory of the top predators. Fat decouples organisms from the ecosystem food web; the ecosystem food web is decoupled from climate by water storage in soil. Because the ecosystem is decoupled from its parts it can possess a goal; goal function equations for ecosystems and for water transfer in soil are given.

2.15.1 Introduction

You cannot serve both God and Mammon
St Matthew : 24

The quotation above illustrates the religious hypothesis that a system can have only a single goal at any one time. Any single goal can be described mathematically by a single goal function. Yet the world is obviously complex, with many goal orientated processes going on simultaneously. How can we bridge these two views of reality? We use the example of the soil as a system which appears to possess more than one goal function to examine general principles of system goals and to show the value of these principles to system modelling. The objective of this paper is to define the conditions under which the ecosystem (Cousins 1990) can possess a goal and to identify candidate goal functions for the ecosystem thus defined. It is shown that real world separation of the dynamics of systems is necessary to allow a system goal to operate. The soil provides

F. Müller M. Leupelt (Eds.)
Eco Targets, Goal Functions, and Orientors

this means of separation of the ecosystem from the global climate referred to as the Earth system in the title, and so allows the ecosystem to possess a goal function.

2.15.2 Systems and Goal Functions

If we take an arbitrary area on the Earth's surface and call that area *The System*, then we can call the remaining larger area of the Earth, the System's Environment. The line enclosing the system, as we have defined it, is the System Boundary. It is unclear in this case whether the system as a whole has a goal or indeed whether it has several goals because the system has been arbitrarily, perhaps even randomly, constructed. However there may be other ways of more carefully choosing the boundary position such that only a single goal is exhibited by the system as a whole. Thus it is proposed that the search for goal functions is directly linked to where system boundaries are placed and therefore to the task of system definition. Similarly the search for the goal function of the ecosystem is directly linked to the location of ecological system boundaries and to the definition of what an ecosystem *is*.

The presently dominant definition of ecosystem (Lindeman 1942), that it contains biotic and abiotic (i.e. everything) and exists at every scale from microscopic to planetary, is not constructive for the identity of goal functions since perhaps ' every' goal function could be included under these conditions. In short, the meaning of the word 'system' within the word 'ecosystem' is totally subjective allowing the ecologist to draw a boundary in whatever arbitrary or informed way he or she may choose.

The alternative approach as suggested above is to invert this problem and construct system boundaries such that they are defined by a single goal, this single goal being characteristic of each system and the boundary enclosing the domain of action of the goal, where the goal operates as a system attractor.

A further concept is needed to deal with the reality that there are many simultaneous goals being pursued in the physical space occupied by a system. Because of containment, that is, ecosystems contain organisms which contain cells and so on, the ecosystem goal may differ to the goal of the organism and of the cell. The concept of containment resolves this particular problem, by showing there may be multiple goal functions active within the space occupied by the ecosystem, while the ecosystem itself has a single goal. This may be generalized as the whole has a single goal function but the whole is made up of parts which each have a single goal function and so on in a hierarchy.

The problem that remains is how can these separate goals interact and yet be preserved as separate goals. Simon (1973) provides a description of what is required which he defines as semi-autonomy. Thus in a hierarchy of contained systems, each system is semi-independent of the levels above and below, that is, independent within certain thresholds of behaviour.

The identity of a hierarchy is at least as controversial as the identity of system boundaries (Allen and Starr 1982). However goal functions may play a similar role in the non-arbitrary definition of hierarchy as they have for system boundary definition above. Thus we may say that in the real world, nature is constructed by a hierarchy of systems where each system is in possession of a single goal function and is semi-autonomous from the large system in which it is contained and is itself made up of parts which are single goal function semi-autonomous systems, and so on. The identification of single goal functions and their domains thus defines objective hierarchies of structure in nature.

In this paper we put some flesh on these theoretical bones using soil as focus but with the objective of defining the ecosystem goal function. The hydrological properties of soil are central to the soil's role as a growth medium for plants and we begin there.

2.15.3 Hydrological Systems

Let us consider the movement of water as part of the global water cycle. Overall water minimises its energy level both in the transition between phases and within each phase; sorption of water in soils minimises energy also. Water at the surface is energised by solar energy causing a phase change to water vapour. In this gas phase water moves to regions of lowest pressure which we take to be the goal of all air flow. However when the water vapour is returned to the liquid phase as rain, its goal is to move downwards in response to gravity. This goal is expressed by directed movement down the maximum slope available on the land surface. The goal of water minimising its potential energy can be used to define the boundary of the waterflow system. This is identified as the watershed and the area over which water travels is the catchment. Thus the goal is achieved in catchments by the movement of water through a network of drainage channels to the river and ultimately to the river mouth. The rate at which this occurs is due to the characteristics of the land surface e.g. soils and vegetation.

It is important to observe that the catchment boundary is uniquely defined by the water flow goal and the catchment has an intrinsic scale. That is to say, the size of the catchment is independent of the observer and the system defined from its single goal function is independent of the observer also.

Finally it is also relevant to note that the intrinsic scale of the system changed when the water changed phase from liquid to gas or vice versa, and the scale change was determined by the change in goal. At a more general level this phase change may be thought of as a change in the currency of the flows which are proceeding to the goal of each system.

We can now examine a second relevant system, the ecosystem, in which a completely different currency is transferred to the ecosystem goal.

2.15.4 Ecological Systems

We have already touched on the problems that arise in identifying ecosystem goals due to the highly subjective approach to the definition of ecosystem proposed by the founding fathers of ecology (Tansley 1935; Lindeman 1942). Here we take the hypothesis that nature is structured as nested (contained) domains of single goal systems. We search for a single goal function which describes the key properties of ecosystems and from which to define objective ecosystem boundaries which correspond to the domain of that goal.

An approach to this problem is provided by McNaughton et al. (1989) 'Ecosystems are structurally organised as food webs'. Thus we can reframe the ecosystem goal by seeking the goal function of food webs. When we do this a solution analogous to the watershed is produced.

Cousins (1990, 1996) describes ecosystems as objects defined by energy flowing through the food web to the top predator. These objects are bounded by the territory edge of the top predator, such that if solar energy (photons) fall on one side of the boundary the energy will tend to move to one top predator, and if it falls on the other side it will flow to the other top predator. Little of the energy reaches the top predator but is consumed and respired by intermediate organisms in the food web. In this framework ecosystems as functioning objects only contain live organisms and food items. The abiotic is specifically excluded and assigned to some other (abiotic) goal orientated system, e.g. the watershed.

Before leaving this question of how to view ecosystems, their boundaries and their goal functions, we note that it has been traditional to study ecology 'at the ecosystem level' precisely as the study of the abiotic, or as the study of nature using energy and materials as currencies. Population ecology has adopted live organisms as its currency but studies tend to be limited to only a subset of species in an ecosystem. However as Martinez (1995) observes, there is nothing unique about the ecosystem level in its use of energy and materials. Cells use energy and pass materials across their boundaries; single organisms and therefore populations do also. What is unique about ecosystems is that they are structured by feeding interactions; food is their currency.

The behaviour and goal function of ecosystems is in many ways similar to catchment hydrology. Catchments trap incoming rainfall and ecosystems trap incoming radiation. In catchments water is transferred through a hierarchical network of drainage channels to the river and in ecosystems energy is transferred through hierarchical sequences of organisms to the top predator (a food web). The boundary of the catchment is the watershed and the boundary of an ecosystem is the photonshed, in which the photonshed is defined by the territory of the top predator (its feeding area), and that within a photonshed the energy from intercepted radiation is statistically most likely to be directed toward the top predator.

2.15.5 Soils as the Interface

Biological and hydrological systems interact because they are not totally autonomous. So that each system impinges on the ability of the other system to achieve its own goal function.

Table 2.15.1. Mean values of available water for some common British soil profiles (adapted from Hall *et al.* 1977 and Mackney *et al.* 1983)

Soil series	Parent material	Depth of integration (cm)	Total available water (mm) (0.05-15 bar)	Easily available water (mm) (0.05-2 bar)
Aberford	Permian, Jurassic and Eocene limestone	60	110	70
Ardington	Cretaceous glauconitic sand, loam and clay	100	150	100
Bardsey	Carboniferous mudstone with interbedded sandstone	100	130	80
Bridgnorth	Permo-Triassic and Carboniferous reddish sandstone	80	85-105*	65-90*
Bromsgrove	Permo-Triassic and Carboniferous sandstone and siltstone	80	105	75
Clifton	Reddish till	100	150	100
Crewe	Reddish Glaciolacustrine drift and till	100	140	75
Denchworth	Jurassic and Cretaceous clay	100	160	82
Eardiston	Devonian and Permo-Triassic sandstone	60	105	70
Elmton	Jurassic limestone	30	65	40
Flint	Reddish till	100	120	70
Gresham	Aeolian drift and till	100	140	80
Marcham	Jurassic limestone	35	45	30
Newport	Glaciofluvial drift	100	75-150*	65-130*
Oak	Reddish till	100	130	65
Ragdale	Chalky till	100	130	75
Salop	Reddish till	100	135	80
Sherborne	Jurassic limestone and clay	35	55	30
Wick- Arrow-Quorndon	Glaciofluvial drift	100	100-200*	70-160*
Wilcocks	Drift from Palaeozoic sandstone, mudstone and shale	100	215	135
Worchester	Permo-Triassic reddish mudstone	100	115	60

*varies greatly according to content of 60-100 μm sand and surface organic matter levels
** (to 100 cm or to rock if shallower)

It is the contention of this paper that soils provide the interface between the goal functions of these two systems; the ecosystem goal function and the Earth system goal function for water distribution.

Because a wide range of soil types and properties occur in nature, the ability of soils to act as an interface depends largely on these properties and their distribution in space. For example, differences in soil types have a number of influences on the movement of water in catchments, including:

- infiltration (soil structure and porosity);
- surface runoff (surface roughness, soil water content);
- lateral flow (hydraulic conductivity);
- evapotranspiration (soil effects on plant growth and distribution).

Some examples of the differences in the physical properties of soil types are given in Tables 2.15.1 and 2.15.2. These examples are illustrative and data for other sites can be found elsewhere (Schlesinger 1985, 1991).

Table 2.15.2. Topsoil porosity classes for textural groupings with a medium packing density (1.4-1.75 g cm^{-3}) (after Hall et al. 1977).

Particle size class	Air capacity (%)	Porosity class
Clay	5.0-9.9	Slightly porous
Sandy clay	-	-
Silty clay	10.0-14.9	Moderately porous
Sandy clay loam	10.0-14.9	Moderately porous
Clay loam	5.0-9.9	Slightly porous
Silty clay loam	5.0-9.9	Slightly porous
Silt loam	10.0-14.9	Moderately porous
Sandy silt loam	10.0-14.9	Moderately porous
Sandy loam	15.0-20.0	Very porous
Loamy sand	>20.0	Extremely porous
Sand	>20.0	Extremely porous

Soil formation and the development of soil properties have been described using a functional relationship first proposed by Jenny (1941):

$$s = f(\text{parent material, climate, organisms, topography, time}) \qquad (2.15.1)$$

in which s is a soil property. It is interesting to note from Equation (2.15.1) that both the biological (*organisms*) and hydrological (*climate*) systems contribute to the development of soils. In doing so, they influence the interface between their goal functions. For example, the biological system influences the hydrological system by:

- extracting and transpiring soil water;

- enhancing soil water retention (through returns of organic material or the effect of roots on soil structure);
- reducing surface runoff (through interception) and thus, enhancing infiltration.

In each of these examples the ability of the hydrological system to achieve its goal function is impaired, whereas the biological system benefits. Table 2.15.3 gives some examples of how different vegetation types influence the organic carbon content of soils.

Table 2.15.3. Examples different soil organic carbon contents under different land uses or vegetation types (after White 1981).

Land use or vegetation type	Organic carbon (t ha^{-1})
Alpine and arctic forest	0.1
Arable farming (cereals)	1-2
Temperate grassland	2-4
Coniferous forest	1.5-3
Deciduous forest	1.5-4
Tropical rainforest (Columbia)	4-5
Tropical rainforest (Africa)	10

2.15.6 Soil Water Goal Function

In its simplest form, soil water movement can be described as follows. Water movement through porous materials is expressed by Darcy's equation:

$$v = -k \text{ grad } \psi \quad (2.15.2)$$

where v is the volume of water per unit time crossing an area A perpendicular to the flow, grad ψ denotes the change in water potential per unit distance in the direction of flow (which may be horizontal or vertical), and k is a constant (at saturation) called the hydraulic conductivity (defined as the reciprocal of the flow resistance).

The main components of the soil water potential (ψ) in head units (h) are the pressure head (p), where -p = suction for soil water under tension, and the gravitational head (z), thus:

$$h = p + z \quad (2.15.3)$$

So that, Equation (2.15.2) becomes:

$$v = -k \text{ grad } h \quad (2.15.4)$$

Thus, as a plant extracts soil water through its roots (by transpiration), the pressure head in the immediate vicinity of the roots declines, and grad h increases. Water will then move toward the rooting zone at a rate that is defined by the size of the pressure gradient (grad h) and by the hydraulic conductivity (k). The value of k varies greatly between different soil types, see Table 2.15.4. Thus, the properties of different soils determine the efficiency of the plant 'water pump'.

Table 2.15.4. Saturated hydraulic conductivity (using the auger hole method) for some selected soils in England and Wales

Soil series	Horizon	Depth (cm)	Clay (%)	Saturated hydraulic conductivity (cm day^{-1})
Bignor	Eb(g)	16-40	23	74
Bignor	2Btg	40-55	30	45
Bignor	Bt(g)	45-61	29	36
Denchworth	Bw(g)	19-38	53	0.5
Everingham	Bg2	37-60	8	183
Evesham	Bw(g)	50-70	75	1.4
Evesham	Bw(g)	32-52	88	6
Evesham	BCgk1	52-76	87	2
Evesham	BCgk2	94-100	85	0.5
Stow	Bg	45-70	38	0.1
Whimple	Bt(g)	27-50	40	2
Whimple	2BCt	70-90	30	0.02
Winsdor	Bg	50-70	45	0.3

2.15.7 Ecosystem Goal Function

We have described ecosystems as the outcomes of interactions within a food web. These interactions occur within the area of the top predator territory. For lions or for Wolves such a territory is of the order of $100km^2$. The net result of feeding interactions is the growth and reproduction of the feeders and when the populations of all species on the territory are observed, the relative abundance of all the organisms on the territory.

Peters (1983) plot of the abundance of different species of different body mass shows a steep decline in abundance of species of progressively larger size. Elton (1927) was the first to observe that it was the faster reproduction and growth rates of small organisms that fuelled the food web in which larger predators ate smaller prey.

The equation for the 'Peters' line is

$$D = 3\,w^{-1} \qquad (2.15.5)$$

where D is the population density of a species in number km^{-2} and w the mean weight in kg.

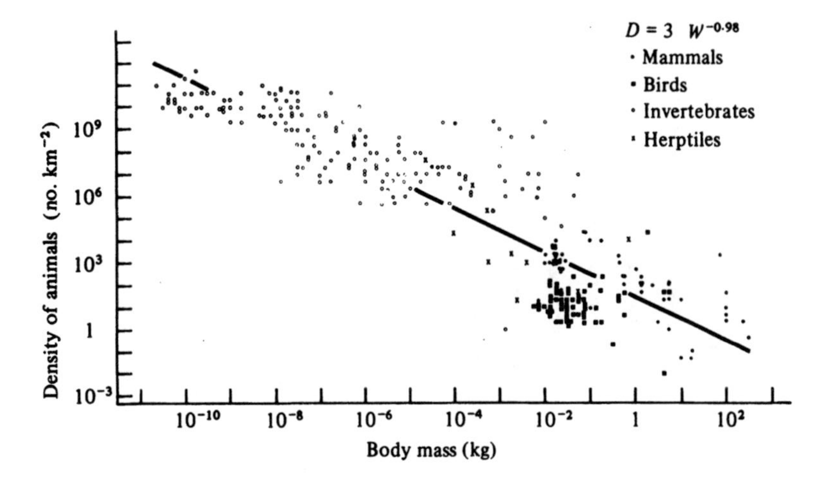

Fig. 2.15.1. The general relationship between animal body size and population density in published studies. With Permission (Peters 1983)

However, species are not the appropriate variable to parameterize the goal function since it is individuals who eat or are eaten in a food web not species. Members of a species may exist at very different sizes within the same ecosystem and can be cannibalistic. The alternative is to multiply equation (2.15.5) by the number of species at that weight to give the number of individuals at any weight. In practice it is easier to measure this quantity directly without identifying species. This approach has been most thoroughly pursued in the marine environment where automatic particle counters can be towed through the ocean. Again smaller organisms are eaten by progressively larger ones as energy passes in the direction of the top predator. The resulting abundance distribution of biomass is described by

$$b\,(w) = k\,w^{-.22} \qquad (2.15.6)$$

Where b, is the biomass of organisms of size w + dw kg, and k is a constant. Platt et al. (1984) found equation (2.15.6) to be a good fit for data collected in the central gyre of the north Pacific ocean.

This distribution of animal number by size of organism is proposed as the attractor to which the ecosystem converges and equation (2.15.6) is proposed as the form of the ecosystem goal function. Equations for the dynamics of this distribution are given elsewhere (Silvert and Platt 1980, Cousins 1985).

The operation of the ecosystem goal function within soil is illustrated in Fig. 2.15.2 showing the size and abundance of soil organisms found on long term experimental agricultural and grazing plots at Rothamstead, UK.

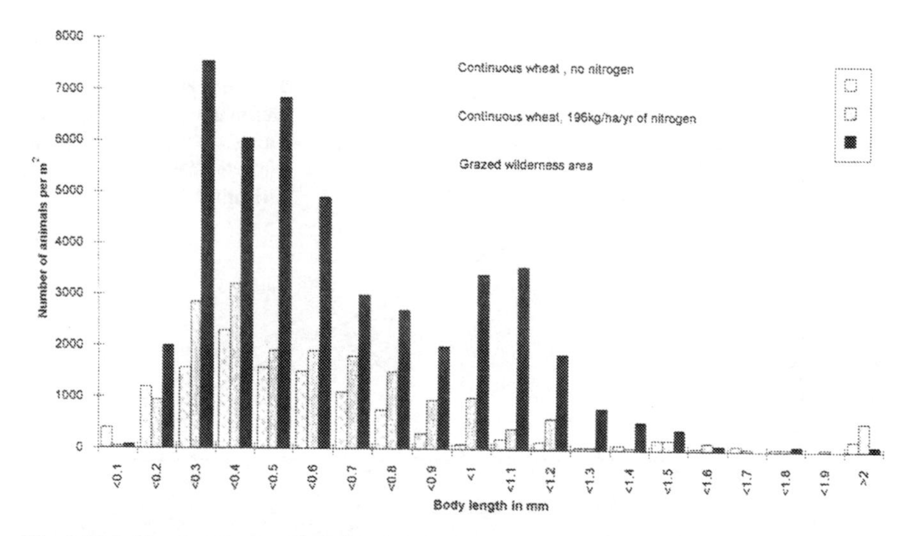

Fig. 2.15.2. The abundance of individuals in body size classes for all arthropods (0-2mm) from soil derived from three different long term farming regimes at Rothamstead. With Permission (Park and Cousins 1995)

Finally, we can ask what relationship the proposed ecosystem goal function has to other goal functions proposed such as minimising entropy loss or maximising exergy.

2.15.8 Negentropy Change in the Food Web

The objective of this section is to illustrate that the proposed ecosystem goal function can be described in thermodynamic terms.

We may consider the operation of a food web over the top predator territory as energy capture via photosynthesis then movement of this energy by the plant, and fungi as well as concentration of energy into large body masses in the animal food chain. In the plant translocation and chemical transformation create locally energy dense states (e.g. seeds as an extreme example; fruiting bodies in fungi).

We may in general consider the operation of the food web and plant growth to create an ecosystem populated by statistically unlikely occurrences of local energy dense states, the organisms (or plant parts) themselves. These locally energy dense states arise within a uniform field of solar energy input to the system. We can calculate the change in entropy which is associated with the aggregation of food into larger particles. Suppose we take a uniform food item and break it into smaller parts and scatter them over an ecosystem. This act produces a change in entropy ΔS where

$$\Delta S = -k \,.\, \ln W \tag{2.15.7}$$

We suppose that the item has been broken into n equal sized particles and scattered over an ecosystem which is itself partitioned into m boxes each the size of the n particles. k is the Boltzmann constant and W is the number of ways of rearranging n particles into m boxes,

$$W = \binom{m}{n} \tag{2.15.8}$$

Thus taking 10^2 particles and scattering them over an ecosystem of 10^6 boxes of the same particle size then,

$$\Delta S = -k\ 10^6! \ / \ [(10^6!-10^2!)\ 10^2!) \tag{2.15.9}$$

$$\Delta S = 1.4 \ .\ 10^{-20}\ \mathrm{J\,K^{-1}} \tag{2.15.10}$$

This can be compared to the energy change due to the oxidation of 1 g glucose

$$\Delta S = 1.4 \times 10^3\ \mathrm{J\,K^{-1}} \tag{2.15.11}$$

However, it is important to contrast the small but calculable change in negentropy caused by the aggregation of animal biomass into larger particles with the inefficiency (and hence thermodynamic irreversibility) of the food web which collect the food together into larger particles.

A reference state of all biomass being distributed evenly in particles of the size of the smallest organism can be compared with the distribution of the same biomass in body sizes according to equation 2.15.6 to provide a measure of entropy change from ecosystem organisation.

2.15.9 Semi-Autonomy and Rules for System Identification

So far we have identified goal functions for the water and ecological systems and from these we have identified the system boundaries defined by the domains of these goals. The remaining issue concerns how these goal directed systems remain sufficiently independent of each other to allow their goals to operate. For ecosystems we have identified the 'whole' as the relative abundance of organisms within a spatially defined food web. Here the components are the individual organisms. The degree to which the organisms are independent of the food web, or are entrained by it, is given by the size of energy store the organism possesses.

Fat stores allow the organism to pursue other activities than just eating. Without those stores every expenditure of energy would have to be balanced by an immediate intake of food to compensate.

A general view of this process is that stores and buffers allow parts to decouple from the whole and therefore allow the parts to possess separate goals. In a hierarchy of contained systems where we focus on ecosystem (as defined above) we now see that it is decoupled from its parts, the organisms, by their energy stores. However a bounded ecosystem is also contained in a larger system which we will call the Earth and its property known as climate. For ecosystems to possess a goal function it is necessary to have some autonomy with respect to climate.

We can now see the importance of soil to the ecosystem goal function. The principle components of climate are rainfall and temperature and these determine the life zones on Earth (Holdridge 1947). Here we have shown that it is through the biological system increasing soil organic matter, that the supply of water to the plant can be moderated. Soil organic matter acts as a water store, decoupling the ecosystem from the climate. This decoupling is not total and acts to smooth out variation and allows the plant to respond to the long term average of temperature and rainfall.

2.15.10 Conclusions

We concluded that systems must have some autonomy in order for them to possess goal directed behaviour. If the system contains parts, then these parts can have their own goals if the parts have some autonomy from the whole. Thus the ecosystem has to be decoupled from the climate of the global system and from its own parts, the organisms. One mechanism to achieve autonomy is storage of the material that links the part and the whole. Fat decouples organisms from the ecosystem food web; the ecosystem food web is decoupled from climate by the possession of water storage in soil organic matter and in mineral soil components.

Although soil is normally treated as a system in its own right, it is shown here to be the interface between two system objects each with their own goal function, given by equations (2.15.4) and (2.15.6), for the water catchment and the food web respectively. The interaction of the two system goal functions is consistent with organisms as 'ecosystem engineers' (Lawton and Jones 1995). The parameters of the soil water movement equation for different parent soils are shown to vary widely and so determine the characteristics of the interface between the two systems.

It is also proposed that system boundaries can be identified by the area of operation of the system goal and that this process can be used also to define objective hierarchies in nature. The use of goal functions may simplify modelling and in this case the existence of hierarchical catchment models may inform energy catchment models of the food web.

The major conclusion of this paper is that soil allows ecosystems to be semi-autonomous and therefore goal directed and described by a goal function. It will be for others to determine how an awareness of these system goals is useful to resource management or helps the expression of human preferences.

Acknowledgements. I thank the editors and organisers of the Salzau meeting for establishing a stimulating workshop which has contributed to the construction of this paper. I would particularly thank participants Yuri Svirezhv and Bernie Patten in discussion and the referees for detailed comments.

References

Allen TFH and Starr TB (1982) Hierarchy: perspectives for ecological complexity. Chicago University Press, Chicago

Cousins SH (1985) The trophic continuum in marine ecosystems: structure and equations for a predictive model. Can Bull fish and Aq Sci 213:76-93

Cousins SH (1990) Countable ecosystems deriving from a new food web entity. Oikos 57:270-275

Cousins SH (1996) Food webs: from the Lindeman paradigm to a taxonomic general theory of ecology. In: Polis GA and Winemiller KO (eds) Food webs: integration of patterns and dynamics. Chapman and Hall, New York

Hall DGM, Reeve MJ, Thomasson AJ and Wright VF (1977) Water retention, porosity and density of field soils. Soil Survey Technical Monograph No. 9. Soil Survey of England and Wales, Harpenden

Holdridge LR (1947) Determination of world plant formulations from simple climatic data. Science 105:367-368

Jenny H (1941) Factors of soil formation. McGraw-Hill, New York

Lawton JH and Jones CG (1995) Linking species and ecosystems: organisms as engineers. In: Jones CG and Lawton JH (eds) Linking species and ecosystems. Chapman & Hall, New York

Lindeman, RL (1942) The trophic dynamic aspect of ecology. Ecology 23:399-418

Mackney D, Hodgson JM, Hollis JM and Staines SJ (1983) Legend for the 1:250,000 soil map of England and Wales. Soil Survey of England and Wales, Harpenden

Martinez ND (1995) Unifying ecological subdisciplines with ecosystem food webs. In: Jones CG and Lawton JH (eds) Linking species and Ecosystems. Chapman and Hall, New York

McNaughton SJ, Oesterheld M, Frank DA and Williams KJ (1989) Ecosystem patterns in primary productivity and herbivory in terrestrial habitats. Nature 341:142-144

Park J and Cousins SH (1995) Soil biological health and agro-ecological change. Ag Ecosys Env 56:137-148

Platt T, Lewis M and Geider R (1984) Thermodynamics of the pelagic ecosystem: Elementary closure conditions for biological production in the open ocean. In: Fasham MJR (ed) Flows of energy and materials in marine ecosystems: theory and practice. Plenum Press, London

Peters RH (1983) The ecological implications of body size. Cambridge Univ Press, Cambridge

Schlesinger WH (1985) Changes in soil carbon and associated properties with disturbance and recovery. In: Traflka JT, Reichle DE (eds) The carbon cycle: a global analysis. Springer-Verlag, New York

Schlesinger WH (1991) Biogeochemistry: an analysis of global change. Academic press, New York

Silvert W and Platt T (1980) Dynamic energy flow model of the particle size distribution in pelagic ecosystems. In: Kerfoot WC (ed) Evolution and ecology of zooplankton communities. University Press of New England, New Hampshire
Simon HA (1973) The organisation of complex systems. In: Pattee HH (ed) Hierarchy theory. Braziller, New York
Tansley AG (1935) The use and abuse of vegetational concepts and terms. Ecology 16:284-307
White RE (1981) Introduction to the principles and practices of soil science. Blackwell Scientific, Oxford

2.16 The Physical Basis of Ecological Goal Functions – An Integrative Discussion

Felix Müller and Brian Fath

The preceding articles have illuminated many different aspects of the scientific orientor and goal function principles. They have analyzed different optimizational approaches, and they have successfully illustrated the evidence of general developmental ecosystem tendencies. Therefore, some important pieces of mosaic, depicting parts of the aspired theoretical pattern have been covered. These elements will be used as starting points for the following summaries and interpretations. In order to gain the necessary overview of the elements, a short review of the papers will be helpful to introduce the subsequent argumentation.

A Short Overview

In the first article of this section, Bossel (Sect. 2.2) provides a general introduction to orientor theory and to the technical principles of multidimensional value orientations. He explains the concept of basic orientors and demonstrates their emergence on the systems level a as result of computer simulation experiments. It is shown that the basic orientor approach allows comprehensive assessments of the fitness, performance, and viability of different types of systems. Following this systems analytical introduction, Bröring and Wiegleb (Sect. 2.3) give a detailed bio-ecological report on some general concepts of succession, which are applied to structural communities as well as functional ecosystemic phenomena. Basing upon terminological discussions the fundamental succession mechanisms are described, the recent state of succession theory is illuminated critically, and some examples are used to stress the corresponding methodological problems. Bröring and Wiegleb state that ecosystem development is accompanied with a high uncertainty, particularly on the community level. They complain that an orderly, predictable sequence of species often is hard to find due to rare events and unknown interrelationships. This statement may be based on scale problems in their quoted case studies, on the observed developmental direction (retrogression) and on the analyzed parameter set. Nevertheless, we have to be aware that a prognosis of the specific abundances of specific species cannot be a general goal of the orientor concept because only in a few cases we will be able to

F. Müller M. Leupelt (Eds.)
Eco Targets, Goal Functions, and Orientors

measure and integrate the necessary aut-ecological information, and because the high complexity of ecological systems causes extraordinary systemic uncertainties (Breckling 1992).

In Sect. 2.4 Jørgensen and Nielsen give an introducing outline of the thermodynamic concept of exergy and its utilization as a goal function in structural dynamic modeling. The authors describe the connection between exergy and evolution, which is documented in the "fourth law of thermodynamics" that is used as a point of departure for the quantification of the information embodied in the genes as an ecosystemic goal function. An application of the exergy concept in benthic communities can be found in the paper of Marques et al. (Sect. 2.5). Here exergy is used to model the change of species compositions as a thermodynamic optimization process: In whatever situation the system operates, the development will always attempt to follow a local maximization of the stored exergy. This approach has been validated by measured community structures. Marques et al. discuss the correlation between exergy, specific exergy and biodiversity, and they stress the potential significance of genetic characteristics for the assessment of ecosystem qualities. In a third thermodynamic statement, Svirezhev (Sect. 2.6) presents a theoretical, fundamental article on the potential applications of thermodynamics to ecosystem theory. He additionally develops an abstract characterization of anthropogenic inputs, which is focused on processes of entropy production in the form of an "entropy pump hypothesis": Only natural ecosystems can reach a dynamic equilibrium with the natural external factors, such as climate or hydrology. Comparing calculated, derived, and estimated natural values of entropy production with measured degrees leads to an evaluation of the environmental stress the observed system has to bear. Besides this applicable approach, Svirezhev discusses the interactions of entropy, exergy, information and stability. In a last paper concerned with thermodynamics, Jørgensen and Nielsen (Sect. 2.7) give an overview on the interrelations between different ecosystem orientors. A number of potential goal functions is described, such as emergy, exergy, ascendency, overhead, ratio of direct vs. indirect effect, and structural exergy. It is shown that these ecosystem indicators are correlated to a very high degree.

Looking at ecosystems from a network theoretical approach, Patten (Sect. 2.8) provides a holistic analysis and derives 20 major ecosystem properties which are presented as directional tendencies, that he denominates developmental, organizational, and evolutionary orientors. As a whole, Patten's ecosystem features form an integrative pattern of network-oriented tendencies in the self-organized development of ecosystems. Starting from a similar point of view, Fath and Patten (Sect. 2.9) analyze the property 'network synergism' that describes the development of holistic positive effects that emerge from the linkage of direct and indirect interactions between the ecosystem agents. They introduce an input-output analysis based methodology that measures the total integral utility of a system. It is shown how this feature emerges in simple systems. Using complex matrix models the utility function is interrelated with other proposed goal functions. A third paper founded on networks is presented in Sect. 2.10. Ulanowicz discusses Popper's notion "propensity" and articulates a theory of ecosystem development which is based upon probabilistic indices, borrowed from information theory. He explains the holistic ecosystem feature ascendancy as a flow-averaged, ecosystem level propensity for activity. Ulanowicz states that ascendency can serve as an orientor which has an observed propensity to increase over time.

In Sect. 2.11 Bass presents an additional approach to ecosystem dynamics. He interprets Holling's Figure-Eight Model (Holling 1986), which seems to be contradicting the goal function principle at first glance, especially due to the concept of an immanent stage of creative destruction. Bass determines two orientations with respect to Holling's model: the optimization of exergy consumption (Schneider and Kay 1994a,b) and exergy storage (Jørgensen 1992), and the emergence of a self-organized critical state near maturation which might be coinciding with a maximized distance from thermodynamic equilibrium. Consequently, an additional focus has to be put on the dynamics of orientors, which develop in the goal function manner for a long while but can break down in phases of the system's self-adaptation.

In the end of this section, four case studies demonstrate relevant concepts and methods of ecosystem analysis to find and evaluate orientors. Furthermore, these studies prove the evidence of developmental trends. Kutsch et al. (Sect. 2.12) present an empirical test of the significance of ecosystem properties referring to Odum's maturity concept on three ecological levels of organization (ecosystems, plant communities, microflora) in 6 terrestrial ecosystems. Baird shows 11 examples of coastal network analyses in Sect. 2.13, demonstrating a methodology to derive the respective systems level properties. Interrelating these features Baird resumes that there will not be one universal property which can be satisfactory used to describe ecosystem states as entities, but a number of them, viewed in relation to each other. A third empirical aspect is presented in Sect. 2.14 where Nielsen et al. are reviewing the current state of structural dynamic models, their applications and techniques. This presentation demonstrates the demand for goal functions, if the development of the community has to be described by models in accordance with ecosystem physiological values. Finally Cousins and Rounsevell (Sect. 2.15) give an example for the integration of soil water and carbon balances. Both subsystems are restricted to specific areas (watershed vs. photonshed), both are bounded in hierarchies of organizations, and both comprise of different orientors on different scales.

Problems of Terminology

Reading through the texts, as a first point of discussion terminological problems become obvious. As Bröring and Wiegleb (Sect. 2.3) state the "recent succession theory is not yet connected properly to the terminology of modern systems theory." Therefore, the vocabulary of structural changes (Gleason's succession) is not adjusted to the modern terminology of ecosystem physiology (Odum's maturation). In this situation it is feasible to follow Patten's pragmatic attitude (Sect. 2.8): He unifies all terms for targets, goal function, and orientors and stresses what these extremum variables and principles have in common: They are "*directional criteria* accounting for systems being pushed (by "forcing functions") or pulled (by "attractors") toward certain configurations, either with or without goals.". As members of this class of system characteristics the three terms "orientor", "goal function", and "attractor" should be clearly defined to facilitate future discussion.

According to Bossel (Sect. 2.2) the term *orientor* is "used to denote normative concepts that direct the behavior and the development of systems in general. In the social context, values and norms, objectives and goals are important orientors." Orientors are aspects, notions, properties, or dimensions of systems. Orientors can be used as criteria to describe and evaluate a system's developmental stage (Bossel 1992, 1994). The corresponding ecological state variables will be called orientors in this text, reflecting the fact that their dynamics seems to be oriented toward certain attractor points.

Goal functions are technical counterparts of orientors in ecological modeling. They are mathematical functions that describe the direction of ecosystem development as a consequence of its self-organizing ability. This capacity enables the system to meet perturbations by directive reactions which can be described by goal functions (Jørgensen and Nielsen in Sect. 2.7). Marques et al. (Sect. 2.5) define goal functions like this: "In ecological models goal functions are assumed to measure given properties or tendencies of ecosystems, emerging as a result of self-organization processes in their development". They are "suitable measures of system oriented characteristics for natural tendencies of ecosystem development, and good ecological quality indicators." In order to stress the teleological restrictions that are interrelated with this notion, Jørgensen and Nielsen write in Sect. 2.4 that "the term goal function should solely be applied in the modeling context, while the term ecological indicator is more appropriate to discuss the propensities that characterize the development of ecosystems." In Sect. 2.11 Bass introduces a third concept of directing developmental features. He defines attractors as points in the state space where the system is drawn to. This directiveness in the trajectory of an orientor is especially evident after a perturbation. Attractors are points of stable equilibrium in the state space.

Neither orientors nor goal functions or attractors are attained on the base of special purposes in non-human systems. Therefore, all three terms should be used without any teleological background (see Barkmann et al. Sect. 3.1). Much more, the respective developments are observable consequences of self-organization processes, reflecting the regular change of systems that are moving away from the thermodynamic equilibrium state. In this context the term "*propensity*" which has been introduced into the developmental debate by Ulanowicz (Sect. 2.10) characterizes the uncertainties which are connected with ecological dynamics: "The system is not driven toward a certain endpoint. Rather, the probabilistic orientors guide the system in a vague direction" (Ulanowicz, Sect. 2.10). The orientation propensity describes the tendency of a certain event to occur under given circumstances, whereby the dynamics of the surroundings (e.g. the site conditions, the history of the investigated case, or the abundant species) have to be included as well as the respective probabilities. Propensities are conditional probabilities.

Orientors and Self-Organization

In many of the papers it has been postulated that the dynamics of orientors is the result of dissipative self-organizing processes, which generally can be characterized by the transformation of microscopic disorder into macroscopically ordered structures (Müller et al. 1997). For such dissipative self-organized systems some general prerequisites and features have been described (Haken 1983; Kay 1984; Ebeling 1989; Schneider and Kay 1994a; Müller 1996a) which have to be considered with regard to the problems of orientor theory. These characteristics are:

1. Openness of the system: The dissipative system must be enabled to exchange energy, matter and information with its environment. For ecological systems this means that inputs as well as outputs of energy and nutrients are basic suppositions for any self-organized development.
2. Import of convertible energy (exergy) in a degradable quantity: The dissipative developmental processes are energy-pumped. They depend on the availability of a "high quality" energy form which can be converted into mechanical work. The quantity of this energy must fall within a window between the minimum supply, necessary for the start of the dissipative process and a maximum input which starts to turn the achieved order into turbulent structures.
3. Internal energy transformations (exergy degradation): The imported exergy is converted in dissipative processes. Throughout these reactions which are energetically connected to ramified reaction chains, the initial exergy is degraded in several succeeding steps. The number of degrading reactions and the connectivity of the relating flows are functions of the state and degree of the self-organized development.
4. Export of non-convertible, non-usable energy (entropy): As one result of the exergy degradation, non-usable energy forms (entropy) are produced. They have to be exported to the environment of the system. The ecological entropy exporting processes are heat transfers, matter transfers such as respiration, transpiration, and losses of ecosystem constituents. The degraded products are reconducted to the next higher hierarchical level in the system's environment.

5. Distance from thermodynamic equilibrium: The self-organized processes create internal gradients in the system. With these gradients, spatio-temporal structures and heterogeneity arise which can be passed over to functional hierarchies. The respective degree of heterogeneity and structurization can be taken as an indicator for the distance to homogeneous distribution patterns or to the state of thermodynamic equilibrium.
6. Nonlinear interactions between the elements of the system: Changes of one state variable in self-organized systems can lead to nonlinear variations of other parameters. As a consequence, different stationary states can be achieved. Thus, the actual state is always dependent on the system's history. This means that all linear descriptions of ecological interaction patterns are extremely small- and short-scaled sections from a totally nonlinear reality. As a regrettable consequence, all predictions that refer to the development of ecosystems are linked with a high "uncertainty of the detail".
7. Strong amplifications in phase transitions: Caused by the nonlinear interactions, strong and destabilizing amplitudes of state variables can appear when the system's trajectory reaches states in the proximity of structural modifications.
8. Internal structural processes: The self-organizing processes develop in a self-referential, self-regulated manner. There is no external regulation which exceeds the limitations of the single processes' degrees of freedom, that arise from hierarchical constraints.
9. Constraints in hierarchies: The coordination in self-organized systems is based on the interactions of partial processes that operate on different scales. During phases near steady states, the interrelationships are dominated by slow processes that operate within large spatial extents. The potential state space of smaller and more rapid entities is enclosed by these constraints.
10. (Meta)stability after small impulses: Disturbances can be buffered by self-organized systems within certain limitations.
11. Historicity and irreversibility: Self-organization is based upon irreversible processes. It can be understood only if the history of the system is known.

All these items are strongly linked with the concept of *ecological gradients*. The thermodynamic non-equilibrium principle (Schneider and Kay 1994a,b) hypothesizes that self-organizing ecological systems build up internal, functional gradients as a consequence of the strong external exergy gradient which is provided by the solar radiation. In the long term, the imported (captured) exergy is transformed into a hierarchical systems of interconnected gradients which imply all structural features that characterize the system's distance from the thermodynamic equilibrium. Thus, all structural ecosystem properties can be comprehended as concentration gradients in space and time. They build up the potentials to carry out mechanical work, chemical reactions or biological interactions. Ecosystem function can therefore be defined as the general characteristic of the system's gradient dynamics. The above listed features are valid in all dissipative systems, but living entities are fitted out with some very special features that are suitable to amplify the self-organized process sequences (e.g. reactivity, processing of information, genetic variability, self-replication, ability for evolution, growth, development, hierarchical organization). Therefore, these systems proceed through orientor changes with special intensities. Herewith, the above listed general features must be comprehended as a framework in which all single orientors have to be arranged.

A Schedule of Ecosystem Orientors

The contributions of Chap. 2 propose a variety of state variables and parameters which can be used as orientors, goal functions, or indicators that represent the effects of the above described features of self-organizing processes on the ecosystem level. The respective properties are summarized and structured in Figures 2.16.1 to 2.16.3, each reflecting different focal points of view. Figure 2.16.1 characterizes the *thermodynamic approach* to ecosystem dynamics. In general this access describes the energetic demands of a self-organized ecological development that leads from homogeneous patterns to strongly heterogeneous, gradient dominated systems (Jørgensen 1992, 1997, Müller and Nielsen 1996). In a first group of parameters the energetic feature exergy is described by different criteria, basing on the papers of Jørgensen et al. (Sect. 2.4 and 2.7), Marques et al. (Sect. 2.5), Svirezhev (Sect. 2.6), and Bass (Sect. 2.11). Exergy is defined as the amount of work that a system can perform when it is brought into chemical equilibrium with its environment. It is the convertible, usable energy portion contained in a system with respect to its environment. In ecological systems the solar radiation functions as the primary exergy source. The imported exergy is partially transformed into biochemical energy, and this is transferred and degraded in the numerous ecological reaction chains. The respective hypotheses pronounce that during their maturation ecosystems optimize their ability to capture exergy (Schneider and Kay 1994a,b). Thus, the photosynthetic assimilation capacity is optimized in relation to the site conditions. Along with this development the exergy consumption increases, and the exergy flows through the system rise, in their total quantity (the total energy passed through the system) and their complexity (the number and heterogeneity of individual flows). Throughout these energy transfers the initially high qualitative energy which was convertible into many other energy forms at the time of its import, is degraded. The convertibility is reduced and high energy fractions are lost, e.g. through heat and radiation exports, respiration losses, or transpiration processes.

The more elements are participating in the exergy transfer networks, the higher is the demand for their maintenance, and the more exergy has to be degraded. If the system receives and transforms a surplus exergy input that exceeds the demands for its maintenance, it is enabled to invest the additional energy into further growth (which can be indicated by an increase of biomass or the total system throughput of energy and matter) or further structurization (which can be perceived by a rise of the structural gradients and the biocoenotical complexity). In both cases some fractions of the imported exergy are stored as a convertible reserve, either in biomass, organic matter, or in complex structures. The appropriate exergetic investments into structures can be derived on the base of information theory, because the respective energy fraction is transformed into information (Jørgensen and Nielsen, Sect. 2.4). A reference number for this energy component is the specific exergy (or structural exergy), which is the exergy per unit of biomass. The specific exergy indicates the information embedded in the biomass.

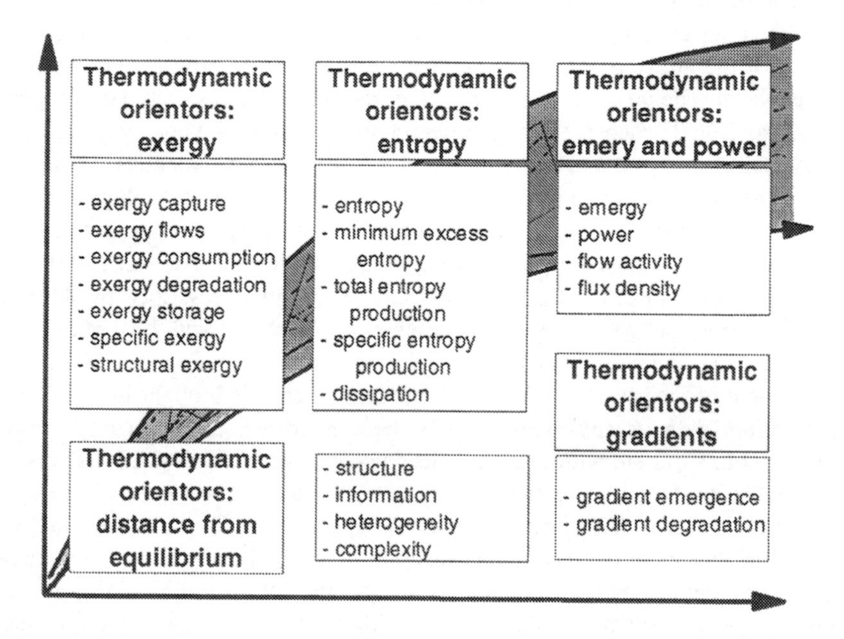

Fig. 2.16.1. A list of proposed thermodynamic orientors. For a better understanding of the basic Figure, see Fig. 2.1.1 in Sect. 2.1.

Exergy can be used to describe the "constructive part" of the developing systems' energy budgets. On the contrary, entropy is an indicator for the cost side of the exergetic structure. To build up the complex pattern, and to enhance the distance from thermodynamic equilibrium, a high amount of energy gets lost via respiration and heat transfer. And also for the maintenance of the complex structure, a significant energy fraction has to be degraded in order to support the complex life processes. The sum of these degraded energy portions is the system's entropy production. Therefore, an optimization of exergy degradation, gradient dissipation or exergy consumption must lead to local optima of entropy production. . .

As ecosystems often become more and more diverse and as the ecosystem space becomes more and more filled with active biological units, the total entropy production will rise in correlation with the complexity of the system. Nevertheless, adaptational processes may lead to ever better adjustments of directly linked processes or elements. This will be correlated with reduced energy and matter losses. Thus, the specific entropy production, which is restricted to one process or one transfer reaction, will decrease, although – as a function of the system's complexity – the total amount of entropy production rises throughout the development. As Svirezhev shows in Sect. 2.6, this energetic adjustment is restricted to naturally evolving ecosystems. Anthropogenic inputs lead to decreasing efficiencies, and therefore environmental degradation, dissipation, and the entropy production increases as a consequence of human disturbances.

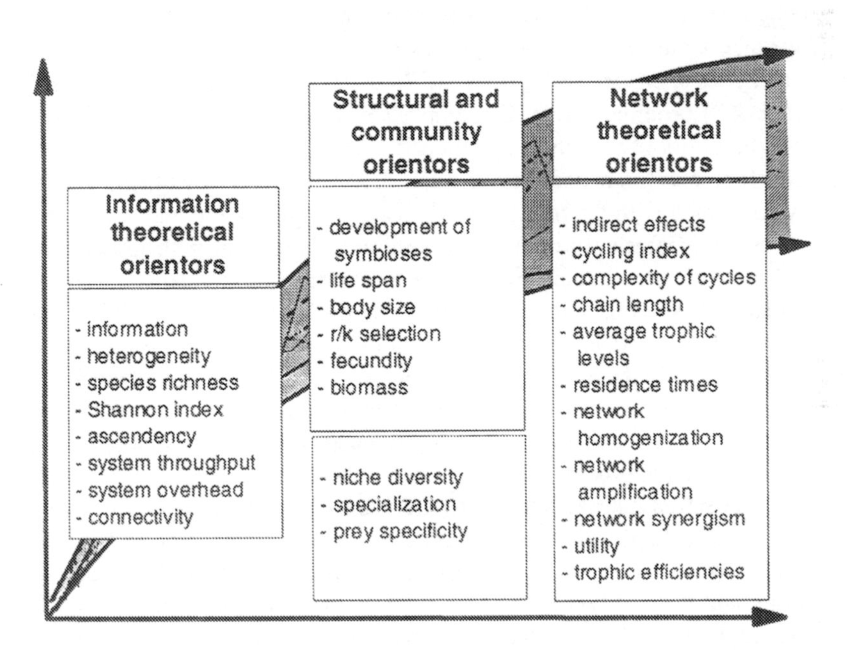

Fig. 2.16.2. A list of proposed information theoretical, structural and network theoretical orientors

Similar thermodynamic ecosystem features have been developed by Lotka (1925) and Odum (1983). They are based on the increase of flux densities and activities throughout natural ecosystem developments, postulating a maximum of flows of usable energy in mature ecosystems (maximum power principle) as it has been described above. Furthermore, Odum (1983) postulates the emergy-maximization principle (see Jørgensen and Nielsen Sect. 2.7): Ecosystems tend toward states where the "energy quality", represented by solar energy equivalents that are incorporated in the organisms, is optimized. Following these hypotheses the increasing complexity of organismic flows and the growing total amount of materials and energy flowing in the system are subjects of extremum principles.

Many more orientors are tightly connected with the thermodynamic variables. As Jørgensen and Nielsen (Sect. 2.4) and Svirezhev (Sect. 2.6) show, e.g., there are very strict mathematical connections between thermodynamics and *information* theory. Both approaches can be used to describe the heterogeneity of a system, which represents the distance from thermodynamic ground and which can be observed on the base of the complexity of the respective systems of gradients. Besides the solely structural variables, Ulanowicz's system indicator ascendency and its corresponding characteristics (see Sect. 2.10), refer to the variability of flows as well as the total amount of energy or matter that flows through the biological networks. Ascendency therefore is a holistic ecosystem characteristic, reflecting functional qualities as well as structural features.

If we take a closer look at the systems of flows and storages, a multitude of evolving *network properties*, such as the 20 major ecosystem features from Patten (Sect. 2.8) become visible. One important result of network analysis is the fact that the utility of the whole system arises throughout an undisturbed ecosystem development (Fath and Patten Sect. 2.9). Network synergism, which is strictly connected with the principles of network amplification and homogenization, with the increasing dominance of indirect effects and with growing residence times, offers an interesting relationship to modern, mutualistic aspects of evolutionary processes, such as the concept of Weber et al. (1989). These authors take the ecosystem level as an evolutionary unit, and they propose that new organisms will be integrated into this system, if they help reducing the losses of the whole and if they take part in an amplification of system-internal flows. Otherwise an extension of the new organism will not take place. Thus this mutualistic selection process, which can only take place in phases of dominating biological selection mechanisms, when there are no dominating physical events, leads to the structural and thermodynamic orientation we have described on the preceding pages. As consequences from this network theoretical concept, specialization and niche diversity must increase in succession dynamics until a site specific, maximal value is attained. In parallel, the systems will accumulate biomass, symbioses will become more and more significant, and the life spans of the organisms can be prolonged. r-strategists will be relieved by k-strategists (Kutsch et al. Sect. 2.12), and a much broader system of niches will be available.

These structural changes are accompanied by interesting *functional modifications* of the systems (see Figure 2.16.3). The ecophysiological features of this figure can be easily attributed to the thermodynamic variables. The reduction of loss (of scarce resources) refers to the relative minimization of entropy flows, while the storage and biomass optimization can be seen in connection with Jørgensen's exergy storage theory. Transpiration and respiration are efficient mechanisms to export the excess entropy, thus, they are principle subprocesses of gradient degradation. Analyzing the succeeding systems from an abstract systems-analytical point-of-view will lead to the features that have also been mentioned in the beginning: The complexity increases, the number of hierarchical levels rises, the autonomy of the systems gets higher, the systems attain a holistic determination with a maximum distributed control, the coevolutionary processes are enhanced, and as a consequence of the increasing numbers of species the redundancy is increased.

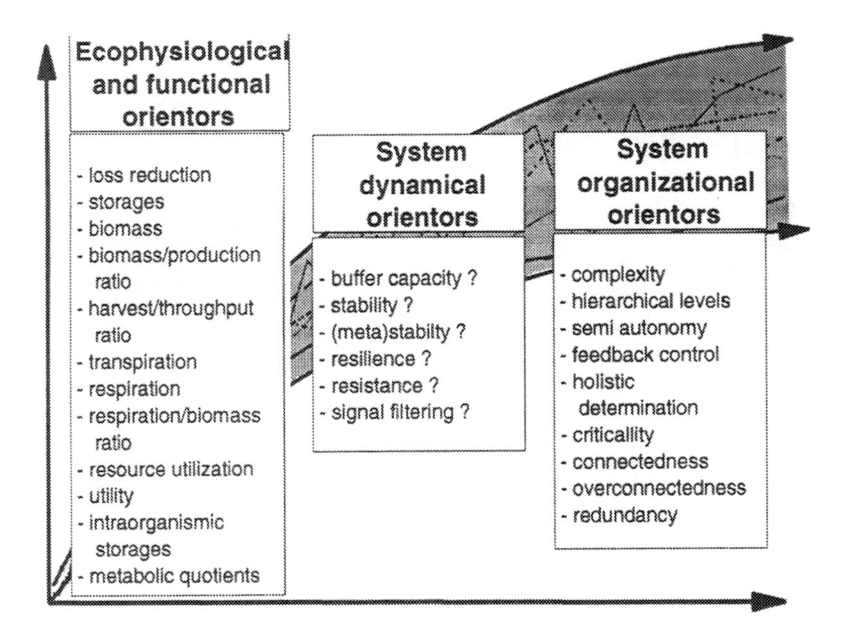

Fig. 2.16.3. A list of functional, system dynamical and organizational orientors. The proposed stability attributes are indicated by question marks because the significance of their optimization throughout succession and maturation is debatable.

Orientor Dynamics in Long-Term Ecosystem Development

On the other hand, in a climax stage of perfect maturity, the degree of the self-organized criticality is getting higher, too. The captured exergy is totally utilized for the maintenance of the existing structures. Thus, the energetic reserves for further structural development decrease. The adaptability of mature systems, that have been structured for long times in similar external conditions may be low, and the high connectivity may turn into overconnectedness (Holling 1986). Therefore, as Bass describes in Sect. 2.11, the stability characteristics may not follow the orientor functions of the other parameters. Schneider and Kay (1994a,b) support this concept. They have formulated the hypothesis that the tendency to return to thermodynamic equilibrium is the higher, the further a system has been moved away from thermodynamic equilibrium. Holling (1986) also points out that the longer the destructive shift has been delayed the greater the destruction will be; the longer the system remains in the critical (mature) state, the larger will be the consequences of major disturbances. The stored exergy in this case of creative destruction serves as fuel for an extensive renewal in a restarting exploitation phase. Therefore, a high maturity is always connected with a high risk.

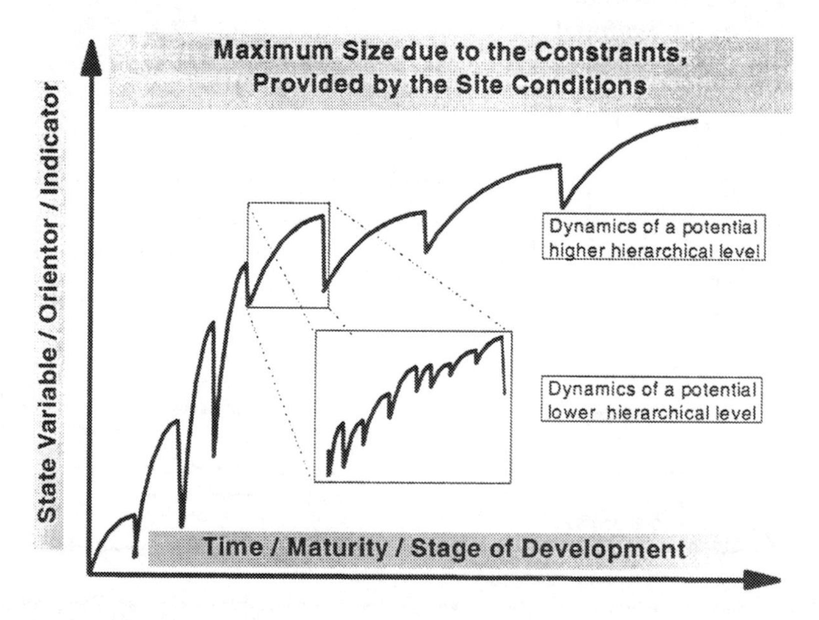

Fig. 2.16.4. Sketch of a nested hierarchical organization of orientors which are developing on different temporal scales. The resulting curves consist of many sequential "Holling cycles" on the respective lower levels. Thus there is self-similar behavior on different levels-of-organization such as ecosystem dynamics, ecosystem succession, or ecosystem evolution.

As Bass (Sect. 2.11) stresses these temporal characteristics can be found on many different scales. There is a sequence of dynamics which follows the general principles of hierarchy theory (Allen and Starr 1982; O'Neill et al. 1986; Müller 1992). Bass summarizes these aspects: "An ecosystem is comprised of several smaller attractors, perhaps semistable attractors, of thermodynamic risk, being destabilized at different times. Although a smaller attractor may be destabilized, it is subsumed by a larger attractor which remains stable. Essentially, maintaining the system at the conservation stage changes this larger attractor into a semistable attractor, but the system cannot remain at this attractor indefinitely, thus the damage occurs on a much larger scale." If that disturbance occurs, the regulating steady state based hierarchy of signal transfers is broken, and a new orienting phase can start. The orienting functions are actualized and newly defined. If we subsume this hierarchical aspect, and if we comprehend emergence as a consequence of self-organization (Müller 1996a; Müller et al. 1997), it becomes obvious that the state variables that have been discussed before, are functioning as emergent properties property (if they are based on processual interactions between the parts of the system) or collective properties:collective (if the sum of the parts is equal to the quantity of the whole) which are suitable to partially describe the system's behavior. Therefore, the system itself constructs a regulating, autocatalytic hierachical level. Following hierarchy theory, this step must lead to the emergence of constraints which lead system dynamics and which determine the respective orientors. The degrees of freedom are restricted self-referentially when a new self-organized level of control is built up.

Problems of Orientor Aggregations

The remaining question is: "What will be the correct indicator set of ecosystem behavior on what level-of-description?" In Figures 2.16.1 to 2.16.3 many different potential orientors, goal functions, and indicators are listed. In spite of the fact that some authors believe that they have found the one parameter which can be used as the only holistic ecosystem property, we will stick to the idea of a multidimensional systems characterization. Of course many different arrangements of ecosystem properties are possible. E.g. the most basic structure can be taken from the features and prerequisites of self-organization that have been summarized in the first half of this paper. Although, the best answer to the question for an optimal indicator selection will be the (re)-question "What are the specific indicanda?", three approaches of structuring ecological orientors are demonstrated in the following Sections. At first we refer to Schneider and Kay (1994a) who take a thermodynamic, gradient oriented position and who propose the following ecosystem properties as fundamental orientors to describe the degree of organization and maturity in ecological systems:

1. Exergy capture
2. Energy flow activity
3. Cycling of energy and materials
4. Average trophic structure structure
5. Respiration and transpiration
6. Biomass
7. Types of organisms

Similar high-level orientor groups have been proposed by Müller (1996b) as aggregated indications for the functionality of ecosystems. They include properties such as

1. Exergy capture

2. Energetic and material flow densities in the system
3. Cycling of energy and matter
4. Storage capacities
5. Matter retention (Loss reduction)
6. Respiration and transpiration
7. Diversity and heterogeneity
8. Organization and hierarchy (signal filtering and buffering capacity)

The third approach refers to Bossel's basic orientors. Following this access, the viability of systems is strongly based on the properties of the system itself and the features of its environment. In viable systems, structures and functions of the systems therefore have to be adapted to the predominant environmental conditions. Bossel (Sect. 2.2) emphasizes 6 classes of environmental challenges which effect the most fundamental conditions of the system's behavior (normal environmental state, scarce resources, variety, variability, change, and other systems). Following the orientor theory . (Bossel 1992, 1994), the corresponding systems have to develop adequate responses to the dominating environmental properties. These fundamental criteria are denoted as 'basic orientors'. They are identical for all complex adaptive systems. The basic orientors are existence, effectiveness, freedom of action, security, adaptability, and coexistence.

Table 2.16.1 Basic orientors and some exemplary corresponding ecosystem properties

Environmental challenges	Basic orientors	Ecosystem properties
Normal environmental state	Existence	(Meta)Stability
		Resilience
Scarce resources	Effectiveness	Cycling
		Loss reduction
Variety	Freedom of action	Heterogeneity
		Diversity
Variability	Security	Redundancy
		Storage
Change	Adaptability	Genetic diversity
		Patch dynamics
Other systems	Coexistence	Landscape gradients
		Ecotone structures

The construction of Table 2.16.1, which is situated on rather a high level of systems performance evaluation, is a much more difficult task than this may appear at first glance. The reason is simple: While the ecosystem theoretical orientors have been selected, attempting to remain on an ontological, descriptive level, Bossel's orientors are arranged on the base of a specific axiological purpose. They have been defined to evaluate the fitness of systems which are situated in an anthropogenic environment, where the conscious setting and following of goals is a central part of systems management (see Chap. 1). The basic orientors are teleological entities and they are allowed to be. Thus, this approach represents a good starting point to discuss the communication between ecological orientors and human goals (Chap. 4) and to struggle about teleological aspects from the point of view of philosophy of science (Chap. 3).

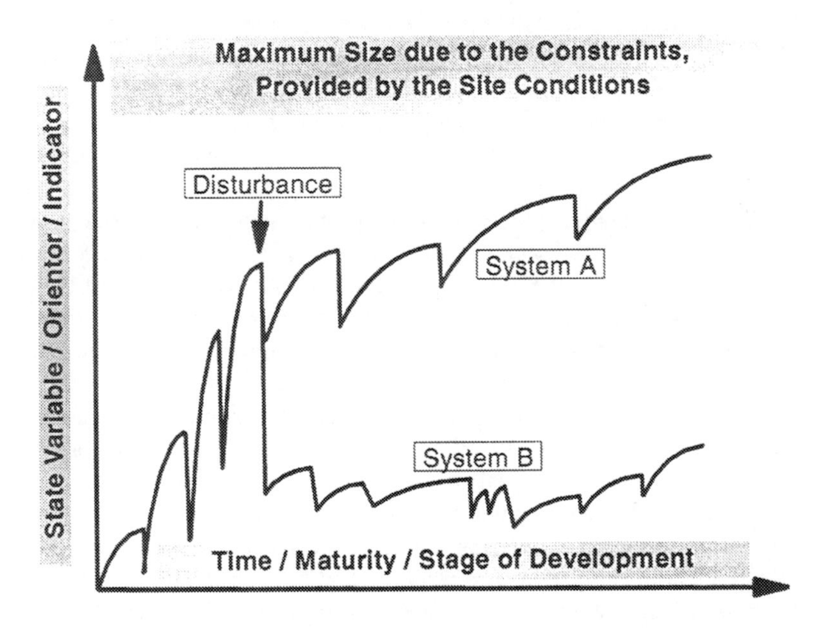

Fig. 2.16.5. Two hypothetical developments of an orientor throughout systems dynamics after a perturbation. While the "healthy" ecosystem (System A) is able to develop into the maturing direction after a perturbation, the stressed system (B) does not follow the natural developmental trend.

Conclusions

These last points will be discussed in the following sections. But before we have to come back to the initial questions and propose some answers as conclusions from this second section:

In fact, during the development of ecosystems there are properties which are regularly optimized. The examples, case studies, and derivations could not provide evident hints that the orientor principle is a non-valid scientific concept. Although some doubts were discussed in context of succession and the prognosis of species abundances the generality of the optimizing principle has become obvious. The case studies show that the orientors can be used to characterize, classify and distinguish different ecosystems. Their quantitative differences are high enough to see the differences between even directly neighboring ecosystems, and thus a group of orientors in fact can form a suitable tool for a holistic environmental indication. Ecosystem science and theory provide the necessary information to draw a clear connecting line between the indicators and the respective indicanda. As these are holistic issues, both ecological object classes, ecosystem functions and ecosystem structures are described. It is also evident that many of the proposed properties are complementary items. They differ due to the specific aspect used as the ecosystems are observed. And as the fundamental processes are those of biological self-organization processes, we must not wonder that similar results are found when different orientors are investigated. Thus, in fact, the orientors describe a potential for self-organization. They refer to developmental processes, ecological dynamics, and change. Therefore, the temporal parameter modifications should be taken into account —as a fundamental for any evaluation procedure— with much more significance than the final state that might represent a brittle configuration at the edge of a creative destruction. Furthermore, there is no question, that these concepts must be introduced into modern strategies of ecosystem protection. And, of course, ecological orientors are a good basis for finding usable indicators for ecosystem health, ecological integrity and for the ecological aspects of sustainability.

Besides these positive notions, a number of questions have been discussed that are not yet answered satisfactory: What are the prerequisites from philosophy of science for the utilization of goal functions and orientors? How can ecological concepts be applied in environmental management? How can we derive ecosystem indicators form orientors? How many and what indicators do we need to describe the whole ecosystem satisfactory? How can these indicators be aggregated and on which scale do they operate? These points will be objects of intensive future research, which is necessary to answer the important questions that will be asked in the following paragraphs.

References

Allen THF and Starr TB (1982) Hierarchy - Perspectives for ecological complexity. The University of Chicago Press, Chicago

Bossel H (1992) Real-structure process descriptions as the basis of understanding ecosystems and their developments. Ecological Modelling 63:261-276

Bossel H (1994) Modelling and simulation. AK Peters, Wellesley MA and Vieweg, Wiesbaden

Breckling B (1992) Uniqueness of ecosystems versus generalizability in ecology. Ecological Modelling 63:13-28

Ebeling W (1989) Chaos - Ordnung - Information. Lizenz (-ausg.), Frankfurt/Main

Haken H (1983) Synergetics, an introduction. Springer-Verlag, Berlin Heidelberg New York

Holling CS (1986) The resilience of terrestrial ecosystems: Local surprise and global change. In: Clark WM and Munn RE (eds) Sustainable development of the bioshpere. Oxford, pp 292-320

Jørgensen SE (1992) Integration of ecosystem theories: A pattern. Kluwer Academic Press, Dortrecht Boston London

Jørgensen SE (1997) Thermodynamik offener Systeme. In: Fränzle O, Müller F and Schröder W (eds) Handbuch der Ökosystemforschung, Sect. III-1.6, ecomed, Landsberg

Kay JJ (1983) Self-organization and the thermodynamics of living systems: A paradigm. Ph.D. Thesis, University of Waterloo, Canada

Müller F (1992) Hierarchical approaches to ecosystem theory. Ecolog Modelling 63:215-242

Müller F (1996a) Emergent properties of ecosystems - consequences of self-organizing processes? Senckenbergiana maritima 27(6):151-168

Müller F (1996b) Ableitung von integrativen Indikatoren zur Bewertung von Ökosystem-Zuständen für die Umweltökonomischen Gesamtrechnungen. Studie für das Statistische Bundesamt, Wiesbaden

Müller F, Breckling B, Bredemeier M, Grimm V, Malchow H, Nielsen SN and Reiche E-W (1997) Ökosystemare Selbstorganisation. In: Fränzle O, Müller F and Schröder W (eds) Handbuch der Ökosystemforschung, Sect. III-2.4, ecomed, Landsberg

Müller F and Nielsen SN (1996) Thermodynamische Systemauffassungen in der Ökologie. In: Mathes K, Breckling B and Ekschmitt K (eds) Systemtheorie in der Ökologie. ecomed, Landsberg, pp 45-62

O'Neill RV, De Angelis DL, Waide JB and Allen THF (1986) A hierarchical concept of ecosystems. Monographs in Population Ecology 23, Princeton University Press, Princeton

Schneider ED and Kay JJ (1994a) Life as a manifestation of the second law of thermodynamics. Math Comput Modelling 19:25-48

Schneider ED and Kay JJ (1994b) Complexity and thermodynamics: Towards a new ecology. Futures 26:626-647

Weber BH, Depew DJ, Dyke C, Salthe SN, Schneider ED, Ulanowicz RE and Wicken JS (1989) Evolution in thermodynamic perspective: An ecological approach. Biology and Philosophy 4:373-405

Chapter 3

The Philosophical Basis: Aspects from Evolution Theory and Philosophy of Science

3.1 Introduction: Philosophical Aspects of Goal Functions

Jan Barkmann, Broder Breckling, Thomas Potthast and Jens Badura

Introduction

Questions of directionality in nature or about goals in the developmental processes of natural entities have been raised by philosophers of nature since the origin of philosophical reasoning, mainly under the flagword of teleology. Natural science has played an important role in addressing the question from its own perspective. In this century, ecology, evolutionary biology, and systems theories have shaped the general picture of what could be termed trends, directions, or goals within any spatio-temporal change of ecological units. However, there is a strong reciprocal interaction between the philosophy of nature, the philosophy of science, epistemology, natural sciences, and our common phenomenological experience of change in nature. This leads to a multitude of speculations about the underlying reasons and mechanisms.

Few people would deny that they perceive temporal trends, directed development, or some kind of adaptive evolution in living nature. In spite of its seemingly ubiquitous character, the idea of directionality in nature has shown to belong to the most recalcitrant scientific problems. Many ecologists as well as philosophers of science press for a rigorous analysis of the conceptual preconditions and the paradigmatic limitations of any new concept, e.g. the goal function concept, which is set out to tackle aspects of directionality in ecology.

In this chapter, some of the basic questions and underlying assumptions of goal functions are discussed. While the second chapter of this book asks "What are goal functions and what is their physical basis?", this introduction and the following papers focus on the relation of the goal function concept to epistemological considerations and social constraints, both of which are traditionally located in the realm of philosophy of science. The philosophical implications of a goal function concept extend beyond its relation to teleological thinking. The analysis has to employ the philosophical disciplines of ontology (to clarify the answers to "What are goal functions?"), epistemology ("What can we know of goal functions?"), the history and sociology of science ("To which extent and by which means is the construction of the goal function concept subject to historical

F. Müller M. Leupelt (Eds.)
Eco Targets, Goal Functions, and Orientors

constraints or social bias?"), and ethics ("If possible at all, in which way should goal functions guide human action?").

After clarifying the conceptual sources and the implications of teleology, this introduction analyzes the epistemological status of goal functions. Problems with an unsettled epistemological status are not unique to the goal function concept. However, they require attention especially in the field of ecology where the validity of derived hypotheses is not easily testable (questions of empirical evidence for patterns and processes of directed change are not the topic of this chapter, they are discussed by Bröring and Wiegleb (Sect. 2.3). Similarly, the relation of ecological concepts to the ideological framework at the time of their invention has to be questioned with utmost rigor. When it is sought to identify goal functions that operate in social and in ecological systems alike, special attention has to be paid to trace and to analyze basic theoretical presumptions and ideological substructures. Whether such analogies or homologies exist or not, the goal function concept may transport some hidden prescriptive meaning instead of supplying a pure description that is appropriate to capture the specificity of the course of social or ecological developments.

Goal Functions and Aristotelian Teleology

Contemporary versions of the idea of directionality in nature can be traced back to their origins in early Greek philosophy. Directed movement is one of the traditional topoi of classical western philosophy extensively discussed in Aristotelian physics and metaphysics. Critics of the goal function approach argue that talking about 'goal' functions implicitly attributes intentions into natural processes, which spring from Aristotle's teleological account of nature: The goal function concept deals with the difference between a given state of a system and an optimized (or maximized) state. "Optimal states" of biological systems are often identified by reference to theoretically devised measures of the capacity of the system to self-organize (Müller et al. 1997). Thus the notion of an internal goal or aim ecological systems tend to move towards resembles the contested general idea of an internal directionality in nature.

In traditional Aristotelian ontology, presented in his "books of physics" (Aristoteles 1987), a predetermined state (goal, *telos*) is thought to cause any object, regardless of being living or non-living, to move towards its adequate state by final causation (*causa finalis*). This final cause is complemented by an efficient cause (*causa efficiens*), which is very similar to the forcing factors of modern causation concepts. Both forms of causation are simultaneously acting, and so closely intertwined that they can be regarded as expressions of *one* formal cause (*causa formalis*). Together they drive a system towards its predetermined state, the *telos*, which is thought to reside in the object itself. The material substrate of the transformations of a system or an object is attributed to its own causation – material causation (*causa materialis*) in the Aristotelian account. But matter itself is merely regarded as "stuff" (*hyle*), which is subjected to the formal (and forming) causes, and thus subsisting in a latent mode. Matter represents a *potentiality* to be formed. It is brought into existence as an actuality – that means "formed" to an object or system – by the joint operation of final and efficient cause. The formal causes also determine the evolution of the resulting entities. In short, formal causes transfer potential objects to actual objects and are responsible for the movement of the resulting objects towards their destination. Aristotle calls the all-important organizing activity of formal causation *entelechy*. He epitomizes the entelechetic principle by the directed and purposeful development of organisms (ontogeny): Organisms possess a decisive principle in themselves that organize their development, which is their *telos*.

The Aristotelian example of ontogenic development might have some appeal today by re-interpreting developmental expression of genetic *information* as the action of *formal self-organizing* causes (cf. Kauffman 1993). However, any idea proposing a purposeful or, even more problematic, morally good *telos* in nature, must firmly be rejected in the realm of natural science for methodological and epistemological reasons. At the heart of the perpetuating question is the status of final causation and the related teleological explanation for apparently directed development. Contemporary research on classic ontology stresses, however, that within ancient Greek philosophy, teleology and finality did not carry the connotation of "higher" intentions and divine determination or planning (Gloy 1995). These notions are additions of the Christian reception of Aristotle in the Middle Ages.

With the propagation of the mechanistic world view during the scientific revolution of the 18th and 19th century, the Aristotelian account had finally been discarded for scientific use. The idea of a purposeful goal (*telos*), inherent to a system and determining its changes, is inaccessible to scientific methods. Moreover, the perspective of Aristotle's notion of teleology is in principle incommensurable to empirical questions of modern ecology as a natural science. The paradigmatic background (Kuhn 1988) of modern science cannot be equated with Aristotelian 'biology', which is part of his "physics", the latter including his view of living as well as non-living objects (Engels 1982).

Even bearing in mind that ancient Greek philosophers of nature did not speak of "optimization" in any contemporary sense of the word, some ecologists seem to draw more or less hidden, and mostly unreflected analogies. These analogies between the Aristotelian account of movement and development on one side, and goal-directed concepts for biological development on the other are close enough to raise concern about a relapse to pre-scientific, teleological thinking. A goal function, e.g., can be used in ecological modeling as a forcing function for directed system processes and development. As many biological developments are thought of as self-organizing activities, the "goal" resides "in" the system itself. This way, it seems to provide a veritable "*telos*" for the system. If the model is regarded as a hypothesis how "real world" ecosystems work, teleological thinking re-enters our description of nature.

Neither ecosystems nor evolutionary processes possess goals in themselves. Scientifically, we can only observe the occurrence of certain developmental tendencies or trends. These are, however, often occluded because contradicting evolutionary or successional trends act simultaneously within one single system. Adaptive evolution or regular succession would not be possible, though, if there were no long term system constraints leading to certain (thermodynamically preferred) tendencies in ecosystem development and phylogeny. In the perspective of ecosystem ecology, this seemingly directional system behavior has its ultimate cause in the thermodynamics of open systems.

However, evolutionary trends can be interpreted in retrospective only. In this *post hoc* view it appears to an observer *as if* a certain development follows a particular goal. A strict distinction between process and this "as-if goal" is not useful because evolution is self-organizing, and the direction of further development changes together with the state of the process relative to the evolutionary flux in the environment. Thus, it makes sense to talk about "goals" in natural systems at best in a metaphorical sense. Otherwise the term may imply intentional structure within ecological systems. Nature in the scientific perspective does not and cannot possess intended goals, which could be derived from the observation of biogeochemical or thermodynamic processes. Therefore, the usage of the word should be restricted to well-defined circumstances to avoid teleological misunderstandings.

Epistemology

It is one of the central questions raised over and over in the course of the history of ecology if ecosystems exist as 'real' units in nature or not (Trepl 1988; Golley 1992). Opposing epistemological viewpoints concerning the status of ecosystems can be demarcated by the notion of realists and constructivists. Taking either one or the other of these positions also implies differing attitudes toward the idea of goal functions. The outline of epistemological questions presented here is a simplification to help sketching the problem. In the following section we are not addressing ontological questions about how nature *as such* really is (the famous Kantian "Ding an sich"). Realist and constructivist views are discussed *only* within the epistemological framework of natural science. This means that both positions are grounded in a broader sense of *methodological* realism. Proper scientific methodology is opposed to any ontology of naïve realism in which natural sciences are thought to provide a "real" or "true" picture of nature free from any theoretical, methodological, and ontological presuppositions.

Realist view

The definition of an ecosystem as a collection of biological and geoclimatical objects linked by interrelating processes to form a distinct subsection of the biosphere is a common feature of many definitions for "field use". It is acknowledged that ecosystems may be hard to delimitate, but in this view they do nevertheless exist as units, which can at least in principle be found empirically. They are neither simple assemblages of living and non-living objects nor collective superorganisms of Clementian proportions (Clements 1916; cf. Golley 1992), but something in-between exhibiting characteristics of a semi-open system.

Ecosystems exhibit developmental trends, which are abstracted from observational data. Described in a theoretically consistent manner, it is inferred that they undergo *succession*. Temporally explicit hypotheses explain some phenomena essential to the successional development. The Holling Cycle (Holling 1986) or the trend to increasing exergy storage and exergy degradation represent such hypotheses (e.g. Jørgensen Sect. 2.4; Bass Sect. 2.11). Using a suitable quantification, a goal function may be constructed, which incorporates the physicochemical and the mathematical essence of these tendencies. Thus, tendencies are in principle reducible to the action of natural law ("efficient cause"). For instance, the tendency of increasing exergy storage could be derived from the thermodynamics of open systems far from thermodynamic equilibrium (Schneider and Kay 1994).

From an explanatory point of view, the goal function can be regarded as the driving factor of ecosystem change: it is an expression of the self-organizing capacity of the system in question. It seems to acquire an epistemological status on par with the observed changes because it appears as real as the developments it causes. Thus, epistemologically informed methodological realists observe goal functions at work in ecosystems. Non-scientific forms of causation or intentional spirits in nature need not be employed. Thermodynamics protects the realist position against the teleology charge.

Constructivist View

While the teleology charge can be fended off within the realist view, there are valid reasons to withhold ecosystems and goal functions the empirical status of natural entities. The original ecosystem definition (Tansley 1935) as well as more recent work on system theory and on the observer-dependence of ecological phenomena express caution against any identification of ecosystems with "real world" entities - common usage to the contrary notwithstanding (e.g. Lenk 1980; Allen and Starr 1982; Bröring and Wiegleb 1990). No explanatory power is lost, however, when a more abstract, constructivist view is taken. Any metaphysical assumptions on the relation between concept and reality are rejected.

The ecosystem concept provides a framework for the collection and interpretation of ecological data. It is a highly abstracted *model* for the local interaction of biotic and abiotic agents. Its introduction to the biological sciences was not arbitrary. It served to integrate the phenomena while "preserving" them (cf. Aristoteles, above). Instead of inferring "goal functions in ecosystems" from observed spatio-temporal patterns, these observations are used to reconstruct tendencies in ecosystem *models*. Depending on empirical and theoretical support, purely heuristic or conceptually well-based goal functions can be used to determine the behavior of the ecosystem model. Confining goal functions to models has the advantage to restrict speech of "goals" in the context of ecology. Scientific models are purposefully constructed, and goal functions are devised to capture directional tendencies in the models. As it is not claimed that goal functions exist in nature, the charge of teleological thinking and of (naïve) realism or unscientific ontology is not substantiated. To reduce the danger of misunderstandings, we suggest that authors discussing ecological results, especially those who deal with goal functions, explicitly state their epistemological affiliations.

Historic and Social Considerations

Many phenomena of the natural sphere are paralleled by comparable societal phenomena. Conceptual parallels translate to parallel modeling approaches. Not surprisingly, competition related with scarcity turns out to be a unifying concept applicable to explain many ecological and economical phenomena alike (Trepl 1994). Evolutionary change is another potentially unifying concept that is detected in markets, the succession of scientific knowledge as well as in ecosystems (cf., e.g., Saviotti and Metcalfe 1991).

The application of a unifying concept reaffirms those who claim that ecological and societal systems share some structural properties leading to partly homologous phenomena. At the same time, this application reaffirms those who suspect that ecology has completely fallen prey to the ideological framework of market society.

Schwarz and Trepl (Sect. 3.2) investigate the interdependence between different concepts of "nature" and concurrent ideological developments to elucidate the contextual framework in which the goal function concept came up. Their essay identifies three concepts of nature – romantic holism, bourgeois individualism, and democratic technocracy – that correspond to differing schools of thought in ecology. This analysis helps to identify some of the basic theoretical presuppositions of ecological thinking with more general types of modern Western thought.

Metzner (Sect. 3.3) poses a similar question from the point of view of an environmental sociologist. Unlike Schwarz and Trepl (Sect. 3.2) he is not aiming at a reconstruction of the concepts of ecological science, instead he highlights the way how environmental issues enter public discourse. Again, there are two opposing positions on the relation between society and science. In the first case, ecologists discover problems caused by an alteration of ecological systems. Now discovered, these problems may gather attention in society as a precondition to be eventually solved. In the second case, the state of the environment may not change significantly at all. Only the cultural conditions within society are altered, and a different pattern of problem-construction is employed. Metzner does not support a particular one of the epistemological distinct positions but opts for a pragmatic approach that takes "real world problems" seriously while including the concerned population as far as possible in the environmental discourse. Right here, ecologically constructed goal functions have the potential to communicate information about the state of the environment. Being part of scientific models, they can help to balance and integrate society and environmental processes in a better way. If one attempts to address the public awareness directly, the goal functions concept appears to posses less symbolic and metaphorical force as "ecological integrity" or "environmental health" arguments, Metzner holds: they are charged more strongly with normative connotations and provide "orientation" rather than "knowledge".

From the epistemological perspective, however, there is a caveat to pragmatic solutions with mainly strategic intentions. On the one hand, the range of possible empirical models is not completely arbitrary. The applicability has to be empirically assessed. On the other hand, as talking about "real world problem" and possible solutions, one has to be aware that confusing any practicable model with reality is part of the unsound ontology of naïve realism.

Ethical Considerations

While Metzner's paper touches upon the mechanisms necessary to introduce scientific findings into public discourse efficiently, Theobald (Sect. 3.4) focuses on the question, how environmental goals ("eco targets") can be anchored in the ethical system of economically active individuals. The exclusively institutional approach to set eco targets by environmental legislation is not effective in the long run. It does neither provide genuinely ethical reasons why the environment should be protected, nor does it stimulate an active co-responsibility of the individual to reach the targets. Theobald's main point is that talking about environmental ethics only makes (practical) sense if it is strongly connected with reflecting the ethics of economy. To induce environmentally responsible behavior, he propagates a subjectivistic approach that does not search for "objective values" in nature, but takes the "subjective dismay" with the state of the environment seriously. In pluralistic societies, subjective arguments are "irrefutable" —but they are not easily forged into binding norms for others.

With the discussion of some fundamental aspects of ethical considerations the papers of this chapter set the stage for more applied ethical problems that occur when natural and human orientors have to be combined (Chap. 4) or when practical consequences for ecosystem management are to be drawn from the interpretation of goal functions as eco targets (Chap. 5). Bossel (Sect. 4.2), for instance, points to the necessity of an explicit ethical framework to guide developmental processes. The goals cannot be imported from nature because *Nature's* terms "would not correspond to human goals." Values, being mental constructs, cannot be found in nature. The concept of sustainable development provides such a framework for human centered goals to attain a balanced satisfaction of fundamental "interests" of natural systems. On the other side to monitor sustainable development, indicators and standards for sustainability have to be available. Jüdes (Sect. 4.3) claims that unifying goal functions can be used to develop these measures.

The following papers will work out in some detail specific insights to historical, philosophical, sociological, and ethical aspects of the goal function concept in ecology and its potential practical implications. Considering the workshop discussions, we are confident that the goal function concept in its ecological dimension can be significantly specified, conceptually sharpened and freed from common misunderstandings by sound philosophical arguments. On the other hand, the topic of goal functions implies a challenge for philosophical, epistemological and ethical reasoning to cope with the development of fundamental terms used in the natural sciences. The workshop discussions on this topic brought about some approaches of this interdisciplinary discourse.

References

Allen TFH and Starr TB (1982) Hierarchy - perspectives for ecological complexity. University of Chicago Press, Chicago London

Aristoteles (1987) Physik 2nd edn. Meiner, Hamburg

Bröring U and Wiegleb G (1990) Wissenschaftlicher Naturschutz oder ökologische Grundlagenforschung? Natur und Landschaft 65(6): 283-292.

Clements FE (1916) Plant succession – an analysis of the development of vegetation. Carnegie Institution Publ 242, Washington D.C.

Engels E-M (1982) Die Teleologie des Lebendigen. Kritische Überlegungen zur Neuformulierung des Teleologieproblems in der angloamerikanischen Wissenschaftstheorie. Eine historisch-systematische Untersuchung. Duncker & Humblot, Berlin

Fränzle O, Müller F and Schröder W (1997) Handbuch der Umweltwissenschaften: Grundlagen und Anwendungen der Ökosystemforschung. ecomed, Landsberg am Lech

Gloy K (1995) Das Verständnis der Natur. Geschichte des wissenschaftlichen Denkens. Beck, München

Golley FB (1993) A history of the ecosystem concept in ecology – More than the sum of the parts. Yale University Press, New Haven CT London

Jax K, Jones CG, and Pickett STA (1998) The self-identity of ecological units. Oikos (in press)

Kauffman SA (1993) The origins of order – Self-organization and selection in evolution. Oxford University Press, New York Oxford

Kuhn TS (1988) Die Struktur wissenschaftlicher Revolutionen [Orig. 1962: The structure of scientific revolutions]. Suhrkamp (stw 25), Frankfurt/Main

Lenk H (1978) Wissenschaftstheorie und Systemtheorie. Zehn Thesen zu Paradigma und Wissenschaftsprogramm des Systemansatzes. In: Lenk H and Ropohl G (eds) Systemtheorie als Wissenschaftsprogramm. Athenäum Verlag, Königstein/Taunus.

Saviotti PP and Metcalfe JS (eds) (1991) Evolutionary Theories of Economic and Technological Change. Present Status and Future Perspectives. Harwood Academic Publichers, Chur

Trepl L (1988) Gibt es Ökosysteme? Landschaft + Stadt 20(4):176-185

Trepl L (1994) Competition and coexistence: on the historical background in ecology and the influence of economy and social sciences. Ecological Modelling 75/76:99-110

Wiegleb G and Bröring U (1996) The position of epistemological emergentism in ecology. Senckenbergiana maritima 27:179-193

3.2 The Relativity of Orientors: Interdependence of Ecological and Sociopolitical Developments

Astrid E. Schwarz and Ludwig Trepl

Abstract

The discussion in the following contribution rests on two fundamental questions: first, the general question as to the influence of social factors on biological theory and its implications on historical and scientific theory, and second, the question as to the specific conditions in the field of ecology. We will consider the hypothesis according to which the holistic-reductionistic polarity, consisting in the concepts of bourgeois individualism, romantic holism and democratic technocracy, can be resolved by a triadic construction. This triadic model will serve as a basis for the presentation of various examples of ecological discourse from the beginning of the 20th century. Finally, we will consider the question whether the elaboration of a concerted theory in ecology is possible at all.

3.2.1 Theoretical Background

We would like to begin by rephrasing the title as a question: namely, "to which degree are ecological goal functions dependent on social and political development"? Further questions will also be which could serve to define the conditions required to investigate the "relativity of orientors".

The basic question, of course, is whether "pure" ecology, i.e. free of external influence, is conceivable at all. If it is not, then how are ecological concepts rooted in political and ideological contexts as well as in the history of ideas? What position can we ourselves take in analyzing this interplay? What does ideology mean in this context? What role play ecological concepts such as "competition" in this respect? And are such concepts necessary, or could they be avoided, even though everyday experience suggests that our thinking about living nature is determined by such analogies?

In order to see how the theoretical background of ecology is influenced by the general social background of a given period, one needs to take longer historical periods into consideration. By reflecting on historical events, one can see more

F. Müller M. Leupelt (Eds.)
Eco Targets, Goal Functions, and Orientors

clearly the mutual influences that may still be involved in the process of acceptance or rejection of current developments.

A further useful approach is to think about the use of concepts and ideas which emerge both in ecological and sociopolitical contexts and are necessarily combined; such concepts as "the balance of nature", "stability", "complexity", "community" and "competition". The critical reception of concepts in the sciences as well as in the humanities has a long tradition, and problems concerning their transfer have in fact been subjects of discussion for a long time. Early on, philosophers and scientists realized that it was necessary to reflect on concepts underlying their disciplines. Darwin himself pointed out that the notion of "struggle for life" originated in the context of Malthusian demographic theory. Soon after, the notion of competition, promoted to a sociomorphic model, became the object of a veritable campaign with parallels being drawn with the then-prospering liberalism in economics (Peters 1960; Topitsch 1962). In 1875, Friedrich Engels wrote that the Darwinian idea of struggle could be seen as a transfer of Hobbes' theory of the struggle of all against all, of the bourgeois ideology of competition, and finally of the Malthusian theory of a "living Nature" (cf. Pörksen 1978).

However, there was also an inverse transfer. In 1859, the very same year that Darwin published his seminal "On the Origin of the Species by Means of Natural Selection, or the Preservation of Favored Races in the Struggle for Life", Walter Bagehot transferred the notion of "natural selection" back into a political context. According to his theory, relations between nations should follow the principle of "natural selection"; consequently he regarded war and repression as legitimate instruments of progress (cf. Pörksen 1978). The case of Walter Bagehot was no exception; it perfectly suited the 'common sense' assumptions concerning the competitive relations between individuals according to liberal theory, and later, between nations according to imperialistic conservatism.

These are well-known facts and it is relatively common for scientists and historians to say that the appearance of the concept of competition in biology was a result of social – and specifically economic – experience. Some ecologists believe that problems were caused by this concept in biology, because it was transferred to an empirical field in which this concept did not fit. If ecologists were to adhere to these facts, the concept of competition would lose its dominance. But this is a problematic view. One necessary consequence of this way of thinking – and this is the crucial point – is that ecology could exist as a 'pure' science, i.e. free of external influence (Trepl 1994a).

However, it seems clear that it is very difficult to maintain the claim of 'pure science' at all. On the contrary, one should ask instead:

Are there such things as 'natural facts' that are independent of the structure which social systems and their subsystems – especially economics – hold to be the very structure of reality? And is it not precisely this structure that is reflected in the various philosophies or ideologies?

3.2.1.1 Epistemological Approach

The following analyses rest on the assumption that nature is the product of the same general structure of thought as society and that, therefore, scientific concepts arise in a context with a necessary conceptual background; in other words, there are not just empirical objects waiting to be found.

First argument: It is inevitable to think some things, because other things inevitably make this necessary.

We have seen that the concept of competition brought about a new way of thinking about nature. In this connection, the major change was that organisms no longer followed a 'natural purpose', but became associated with a non-teleological one that followed the principle of causality. In order to think of life in non-teleological terms, the theory of evolution needed the concept of the destruction of life (negative selection), and finally competition precisely made destruction inevitable. The other factors of destruction (predation, physical factors etc.) may not necessarily appear, but in this case competition inevitably became the destructive factor, for there is a surplus production of offspring. This has been observed, since human requirements have grown boundlessly. This surplus production must be taken into consideration and is not just an empirical finding, because even a minimally negative average production would result in extinction; living organisms without a surplus production of offspring could not perpetuate themselves. Thus, competition inevitably appears and therefore destruction: i.e. competition is necessary for non-teleological living organisms inevitably to be conceived (Trepl 1994a).

Second argument: We cannot think about nature at all without reference to certain concepts; for example, the historical and non-teleological character of living beings.

Neither horses with wings nor Nasobems exist in nature. However, it can be imagined that such animals exist. *Nasobema lyricum* (Fig. 3.2.1), for example, belongs to a very popular systematic group, called the *Rhinogradentia*, which has been described in the zoological literature (Stümke 1989). Systematic, anatomy and physiology of the animal seems to be largely known, although some ethological questions remain open. Yet no one has observed these animals in the wild: there are no empirical facts. On the other hand, living beings not subject to history are unthinkable for us. In the 18th century, however, it was impossible to think that living beings had a history. Their essence was thought to have been created at the beginning of the world, it was something given, unchangeable. No empirical finding would have caused any doubt of this certainty. The facts, therefore, had to be interpreted in such a way as to be in accordance with the certainty that living beings are not subject to historicity.

Another example for the imperative role of concepts is that without the concept of abstract space or the concept of causality, nature is not conceivable for us at all.

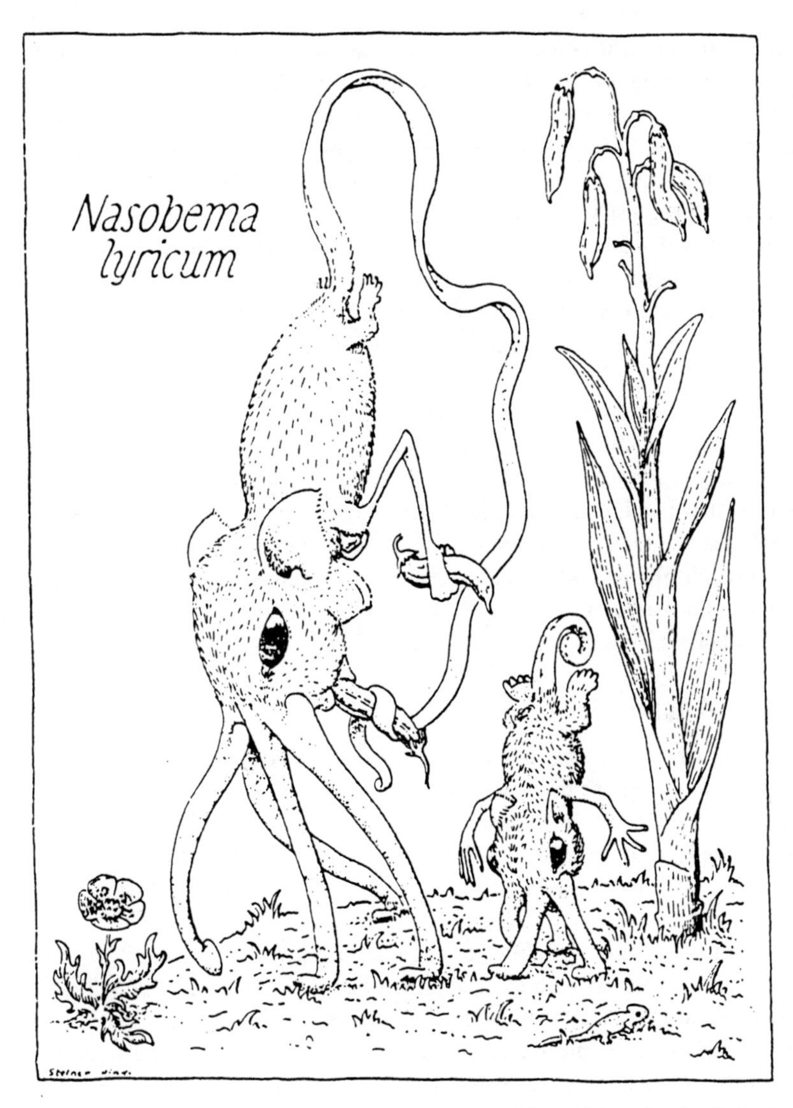

Fig. 3.2.1. Nasobema lyricum, best known representative of the suborder Polyrrhina (source: Stümke H (1989) Bau und Leben der Rhinogradentia. Fig. 13, p 67; Fischer-Verlag, Stuttgart)

There are theories that explain this as an effect of economics, specifically of the necessity to postulate the invariance of the physical qualities of commodities in the act of exchange as the principle of social synthesis (e.g. Sohn-Rethel 1978). Finally, it is this principle which allows a society to act coherently and therefore constitutes society and, with it, any principle has to interpret reality.

Later on, we will take up this central role of economics and give some examples of the nexus between economics and views of nature.

Another more sociological consequence of the second argument is the fact that these concepts of nature are of different relevance for different social groups and therefore produce different ideological reactions. Even certain historical developments must produce political opposition, because they involve setting up a certain position. This opposition needs to be legitimated by an ideology. At the same time, if there is 'progress', this will create disadvantages for some sections of society, and this will lead to 'conservatism'. There will necessarily be thinking along the lines of conservative ideology somewhere in society, but not everyone must think in this way. On the other hand, absolutely everybody must think that nothing happens in nature without a cause or spatial reference.

To summarize: we would say that to progress on the level of theory, it is of interest to consider past historical developments. On the one hand, this may lead to the intuition to see that the same mutual influences which can be clearly seen in the past, still play a role in the acceptance or rejection of current developments today. On the other hand, consideration of the historical background can show us which aspects lead to the congruence of scientific developments and social processes. An approach like this which calls attention to given structures of discourse can show what is possible, impossible or probable (if not necessary) for an ecologist to think or believe.

3.2.1.2 Historical Approach

In the following, we will present a history of ideas of ecology based on three different views of nature. These different "Natures" will be described in advance by mapping out their different epistemological fields. Then we will exemplify the position of some ecological theories in this system of "epistemes" (Foucault 1989). To do, we will make use of two approaches to history, one historiographical, the other structural.

The former provides findings about concrete links between facts about persons and/or institutions connected with a certain historical period, based principally on detailed research on historical sources. In this manner, the Industrial Revolution, the Romantic crisis and its aftermath, and the detachment of transcendental philosophy from the philosophy of history – which will be discussed in more detail later on – could be described as follows: In 1848, physical-mechanical reductionism was not at all a new idea in biology, but rather a counter-reaction to the crisis in the philosophy of nature as well as in the philosophy of the State, accompanied by an intensified socio-political crisis. The works of Virchow show this growing intensity of new ideas and interests, which accelerated the invention of instruments to analyze the crisis that characterized the social and political scenery in the early 1840s (Mazzolini 1988). Clearly, Virchow needed a change in the sciences to coincide with the political and social changes. He believed that a radical change of his own view of society could lead to a change in the processes of science itself. Such ideas can also be found in the works of Virchow's contemporaries who were less outspoken and therefore less understood. According to the historian

Everett Mendelsohn, these ideas will not be understood adequately unless their publication are integrated within the intellectual climate of the time (Mendelsohn 1974).

Structural historiography goes well beyond this viewpoint: it emphasizes not a history of development – for instance of the reductionism-based sciences – but the description of a specific epistemological field in a specific period. It describes the knowledge that was important to formalize the empirical and unreflected fund. Another point is to show in what form problems, concepts and topics defined in a specific field, appear in a scientific or a political discourse (Foucault 1969).

3.2.2 Three Concepts of Nature

We will now describe the particular epistemic field of the three concepts of nature in the model. This model was advanced by the philosopher Odo Marquard. He situated his three "Natures" – Nature as wholeness, Nature as instinct and Nature as control – in a transitional period between transcendental philosophy and the philosophy of history (Marquard 1987). This period has already been mentioned in the brief historiographical example of Virchow. At the same time this was the period in which the discipline of ecology was founded.

However, the idea of our triadic construction will not be limited to this period. It is our hypothesis that the development of theory is still going on within this triadic construction. The three concepts that constitute this triad: Romantic holism, Bourgeois individualism and Democratic technocracy, correspond in this order to the Marquardian "Natures" – Nature as wholeness, Nature as instinct and Nature as control.

3.2.2.1 Bourgeois Individualism – Nature as Instinct

The separation of state and society and the strengthening of the latter is a driving force of the liberal ideal. The society constituted of individuals is considered as an absolute value and is opposed to the state. Of course, the state maintains a regulating function, because the society would otherwise fall into a state of maximal disorder and destroy itself. The inevitable condition of the dynamics that determine state and individual equally is that self-preservation and self-assertion are always related to the individual acting alone. Human beings are considered aggressive and warlike by nature, but cannot be condemned, because they are just following the Law of Nature. Thus it happens that "during the time men live without a common Power to keep them all in awe, they are in that condition, which is called Warre", "where every man is enemy to every man", a "condition of Warre one against another". This description originates from Thomas Hobbes, who decisively influenced the individualistic theory of society (Esfeld 1995).

Thus, each is the autonomous creator of their own circumstances, and each is allowed and obliged to act in accordance with their own selfish interests. But in doing so, one necessarily clashes with the interests of others. The former harmonious cooperation is replaced by an active element: struggle. But there is also a passive element in this configuration and that is the subjection of individuals to environmental constraints. This oppositional relation leads to an open-ended, aimless evolutionary process. As a result of this thinking, 'progress' exists (Trepl 1984b).

3.2.2.2 Romantic Holism – Nature as Wholeness

The attributes of the Nature of Romanticism are harmony, expediency, the rational-historical and organic well-being. Nature is considered as the 'better half' and is absolutely purposeful, being conceived as an organism. The same organic idea can be found again in the relation of the individual to the state. Human beings must operate like institutions, and institutions like persons - person and institution being mutually interchangeable (Herder; cf. Greiffenhagen 1986). Externally, the person is of course subordinated to the state, but at the same time the person is inwardly the expression of an organic totality. In any case, the person is considered as a whole. The solution of conflicting interests in society, whether between persons or institutions, is completely state controlled, and generally the authority of the state takes precedence over social interests. All in all, the concept of the conservative state is characterized by the attempt to revoke the liberal distinction between state and society or at least to modify it in such a way that the state gains more importance than the concept of the liberal state concedes. On the economic level, there are no individual entrepreneurs who follow the principle of competition: instead, hierarchical ordering of traditional professions, prescribed in function and size, control the economic life. Finally, the organic theory of the state can not be understood without the liberal respectively rationalist theory of state against which it is objected.

3.2.2.3 Democratic Technocracy – Nature of Control

The main objective is to control nature and to make it predictable and technologically susceptible. This control of perceivable nature should result from the application of the laws of nature. Nature is therefore imagined as something that follows the rules and laws of rationality; indeed, Nature itself is the totality of these laws. All phenomena must follow these laws if they are to be thought of in causal connections. The relation between the state and society must be a strongly rational and institutionalized one. At the economic level, progress is subject to a democratic process of planning that works in the interest of economic continuity (Mütze 1989). Mechanization and a bureaucratization of social moderation appears. The dynamic, individual entrepreneur is no longer responsible for technical innovations and inventions: instead, this role is taken over by a group of professional specialists that has its seat for example in a capitalist trust – or more often – in a state. Although interventions are still possible, for example in trust capitalism or state socialism, we can no longer easily change the very institutions which we have created. Individual and society exist as realities, but the whole can be established only on the basis of the individual. The individual in such a society takes on a double role: he is at the same time the indivisible, singular, autonomous whole, but also reduced to an "abstract-individual" of his society, an "anonymous individual" (Watkins 1953).

3.2.3 The Concepts and their Correlation with Ecological Theories

According to us, the interplay of these three alternative concepts in ecology is indissociably linked with the sociopolitical conditions of the same period. Furthermore, we claim that although several concepts can of course co-exist at the same time, in most cases one of the concepts dominates the others: namely the one which coincides the best with the mainstream bias in the society.

The bourgeois individualistic concept was the main orientation of early ecology. In the late 19th century, an important founder of ecology, Eugen Warming (1896), following Darwin, emphasized competition as the main relation between the members of a plant community. Consequently, the communities were seen as accumulations of individuals and these individuals primarily had an antagonistic or competitive relationship with one another. This led to a view of the biological community in which individuals or species only occurred together at the same place, basically determined by their individual biological capabilities. This corresponded structurally to the situation in society, according to which individuals are bound together in societies by contract, by virtue of their own free will. In liberal social scientific theories, the contradiction between the active component of this configuration – the struggle of individuals – and the passive component – the subjection of individuals to environmental constraints – was resolved in an open-ended, aimless evolutionary process. This is called progress, an issue we have already discussed before. This open-ended, progressive character of Darwinian theory, and later of Gleason's theory of ecological succession (Gleason 1926), was given a conservative twist in mainstream ecology at the time. The biotic community was not conceived primarily in terms of competition between individual species, but rather of species adaptation, with all species being interrelated and mutually interdependent. As with conservative social philosophy, the basic configuration was the "super-organism". It begins with the whole, then the parts do their "work" or duty in a teleological, functionalistic way.

Out of this mixture of liberal progressive and conservative elements, the holistic, "organismic" concept finally emerged during the first half of the 20th century. Paralleling this development, individualistic concepts and, later on, population-oriented schools of ecological thought were also worked out, but could not become dominant, because the leading ideology in society was neither the liberalistic one nor that of a democratic technocracy. Therefore we have to focus on the conservative model.

Its view was that of a super-organism, an individual of higher order with cooperative elements. "Like other but simpler organisms, each climax not only has its own growth and development in terms of primary and secondary succession, but it has also evolved out of a preceding climax. In other words, it possesses an ontogeny and phylogeny that can be quantitatively and experimentally studied, much as with the individuals and species of plants and animals" (Clements 1916, p 181). This was precisely in accordance with the conservative, holistic ideology common at the time (e.g. Schnädelbach 1983; Müller 1996). It is a characteristic of conservative ideologies to conceive of a society in which the individuals function for the benefit of the whole, regardless whether their relations with one another are of competitive character or not, as in relations of cooperative and mutual benefit.

One can say that this is the necessary and inevitable ideological reaction to the bourgeois conception of human beings that necessarily entails the resurrection of the organic scientific idea. As we have already mentioned, the conservative state theory is characterized by the effort to remove the liberal distinction between state and society, or at least to modify it in such a way that the community (Gemeinschaft), which merges state and society in an inseparable unity, regains a more important role than the liberal theory concedes. "Thus the social organism in which the power of the state is the most highly concentrated, while possessing the highest degree of autonomy, self-determination and freedom of action must be regarded as the relatively more highly developed one". (Lilienfeld 1873, p 83)

In ecological theories of the romantic-holistic type, we can identify similar structures which emerged at about the same time: "Every oyster bed is to some extent a community (Gemeinde) of living beings, a selection of species and a sum of individuals that just in this place meet all the conditions for their development"; "All of the living members of a community of life counterbalance with their organization the physical conditions of their biocoenosis" (Möbius 1877, p 74, p 78).

The relation between the organs of an organism and those between organisms or persons and their biocoenosis or state is described in a similar manner as in contemporary sociological theories of the time. Since then, this conservative image (Denkfigur) has come up again and again. For example, it can be easily seen that modern terms like ‘community health’, ‘ecosystem health’, ‘maturity’, and others, fit in conservative thinking, because they are based on notions of super- or hyper-individual units.

In the 1920s and 30s, this idea of super-individual units was dominant in limnological ecology, one important exponent of which was Richard Woltereck, who, together with Thienemann, was one of the "leading hydrobiologists in Germany" (Zirnstein 1987).

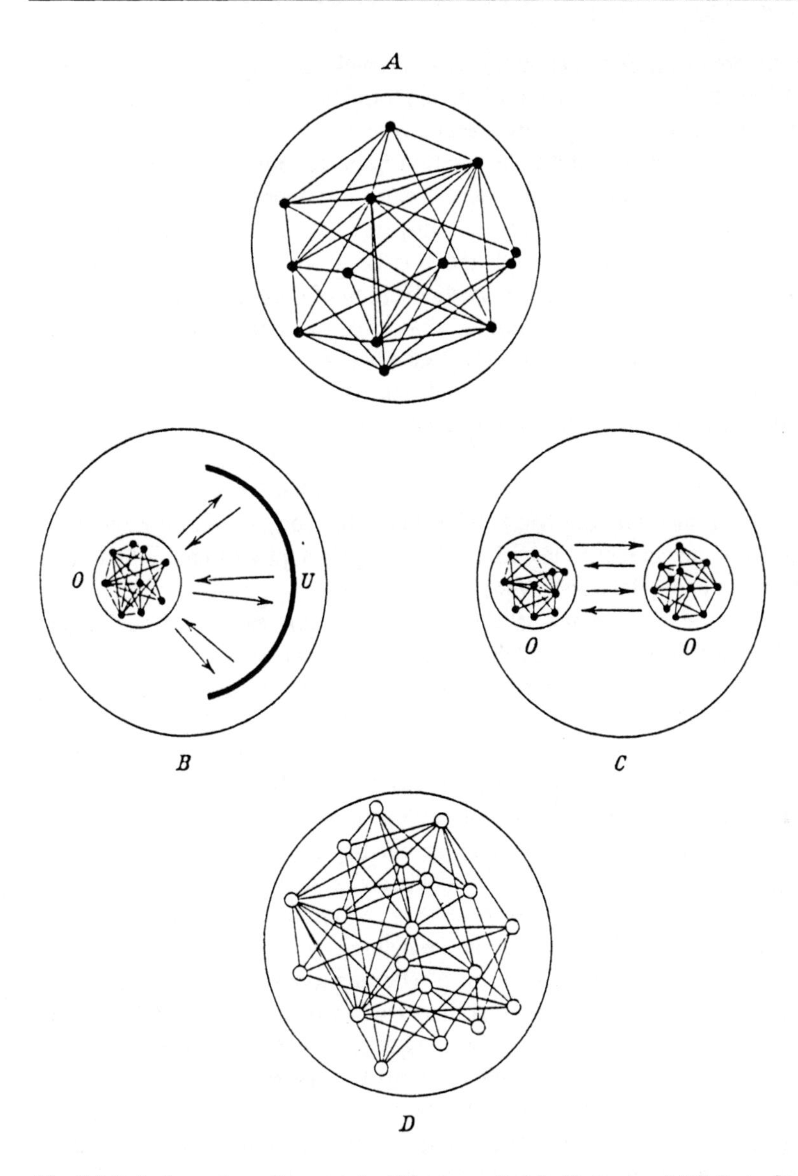

Fig. 3.2.2. **A** shows the units present within a so-called "self-structure" ("Selbstgefüge"). **B** and **C** respectively show the relation between organisms, and with the environment for two organisms that form a pair or a symbiosis. **D** represents individuals that are integrated into a colony or a biocoenosis (source: Woltereck R (1940) Ontologie des Lebendigen. Fig. 113, p 209; Enke-Verlag, Stuttgart)

Part A of Fig. 3.2.2 shows the units present within a so-called "self-structure" "Selbstgefüge" (Woltereck 1940). These units are all interconnected without exception, as well as being joined together in a complex whole. "Self-structure" can mean an organism or a "person" – as Woltereck also says – or a single cell. In part D, the isolated circles represent individuals that are integrated into a colony or a biocoenosis, merged by "relations and qualities of wholeness". Fig. 3.2.2 B and C respectively show the relation between organisms, and with the environment for two organisms that form a pair or a symbiosis. Formally, these so-called "one-person structures" and "multi-person structures" do not differ, or the difference is reducible to filled and non-filled circles. Woltereck argued insistently against an isolation and individualization of these structures :

"The idea of chance plus selection as creators of diversity and purposeful order of living beings will in the future be put down as one of the oddest aberrations of the human 'causal instinct' (Kausaltrieb). Today many of us again sense the ordered, systematic, harmony of the cosmos and the living world not out of a religious or sentimental vagueness, but realistically on the basis of a refined sense for balance, style and rhythm." Obviously, when one compares Woltereck's political and religious views with his philosophy of biology there are parallels and one meets clear indications that his philosophy of nature had taken a turn towards holism (Harwood 1996).

In the 1930s, one reaction against this picture of Nature as wholeness that was not an individualistic concept was the so-called third position. Several authors claimed to have worked it out, one of the most prominent being Ludwig von Bertalanffy.

This concept turned out to be very successful in ecology, because it finally led to the early ecosystem theories (Schwarz 1996). In 1929 Bertalanffy wrote: "Our conception is that of a theory about the system in an inertial state" and "if the organism is a system in an inertial state, as our law expresses it, the metabolic processes generally have to follow the established system; the ever progressing findings must replace the general expression of 'a system in an inertial state' by a more and more detailed knowing about the nature of this system and its chemical, osmotic, fermentive system conditions." (Bertalanffy 1929, p 95, p 97)

To which economical view does this theory correspond structurally? The old organismic concept gives the view of an organism whose growth must not be encroached upon. The ecosystem theory, however, gives also the view of a whole but differs in several important points: its 'whole' can be controlled and, in the long run, constructed. It is the picture of an economy as a single large-scale enterprise, in which the primary relation between the elements is not competition but cooperative division of labor. This is made very clear in the case of the current political slogan 'global ecosystems management' (Hauff 1987). One economic model that refers to similar characteristics is that of the trust capitalism that in turn fits with the ideas of democratic technocracy. As we have already explained, according to this idea, economy can become more systematic, more rational and therefore more stable if one creates large companies. These structures guarantee innovation and progress on the one hand, and a certain stability on the other hand.

Thus, one can ask whether the relation between 'ecological goal functions' and social and political development is a question of degree? In other words, is ecology a natural science at all, or nothing but a facade for different ideologies, whether Romantic holistic, bourgeois individualistic or democratic technocratic?

We believe neither in the former nor in the latter. Yet one has to admit that natural scientific facts, like all facts, generally appear as a result of such projections. This is neither a question of degree, because there exists no 'pure natural fact' and therefore no 'impure' fact. Nor can we solve the complex problem of the relation between reality and cognition by stating that the natural sciences are facades for different ideologies. Human beings are limited by the conceptual constraints imposed by their societies, and also by the paradigms of scientific communities. In the 1930s, Ludwik Fleck, a physician and microbiologist, said that truth is the resolution of problems according to a certain style of thought (Fleck 1983). It is neither relative nor even subjective in the popular sense of these words. It is always or almost always completely determined and defined by a style of thought. There is no observation that is not biased by an orienting and restrictive direction of thought. One can never say that the same is true for A and false for B; if A and B belong to the same common thought, for the idea is either true or false for both. But if they belong to different communities of thought, then they are simply not talking about the same idea. What makes the situation so confusing, then, is that there is apparently not just one style of thought in ecology. But it is not possible to create a unique, one and only style of thought in ecology. Scientists belong to different social groups and therefore to different styles of thought, and the sciences follow democratic structures. Clearly, much more critical thinking has to be brought to bear on these questions.

3.2.4 Conclusions

Many examples show that there is narrow connection between social factors and the development of ecological theories. The holistic-reductionistic polarity in ecological and social theory development will be abandoned here in favor of a triadic model consisting of three concepts of Nature: Romantic holism, bourgeois individualism and technocratic democracy. These three views have their roots in the 19th century, specifically in the transitional period between transcendental philosophy and philosophy of history, and they characterized the formative stages of the discipline of ecology. The link between socioeconomic and ecological theory in the framework of this triadic articulation will be demonstrated by examples from the beginning of the 20th century. It has been shown that, in ecology, different styles of thought coexist at the same time, though with varying degrees of dominance. Given this state of affairs, no concerted theory development in ecology is to be expected.

References

Bertalanffy L von (1929) Vorschlag zweier sehr allgemeiner biologischer Gesetze. Studien über theoretische Biologie, Part 3. Biol Zb 49:83-1

Clements FE (1916) Plant Succession. An Analyses of the Development of Vegetation. Carnegie Institution of Washington, Publ. No 242

Esfeld M (1995) Mechanismus und Subjektivität in der Philosophie von Thomas Hobbes. Fromann-Holzboog Verlag, Stuttgart, Bad Cannstatt

Fleck L (1983) Erfahrung und Tatsache. Suhrkamp, Frankfurt/Main

Foucault M (1989) Die Ordnung der Dinge. Suhrkamp, Frankfurt/Main

Foucault M (1969) L'Archéologie du Savoir. Éditions Gallimard, Paris

Gleason HA (1926) The Individualistic Concept of the Plant Association. Bull Torr Bot Club 53:7-26

Greiffenhagen M (1986) Das Dilemma des Konservatismus in Deutschland. Suhrkamp, Frankfurt/Main

Harwood J. (1996) Weimar Culture and Biological Theory: A Study of Richard Woltereck (1877 - 1944). History of Science 34:347-377

Hauff V (ed) (1987) Unsere gemeinsame Zukunft: der Brundtland Bericht. Eggenkamp, Greven

Lilienfeld P von (1873) Gedanken über die Socialwissenschaft der Zukunft. Die menschliche Gesellschaft als realer Organismus. Vol. 1. Behre, Mitau

Marquard O (1987) Transzendentaler Idealismus, Romantische Naturphilosophie, Psychoanalyse. Verlag für Philosophie Jürgen Dinter, Köln

Mazzolini RG (1988) Politisch-biologische Analogien im Frühwerk Rudolf Virchows. Basilisken-Presse, Marburg

McIntosh RP (1976) Ecology since 1900. In: Taylor BJ, White TJ (eds) Issues and Ideas in America. University of Oklahoma Press, Oklahoma, pp 353-372

Mendelsohn E (1974) Revolution und Reduktion: die Soziologie methodologischer und philosophischer Interessen in der Biologie des 19. Jahrhunderts. In: Weingart P (ed) Wissenschaftssoziologie, vol. 2. Fischer Athenäum, Frankfurt/Main, pp 241-261

Möbius KA (1877) Die Auster und die Austernwirtschaft. Hempel & Parey, Berlin, pp 72-87

Müller K (1996) Allgemeine Systemtheorie. Geschichte, Methodologie und sozialwissenschaftliche Heuristik eines Wissenschaftsprogramms. Westdeutscher Verlag, Opladen

Mütze S (1989) Strukturwandel und Wirtschaftsentwicklung bei J.A. Schumpeter: Kritische Würdigung und Alternativen. Lang, Frankfurt/Main

Peters HM (1960) Soziomorphe Modelle in der Biologie. Ratio 3:22-37

Pörksen U (1978) Zur Metaphorik der naturwissenschaftlichen Sprache - dargestellt am Beispiel Goethes, Darwins und Freuds. Neue Rundschau 89:64-82

Schnädelbach H (1983) Philosophie in Deutschland 1831 - 1933. Suhrkamp, Frankfurt/Main

Schwarz AE (1996) Aus Gestalten werden Systeme: Frühe Systemtheorie in der Biologie. In: Mathes K, Breckling B, Ekschmitt K (eds) Systemtheorie in der Ökologie. Ecomed, Landsberg, pp 35-44

Sohn-Rethel A (1978) Warenform und Denkform. Suhrkamp, Frankfurt/Main

Stümke H (1989) Bau und Leben der Rhinogradentia. Fischer, Stuttgart

Topitsch E (1962) Das Verhältnis zwischen Sozial- und Naturwissenschaften. Dialectica 16:211-231

Trepl L (1994a) Competition and Coexistence: on the Historical Background in Ecology and the Influence of Economy and Social Sciences. Ecol Modelling 75/76:99-110

Trepl L (1994b) Holism and Reductionism in Ecology: Technical, Political, and Ideological Implications. CNS 5:13-31

Warming E (1896) Lehrbuch der ökologischen Pflanzengeographie. Eine Einführung in die Kenntnis der Pflanzenvereine. Berlin

Watkins JWN (1953) Ideal Types and Historical Explanation. Brit J Philos Sci 3:22-43

Woltereck R (1940) Philosophie der lebendigen Wirklichkeit. Grundzüge einer allgemeinen Biologie, 1st ed. 1932, Ferdinand Enke, Stuttgart

Zirnstein G (1987) Aus dem Leben und Wirken des Leipziger Zoologen Richard Woltereck (1877-1944). NTM-Schriftenr Gesch Naturw, Technik, Med 24:113-120

3.3 Constructions of Environmental Issues in Scientific and Public Discourse

Andreas Metzner

Abstract

We have two possibilities to explain why there is continuing change on the public agenda of environmental issues. Either the changes are viewed as reactions of social actors to real alterations of the environment, or they are viewed as reactions to alternating cultural conditions. The willingness of society to recognize and solve environmental problems depends upon the claims-making activities of differently motivated social actors in public discourses. The sciences produce cognitive and interpretative claims, focusing either on pure description or giving evidence for the need to act. The deconstruction of scientific claims may lead to a relativistic overall devaluation of knowledge needed to encounter environmental problems.

A model is introduced that integrates socio-ecological approaches with cultural studies oriented ways to treat societies and their environment. The necessity to built complex procedures of eco-social regulation increases historically along with the growing complexity and differentiation of societies. Society has to determine environmental limits and goals, which are not directly observable, but need to be reconstructed and agreed upon. If it is the task to mediate environmental problems, the more general concepts of ecosystem integrity or environmental health perform better than the rather specific goal function concepts.

3.3.1 Introduction

Environmental problems do not establish themselves on the public agenda. "They must be 'constructed' by individuals or organizations who define pollution or some other objective condition as worrisome and seek to do something about it" (Hannigan 1995, p 2). In this regard, environmental problems are not very different from other social problems such as child abuse, juvenile crime or AIDS. But there are, however, a few notable differences: Social problems derive much of their rhetorical power from moral rather than from factual arguments.

F. Müller M. Leupelt (Eds.)
Eco Targets, Goal Functions, and Orientors

Environmental problems have a more imposing physical basis than social problems, which are rather rooted in personal troubles growing into public issues (Hannigan 1995, p 38).

A good example for the interaction of physical and social aspects is the 'Love Canal' case, a settlement built on top of a hazardous waste-dump in New York State (Fowlkes and Miller 1987; Mazur 1991). Toxic vapors emerged from the subsoil and entered into people's homes, but as long as health defects were not apparent, and not connected with the toxic wastes, nothing happened. There was neither a demand for action, nor were there any protests—in other words: there was no *problem*. The difference between a scientific-objectivistic and a constructivistic, cultural-studies-oriented concept of a problem becomes quite apparent here. The natural scientist gives an objectivistic description of the facts, determines the type and concentration of the toxins in the soil, and in the bodies of the persons affected, and grasps *this* as the problem. The social scientist, however, endeavors to treat the problem by investigating the action-leading-concepts of human participants, the state of consciousness of the individual subjects, and the motives of interacting social participants. "Social problems are what people think they are" (Spector and Kitsuse 1987, p 73).

In principle, we have two possibilities to explain why there is continuing change on the agenda of environmental issues. Either the changes are viewed as reactions to the alterations of the environment, or they are viewed as reactions to alternating public attitudes (Ungar 1992). The first approach concentrates on *problems of eco-social metabolism and societal (re)production,* the other on *phenomena of social construction and perception of environmental issues.* The social sciences orientate to either of these leading themes; they do exist simultaneously inside (environmental) sociology (Dunlap 1993; Dunlap and Catton 1994).

The development of methods of environmental management and some basic rules for using environmental resources according to the principles of sustainable development (Constanza, Daly and Bartholomew 1991; Huber 1995, pp 49; WCED 1987, pp 44-60; Vellinga, de Groot and Klein 1994) are examples for the first main leading theme. In order to identify the necessary requirements and strategies of improvement, this approach poses the following questions: What are the social causes of environmental changes leading to degradation, and what are the social effects of the emerging environmental problems? The difference in attention paid to anthropogenic climate changes in contrast to the low attention paid to the problem of worldwide soil erosion exemplifies the importance of the second main leading theme: Why are certain environmental phenomena diagnosed as problematic and brought towards a solution, while others – factually not less 'problematic' or 'risky' events – are hardly or not at all noticed?

3.3.2 Understanding Environmental Problems and Technological Risks

3.3.2.1 The Constructivistic Approach and its Critics

The constructivistic approach comprehends "risks" as constructs of social communication and explains "the increase of environmental and technological risks" through cultural processes of change. Douglas and Wildavsky (1988, pp 10) hold that the balance of power between the central cultural institutions (including the market and the hierarchy) and the "sect" as the socially peripheral subculture has shifted so eminently that the risk-aversions of subcultural, social movements (with "egalitarian" life-styles) have turned into the prevalent subject-and thereby predominant reality—for the developed industrial societies.

Because the approach has freed risk assessment from of the grip of purely technological-scientific and psychological handling, it has its merits. It needs to be mentioned, however, that it merely deals with risk *constitution*, not with (social and industrial) risk *production*. Therefore, it (mis)understands ecological criticism as anti-industrial and anti-modernistic. The objective contents of environmental problems is simply neglected. The ecological problems do not appear as a threat to "society", but the environmental movement does, which challenges the established institutional order and threatens its beneficent functioning.

In spite of all psychological theories that try to reduce the characteristic of the perception of environmental problems to patterns of individual psychology (Jungermann and Slovic 1993), it remains unexplained why a dispersion of divergent risk evaluations and different assessments of endangered (environmental) goods – spread over the whole population – are found. Sociological explanations of this observation are referring to divergent utilization interests of the participants, to disparate states of knowledge, to different frames of rationality, and to deviating patterns of cultural values.

Its (exclusive) explanation claim of the constructivistic approach closes the gap left by risk psychology. Risk psychology provides an understanding of individual differences in recognizing and accepting risks, but it remains to be answered, why a variance of certain patterns of risk preferences and risk aversions can be observed within the population. The answer is thought to be found in the different sociocultural contexts, in which individuals live, and in the orientation of the thinking and acting of individuals towards institutions (Thompson, Ellis and Wildavsky 1990). The different biophysical conditions of the individuals' environment and the latter's ecological alteration do not play a role, not even as a constitutive element within a multi-factorial explanation model of particular risk cognition.

Sociologically, it is by all means imperative to inquire into the structures of the material production of the "risk-society" (Beck 1986). Instead, culture is presented as the only explanation. The (rhetorical) question "are dangers really increasing or are we more afraid?" (Douglas and Wildavsky 1988, p 1), can only be answered in one way: "We are more afraid!". The phenomenon of "growing environmental and technological risks", which requires further explication, is merely diagnosed as the *result* of growing sensibility towards these risks, *caused* by processes of cultural change. Thus, Douglas and Wildavsky represent a radical thesis: cultural processes of change put the individual into a state of uncertainty. The individual looks for an object to project its anguish onto and finds it in environmental and technological risks. This thesis implicates that the (environmental) anxieties are actually misleading and unfounded. This approach is decisionistic in so far, as processes of the industrial-technological (re)production and formation of society cannot be investigated. They are understood as something *extra-societal* – similar to the factors of the "natural" environment. According to the dogma "social matters can be explained only by social phenomena", they are no longer to be comprehended as subjects of social sciences research.[1]

Rejecting this mono-dimensional approach, I prefer a combined, multiple approach. The perception of technological risks and environmental problems, of social risk-preferences and -aversions may depend on

- socio-cultural changes (e.g., post-materialism, pluralization of life-styles);
- scientific-technical progress in the detection and measurement of potentially harmful substances,
- growing technical capacities of diagnosing and treating environmental and technological problems;
- technological alternatives for a more sustainable utilization of nature;
- changes in the state of the environment or in the intensity of risks, be they observable, effective or apparent.

3.3.2.2 The Realistic Approach and its Critics

The realistic approach comprehends "risks" as objective elements of interaction between nature and society. It explains the "increase of environmental and technological risks" through the intensification and extensiveness of nature's utilization. According to Dunlap (1993) the environment has to fulfill three essential functions for society, namely: a) to provide it with resources, b) to absorb its refuse, and c) to serve as "living-space" and habitat of man. If an environment is used by one function, the other two are impaired. *Usufructuary competition* and

[1] This exclusion of a substantial research-perspective takes place with different accentuations also in the systems-theoretical treatment of environmental, technological and risk problems in so far as the analysis runs under the communications-theoretical premises of the sociology of Luhmann (1984, 1986, 1991). The disadvantage of this methodological exclusive approach, that assigns itself lately as "constructivistic systems theory" (Japp and Krohn 1996), lies in the fact, that the possibilities to intensify an inter- and trans-disciplinary research – as they are offered in the context of the "general systems theory" – are neglected in favor of a pure "sociologism".

social conflicts may arise. The ecological systems are affected by pressures, which may in extreme cases disrupt the capacity of the ecosystem to regenerate and function. The simultaneous expansion of all three functions leads to a transgression of the global carrying capacity and of the ability of the environment "to withstand" the stress. Dunlap (1993) infers that an *original* ecological problem of the society subsists, manifesting itself in increasing practical problems and growing risks fulfilling the three basic functions. This problem will remain incomprehensible, if it is understood merely in the framework of the "construction of social problems" as a *genuine* phenomenon of social cognition and communication. The growing social attention towards environmental and technological risks is therefore in essence a responsive reaction to their actual increase.

A constructive critique of Dunlap's statements has to make – amongst others – ecological points because his description of nature lacks an elaborated ecosystemic argument. His view does not accommodate the fundamentally dynamic nature of ecological phenomena. The distinction between ecologically harmless utilization of resources or smaller scale "disturbances" and irreversible damages remains systematically unexplained. Nor does Dunlap's approach consider positive interactions between different types of the utilization of nature, for instance between adapted agriculture and biodiversity.

Roughly speaking, it is the strength of Dunlap's approach to develop footholds for integrated ecological and sociological problem analyses and crisis management strategies. The weak point is its inability to address aspects of the communicative processing of perceived problems. He does not take up the following questions: Which socio-economic achievements produce relevant material and energy flows? Which functional prerequisites are found in supra-technological infrastructure-systems (Mayntz and Hughes 1988; Gras 1993)? Which institutions and media are needed for the control of these systems (Ayres and Simonis 1992)? In connection with a model of functional societal differentiation (Luhmann 1984, 1986, 1991), these questions, however, have to be answered, before another question can be tackled: Why are certain environmental problems intensely heeded while others – sometimes "objectively" equally important ones – are neglected?

3.3.3 Claims-making activities: assembling, presenting and contesting environmental issues

Hanningan (1995) presents an analytical frame that allows analyzing processes of the construction of environmental issues related to the different functions of sciences, the mass media and politics in the field of public discourse. According to Hannigan, the willingness of society to recognize and solve environmental problems rests primarily upon the claims-making activities of a handful of 'issue entrepreneurs' in science, in the mass media and politics. He identifies three key tasks in the definition of any environmental problem: assembling, presenting and contesting, each of which carries its own activities, opportunities and pitfalls.

Table 3.3.1. Key Tasks in Constructing Environmental Problems (Hannigan 1995, p 42)

	Assembling	Presenting	Contesting
Primary activities	• discovering the problem • naming the problem • determining the basis of the claim • establishing parameters	• commanding attention • legitimating the claim	• invoking action • mobilizing support • defending ownership
Central forum	• Science	• Mass media	• Politics
Predominant 'layer of proof'	• Scientific	• Moral	• Legal
Predominant scientific role(s)	• Trend spotter • Theory tester	• Communicator	• Applied policy analyst
Potential pitfalls	• lack of clarity • ambiguity • conflicting scientific evidence	• low visibility • declining novelty	• co-optation • issue fatigue • countervailing claims
Strategies for success	• creating an experiential focus • streamlining knowledge claims • scientific division of labour	• linkage to popular issues and causes • use of dramatic verbal and visual imagery • rhetorical tactics and strategies	• networking • developing technical expertise • opening policy windows

When scientists engage in claims-making activities, they are mainly handicapped by a combination of scholarly caution, excessive use of technical jargon and inexperience in handling the media (Hannigan 1995, p 44). In order to fill this gap, popularizers are needed – individuals, such as Jeremy Rifkin, or organizations, such as Greenpeace. Accordingly, the success of the claims-making activity by Greenpeace lies not so much in its ability to construct entirely new environmental problems, but rather in its genius to select, frame, and elaborate scientific interpretations that might otherwise have gone unnoticed or would have been deliberately glossed over (Hannigan 1995, p 44, in reference to Hansen 1993, p 171).

In sum, Hannigan (1995, pp 54) identifies six factors necessary for the successful construction of an environmental problem:

1. Scientific authority for the validation of claims.
2. Existence of 'popularizers' who can bridge environmentalism and science.
3. Media attention, where the problem is 'framed' as novel and important.
4. Dramatization of the problem in symbolic and visual terms.
5. Economic incentives for taking positive action.
6. Emergence of an institutional sponsor who can ensure both legitimacy and continuity.

The success of environmental movements is one facet of the discrepancy between growing environmental consciousness by successful problem construction and continuing, ecologically harmful behavior (Schluchter and Metzner 1996). Their success is considerable less noteworthy in getting their policies institutionalized, especially, where these policies might require the reallocation of resources away from potent capital interests and inflexible bureaucratic actors (Hannigan 1995, p 49; in contrast to Maxeiner and Miersch 1996).

3.3.4 Science, Politics, and the Public

Starting from the observation that hardly any environmental problem, e.g., acid rain, loss of biodiversity, or dioxin poisoning, does not have its origin in a body of scientific research, the orthodox way how science sees itself is scrutinized. The task of traditional science seems to be the search for truth with the goal to obtain a clear *reflection of nature*, as free as possible from any social and subjective influences that might distort the 'facts'. Nevertheless, the sciences are not merely a medium to which one refers in order to construct environmental problems. Frequently, they become the target of environmental claims themselves, for instance, in the case of the potentially harmful effects of genetic engineering. A paradoxical situation originates, because environmental argumentation has to presuppose the credibility of scientific expertise when defining these problems. When fighting against science-based industrial projects, however, they are forced to doubt the technocratic rationality of science by referring to 'social rationality' or folk wisdom (Perrow 1984).

Science has become dependent on social and political processes. Whatever is on the political agenda gets researched, since scientific knowledge is produced, where research funds are available. The consequences of these developments have two symptoms: the *politicalization of science* and the *scientification of politics* (Weingart 1983). One can assume that the scientification of politics is less problematic for the scientific community than the politicalization of science because politics may improve through better expertise. But the conduct of politics – that is mediation of problems, balancing of interests, etc. – is more difficult when scientific experts claim the objective truth of their findings. Furthermore, the credibility of science is undermined when experts present different results dealing with the same issue. Consent in the scientific community and the lack of scientific controversy influence the public credibility of science profoundly. This way, a certain pressure emerges to reduce the normal and important scientific controversy in favor of more consensual and coherent statements on part of the particular scientific community in order to gain public resonance and political support. The relationship of scientification and politicalization carries many more critical tendencies, which cannot be explicated here.

To gain a better understanding of science's role in modern societies, it is useful to make a distinction between *cognitive* and *interpretive claims*, corresponding with the concepts of 'instructional' and 'operational knowledge', which means the difference between "knowing how to do" and "knowing what to do". On the one hand, there is the pure description of a situation, e.g., the charge of fresh water ecosystems with sulfuric acids. On the other hand, there is the purpose to give evidence for the need to act (often mediated with adequate semantic constructions like a 'critical load' of an ecosystem that implies the need not to exceed its limits) (Alcamo 1984). The activity of scientists does not end, however, with making knowledge claims. They also routinely construct *ignorance claims* highlighting 'gaps' in the available knowledge in order to promote the necessity of further research, or to retard unwanted or disturbing political action because not enough hard data exist to justify legislative or administrative activity (Hannigan 1995, p 77).

Table 3.3.2. Uncertainties regarding processes of global climatic change

Uncertainty regarding the mechanisms of the effect. Does the sea-level rise? Will there be more storms?
Uncertainty regarding anthropogenic causation. Does the variance perhaps lie within the range of natural fluctuations? Are, e.g., the effects of a volcanic eruption bigger?
Uncertainties related to the time scales of prognosted changes and to the localization in space. Are they detailed or rough, what is the speed of the delocalization of eco-zones on the earth's surface, where will extreme weather events become more frequently?
Uncertainty regarding the socioeconomic and political repercussions. Will there be crop shortages? Will there be mass migrations and conflicts about eco-refugees? Do higher energy prices have a negative effect on the growth of the gross national product, or do they trigger a technical breakthrough in the direction of higher energy and resource productivity?
Uncertainty regarding the efficiency of countermeasures. Does a 25 % reduction of CO_2-production have any considerable climatic effect and when does it occur? Are the expenses of the reduction greater than the costs of the prevented dangers?

The uncertainties of forecasting the effects of global change are a good example in case (Bechmann et al. 1996). Concerning these uncertainties, we have to be aware of the phenomenon, that science has undergone a metamorphosis from a 'classic' (mechanistic) to a 'transclassic' (postmechanistic) science (Prigogine and Stengers 1981; Krohn, Küppers and Paslack 1987). In order to investigate this complex matter of real world phenomena, it is not sufficient anymore to 'construct' (mono)causal linkages between cause and effect under the 'ideal' circumstances of the scientific laboratory. Instead, it is necessary to work with models including a high number of factors and parameters. The models show different kinds of dynamic behavior according to alternating starting points and conditions, and according to different algorithms describing different potential causal linkages. There is no other – or better – way to verify the prognostic capacity of such models than testing the validity of underlying assumptions and by simulating past or present processes that can be compared with empirical data.

Regarding the growing necessity to deal with the effects, problems, and risks emerging from the relationship between human societies and their ecology, science loses more and more the character of a pure basic science and rather gains some characteristics of applied science (such as medical and engineering science). Furthermore, the very idea of science itself has changed. In former times, science was thought to produce media for the increasing human domination of nature. Today science has to deal already with the problems of man's successful appropriation of nature – it has turned into a *reflexive science* (Beck, Giddens and Lash 1994; Lash, Szerszinsky and Wynne 1996). Thus science works in the context of a *reflexive industrialization*, which needs to master the charges and after-effects of the conventional modernization (Mol 1995).

The acceleration of the adaptive processes of social institutions (related to technological innovations, economical restructuring, cultural changes, and environmental alterations) poses a great challenge to politics. Reactive politics that intervenes only when manifest damages occur, is systematically late. A proactive politics could be effective in time, but it is forced to work with uncertain hypotheses regarding the problems or corrective measures. According to the *principle of precaution*, even scientific uncertainty is no excuse for the delay of protective measures concerning ecologically destructive processes. To be useful as reference points for social practice, it is not necessary that models are scientifically complete in detail. In view of the ever lasting hypothetical character of all scientific knowledge, as Critical Rationalism has pointed out (Popper 1982), completely detailed models are impossible as it is. As (even temporary) points of reference it is sufficient that the models be more coherent and covering more observations and experiences than concurrent action-leading cognitive maps (Fischer-Kowalski 1991).

3.3.4.1 Scientific Advertisement

In politics top down and bottom up oriented approaches are linked to different world views. One side is convinced "we need more democracy" because "the people know best" how to attain a growing public welfare. The other side underlines the necessity of an institution, capable to integrate vested interests and to force decisions when necessary against resistance.

The advisory function of science has to change while reacting either to 'top down' or 'bottom up' approaches. Several media can be used to implement political decisions, to formulate political targets, or to influence processes of decision-making. Of these, *money*, *power*, and *knowledge* are three most important ones. Knowledge is the specific medium of science. As on other markets, the impact of ecological knowledge depends on demand as well as on the way this medium is offered. Furthermore, there is competition between ecologists and – often publicly better-established – economists in advertising their political ideas in the decision-making process. Conflict arises, for instance, in the debate over the need for economic adaptation to the ongoing 'globalization' of the world market. This debate distracts attention from discourses that pled for ecologically inspired future scenarios, such as "Sustainable Netherlands" (van Brakel and Builtenkamp 1992) or "Zukunftsfähiges Deutschland" ('Future Oriented Germany') (BUND and MISEREOR 1996).

The course of the concurrent debates suggests to make ecological knowledge more attractive. To strengthen its position, it is of growing importance to introduce competencies of ecology into interdisciplinary working groups and projects. To demonstrate the positive effects on ecological knowledge, social and economical advantages or positive effects to the quality of human life should be mentioned frequently. Ecology as a scientific discipline should in each case try to satisfy the top down as well as the bottom up induced demand for advertising and expertise. Ecology should provide environmental knowledge understandable for all participants in public discourse and lucid management schemes. Its models should be far-reaching and complete to enable overview and orientation.

3.3.4.2 Shared Responsibility and the Impact of the Sciences

Natural and social sciences should take equal responsibility for the alteration of ecologically relevant social practices. Not only the claims making activities of objectivistic sciences are responsible for environmentally destabilizing human actions. By pointing to the necessary relativity of such scientific truth claims, or by deconstructing the social genesis and social functions of these claims, subjectivist and constructivistic studies may invalidate possibly correct and practically important knowledge. On the other hand, the deconstruction of scientific knowledge shares responsibility for the validation of the incorrect. Take the quarrel about global environmental changes. Several authors stress the uncertainty of scientific data, models and prognoses (Bailey 1995; Wildavsky 1995). Their work would be misleading, however, if (ordinary) people would think "OK then, scientists construct their knowledge as they want", and deny the need for any action. In this context we have to recognize that the own interests of scientists are a decisive factor as the need of the acquisition of research-funds has become overwhelming. This effect has been reported, e.g., in the case of the climate research community and its 'construct': the greenhouse-effect. The ideology-critical motive to deconstruct knowledge claims, however, can lead to the unintended consequence of increasingly opportunistic use of expertise and research. The uncontrollable use of science by vested interest is nothing else than the direct opposite of the original motive.

3.3.5 Modeling Ecology and Society

In general, models striving to integrate natural and social processes cannot consist exclusively of simple cause and effect relations. Between a physical event (or condition) and its social effect (or consequence) lies the *contingency* of human action and perception. The notion of contingency refers to the distinction of animal *behavior* and human *action*[2], as it was established by Max Weber, several anthropologists, and in social systems theory by Parsons followed by Luhmann[3]. Contingency means that it is possible for humans to act in various ways rather than having to act in one way.

Questioning the relationship of social and natural systems, we have to recognize two prejudgements. First, social systems are following other, self-generated and symbolically organized laws than natural systems. Second, social systems, nevertheless, depend upon ongoing processes of interaction with natural systems. They are emergent entities built up of populations of human organisms that have to be stabilized through the metabolism of society and environment. Following these prejudgements, we ask: How far are social systems heteronomous ('material' entities) and how far are they autonomous ('cultural' entities) related to their biophysical environment?

3.3.5.1 Symbolical and Material Dimensions of Social Systems

The concept of "socio-anthropogenesis" (Leakey 1993; Löther 1988) signifies that the evolutionary process of anthropogenesis and the evolvement of a society capable of sociocultural evolution happened concurrently and interdependently. Every "animal-society", such as "monkey-societies", lives as a natural unit in a natural environment. In the transition towards human society, these two dimensions do not vanish but additionally become symbolically constituted. Societies are no longer merely real existing populations in real existing environments. They become symbolically existing societies (described by themselves) in a symbolically existing (socially described) environment. Human society and ecological environment exist in a physical-natural as well as in a cultural-symbolical sense.

Neglecting these interdependencies, e.g., by postulating the independence of the autopoiesis of sense-processing systems (such as social systems or systems of individual consciousness) from human actors, which are viewed merely as parts

[2] We are using this distinctive concept further on, well knowing that in the light of recent primate research, especially related to questions of the capacity to lern, use and generate elements of symbolic (sign) language, and to the capacity to traditionalize or enculturate techniques of how to use and make tools, a sharp diversion between animals and humans no longer exists.

[3] The term 'double contingency' as introduced by Luhmann indicates that there is neither a fixed action (from ego) nor a re-action scheme (from alter-ego). The reduction of the complexity of the given possibilities to (re)act works by means of (mutual) expectations, condensating to norms, values and institutions, stabilizing clusters of corresponding actions that generate (via order-from-chaos-principles) coherent social systems (Luhman 1984).

of their environment, like Luhmann (1984) does, leads on the one hand to mystifications, such as a re-ontologization of systems-theory, and gives up an analytical definition of 'system' in favor of an idealistically biased ontological viewpoint. On the other hand it leads to a sharp division between social and ecological system conditions of human existence: They cannot be analyzed as interdependent complements, and the integral unity of the human actor fades away. Rather, 'social systems' are social constructions, born out of communicative processes between human actors, serving their orientational needs.

3.3.5.2 Societies and their Ecology: How to Describe Divergencies and Convergencies?

In the following model of interactions and relations between human society and ecological environment (Fig. 3.3.1), I plead for the concept of a two-stage system of human societies, which does not comprehend the latter merely as sociocultural units, but also comprehends them in a naturalizing way as ecological entities (Metzner 1993, 1994).

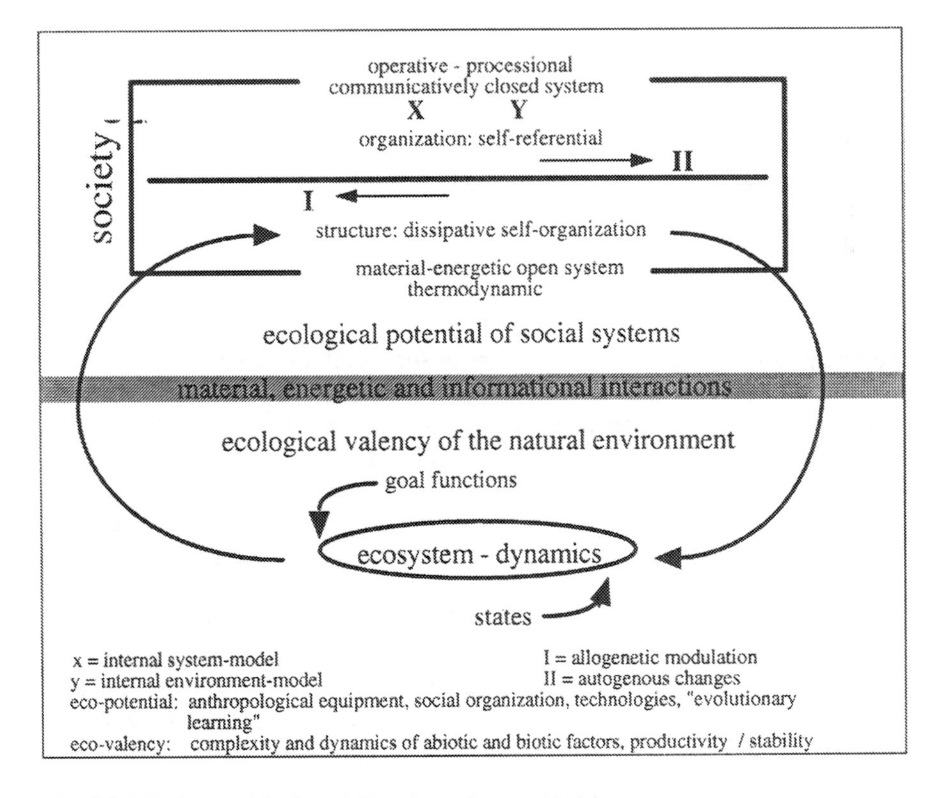

Fig. 3.3.1. Ecology and Society: A Two-Stage Systems Model

Because they are materially and energetically open systems, human societies are *directly* (retroactively) affected by (natural or self-induced) environmental changes. They are, however, as *operatively* closed systems, only *indirectly* related to their ecological environment. They are able to notice, pass over, or suppress environmental changes judging by their own affectedness. They can deny these phenomena or interpret them as problematic or risky. They can preserve the 'tried and tested', or accomplish learning processes capable to reorganize economic, political or scientific behavior. When learning processes and behavioral changes take place, the effectiveness and the consequences of human actions change as well: the substantial connection between environment and society is affected.

All living things are elements of ecological systems, they have a metabolism and consume exergy. The *co-evolution of species*, e.g., of the different populations of one habitat, generates continuing dynamics towards an optimization of 'their' ecosystem. This optimization is characterized by two qualities:

1. metastability (a high probability of the populations' continuing existence) and
2. a configuration of the flow of substances and energies that results in an actual closing of the circuits, and in an energetic intensification of the utilization-cascades.

Only human beings have, in the course of their social development, left this path of co-evolution. But this does not change man's integration into the structural organization of ecological systems.

3.3.6 Reconstructing Issues: Ecological Limits and Goal Functions

To return to a 'metastable' development of society, it appears necessary to find ecological limits to human actions. Integrated goal functions may help identifying these limits. Procedures regulating the population density, the utilization of ecological resources, and limiting non-intended anthropogenic alterations of the environment can be found in all cultures (Weichhart 1989). In an "eco-functional" sense, these procedures represent attempts to install self-selected limits to the interaction with nature, before nature itself manifests limits, for instance, when the carrying capacity is exceeded. If critical limits are surpassed, ecological chains of effects are activated that may lead to catastrophic dynamics of unknown proportions. Returning to non-critical conditions may be impossible without paying a high price in terms of human life, public health or common wealth. Besides this *preventive function,* the regulation procedures have an *optimizing function*, namely to use ecological resources efficiently, durably and sustainably.

To work out suitable regulation procedures, it is necessary to integrate the particular performance and development of ecological systems conceptually into the communicative structure of socio-cultural institutions. Society has to encode its perception of its own state of well-being and goal-attainment, along with an adequate perception of environmental states, limits of appropriation, and effective allocational principles. On grounds of internal contradictions between different societal subsystems, however, an adequate and up-to-date system of socio-ecological regulation procedures is missing. The question is whether an integrated goal function concept could be a valuable contribution to this challenge.

3.3.6.1 How to Determine Limits: Biophysical or Sociocultural?

The analysis of the construction of environmental issues revealed so far that from natural science as well as from social science approaches valuable insights can be gained. Instead of employing the approaches in an isolated fashion, a complementary use of the approaches is required. Still, it has to be analyzed in detail, how ecological limits, e.g., to anthropogenic emissions of greenhouse gases, can be determined. As part of the *natural* greenhouse effect, carbon dioxide is considered a good thing because it is an crucial substance in the climatic life support system. As part of the *anthropogenic* greenhouse effect, carbon dioxide turns into the opposite. But what is the point of reference for an emission reduction goal? Is it the 'natural' concentration of CO_2 in the atmosphere (yet, 2, 10 or 100 thousand years ago), the (maximum) emission quantity that is considered 'neutral to the climate', the economically optimal ratio of marginal utility/marginal costs, the emission quantity that is necessary for mankind to survive, or a concentration that is deemed "optimal" (perhaps somewhat more than during the last ice-age but less than in the following warm-age)?

When our resources come closer to exhaustion or when they are increasingly threatened, we will be fighting over the remaining ones. The definition of limits to the utilization of nature has to reflect these social conflicts. But human society itself consists of population units and social systems that are differentially affected by resource shortages or by other environmental problems. Furthermore, issues are perceived and processed differently according to the interests, ratio, knowledge, and preferences of the individual sub-culture within society. Thus, it is even difficult to say what would constitute substantial ecological limitations to social reproduction:

- Shall the existential minimum of human dependency upon ecological resources serve as the yardstick, or shall the best possible accommodation of man's needs prevail?
- Shall the maximum stability or integrity of ecological systems be the criterion, or shall a boundary line be drawn, where ecological systems can be maximally utilized without being endangered?

- Shall the risks and damages to the environment potentially affecting humankind be as small as possible, or shall dealing with the risks and damages be economically optimized?

For many environmental questions different approaches and answers are possible, and often they are controversial (cf. Marticke, this vol.). In principle the same is true about goals for the quality of life, concepts of ecological balance or their underlying models. This touches on the complications of the *naturalistic fallacy*. Is it possible to deduct from 'being' a 'should be', from the factual the normative, from the real the aspired, from the conditions of nature the conditions of culture? The fact that we are dealing always with cultural descriptions has to be taken into account. Our concepts of nature as well as our concepts of man and human culture are symbolical-linguistic units within communication processes and as such 'social constructions'.[4] Every concept of nature has a *cognitive dimension*—it incorporates manifestations and principles of effect—as well as a *normative dimension*—by which nature is being judged beautiful or emaciated, violent or harmonious, and a *practical dimension*, that encodes (im)possible utilization forms, necessary restrictions to utilization, the value of ecological resources, useful or harmful substances. The practical dimension implies what to do in order to take care for a culturally altered 'natural' landscape, and to sustain its productivity in human terms.

3.3.6.2 How to Build Integrative Goal Functions?

The argumentative effectivity of goal functions, such as 'exergy' or 'ascendancy', or of more explicitly normative concepts, such as 'ecosystem integrity' or 'health', is quite different regarding their usefulness in science, their applicability in the management processes, or their quality in public discourse (cf. de Jong, this vol.). Additionally, the new goal functions concept has to demonstrate that it mediates better (with less costs; providing a more accurate basis for decisions) than competing approaches. What could be the function of goal functions in the negotiation of environmental alternatives and strategies? Environmental policy objectives are constructs, negotiated at the interface of science and politics, and stabilized in public discourse. The significance of deducing goal functions from dynamic structural models of ecological systems is not purely cognitive, because they contain interpretative claims to promote objectivistic environmental policy goals. Should this operation succeed, it would have the advantage, that protection goals and principles for the 'right' dealing with different ecosystems could be determined. They would be beyond the dispute of different interest groups. This path, however, is paved with several complications.

[4] Even the discrimination of conclusions from the factual to the normative isn't indeed a logical fixum or 'law' but a construction of a particular (modern) culture that has foregoing divided these two spheres of world's reality.

In scientific modeling, goal functions are an instrument to describe system dynamics. It is one of the purposes of goal functions, to increase the prognostic capacities of models, especially, of ecosystem models. The transfer of this kind of elaborated scientific knowledge into practically useful management schemes generates a tendency to convert *descriptive tools* into *prescriptive rules*. This tendency is particularly worrysome because the definitions of goals or targets for individuals, for social and ecological systems are simultaneously affected by decisive cultural concepts. This is observable, e.g., in form of the diffusion of categories and ideas from physics and biology to economy and sociology, especially, whereas they appear in more general approaches, such as systems theory or synergetics (cf. Schwarz and Trepl, this vol.).

The idea of a legitimate integrative approach stands and falls with the proof that goals or goal functions exist for human individuals, social, economical and political systems, which are somehow compatible to ecosystemic goal functions. The following overview indicates some similarities.

- human individuals try[5] - to keep alive, - to produce offspring, - to acquire a lot of goods, - to satisfy themselves (or each other) by sublimation and cultivation of abilities of self-expression (e.g., piano or cooking), or - to transcend of the mere boundaries of existence;
- social systems try - to prolong their self-reproduction,..., or - to free humankind from every burden by reaching (in Hegelian/Marxian terms) the 'realm of freedom' surpassing the 'realm of necessity', which is to be understood as a relict of natures' domination of man;
- economical systems try - to reproduce an economical unit, e.g., an enterprise, - to make profit, - to determinate a market, - to change financial in political power in order to determine (world) markets and to limit unwanted political influences or better: to control them;
- political systems try - to sustain their power, - to maintain a given order, - to administrate or solve problems, - to change from the domination of man to the administration of things, or - to make progress related to justice, democracy, welfare, etc.

Societies are not homogeneous units, however. Thus, a mediation of ecological goal functions and the goals of human societies has to recognize the functional and structural differentiation of modern societies (Parsons 1975, 1996). Societies consist of several sub-systems, such as the economical, political, scientific, and the juridical system, which have differentiated, heterogeneous and conflicting goals while following their particular operational logic (Luhmann 1984). The different sub-systems generate varying preferences and allow actors to interpret the targets of the social unit. Society does not have a given and accepted system-intrinsical goal definition.

[5] This overview works without regard to the philosophical necessary distinctions between 'teleonomy' (non-intentional processes as for instance adaptation, that lead to effects that are to describe as goal-attainment) and teleology (intentional goal-attainment).

The second problem of a unified goal function concept emerges on the level of organizations that have to integrate targeting conflicts. They are frequently forced to perform internal reforms by redefining their targets because their social environment changes. Profit maximization, for instance, is obviously a goal function of enterprises under conditions of a capitalistic economical system. But economic actors have to follow several, sometimes conflicting goals in realizing this target, e.g., to defend a market share against competitors, which may reduce short term profit.

The third and perhaps main problem deals with the contradictions emerging from of the different (but nested) reference systems of goal functions, namely: individuals < > populations < > societies < > ecosystems. The first difference is well known in evolutionary theory and sociobiology: reproductive advantages for the species can be of disadvantage to the individual (Wilson 1975). It is as well known in social science, where classical economy and liberalism hold that economic egoism results in disadvantages to certain individuals, but still leads to public benefits. Ecosystems finally, which are composed of a huge number of different species, are as well a problem with their varying population numbers and community organizations: an advantage for one species must not be (per se) an advantage for the stability of the whole system of coevolving species. The situation becomes even more complicated if the competitive and cooperative interactions of all coexisting human societies in relation to the global ecological system are considered.

Fourth, we have to consider the difference of intentionally defined goals and the non-intentional development of goal functions. Besides all parallels, as self-conscious individuals we can define the goals of ourselves, and – in principle – also of our societies and their institutional sub-systems and organizations. Natural systems may follow goal functions or perform an evolution that can be described as goal-attaining teleonomy, but social systems are in principle units, that are able – up to some degree – to enforce intended and directed developments, which can be described as real teleology. However, it makes sense to speak of sociocultural or societal evolution as an analytical tool in order to investigate not only the parallels but also the differences to biophysical evolution (Giesen and Lau 1981; Bühl 1984; Burns and Dietz 1992a, b). We can try to assimilate an increasing rate of energy and resources, to reduce the consumption of energy and resources, to adapt to the ecological environment, or to determinate the environment by active transformation. Observable strategies are certain combinations of these four main directions of eco-social rationality related to particular cultural ways of life (Metzner 1994). So even if we find (or agree upon) some goal functions common for ecological and social systems, their status would not be any longer one of an objectively given necessity, but a kind of ecologically informed, voluntarily chosen (that is: contingent) strategy. For each of them, other strategies are possible – but not in each case advisable – that have specific advantages and disadvantages.

Selecting appropriate strategies is a process of measuring up alternatives, governed by value judgments. According to this realization we could say that it is in no way better to ask science alone for guidance than using a divining rod to find out the primary goal function out of a list of several relevant items, such as 'exergy', 'ascendancy', 'integrity', 'resilience', 'stability' or 'human happiness'. 'Non-sense', 'prejudice' or 'in-genious' – my divining rod and I, we chose *human happiness*. In the social discourse about which characteristics of nature should be preserved, immutably, we have to deal with different views of nature and the source of its value: functional or aesthetic; anthropocentric or physiocentric. In view of these difficulties, one can justifiably argue that it is impossible to set 'objective' environmental management or policy goals. In this context I do not support an epistemologically radical position, biased either to relativistic constructionism or to positivistic objectivism, but a pragmatically oriented, socio-ecologically reflected realism.

It is as possible and meaningful to build indicators that try to express, e.g., ecosystem health related to the potential to follow goal functions, as of pollution limits that indicate risks to human health or to the utilization of ecological resources. These indicators can be determined in an objectivistic way, and play an important role in the communication and negotiation process within the triangle of sciences, politics and the general public. However, any scientific 'claim' to represent objective truth remains problematic and is not self-evident. Scientific knowledge has to establish itself in interdisciplinary and public discourses, gaining strength from orientational and instructional practicability and effectivity. According to this position, the value of a unifying goal function concept lies rather in its orientational usefulness and not so much in its quantitative utilization. The quantification is of limited value because any complete framing has to integrate goal functions of different classes of parameters that cannot be optimized mathematically in an obvious sense, as for instance, by means of the Pareto-optimality (cf. Gnauck Sect. 5.6). Consequently, value judgments that measure different classes or dimensions, e.g., aesthetical needs, economical needs, preservation of species, security against floods, etc., are not avoidable. There is no way to solve these *targeting conflicts* using only arithmetical means. The conflicts are to be dealt with psycho-socially by negotiation and mediation. In rare cases, when all affected actors agree upon procedures, indicators and the particular integration of all relevant factors, the use of mathematical decision-models is proper and can lead to success.

In terms of the social promotion of environmental knowledge, the goal function concept seems too general or indifferent to advertise politicians, and too complex for public discourse and the mass media. Instead, the more compact terms of eco-systems integrity or health – to which goal functions can be related – do not only sound better but seem to provide that kind of orientational knowledge that is demanded in policy and public discourse.

3.3.7 Conclusion

Recent efforts to encounter the environmental crisis often appear as a kind of 'muddling-through' strategy concentrating on crisis management or the appeasement of conflicts by means of symbolical policy instead of focusing on problem-solving. Consequently, it is decisive to develop orientational approaches and to work out strategies that connect the debates over technology, economy and the environment. We have to look at real world problems and to integrate the concerned population into the discussion about development perspectives, concrete projects and problem-solutions. Besides all mentioned problems, ecological goal functions seems to provide a scientific basis for the evaluation of ecosystem states and developments. They may be highly useful in the continuous process of balancing human (settled) targets with natural (pre-existing, self-organizing) system targets, especially, if they are related to more general models of ecological integrity or health.

It remains a huge theoretical problem, however, how to link analyses of actor oriented studies (working under premises of methodical individualism) and studies based on systems theory that reconstruct functions independently of the action-leading motives of actual persons by lifting the level of abstraction to more general concepts (like equilibrium or homeostasis, stability, resilience, reproduction, self-maintenance, growing and differentiation of complexity, etc.).

Even though interdisciplinary study is necessary and imperative, it is often treated as a mere statement of good will. The various possibilities to interpret and misunderstand specialized expressions (e.g., 'labor' as a category in physics or in the social sciences) poses a serious problem to interdisciplinary work, manifesting itself in short circuited popular theses (e.g., the prolongation of thermodynamics to concepts of 'social entropy'). Another problem is created by the "ontological" way to draw distinctions between scientific, particularly ecological, objects, which becomes increasingly obsolete in face of global ecological problems. The relationship between natural and technological sciences, the social sciences and the humanities needs further clarification. We need to look at things not just from two angels but from four. Programs that are oriented towards unified sciences promoting one scientific view and neglecting others will hardly help. A simple 'as well as' approach, however, is neither satisfactory. Working together in a method- and issue-oriented way points in the right direction (as exemplified by the 'goal functions workshop').

Obviously, one can tackle problems under legal, economical, political, or scientific aspects. In order to promote processes of decision-making, it is important to mediate the use of the different problem concepts and solution styles. It is characteristic for the discourse about environmental problems and technological risks that it is reactively oriented towards negative consequences of industrial projects. It is not oriented towards approaches that try to shape future developments, for instance, by working with expectations and whole scenarios. However, even the re-active approach tries to include social, societal and ecological aspects as well as cultural and psychological ones. However, re-active approaches do characteristically assume that all aspects are quantitatively comparable in order to determine the optimal decision. It is evident, however, that a common measure for all aspects, e.g., by monetarization, is difficult to construct. Each concept that is used to promote social problem solving capacities should be aware that not a contradiction-free system of ideas is in question, but systems of social practice that escape 'pure' abstraction a good part of the way.

References

Alcamo J (1984) Acid rain in Europe: a framework to assist decision making. International Institute for Applied Systems Analysis, Working paper 32, Laxenburg

Ayres RU, Simonis UE (ed) (1992) Industrial Metabolism – Restructuring for Sustainable Development. United Nations University Press, Tokyo

Bailey R (ed) (1995) The True State of The Planet. Free Press, London, Toronto

Bechmann G et al. (1996) Sozialwissenschaftliche Konzepte einer interdisziplinären Klimawirkungsforschung. Forschungsbericht für das BMBF, Forschungszentrum Karlsruhe

Beck U (1986) Risikogesellschaft – Auf dem Weg in eine andere Moderne. Suhrkamp, Frankfurt/Main

Beck U, Giddens A, Lash S (1994) Reflexive modernization: Politics, Tradition and Aesthetics in the Modern Social Order. Polity Press, Cambridge

Brakel M van, Builtenkamp M (1992) Action Plan Sustainable Netherlands – A perspective for changing nothern lifestyles. Friends of the Earth Netherlands, Amsterdam

Bühl WL (1984) Gibt es eine soziale Evolution? Zeitschrift für Politik 31:302-332

BUND, MISEREOR (eds) (1996) Zukunftsfähiges Deutschland. Ein Beitrag zu einer global nachhaltigen Entwicklung. Studie des Wuppertal Instituts für Klima, Umwelt, Energie. Birkhäuser, Basel Boston Berlin

Burns TR, Dietz T (1992a) Technology, Social-Technical Systems, Technological Development, An Evolutionary Perspective. In: Dierkes M, Hoffmann U (eds) Technology at the Outset. Suhrkamp, Frankfurt/Main

Burns TR, Dietz T (1992b) Cultural Evolution: Social Rule Systems, Selection and Human Agency. International Sociology 7(3):259-283

Constanza R, Daly H, Bartholomew JA (1991) Goals, Agenda and Policy Recommandations for Ecological Economics. In: Constanza (ed) (1991) Ecological Economics. The Science and Management of Sustainability. New York

Douglas M, Wildavsky A (1988) Risk and Culture. 2nd ed., Berkley, Los Angeles London

Dunlap RE (1993) From Environmental to Ecological Problems. In: Calhoun C, Ritzer G (eds) Social Problems, McGraw-Hill, NewYork, pp 707-738

Dunlap RE, Catton WR jr. (1994) Toward an ecological sociology: the development, current status, and probable future of environmental sociology. In: D'Antonio WV, Sasaki M, Yonebayashi Y Ecology, Society & the Quality of Social Life, New Brunswick, London, pp 11-31

Fischer-Kowalski M (1991) Was ist ökologisch verträglich? – Über die Schwierigkeiten der Verständigung. In: Geyer A, Getzinger G (ed) Chemie und Gesellschaft – Ansätze zu einer sozial- und umweltverträglichen Chemiepolitik. München, pp 143

Fowlkes MR, Miller PY (1987) Chemicals and Community at Love Canal. In: Johnson BB, Covello VT (ed) The Social and Cultural Construction of Risk – Essays on Risk Selection and Perception. Kluwer Academic Press, Dordrecht, pp 55-78

Giesen B, Lau C (1981) Zur Anwendung darwinistischer Erklärungsstrategien in der Soziologie. In: KZSS 33:229-256

Gras A (1993) Grandeur et Dépendance – Sociologie des macro-systèmes techniques. PUF, Paris

Hannigan JA (1995) Environmental Sociology, a social constructionist perspective. Routledge, London New York

Hansen A (1993) Greenpeace and press coverage of environmental issues. In: Hansen A (ed) The Mass Media and Environmental Issues. Leicester

Huber J (1995) Nachhaltige Entwicklung – Strategien für eine ökologische und soziale Erdpolitik. Sigma, Berlin

Japp KP, Krohn W (1996) Soziale Systeme und ihre ökologischen Selbstbeschreibungen. Zeitschrift für Soziologie 3:207-222

Jungermann H, Slovic P (1993) Charakteristika individueller Risikowahrnehmung. In: Krohn W, Krücken G (eds) Riskante Technologien: Reflexion und Regulation – Einführung in die sozialwissenschaftliche Risikoforschung. Suhrkamp, Frankfurt/Main, pp 79-100

Krohn W, Küppers G, Paslack R (1987) Selbstorganisation – Zur Genese und Entwicklung einer wissenschaftlichen Revolution. In: Schmidt SJ (ed) Der Diskurs des Radikalen Konstruktivismus. Suhrkamp, Frankfurt/Main, pp 441-465

Lash S, Szerszinsky B, Wynne B (1996) Risk, Environment and Modernity, Towards a New Ecology. Sage, London Thousand Oaks New Delhi

Leakey R, Lewin R (1993) Der Ursprung des Menschen. Fischer, Frankfurt/Main

Löther R (ed) (1988) Tiersozietäten und Menschengesellschaften – Philosophische und evolutionsbiologische Aspekte der Soziogenese. Fischer, Jena

Luhmann N (1984) Soziale Systeme. Westdeutscher Verlag, Opladen

Luhmann N (1986) Ökologische Kommunikation. Westdeutscher Verlag, Opladen

Luhmann N (1991) Soziologie des Risikos. De Gruyter, Berlin New York

Maxeiner D, Miersch M (1996) Öko-Optimismus. Metropolitan Verlag, Düsseldorf

Mazur A (1991) Putting Radon and Love Canal on the Public Agenda. In: Couch SR, Kroll-Smith JS (eds) Communities at Risk – Collective Responses to Technological Hazards. Lang, New York San Francisco Bern Frankfurt/Main Paris London, pp 183-203

Metzner A (1993) Probleme sozio-ökologischer Systemtheorie – Natur und Gesellschaft in der Soziologie Luhmanns. Westdeutscher Verlag, Opladen

Metzner A (1994) Offenheit und Geschlossenheit in der Ökologie der Gesellschaft. In: Beckenbach F, Diefenbacher H (eds) Zwischen Entropie und Selbstorganisation: Perspektiven einer ökologischen Ökonomie. Metropolis Verlag, Marburg, pp 349-391

Metzner A (1997) Konstruktion und Realität von Umwelt- und Technik-Risiken – Ansätze sozialwissenschaftlicher Risikoforschung. Zeitschr f angewandte Umweltforschung, in print

Mol APJ (1995) The Refinement of Production, Ecological Modernization Theory and the Chemical industry. Van Arkel, Utrecht

Parsons T (1975) Gesellschaften – Evolutionäre und komparative Perspektiven. Suhrkamp, Frankfurt/Main

Parsons T (1996) Das System moderner Gesellschaften. Juventa-Verlag, Weinheim, p 200

Perrow C (1984) Normal accidents–Living with high-risk technologies. Basic Books, New York

Popper KR (1982) Logik der Forschung, 7th edn.. Mohr, Tübingen

Prigogine I, Stengers I (1981) Dialog mit der Natur – Neue Wege naturwissenschaftlichen Denkens. Piper, München Zürich

Schluchter W, Metzner A (1996) Globalization, Environmental Awareness, and Ecological Behavior shown at the Example of the Federal Republic of Germany. In: Ester P, Schluchter W (eds) Global Development of Societies and Environment. Tilburg Univ Press, Tilburg

Spector M, Kitsuse JI (1987) Constructing Social Problems. Wiley, New York

Thompson M, Ellis R, Wildavsky A (1990) Cultural Theory. Colorado, Oxford

Ungar S (1992) The rise and (relative) decline of global warming as a social problem. The Sociological Quarterly 33:483-501

WCED (World Comission on Environment and Development) (ed) (1987) Our Common Future. Oxford University Press, Oxford NewYork
Weichhart P (1989) Werte und Steuerung von Mensch-Umwelt-Systemen. In: Glaeser B (ed) Humanökologie – Grundlagen präventiver Umweltpolitik. Westdeutscher Verlag, Opladen,
Weingart P (1983) Verwissenschaftlichung der Gesellschaft – Politisierung der Wissenschaft. In: Zeitschrift für Soziologie 12:225pp
Wildavsky A (1995) But is it true? – A Citizen's Guide to Environmental Health and Safety Issues. Harvard University Press, Cambridge, London
Wilson EO (1975) Sociobiology – The New Synthesis. Cambridge London

3.4 Ethics and the Environment. How To Found Political and Socio-Economic Eco Targets

Werner Theobald

Abstract

Modern economic and environmental ethics are preferably conceived as institutional ethics in order to guarantee high effectiveness. The genuine ethical question of how the political and socio-economic eco targets of the institutional arrangements can be individually founded has almost never been answered. It must however be answered to make sure that a sustainable development can take place, because political-structural measures have no effect in the long term if they do not address the subjective willingness of the individuals to translate these measures into actions and to be involved in their creation: The unregulated externalities in economics are often better known to the economic agents than to the state because they continually recreate such externality components with their product and process innovations as well as with inducing forms of consumer behavior. In order to be able to create support here, a corresponding environmental-moral consciousness is demanded of the economic agents themselves, and this can only be founded by individual ethics.

3.4.1 Ethics and the Environment. Two Basic Approaches of Economic Ethics

There is a widespread consensus regarding the belief that the global environmental problems we are confronted with today are consequences of inconsiderately progressing economic growth. At the very least we admit that "the economy" is "centrally involved" in these problems[1]. Environmental ethics therefore has substantially to be economic ethics as well.

In the present economic ethics debate, which has been held with no less enthusiasm than the long-lasting environmental ethics discussion, we can differentiate between two main currents: a so-called "preference-theoretical" and a

[1] Homann, Blome-Drees (1992), p 9

F. Müller M. Leupelt (Eds.)
Eco Targets, Goal Functions, and Orientors

"restriction-theoretical" approach[2]. The former seeks to achieve the (re)structuring of the modern economy via a change in consciousness on the part of the economy's agents (correction of preferences and expansion of the individual's or company's targets). For this purpose they draw on more or less basically founded moral ideals. On the other hand, the latter approach attempts to set in motion an (ecological) change in the economy by altering the economic restrictions and the resulting behavioral incentives. More succinctly: one approach wants to "bring the economy to reason", while the other wants to "bring reason to the economy"[3] (by "reason" we mean here practical-ethical reason).

As often the case with extreme opposites, the "truth must lie in the middle". Preference-theoretical approaches fail to a great extent to recognize the concrete socio-economic conditions under which moral ideas can be shown to advantage – in short: they lack the possibilities of practical application. On the other hand restriction-theoretical approaches with their sights set precisely on the latter predominantly do without the genuine ethical undertaking to establish moral norms of action; in other words, they lack to a certain extent an "ethical substructure". The strength of the one side is at the same time the weakness of the other and vice versa.

As far as I can see, there has been hardly any attempt (with the exception of Ulrich 1994) to convey both (currently) seemingly irreconcilable opposites within the framework of an integrative theory. The necessity of such an undertaking, however, is perfectly obvious: An ethics of economics without "ethical substructure" is basically not "ethics" at all if we accept that the classical feature of ethics is its ability to establish moral norms. (This is why Homann, one of the most important representatives of the restriction-theoretical concept, is consistent in speaking rather of a judicial-political program with normative implications)[4]. An ethics of economics, basically founded but lacking applicability, cannot be an ethics for economics if we posit, like Fichte, that ethics should produce "knowledge that causes action"[5]. If we wish to resolve this dilemma and develop an economic-ethical theory that satisfies both, i.e. the requirements of ethical foundation and normative-practical applicability, then we must attempt to make preference-theoretical approaches refer more to their application or to underpin the restriction-theoretical conceptions with a systematic-ethical foundation. I shall attempt a solution which takes both options into account.

---

[2] Homann, Blome-Drees (1992), 103pp

[3] Kersting (1994), 351.

[4] Homann, Pies (1994).

[5] Pieper (1994), 58.

3.4.2 The Problem of Responsibility

One of the key concepts of environmental ethics is that of responsibility; this has been the case at the very latest since the appearance of Jonas' much noted work "The Principle of Responsibility"[6]. As Bayertz has stressed, the concept of responsibility depends on three major conditions: it presupposes causality, intentionality and individuality[7]. This means, in order for someone to be made responsible for something (e.g. the bad consequences of an action), he or she must in principle be able to be identified as the originator or responsible party (individual attributionality). Furthermore, he or she must be or have been able to have had a causal influence on the occurrence (causality), and also to have intended what is attributed to him or her as the responsible party (intentionality)[8]. These three criteria are valid for the classical concept of responsibility, which is generally retrospective (subsequent pin-pointing of a "guilty party" for the negative consequences of the actions). Likewise, the conditions have to be implemented – albeit within limits – for the modern prospective or preventative responsibility model if talk of "responsibility" in this connection is also to be clearly understood. Above all this is valid for the third condition "individuality", since practically speaking it makes little sense to say, for example, that "humanity" has a responsibility of prevention regarding nature.

Global environmental changes, such as loss of biodiversity, the hole in the ozone layer, climate shifts, etc., simply cannot be attributed to individuals. Individual units, be they individual persons, individual companies or individual business concerns, contribute relatively little to the hole in the ozone layer - so little in fact as to be negligible. From this, we could conclude that it makes little sense to tackle the awareness of the individual—as recommended by the preference-theoretical model—in order to have an efficient influence on events. Still, global scale changes are induced by the cumulative sum of all the small, negligible contributions. Before we are too hasty in giving preference to an exclusively restriction-theoretical therapy concept, we need to consider the following analogy.

In democratic elections, the causal influence of the individual on the end result is extremely, if not negligibly small. Yet, in Germany and other western democracies there is an election turnout of seventy per cent on average. What is (nevertheless) effective here – and only in this way can democracy function properly – is clearly the "democratic attitude" of the citizen, which expresses itself in wanting to vote in the election (or – depending on the assessment – having to vote) even when the factual influence of the single vote is so small. Whoever votes wants to make this influence count and this "wanting" is based on

---

[6] Jonas (1979).

[7] Bayertz (1995).

[8] Although, we should also ask if responsibility should not also be carried for unintended results which one brought about oneself.

the individual awareness of co-responsibility for the state or democracy in which one lives.

Correspondingly, there is a co-responsibility of the individual for the hole in the ozone layer, the climate shifts, etc. As long as the development of an awareness of co-responsibility regarding environmental damage, which draws with it corresponding behavior, is not more widely spread, institutional arrangements are the only effective measures as demanded by the restriction-theoretical approach and exerted by judicial-political measures, such as economic incentives or legal sanctions. Still, we must bring people to co-responsibility with regard to environmental problems: It is a crucial for the function and maintenance of an individual environmental attitude as it is for the democratic attitude. In the long term it is its single most important foundation. Without the subjective willingness of the individuals to translate political-structural measures into action, these are unlikely to have long lasting effects[9].

3.4.3 Eco Targets and the Principle of Self-Responsibility

On security grounds we should first look for solutions at the institutional level. It should certainly not be left more or less to chance as to where those moral ideas come from, which form the necessary condition to create institutional arrangements and to set *ecological targets*. Leading representatives of institutionalism, in reference to the above, name as their moral source the "heads and hearts of individuals"[10] or rely on the power of culture (education of awareness via scientists, intellectuals, the media etc.). We should be able— equally on security grounds—to fall back on systematically founded arguments however. This is exactly what traditional environmental ethics is attempting in its efforts to establish ecological norms of action.

The above-mentioned work by Jonas[11] represents one of the most committed attempts of this kind. It is also one of the most controversial due to its intention to establish the necessity for preventative responsibility for nature objectively, i.e. in nature's own self-esteem. Birnbacher amongst others has shown recently that every attempt to establish assumed values concerning objective structures and states of nature and the environment must fail[12]. In environmental-moral approaches and conceptions the proportion of subjective factors in norm-setting is necessarily high; we can try to reduce this but we will not be able to eliminate it completely. Nevertheless, according to Birnbacher, ecological ethics can show intersubjectively accepted arguments and forms of argument or develop plausi-

[9] SRU (1994), 25.
[10] Homann, Pies (1994), 11.
[11] Jonas (1979).
[12] Birnbacher (1991), 297.

bility arguments, which make one inclined to agree to a moral principle[13]. This way, environmental morals progresses towards the generation of environmental-moral norms by systematic argumentation.

Birnbacher suggests the following solution (translated by the author): "It is considerably less problematic to believe that assumed values concerning the subjective well-being or suffering of beings who are capable of a sense of well-being or suffering can be understood by everyone than assumed values concerning objective structures and conditions: With regard to the moral relevance of well-being, satisfaction of needs, freedom from fear and suffering of conscious beings, a consensus can be much more easily reached than with regard to moral relevance of any other natural values"[14] —for example, an "intrinsic value of nature" which (as with Jonas) is made the basis for establishing environment-protecting and environment-sustaining action.

By reverting to this "subjective dismay" we gain an argumentative starting point in order to systematically anchor the development of environmental-moral attitudes in the "heads and hearts of the individuals", and not to (more or less) leave this to chance. A subjectivist axiology of this kind, furthermore, has the advantage of overcoming the very obstacle that is quoted by the restriction-theoretical approach as the main reason for their abstinence from a genuine ethical establishment of moral norms: the modern pluralism or relativism of values[15]. Reverting to "subjective dismay" it brings into play exactly what principally lies beyond all possibilities seen in perspective, something that is to a certain extent "irrefutable".

Such a procedure has recently been criticized in various respects. On the one hand, it is accused of making the value of nature and its partial systems dependent on the changing social preferences and that it is not possible to foresee whether these preferences will in future be well-disposed towards nature and its values in the same manner[16]. On the other hand, this procedure has to take a good look at all arguments which are directed against any anthropocentric approach[17]. Birnbacher has to a great extent defeated this critique by pointing out the necessary differentiation of subjective preferences into "demand values" and "transformative values"[18]. So it is not always in ones best interests if ones preferences are satisfied as they are, and as they show themselves in the demand for goods. In addition to these „demand values", "transformative values" exist which qualitatively change preference structures and make them richer. The experience of nature in particular, which is not totally deformed by culture, can do this. A

[13]Corresponding to this strategical argument see Nortons "Convergence Hypothesis" (Norton 1991).
[14]Birnbacher (1991), 297f.
[15]Cf. e.g. Homann, Blome-Drees (1992), 15.
[16]Sagoff (1974), 224.
[17]Spaemann (1980), 197.
[18]Birnbacher (1991), 298f.

lot of studies show that in future these non-material values will become considerably more significant as material welfare increases.

Another point of criticism however is in my opinion more significant: with reference to the acceptance- and agreement-causing premises of the avoidance of suffering, there is still no concrete (environmental)moral norm being established. Here we need the imposition of further premises such as the Golden Rule, which makes it understandable why I, objecting to the possibility of my own suffering, should not cause any other conscious organism to suffer or we need the acceptance of an irrefutable, to some extent self-evident "intrinsic value" of suffering. This acceptance is the basis of pathocentrism[19].

The only possibility to rationally establish (environmental)moral norms on the basis of the posited subjective approach might be the development of arguments in consistent continuation of its subject-centered components. These arguments make clear that mankind damages itself when it causes another conscious being to suffer or damages nature in general—also in respect to its abiotic component. Only in this way, in the *consciousness of self-responsibility*, does one come to a basic moral principle according to which environmentally conscious behavior—also and particularly under pluralistic value conditions—can indeed be formed in an effective manner.

3.4.4 The Significance of Institutional Arrangements

If such considerations regarding the systematic founding of ecological norms of action are not supposed to be just "a game of philosophical marbles" and intend to lead to effective results[20], then they need to be imbedded in institutional structures. As long as a corresponding environmental-moral conscious does not yet have blanket coverage, nature will be threatened by overexploitation—above all by economic activities: The environmentally conscious entrepreneur either runs the risk of suffering financial ruin, or returns to exploitative behavior. This dilemma results from the consciously implemented, highly competitive structure of modern market economies. Without institutional regulations, which for example impose the installation of filtering systems by law for all possible economic agents, there is the risk of exploitation by all those competitors who fail to fulfill such measures—for whatever reason[21]. In order to counteract this, the environmental-morally conscious entrepreneur, in the case of environmental restrictions not yet reflected in the judicial-political framework, has to become a "political entrepreneur" who provides for a sufficient internalization of the still unregulated externalities via legislation[22]. "The" entrepreneur is of course a

[19]Birnbacher (1991), 282.
[20]Bayertz (1988), 7.
[21]Homann, Pies (1994).
[22]SRU (1994), 78.

simplified, theoretical construct, which was mostly used to formally describe economic action. In the economic reality one has generally to deal with individual decision-makers in companies – thus concrete, differentiated responsibility models which are in a position to provide the systematic link between individual, collective and corporate responsibility, are necessary in practice[23].

This is not (yet) possible without political-institutional solutions, which in a legally differentiated manner take into account the organizational structures of modern economics. Correspondingly, the institution-ethicists derive the thesis that systematic place of morals in market economics is the judicial-political framework, this thesis is valid only with reference to the applicability "of morals" or —more precisely—with regard to the incentives to moral behavior under dilemmatic competitive conditions. The situation is profoundly altered if an environmental-moral consciousness becomes more widely spread among the economic agents (a high environmental awareness of the consumers for example offers the prospect of success of so-called "competitive strategies" which, with regard to the environmental friendliness of products, gives the manufacturer of such products competitive advantages). The thesis makes no sense with regard to the genuinely ethical undertaking of the foundation of morals; for, what is present as unregulated externalities in economics, is known better to the economic agents than to the state because the agents continually recreate such externality components with their product and process innovations as well as with *their self-induced* forms of consumer behavior[24]. In order to be able to create support here, a corresponding environmental-moral consciousness is demanded of the economic agents themselves, and this can only be established by individual ethics in the manner presented.

References

Bayertz K (ed) (1988) Ökologische Ethik. Schnell & Steiner, München Zürich

Bayertz K (ed) (1995) Verantwortung: Prinzip oder Problem? Wissenschaftliche Buchgesellschaft, Darmstadt

Birnbacher D (1991) Mensch und Natur. Grundzüge der ökologischen Ethik. In: Bayertz K (ed) Praktische Philosophie. Grundorientierungen angewandter Ethik. Rowohlt, Reinbek

Homann K, Blome-Drees F (1992) Wirtschafts- und Unternehmensethik. Vandenhoeck & Ruprecht, Göttingen

Homann K, Pies I (1994) Wirtschaftsethik in der Moderne: Zur ökonomischen Theorie der Moral. Ethik und Sozialwissenschaften 5:3-12

Jonas H (1979) Das Prinzip Verantwortung. Versuch einer Ethik für die technologische Zivilisation. Suhrkamp, Frankfurt/Main

Kersting W (1994) Probleme der Wirtschaftsethik. Zeitschrift f phil Forschung 48:350-371

Korff W (1992) Unternehmensethik und marktwirtschaftliche Ordnung. Zeitschrift Interne Revision 27 (1):1-16

[23] Cf. Lenk, Maring (1995).

[24] Korff (1992), SRU (1994).

Lenk H, Maring M (1995) Wer soll Verantwortung tragen? Probleme der Verantwortungsverteilung in komplexen (soziotechnischen-sozioökonomischen) Systemen. In: Bayertz K (ed) Verantwortung: Prinzip oder Problem? Wissenschaftliche Buchgesellschaft, Darmstadt, pp 241-286

Norton BG (1991) Toward Unity Among Environmentalists. Oxford University Press, New York

Pieper A (1994) Einführung in die Ethik. Francke, Tübingen Basel

Sagoff M (1974) On Preserving the Natural Environment. Yale Law Journal 84:205-265

Spaemann R (1980) Technische Eingriffe in die Natur als Problem der politischen Ethik. In: Birnbacher D (ed) Ökologie und Ethik. Reclam, Stuttgart, pp 180-206

SRU (1994) (Der Rat der Sachverständigen für Umweltfragen) Umweltgutachten 1994. Metzler-Poeschel, Stuttgart

Ulrich P (1994) Integrative Wirtschaftsethik als kritische Institutionenethik. Wider die normative Überhöhung der Sachzwänge des Wirtschaftssystems. Beiträge und Berichte des Instituts für Wirtschaftsethik der Hochschule St. Gallen für Wirtschaft-, Rechts- und Sozialwissenschaften 62:5-42

3.5 Teleology and Goal Functions – Which are the Concepts of Optimality and Efficiency in Evolutionary Biology

Wolfgang Deppert

Abstract

If everything happens strictly according to natural law, which meet extremum principles, in what way can the possibility for life be characterized, so that the optimization processes of evolution can take place? Is it legitimate to 'enlarge' the natural laws by certain laws of conservation? The question then arises of which are the new conservation quantities that are introduced by life itself?

The concept of genidentity, which must not be confused with the biological concept of genes, is introduced and used to characterize the interface between animate and inanimate systems by the principle of conservation of genidentity. It becomes clear that animate systems can differ in the way and in how reliably they achieve their goal of self-preservation. The abundance of possibilities to be – or not to be – able to reach this goal offers the necessary scope, in which the notion of a postulated assumed optimization in the theory of evolution is conceivable. The conservation principle of genidentity cannot be explained by physics as a principle of a supposed reality, or of what is consciously experienced. Consequently, there is a difficulty of principle to fully integrate evolutionary theories of nature into physical reductionism.

The conservation principle of genidentical systems brings about the possibility of evolutionary optimization by ranking these systems. An optimal lifespan of individual genidentical systems refers to the conservation principle of genidentical systems on a second supra-individual level (the species). The optimization of the growth of a species needs the conservation of a genidentical system on a third level (symbiotic systems). The ranking of genidentical systems onto ever higher levels - so that the higher conservation principles always impose restrictions on the ones below - would come to an end when the minimization of raw materials and energy consumption limits all possible and available resources.

Since the spectrum extending between opposite goals of optimization lies within the range of possible means of optimization, supposedly evolutionary goals of optimization are always attributed to nature by the observer.

F. Müller M. Leupelt (Eds.)
Eco Targets, Goal Functions, and Orientors

3.5.1 Introduction

It is an ancient idea that the behavior of animate beings is directed at certain goals. These goals are constant for significant periods, or even during the being's entire life. Coining the *etelechy* concept, this idea prompted Aristotle to define *soul* in a way that it is identical with the totality of all developmental steps a being takes from birth to death. Etelechy has also been identified with an intrinsic goal directing all these steps. This teleological definition of life trying to attain its intrinsic goal has been subjected to numerous paradigmatic shifts in the past. It transmuted to an approach so exclusively causal that anyone who seriously holds that natural processes are predestined by their goals is looked at very suspiciously. Nevertheless, in the life sciences, especially, when dealing with human beings, teleologically inspired terms, such as 'development', 'adaptation', or teleological statements, such as 'it is the function of the lung to supply oxygen for the blood' and 'the immune system provides for the health of the organism' etc., are abundant.

Employing system theoretical considerations, we can translate teleological (also called 'finalistic') into purely causal statements – to be sure, this works the other way as well. Whether to prefer the causal or the finalistic mode of interpretation, everyone has to judge by the standards of her or his own personal metaphysics – which you cannot escape from anyway. Yet, there is a strong propensity nowadays for coining new teleological terms, especially, in those villages of the scientific community that contemplate (in part esoterically transfigured) models of the self-organization of nature. The title concepts of this book belong to this new class of teleological terms. Certainly, one has to caution against hidden mystifications that lead to obscure terminology, such as the 'order parameters' and 'slaved modes' terms introduced by Haken.

Another kind of teleological thinking has to be considered much more thoroughly. Human beings touch upon this dimension when they inquire into ultimate questions of purpose, and into ethical problems. According to Humes' Law that is intransgressible, humans necessarily argue teleologically because answers to these questions can never be derived from any form of causal description. The following paper demonstrates how causally irreducable, teleological definitions of this type enter evolutionary descriptions of nature.

3.5.2 Is the Deduction of Natural Laws from Extremum Principles Compatible with the Idea of Evolutionary Optimization?

In his *Theodicy*, Leibniz (1710) vindicated the existence of evil in the world by saying that in comparison to other possible worlds our world possesses the least evil, otherwise God would not have created it. If we agree with Leibniz's idea, ours is the best of all possible worlds in spite of all its flaws[1], since only such a unique, extremely positive determination of the existing world is conceivable with the grace, kindness and reason of God. If this is transferred to the functioning of nature, it means that natural laws must comply with extremum principles, i.e., if one allows a movement to take any possible course, then it should be possible to uniquely ascertain the subsequent movement by using the correct extremum principle.[2] Indeed, the introduction of extremum principles into theoretical physics was very successful, for instance, in the different formulations of the principle of smallest effect, from which the basic equations of physics can be derived.[3]

According to this view of natural behavior determined by natural laws, it is assumed that everything that happens, happens optimally in two respects: first, everything happens strictly according to natural laws, i.e. behavior that deviates from these laws does not exist; and second, the natural laws themselves meet extremum principles. If everything is thus fixed in time and space, one has to ask what do evolutionary theorists mean when they say that, in the course of evolution, life forms have adapted to so-called objective reality[4], as if they had not

[1] Leibniz says: "gäbe es nicht die beste (optimum) aller möglichen Welten, dann hätte Gott überhaupt keine erschaffen. 'Welt' nenne ich hier die ganze Folge und das ganze Beieinander aller bestehenden Dinge, damit man nicht sagen kann, mehrere Welten könnten zu verschiedener Zeit und an verschiedenen Orten bestehen. Man muß sie insgesamt für eine Welt rechnen, oder, wie man will, für ein U n i v e r s u m. Erfüllte man jede Zeit und jeden Ort; es bleibt dennoch wahr, daß man sie auf unendlich viele Arten hätte erfüllen können und daß es unendlich viel mögliche Welten gibt, von denen Gott mit Notwendigkeit die beste erwählt hat, da er nichts ohne höchste Vernunft tut." (Leibniz 1710/1968, p 101)

[2] On the importance of extremum principles cf. Nagel 1961, p 407: "These principles assert that the actual development of a system proceeds in such a manner as to minimise or maximise some magnitude which represents the possible configurations of the system." In a footnote Nagel adds: "It can in fact be shown that, when certain very general conditions are satisfied, all quantitative laws can be given an 'extremal' formulation."

[3] The extremum principle, which is mostly used within modern physics, is the Hamiltonian principle. It can be regarded as a special form of the principle of the smallest effect. If you characterize a special physical system by means of its Lagrangian function, then the Hamiltonian principle demands, that the time-integral of the Lagrangian, which is taken between two points of time, has to be an extremum. This demand allows us to deduce the Euler-Lagrangian differential equations, which determine the behavior of the system. This procedure is used until today for determining the basic differential equations of the realm of physics, such as the Schrödinger equation, the Dirac equation, the field equations of the theory of General Relativity or of the Cartan-Hehlian theory of torsion. With respect to the classical theory of the Hamiltonian principle cf., e.g., Goldstein (1950) or Landau (1966).

[4] The term 'objective reality' is used as a collective term, denoting different descriptions with the same metaphysical meaning, as they are extensively found within the literature of the theoreticists of biological evolution and from which I give here some examples: "außersubjektive Wirklichkeit" (Lorenz 1973, p 11), ">>objektive<< Wirklichkeit" (ibidem), "reale Außenwelt" (ibidem, p 12), "Außenwelt" (ibidem, p 16), "reale Welt" (ibidem), "Wirklichkeit" (ibidem), "Welt" (ibidem), "außersubjektive Welt" (ibid), "reale Wirklichkeit" (ibidem), "äußere Wirklichkeit" (ibidem), "das An-sich-Bestehende" (ibidem, p 17), "reale Welt, unabhängig von Wahrnehmung und Bewußtsein" (Vollmer 1975, p 28), "objektive Wirklichkeit" (ibidem, p 29),

already been optimally adapted. It is true that evolutionary theorists have always believed that evolutionary development is completely governed by natural laws. Since the term 'objective reality' is certainly used to mean a reality determined by the natural laws, this concept of adaptation in the theory of evolution is used in a contradictory fashion and must be first clarified before a discussion can take place about the way in which concepts like optimality and efficiency can be applied to biological processes.

Natural laws may be deduced within a range of possible worlds with the help of extremum principles, and on top of this, all actions follow natural laws in an optimal manner. The question of the optimal organization of life-forms is, thus, only meaningful, if the potential state space of reality enlarges with the emergence of each new life-form. Certainly in this case, it is not a question of optimal behavior in relation to the existing natural laws, since this behavior, on the basis of the definition of natural laws and their functions, has always had to be optimal.

The aims of evolutionary optimization are not 'objective' in the sense, that they could exclusively be traced back to natural laws.[5] Are the aims of evolutionary optimization settled by the observers of evolution themselves and pushed underneath nature afterwards?

To show this explicitly, one has now to ask in what way the possibility spaces for life, in which the optimization processes of evolution are to be imagined, can be characterized.

3.5.3 The Concept of Retaining Individual Genidentity Creates New Scope for Optimization.

Instead by extremum principles, the natural laws can also be characterized by postulating certain laws of conservation of measurable quantities such as energy, momentum, mass, charge, quantum number etc. This reformulation is possible because the extremum principles identify quantities that cannot change during observation. These quantities are the aforementioned conservation quantities.[6]

"objektive Realität" (ibidem), "das, was wir in seiner 'realen Existenz' zu erforschen trachten", (Wuketits 1978, p 26), "etwas (also die Wirklichkeit), das vor und unabhängig von der Erkenntnis existent ist" (ibidem).

[5] This consequence follows only in that case in which the laws of nature are exclusively interpreted as cosmic laws, unrestrictedly applicable over the whole physical world. If, however, the notion of laws of nature get expanded in such a way, that laws are also defined as natural laws which are restricted with respect to that domain of applicability, which characterizes a single living object, then it is, of course, thinkable, that there are laws of nature that include the principle of preservation of genidentity or that are deducible from it (cf. Deppert 1989, 1992a and 1993).

[6] Cf. for instance Weyl 1931, p 195. Here Weyl, among other things, explains: „Überhaupt kann man es als eine Regel aufstellen, daß jede Invarianzeigenschaft vom Charakter der allgemeinen Relativität, die eine willkürliche Funktion einschließt, zu einem differentiellen Erhaltungssatz Anlaß gibt." From this we can deduce the converse that every property of preservation, which

Thus, one could also ask what new conservation quantities are introduced by life itself.

The conservation quantities describe physically closed, i.e., a system has a certain total energy, total momentum, total mass, total charge and for instance, a total number of baryons. Systems that can be described completely by such conservation quantities do not differ from each other if they agree in all these quantities: they do not have any identity of their own. Thus, in quantum physics one speaks of indistinguishable particles if, in accordance with the theory, their behavior can be uniquely described by certain quantum numbers (these are in fact the conservation quantities of quantum physics). The fact that elementary particles cannot be differentiated is also paralleled in their composition, in how they are made up of atoms, ions or molecules. Accordingly, neither systems comprised of elementary particles, atoms, or molecules are equipped with an identity of their own, since any two systems having the same conservation quantities cannot be differentiated. Hydrogen atoms or water molecules do not have a history.

The situation is different, however, concerning macroscopic objects.[7] They have an ascertainable duration in time, and are subject to certain changes, which, nevertheless, allow us to speak of the one and the same object. For example, consider a phonograph record, which during its existence gradually shows traces of use in the form of scratches. Despite these scratches, we are convinced that this is the same record that we once received as a present from a friend years ago. To characterize objects or systems that have a history which are therefore subject to certain irreversible changes during their existence, I would like to use the concept of genidentity. The first idea was introduced by Kurt Lewin, and was taken up by philosophers of science such as Russell, Carnap, Reichenbach, Grünbaum, and Stegmüller. In some cases, they dealt with at considerable length.[8] To elucidate the concept of genidentity some definitions are given here that are often used in the theory of science.

---

may also, of course, be described as a differential property of preservation, characterises an invariable property of nature, i.e., a natural law. Cf. also Deppert 1989, p 223

[7] There is only one possible reductionist explanation for the fact that macroscopic objects are always differentiable objects although they are supposed to be composed exclusively of indistinguishable particles. This explanation goes as follows: particles (fermions, which fulfil the Pauli-Principle, are meant here) can, through electromagnetic interaction, form a variety of stable configurations, which on the basis of the enormously large number of particles in the case of macroscopic objects form a complex space of possible configurations, so that it is extremely unlikely to ever meet two identically formed macroscopic objects.

[8] Cf. Lewin 1922, p 10ff. There Lewin says: "Wir wollen, um Verwechslungen zu vermeiden, die Beziehung, in der Gebilde stehen, die existentiell auseinander hervorgegangen sind, *Genidentität* nennen. Dieser Terminus soll nichts anderes bezeichnen, als die genetische Existentialbeziehung als solche." The notion 'existentiell auseinander hervorgehen' or 'genetische Existentialbeziehung' refers to the existence of one and the same thing in the flow of time if we regard things different even having different positions only in time. I have not used Lewin's division of the concept of genidentity into 'Avalgenidentität', 'Individualgenidentität' and

Carnap (1928, p 170 (§128) and p 219 (§159)) states with reference to Kurt Lewin: "Zwei Weltpunkte ((das sind die Punkte, die durch die Angabe von Raum- und Zeitkoordinaten bestimmt sind)) derselben Weltlinie ((das sind die miteinander verbundenen Weltpunkte)) nennen wir 'genidentisch'; ebenso auch zwei Zustände desselben Dinges." Within the context of the discussion of the causal theory of time, Grünbaum (1973, p 189) explains: "... we shall begin with material objects each of which possesses genidentity (i.e., the kind of sameness that arises from the persistence of an object for a period of time) and whose behavior therefore provides us with genidentical causal chains." Reichenbach uses the concept in several places in his work, he differentiates material genidentity from functional genidentity. In his final work, "The Direction of Time", he says: "... *physical identity* of a thing, also called *genidentity*, must be distinguished from logical identity. An event is logically identical with itself; but when we say that different events are states of the same thing, we employ a relation of genidentity holding between these events. A physical thing is thus a series of events; any two events belonging to this series are called genidentical" (Reichenbach 1956, p 38). Because it is not possible anymore within the framework of quantum theory to define a physical object in this way, Reichenbach treats the problem of 'The Genidentity of Quantum Particles' in a separate chapter in the same book. In this chapter Reichenbach differentiates between *material genidentity* and *functional genidentity*: "For macroscopic objects, we define *material genidentity* in terms of several characteristics, which can be divided into three groups. First, we associate material genidentity with a certain *continuity* of change... Second, it is a characteristic of material objects that the space occupied by one cannot be occupied by another... Third, we find that whenever two material objects exchange their spatial positions this fact is noticeable." (ibid., p 225) "We say that the kinetic energy travels from one ball to the other, and by this usage of words we single out another physical entity, the energy, whose genidentity follows different rules. This is not a material genidentity; it might be called a *functional genidentity*, a genidentity in a wider sense. Also of this functional kind is the genidentity of water waves, which we distinguish from the material genidentity of water particles..." (ibid., p 226) On the basis of this indistinguishability of elementary particles and the postulate of the avoidance of causal anomalies, Reichenbach arrives at the conclusion: "In the atomic domain, material genidentity is completely replaced by functional genidentity." (ibid., p 236) On the assumption that the totality of material existence consists exclusively of such atomic particles, he continues: "... there is no material genidentity at all in the physical world; there is only functional genidentity." (ibid.)

It will be sufficient here to define the idea of the genidentity of a system by its ability to have a history. Genidentity characterizes the sameness of a system that

'Stammgenidentität' although these concepts occur again below in the ranking of genidentical systems.

is retained despite the changes it undergoes in the course of its existence. The features of genidentical systems should therefore be differentiated from intrinsic and random (accidental) features. A system which has no accidental features does not have the ability to have a history and should, therefore, according to the definition of the concept given here, not be described as genidentical. Electrons, e.g., have no history and no genidentity because they lack accidental features.

The concept of genidentity must not be confused with the biological concept of genes, which was introduced into genetics in 1909 to describe certain classes of hereditary features. At the same time, it remains to be seen in what respect a gene may also be considered a special genidentical system. Conversely, it is quite clear that not all genidentical systems are genes.

Where exactly the immutable essence of a genidentical system lies, depends to a great extent on the observer of this system. Opinions about when a genidentical system has lost its genidentity can differ widely. So, for example, it is conceivable that some inhabitants from Königsberg may be of the opinion that the city ceased to exist in the last months of the war, because for them the essence of the city lies in particular buildings, streets and bridges which were destroyed by the war. Some inhabitants of the city of Kaliningrad, however, may wish to believe in a *genius loci* that is connected with the geographical location of old Königsberg, in which Kant taught, and which characterizes the essence of this city for them, so that they have expressed the wish to rename Kaliningrad Königsberg again.[9]

With respect to the example of a record, the ability for having a history is connected with its ability to 'get scratches'. But there are many possibilities for the meaning of the immutable essence of the record. Someone could say that the immutable essence is conserved if one can recognize the recorded acoustic events when playing the record. Another one could argue that the record preserves its identity as long as it is possible at all to play the record, and a third one could perhaps claim, that the sameness of the record is maintained if the label, which is fixed on it, is readable.

The accidental feature of a living system is often regarded as its ability to mature or, more general, as its changeability, whereas the intrinsic feature could be considered as the genetical determination or its historically continuous existence.

As long as such a vague definition of the invariable essence is connected with the concept of genidentity, it will not be suitable for the description of scientific findings. Besides, the concept of genidentity is just as applicable to inanimate as to animate matter. In inanimate things we cannot recognize an impulse for self-

[9] Königsberg is the former capital of the former province of East Prussia. It is the city where Immanuel Kant spent his whole life working. The old part of Königsberg was completely destroyed in 1945. Afterwards this district came under Soviet administration and the city was rebuilt under the name of Kaliningrad. Today Kaliningrad belongs to Russia and a movement has been formed by the citizens for the re-introduction of the name Königsberg.

preservation, (so können wir etwa einer Porzellantasse keinen Antrieb zur Selbsterhaltung unterstellen). The contrary holds for all animate things. Therefore, the concept of genidentity allows a special differentiation between animate and inanimate matter attributing the *principle of conservation of genidentity* to animate genidentical systems. This would not be opposed even by reductionists.[10] If this were done for inanimate genidentical systems, hardly anyone would consider it meaningful, unless he/she advocated a metaphysics that attempted to include all physical action under one vitalistic principle. On the basis of this asymmetry between animate and inanimate genidentical systems, I would like to use the principle of the conservation of genidentity as a defining characteristic of the living realm. It does not depend in this case anymore on what is regarded as the essential properties and what is regarded as the accidental properties of a particular animate genidentical system. What is critical here is only whether we can consider it meaningful to attribute a principle of conservation of genidentity to the genidentical system in question.

If the interface between animate and inanimate systems is marked by the principle of conservation of genidentity, it becomes clear that animate systems can differ in the way and in how confidently they achieve their goal of self-preservation. In this respect we can imagine an improvement or even an optimization of the self-preservation properties of genidentical systems. If, for instance, the only possibility of preserving an individual genidentity exists in a fleeing movement, then those living systems that are able to achieve a higher speed of flight will be more successful. The abundance of possibilities to be or not to be able to reach the goal of self-preservation of genidentical systems in changing situations offers the necessary scope, in which the assumed optimization (survival of the fittest) in the theory of evolution is conceivable.

3.5.4 The Status of the Principle of Conservation of Genidentity

Such a principle of conservation is a new type of law because it does not involve (at least at present) the conservation of quantitatively determinable factors as is the case with the physical conservation of energy, momentum, or of the number of baryons. What is to be conserved here is the temporal and spatial existence of an entire system that consists of an invariant essential core and an accidental part that can change. However, it is not at all certain if the behavior of a genidentical system, of which it is assumed that its behavior may be described by means of the conservation principle of genidentity, actually brings about this conservation because on the basis of its composition, a genidentical system can disintegrate

[10]They would confidently expect to be able to replace such a principle with a causal explanation by proposing laws and assuming adequate peripheral conditions.

and forfeit its genidentity. Thus, the conservation principle of genidentity can only be specified by the expected strategies of self-preservation.
Hence, it is not surprising that in the description of nature provided by physics there is no natural law concerning the conservation of genidentity. It is simply not conceivable because it is exactly the goal of the laws of physics to trace animate systems back to inanimate ones, to which we cannot meaningfully attribute such a principle in turn.

Humans have always perceived themselves as beings who make plans and pursue goals. All desires and goals, and the values associated with them, are related to the conservation principle of genidentity. This reveals itself to be the principle of self-preservation, a principle of the will to survive, which we can discover at work in ourselves. By employing this teleological principle, the possibility of justifying what is supposed-to-be or what is willed is created. If we take the view of evolutionary theorists, then we conceive of ourselves as a product of evolution. In other words we have to explain the existence of our own will and of what is supposed-to-be by means of evolution.

In the previous discussion I have used the conservation principle of genidentity to characterize living systems because I believe that we are able to determine the principle of self-preservation in our lives. And it would certainly be a circular process if we attempt to derive our instinct for self-preservation in an evolutionary sense from the principle of conservation attributed to genidentity. This, however, is a difficulty that is inherent in all evolutionary attempts to win new knowledge about our own cognitive abilities.

All evolutionary descriptions of nature have to take into account that they can only be founded on the principle of conservation of genidentity. Because otherwise it cannot be argued how and why living systems shall change through adaptation and thereby "create" evolution. Since David Hume it has become obvious that one cannot extract criteria for a "should-be" or an "ought-to-be" from the description of an assumed reality. Normative "ought-statements" are of a different categorical kind than the principles used to describe what reality is. Thus, the conservation principle of genidentity as a principle of the "willed" or the "ought-to-be" cannot be explained by physics or the laws of physics even if the result can be described by physics.[11] Consequently, there is a difficulty of principle in attempting to fully integrate evolutionary theories of nature into physical reductionism.[12]

---

[11]David Hume had described his is-ought-distinction the first time in Hume (1740). This dichotomy has been called Hume's law later on.

[12]I have described the particular difficulties of a reductionist programme in the essay "Das Reduktionismusproblem und seine Überwindung" (Deppert 1992b).

3.5.5 How the Conservation Principle of Genidentical Systems Creates the Possibility of Evolutionary Optimization by Ranking those Systems

The conservation principle of genidentity is outlined here to elucidate the possibility of optimizations for the concept of evolution and development. Theoretically, these optimizations may be arranged within a general framework of possible dangers and the survival strategies associated with them, if the survival of an individual is at stake. The range of such possibilities is unlimited and very complex. In our efforts to formulate theories that incorporate all concepts of optimization we are thus forced to admit that such theories depend on notions about the whole realm of what danger means and which survival strategies may be possible.

If the conservation of genidentity was the only goal – all living beings as long as they were not dead – would be successful in an optimal way because the range of possibilities consists of only two states, namely of survival and non-survival. Judging by the conservation of genidentity, we have won only one non-optimal condition: having failed to reach the goal of conserving genidentity. We can speak of optimality only, however, if, apart from the optimal condition, there are also other conditions which can at least be regarded as satisfactory. Ultimately, the superlative term of optimality can only result from a comparison of a number of candidates for this condition.

One could imagine that lifespan might be a possible quantity to be optimized, particularly, since longer lifespans are enabled by better survival strategies. In this case, however, it is impossible to speak of an optimum since an indefinite lifespan eludes any possibility of determination. Moreover, the measure of a lifespan is relative. How are we to compare the lifespan of a mosquito with that of an elephant? The lifespan of a particular organism depends on the species to which it belongs. Indeed, it makes sense only to talk of an optimal lifespan referring to the preservation of a species, because life is not only determined by the conservation principle of *individual* genidentical systems but also by their ability to reproduce. Therefore a second supra-individual level of genidentical systems may be introduced, which in turn follows a teleological conservation principle, namely that of the *preservation of the species*.

The optimization of vital processes in order to preserve the species or second level genidentical systems involves winning an advantage in reproduction for one species over another species competing in the same habitat. Optimization occurs within a framework of possible reproductive advantages subject to the reproductive conditions in that habitat. Thus, for instance, we can speak of an optimal limitation on the lifespan of the individual of a species if reproductive advantages result for the species.

If we bear in mind that no species can live in isolation, i.e., that it needs a profusion of other, very different species to survive, then the hypertrophic growth of a species can destroy its own essential conditions for life. Thus, a further optimization in relation to the conservation of a genidentical system on a third level is conceivable, which I would like to call a *symbiotic system*. And just as the second level conservation principle imposes restrictions on the first level conservation principle with regard to individual lifespan, so it is possible that the numerical growth of a species is restricted by a third level conservation principle. This theoretical deduction of various levels of the conservation principle seems indeed to correspond as well to biological findings as to the interpretation given by evolutionary theorists. So it has been observed, e.g., that the oldest she-wolf in a pack seems to take particular care to limit the number of offspring in the pack. From the point of view of classic evolutionary theory such behavior may be understood as the wolf providing for the preservation of its living space, which can be identified with a symbiotic system as indicated within the theory presented here.[13]

The ranking of genidentical systems could certainly be so that the higher conservation principles always impose restrictions on the ones below. This ranking cannot proceed infinitely in a biosphere, whose material and energy recources are limited, and subject to the conservation laws of physics. This restriction on all levels of teleological conservation principles suggest to determine optimization in terms of the minimization of raw materials and energy consumption.

3.5.6 Supposedly Evolutionary Goals of Optimization are always Attributed to Nature by the Observers

It is not possible to quickly grasp the totality of all possible optimizations. The following qualitative argument which demonstrates the impossibility of empirically determining the results of optimal development to this quantitative argument can be added.

Generally, nature provides contrasting solutions to the same problem, for example, in overcoming a threatening environmental situation. An organism either adapts in a literal sense to a given situation, or it makes itself independent of it or it chooses something between these two extremes. There are animals that assume exactly the temperature of their environment, and there are those that have to a large extent a constant body temperature, which can be maintained independently of their environment by complicated temperature regulation systems. Or there are animals that do have a temperature regulation system of their own, but whose mean value is adapted to that of the environment. Another example of the great variety of possibilities for adaptation are animals that, at the

[13]It might be assumed that the higher the level of conservation principle the fewer we find it represented within the genetic material of a species. I am convinced that the genetic code of human beeings does not include genetic information for symbiotic behavior with respect to nature as it is the case for wolves. Therefore, we have to hope that the intellectual abilities of human will overcome this lack of genetic information by realizing the mutual connections within nature, the largest symbiotic system which is conceivable.

first signs of danger, take flight at high speed, in contrast to those, such as the hedgehog, the snake or the tortoise, that, when in danger, sit tight and retreat into a state where they are practically invulnerable. The spectrum between these two extremely different survival strategies is in turn filled with an abundance of different ways to ensure survival, be they are passive like camouflage, or active like defence, or disappearance into the third dimension.

Thus, a whole spectrum extending between seemingly quite opposite goals of optimization covers a wide range of possible means of optimization. Therefore, the scientist searching for optimization, can only postulate the aim of a given vital process with respect to which this process may optimized.

The restrictions imposed by the conservation laws of physics appear to be an exception to this arbitrariness of the determination of optimization. So, for instance, the measure of efficiency of the energy consumption and work rate of heart cells is well defined, provided that the problems in measuring these values have been solved. If the optimization of the heart function is to be investigated, the aim of the optimization must be determined first, however.[14] The ranking system of genidentical systems can be a systematic guideline here. If, for example, one investigates a particular energy consuming function of the heart under stress, one could investigate a suspected optimality with respect to the different conservation principles of the individual, of the species or of the symbiotic system. In relation to the preservation of the species, an excessively high energy consumption, which leads to fatal collapse, could still be regarded as optimal of the individual. In relation to the preservation of the symbiotic system, we can only characterize those energy conversion processes as optimal that have an optimal level of efficiency in a physically defined sense.

The optimization of energy consumption does not bring us one step closer to a uniquely determined goal of the optimization. At best one can determine – as Gibbs (1978, 1986) does – that the level of efficiency reaches a maximum under a particular strain, or that the so-called 'fast' muscles are less efficient than the 'slow' muscles, as has been demonstrated by Alpert and Mulieri (1982, 1986). Nevertheless, it is completely open which goal of optimization these investigations suggest.

If scientists believe that they have discovered something "optimal", which evolution has presented us with, it is still the human intellect that attributed these seemingly necessary goals of optimization to nature. The like situation accurs if we try to unify goals or goal functions.

By introducing the concept of genidentity of conservation principles I have shown that there are different levels of scope for optimization, I, nevertheless, think that it will not be possible to demonstrate how this knowledge can applied to a specific case. We cannot but bid farewell to the hope of an objective determination. We have to develop our own concepts of "optimal" living conditions for which we have to stand up for.

[14] Cf. also Oster 1989.

References

Alpert NR, Mulieri LA (1982) Myocardial Adaption to Stress from the Viewpoint of Evolution and Development. In: Twarog BM, Levine RJC, Dewey MM (eds) Basic Biology of Muscles: A Comparative Approach. New York, pp 173-188
Alpert NR, Mulieri LA (1986) Determinants of energy utilization in the activated myocardium. Federation Proc 45:2597-2600
Burkhoff D, Schaefer J, Schaffner K, Yue DT (eds) (1993) Myocardial Optimization and Efficiency, Evolutionary Aspects and Philosophy of Science Considerations. Basic Res Cardiol 88, Suppl 2. Steinkopff Verlag, Darmstadt
Carnap R (1928) Der logische Aufbau der Welt. Meiner, Hamburg
Deppert W (1989) Zeit. Die Begründung des Zeitbegriffs, seine notwendige Spaltung und der ganzheitliche Charakter seiner Teile. Steiner, Stuttgart
Deppert W (1992) Das Reduktionismusproblem und seine Überwindung. In: Deppert W, Kliemt H, Lohff B and Schaefer J (eds) Wissenschaftstheorien in der Medizin. De Gruyter, Berlin, pp 275-325
Deppert W (1996) Synergetik als eine enttäuschte Hoffnung auf fruchtbare Interdisziplinarität. Kritik am Hauptartikel von Haken H (1996) Synergetik und Sozialwissenschaften. Ethik und Sozialwissenschaften 7(4)
Deppert W (1993) Wer schlägt den Takt? Öffentlichkeit und Leben zwischen Gleichschritt und individueller Rhythmik. Vortrag, First Bamberg Philosophical Mastercourse, 28 June-30 June 1993, 'The Resurgence of Time', Bamberg
Gibbs CL (1978) Cardiac Energetics. Physiological Reviews 58:174-254
Gibbs CL (1986) Cardiac energetics and the Fenn effect. In: Jacob R, Just H, Holubarsch C (eds) Cardiac Energetics. Basic Mechanism and Clinical Implications. Darmstadt New York, pp 61-68
Goldstein H (1950) Classical Mechanics. Addison-Wesley, Massachusetts
Grünbaum A (1973) Philosophical Problems of Space and Time. 2nd (enlarged) edn. Kluwer Academic Press, Dordrecht Boston
Hume D (1740) A Treatise of Human Nature: Being an Attempt to introduce the experimental Method of Reasonung into Moral Subjects. Book III, Of Morals. London
Landau L, Lifschitz EM (1966) Lehrbuch der theoretischen Physik. Vol. 2, Klassische Feldtheorie. Akademie Verlag, Berlin
Leibniz GW (1710) Die Theodizee. Translated by Buchenau A (1968). Meiner, Hamburg
Lewin K (1922) Der Begriff der Genese in Physik, Biologie und Entwicklungsgeschichte, eine Untersuchung zur vergleichenden Wissenschaftslehre. Berlin
Lewin K (1923) Die zeitliche Geneseordnung. Zeitschrift für Physik 8:62-81
Lorenz K (1973) Die Rückseite des Spiegels. Versuch einer Naturgeschichte menschlichen Erkennens. Piper, München
Nagel E (1961) The Structure of Science: Problems in the Logic of Scientific Explanation. New York
Oster GF, Wilson EO (1989) A Critique of Optimization Theory in Evolutionary Biology. In: Sober E (ed.) Conceptual Issues in Evolutionary Biology. Cambridge, pp 271-287
Reichenbach H (1956) The Direction of Time. University of California Press, Berkeley
Stegmüller W (1983) Probleme und Resultate der Wissenschaftstheorie und Analytischen Philosophie. Vol. 1: Erklärung, Begründung, Kausalität. 2nd edn. Springer, Berlin
Vollmer G (1975) Evolutionäre Erkenntnistheorie. Hirzel, Stuttgart
Weyl H (1931) Gruppentheorie und Quantenmechanik. 2nd revised edn, Hirzel, Leipzig
Wuketits FM (1978) Wissenschaftstheoretische Probleme der modernen Biologie. Duncker und Humblot, Berlin

3.6 Conclusions: A Generalizing Framework for Biological Orientation Orientation

Jens Badura, Broder Breckling, Thomas Potthast and Jan Barkmann

The discussion of epistemological aspects of the goal functions concept should help to achieve conceptual clarity to attempt a unified glance back to the interaction between natural and social systems and their respective goals and tendencies. As a result of the workshop discussions about the philosophical aspects of unifying goal functions, the scheme shown in Fig. 3.5.1 was developed. It was used to point to the necessary distinctions that have to be made as well as to the mode of interaction and interference of the different aspects of goal functions. It is instructive to compare this scheme with the scheme presented by Metzner (Sect. 3.3).

First, using the term "goal function" we need to differentiate between *intentional human goals* for environmental development as a result of societal interactions on the one hand and *trends and tendencies in ecosystem development* as a result of physicochemical and organism interactions on the other.

A second distinction has to be made between properties of *ecological models* as mathematical constructions and the assumed properties of *empirical findings* as directly derived from field data. In the modeling context, the term "goal function" can be used subject to the following condition: Considering that "goal function" is a modeling term describing the properties of an ecological model, we may regard the behavior of the model as a hypothesis for the behavior of underlying ecological process it refers to. We call these developmental trends *ecosystem tendencies*. Nevertheless, we must be aware that ecological models derived from field data are also simplifications and abstractions that approximate developments only to a certain degree. If validations support the model, a goal function or a derived set of goal functions may be considered the modeler's best guess of the ecosystem tendencies. Even after calibration and validation of a model with empirical data, "goal function" is not a term describing a 'real' process within ecosystems. It may only be considered as a property of the applied mathematical formalism to characterize a situation. The definition of a goal functions as a property of "objective reality" should be firmly rejected.

From hierarchy theory (Allen and Starr 1982) it can be derived that each observational level of an ecological system has its own specific properties emerging on the particular level as an outcome of the interactions on the lower levels,

F. Müller M. Leupelt (Eds.)
Eco Targets, Goal Functions, and Orientors

which are modified by the environmental framework of the particular interactions (cf. Wiegleb and Bröring 1996). The specificity of each integration level notwithstanding, we suggest that ecological theory should develop concepts unifying the different qualities of interaction.

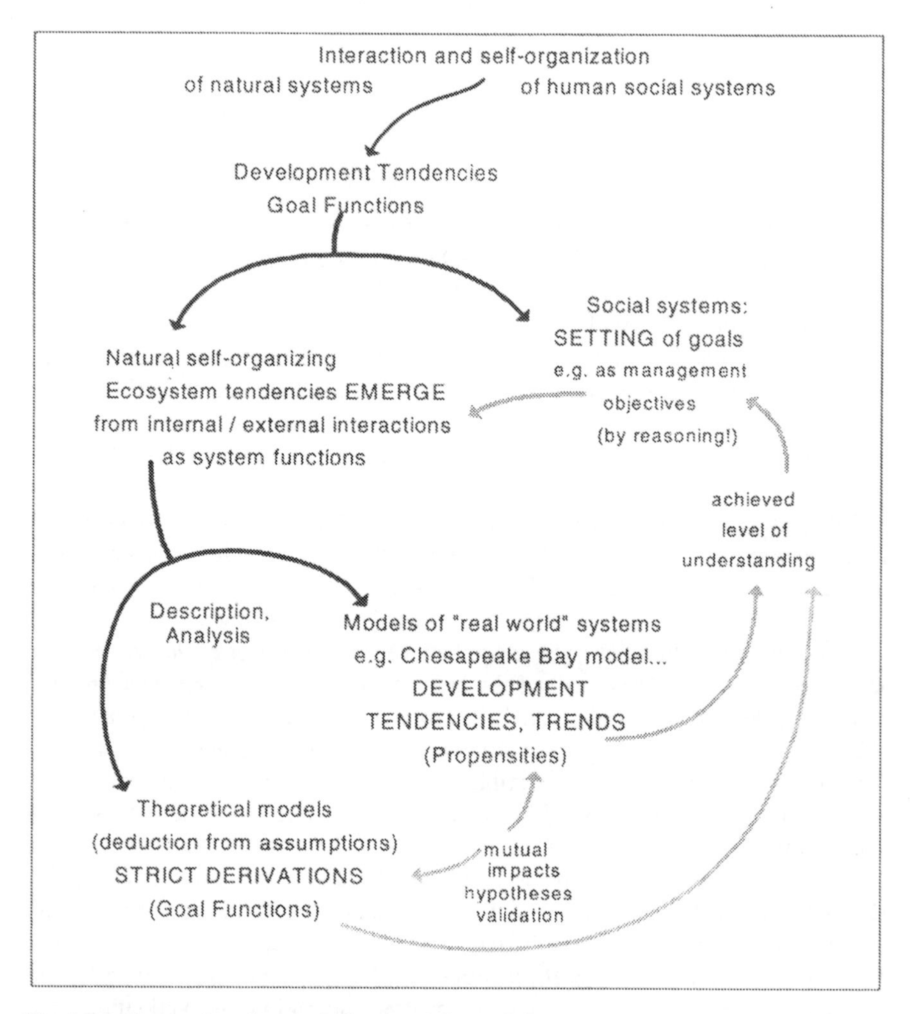

Fig. 3.1.1. Cooperation of ecological and social goal functions

The term goal function addresses natural ecological interactions as well as social phenomena. Both can be understood in terms of self-organization. The involved types of interaction are different, however: Natural systems display emergent system behavior by spontaneous organization. The self-organization depends on the internal interactions of the system elements within their environmental context of external impacts and other external conditions. Social systems have a self-organizing structure intellectually reflected and modified due to intentional reasoning. This means that goals *sensu strictu* can be imposed as consciously defined management objectives within social systems only.

If we look at the characteristics of description and analysis of natural ecological processes, we find another important conceptual distinction, which we can consider as a "bifurcation" between theoretical models on the one side and models on the other, which are designed to depict particular ecological processes. The first are set up to investigate formal implications of certain mathematical assumptions. The latter deal with "real world" systems, using an abstract formalism to depict the essence of what we understand about the investigated ecological processes. In the first case it is justified to use the term *goal function* as a description of a model property. In the second case it is more adequate to use the terms *development tendencies* or *development trends* (propensities) of the processes under consideration.

There is a certain feedback of mutual impacts between both types of models. Theoretical considerations provide a repertoire of potential descriptions, and on the other side receive inspirations from practical problems. Schwarz and Trepl (Sect. 3.2) dwelled on the more implicit influence of social images on the research of and theories about natural self-organizing systems.

Vice versa, the understanding of ecological processes influences the social goal setting process profoundly. This is most obvious in attempts to design sound ecosystem management strategies, which is impossible without at least a fragmentary understanding of the underlying ecological processes.

Theobald (Sect. 3.4) clearly stated the necessity of ethical reflections of the economy, which is driving social systems to undesired or, more hopefully, desired states. Not least, management-induced or natural changes of natural ecosystems constantly pose new challenges to the description and analysis of the ecological systems.

From this point of view, ecosystem management and ecosystem analysis represent an iterative cycle. Understanding the specific feedback process forces us to distinguish properties of natural and social systems, their "goal seeking behavior", and their qualitative differences in order to improve our knowledge about how both work together. The mutual development implications make nature and the human society a common whole, even though they have to be understood considering the specificity of each domain and of each level of integration.

References

Allen TFH and Starr TB (1982) Hierarchy - perspectives for ecological complexity. University of Chicago Press, Chicago London.

Wiegleb G and Bröring U (1996) The position of epistemological emergentism in ecology. Senckenbergiana maritima 27:179-193

Chapter 4

The Diversity of Targets: Problems of Combining Natural and Human Orientors

4.1 Human Targets in Relation to Land Use

Maren Leupelt and Ernst-Walter Reiche

While the third Chapter deals with philosophical aspects of the goal function approach, the fourth Chapter resumes the question "If possible at all, in which way should goal functions guide human action?", focusing on the leading principles of human societies' decision making processes.

In the course of Chap. 4 – taking up the concept of diverging utilization interests – the following questions gain significance: "What are the parallels and differences of natural systems' and human systems' targets? Can we derive binding norms for human society from an ecologically based goal function concept? Could the goal function concept provide indicators for sustainability?"

In the frame of landscape planning and environmental impact assessment it is usual to formulate the assessment based on so called function potentials. They are conceived to illustrate the significance of landscape components for the natural balance. For instance, on the scope of soil protection, soil scientists distinguish between the functions of the biotope, the feedback control systems and the biotic production functions (cf. the self-organization paradigm: Krohn, Küppers and Krug 1992; Mußmann 1995). This classification focuses on the ecological relevance, the landscape balance and the natural balance, while terms such as deposition function, recreation function and location function refer to the suitability of a landscape for human utilization.

Dealing with the concrete description of targets in human systems it is necessary to investigate utilization interests, the emerging conflicts of land use and the resulting land use systems. Accordingly, the potential of utilization characterizes the functioning of landscapes and their parts. Table 4.1.1 shows different utilization-categories and the corresponding utilization targets. Such a conception proceeds on the assumption that the development of society (due to the division of labor) causes a development of lobbies. The different groups of corresponding interests represent special objectives to use the landscape. Their special intentions have to be integrated in the planning process and the resulting conflicts of interests have to be analyzed.

F. Müller M. Leupelt (Eds.)
Eco Targets, Goal Functions, and Orientors

Table 4.1.1 Targets of utilization in different categories (cf. Fränzle et al. 1991)

Category of utilization	Targets of utilization
Agriculture	Production of food; Maintaining quality of food
Forestry	Production of wood
Water-Economy	Providing of drinking-water; water-management in the landscape
Sealing	Providing space for living, traffic and production of goods
Deposition	Waste-disposal
Ecology	Conservation of species and biotopes

There exists a group of special targets for each of the utilization categories. These goals are usually fixed in certain laws, regulations, or decrees. The German food-ensuring-law (1968), for example, stipulates that the supply with agricultural products as well as a satisfactory quality (category 'agriculture') has to be guaranteed. The same applies to the category of water-economy – the supply with drinking water is a target as well as the maintenance of the water quality. The respective law fixes both targets. The contribution of Marticke (Sect. 4.3, this volume) deals with modules of environmental law. He discusses the influence of ecological concepts on binding norms of environmental law.

A specific category in the classification of Table 4.1.1 is the *ecological utilization* of landscape. The target of this category – to use the ecological functioning – is of high importance for human systems as it is a basic foundation of life. It is not only of concern for a single lobby, but for each utilization category itself, at different scales. Ecologists demand the realization of this target of an ecological utilization at different scales of planning. In this context, Dierssen (Sect. 4.7) refers to the objects of nature conservation in Europe. He demands a protection strategy for natural processes as a completion to the present conservation strategy for species and biotopes.

However, we have to state, that the recent concrete (spatial and) environmental planning does not operate on the principle of "ecological goal functions", but bases on the mediation of interests of utilization. That means, each utilization-category competes with the others for the claim of a respective environmental medium (such as water, air, soil, flora and fauna) and for landscape resources. As a consequence, conflicts of utilization emerge – the relevance of these conflicts increases with the dependence of utilization categories on the respective medium. Classical conflicts are, for example, those between agriculture (intensive) and ecology (extensive), or between water management and deposition. As a consequence of these conflicts a general framework developed that embodies general targets concerning the public welfare in laws. However, we have to state that the interests of ecological utilization did have to give way to the pressure of competing lobbies. The process of legislation did not consider adequately ecological needs and the existing protective regulations are not satisfactory and suitable for today's environmental problems.

According to the principle of competing utilization interests —concerning limited resources— institutions for solving these conflicts have developed on different scales. Beside those categories of targets that can be concretely realized – guided by mediation processes on the object-scale (e.g., landscape planning), there exists a lot of targets that are less concrete and belong to a meta-scale (e.g., sustainability). Chap. 4 exemplary discusses the prerequisites for implementing sustainable development on the object-scale. (cf. Bossel Sect. 4.2; Jüdes Sect. 4.4). The sustainability concept is only realizable if the interactions on a global scale – connected with a long-time horizon – are taken into consideration. (Bossel Sect. 4.2; Jüdes Sect. 4.4). However, there exists an increasing number of scientists who cope with the problem of measuring sustainability. They develop parameters and lists of indicators that consider not only economic but also ecological and social factors of sustainable development. Thus, there are promising approaches to transfer those indicators even on a regional scale (object scale) (cf. Bork et al. 1995; Dalsgaard et al. 1995; MISEREOR and BUND 1996; Moldan 1997; Teichert et al. 1997).

The fact of different interacting scales and the issue of a time-horizon that should integrate the demands of future generations, raises some questions:

- In how far will this paradigm ever be realized if the lobby (the future generations) is not able to plead their interests right in time of the decision? Even if the interests of future generations are embodied in laws (Marticke Sect. 4.4): are these laws considered adequately?
- Though there is a tradition of securing long-term interests of general public lasting for centuries, does this tradition still meet the needs of today's society and do the existing instruments (laws, binding norms, decrees) fit to implement a long-term and global ecological idea?

Focusing on a successful environmental protection on the global scale, for example, boycott of tropic timber, it is striking that the initiatives did succeed without laws, but less binding norms such as waiver/disclaimer and claim of boycott. The intensive discussion about CO_2-reduction-programs (climate convention and self-obligations of the industrialized countries) illustrate that there are declarations of intent, but no binding instruments of implementation.

In the context of descriptive indicator sets, we can also find a lack of adequate instruments, for example, the Gross National Product (GNP). The GNP serves as an indicator of economic and socio-political decision making processes. Since this measure does not consider external effects, but to the contrast integrates positively consumption of resources and environmental damages, it is not usable at all to serve as a leading measure of a policy that orientates toward the paradigm of sustainability. To improve the GNP in this sense, the German Statistical Agency (Statistisches Bundesamt 1996) develops indicators for an Ecological National Product (cf. Henderson 1981; Radermacher and Stahmer 1994/1995; van Dieren 1995). In this volume (Sect. 4.6) Radermacher discusses the demands for corresponding indicators as a need for an accounting system.

Further problems concerning the embedding of valid indicators result from the spatial delimitation of landscapes. As generally known, the administration (e.g., offices of environment, water administrative boards) operates with delimitations that do not correspond with the balances of energy or water or metabolic balances as criteria of assessment. Demanding these balances as an aggregation of indicators, they have to be set up at different scales (local scale, regional or global scale) in order to describe the structure and dynamics of the respective system and must not depend on administrative demarcations. If we want to realize eco targets related to ecological principles, the implementation has to be made on the base of ecological system definitions, the relevant time-scale should orientate at the system's lapse of time. Though this problem was recognized decades before (e.g., rivershed of the Rhine), there is no satisfactory solution.

In ecological management human beings and nature are conflicting. On the other hand we know that human beings depend on the natural laws. Confronted with the knowledge that human beings are subject to common natural laws, we have to discuss, for example, if human societies also follow the principle of exergy optimization as nature does. Ruth (Sect. 4.5) focuses on the energetic term of the exergy-theorem and shows methods of calculation for economic systems. In his contribution he stresses that the application of the principle could lead to an increasing efficiency in industrial production processes. Additionally, we have to mention the field of bionic, where engineers copy strategies of nature to improve production processes (cf. Teichert 1997).

The development of the last 100 years obviously shows how far reaching changes in human societies are possible, concerning material and energy flows and their influence on the environment. Most of the information that determines structures and processes in the society is not genetically fixed. That means, it does not follow the principle of hereditary. Therefore, social systems may react more flexibly, but there is an immanent risk of forgetting principles that once were accepted to be system-supporting. The tendency of following short-term utilization-interests in managing nature and landscape seems to prove this thesis.

Summing up the introductory ideas we can formulate the following theses:

- On the local scale the relevant goal functions develop as consequences of mediations between competing utilization interests.
- On the meta-scale there exist many targets (targets of sustainable development), but the instruments of the object-level (concrete spatial relation) have to be modified to fit for the implementation of these targets.
- In relation to evolution of nature, including ecosystems, change of social systems bases on short-term actions. Social systems may forget experience, ecosystems do not.
- On the scientific scale there are still problems concerning the application of the interdisciplinary sustainability approach, as environmental research was the domain of the natural sciences until a few years ago. Today, environmental research also is of high concern for the social sciences, since it trends

towards investigations of the dynamic interaction between societal patterns of acting and natural reaction chains. Thus, there will be solutions for environmental issues not only from a more efficient ecosystem management, but also from a change of societal behavior and an alteration of frame conditions in social systems (cf. Ökoforum 1997).

The following contributions perform different aspects concerning the question: What can social systems derive of goal functions if we want to implement valid instruments for environmental protection that also may consider the targets of sustainability? Sect.s 4.2 and 4.3 focus on the sustainability concept, Sect. 4.4 deals with binding norms of environmental law. Sect.s 4.5. and 4.6 discuss economic aspects of the orientor concept and Sect. 4.7 deals with the targets of nature conservation.

References

Bork H-R, Dalchow D, Kersebaum KC and Stachow U (1995) Regionalanalyse der Auswirkungen veränderter agrarökonomischer Rahmenbedingungen auf Agrarlandschaftsnutzung und Umweltqualitätsziele. Zeitschrift für Kulturtechnik und Landentwicklung 36(4):194-201

Fränzle O, Zölitz-Möller R et al. (1991) Erarbeitung und Erprobung einer Konzeption für die ökologisch orientierte Planung auf der Grundlage der regionalisierenden Umweltbeobachtung am Beispiel Schleswig-Holsteins. Forschungsbericht 10902033, i.A. Umweltbundesamt, Kiel

Hatfield JL and Karlen DL (eds) (1994) Sustainable agriculture systems. Lewis Publishers, Boca Raton, p 316

Henderson H (1981) Politics of the Solar Age. Doubleday

Krohn W, Küppers G and Krug H-J (1992) Organisation. Ein Grundthema der neuzeitlichen Wissenschaft – ungelöst und unabweisbar. In: Krohn W, Krug H-J and Küppers G (eds) Selbstorganisation. Band 3. Duncker & Humblot, Berlin

MISEREOR and BUND (eds) (1996) Zukunftsfähiges Deutschland. Ein Beitrag zu einer global nachhaltigen Entwicklung. Studie des Wuppertal Instituts für Klima, Umwelt und Energie. Birkhäuser Verlag, Basel Boston Berlin

Moldan B (ed) (1997) Sustainability indicators: a report on the project on indicators of sustainable development. Wiley, Chichester

Mußmann F (ed) (1995) Komplexe Natur, komplexe Wissenschaft. Selbstorganisation, Chaos, Komplexität und der Durchbruch des Systemdenkens in den Naturwissenschaften. Leske + Budrich, Opladen

Ökoforum (ed) (1997) Technische Scheuklappen. Memorandum des Ökoforums zur Umweltforschung. Politische Ökologie 52:90-91

Radermacher W and Stahmer C (1994/1995) Vom Umweltsatellitensystem zur Umweltökonomischen Gesamtrechnung: Umweltbezogene Gesamtrechnungen in Deutschland. Zeitschrift für angewandte Umweltforschung 4

Statistisches Bundesamt (1996) Allgemeine Konzepte des Indikatorensystem. Zwischenergebnisse des Forschungsvorhabens: Entwicklung eines Indikatorensystems für den Zustand der Umwelt in der Bundesrepublik Deutschland mit Praxistest für ausgewählte Indikatoren und Bezugsräume. Interne Mitteilung zum Workshop am 29.10.1996 in Wiesbaden, Wiesbaden

van Dieren (ed) (1995) Mit der Natur rechnen. Der neue Club-of-Rome-Bericht: Vom Bruttosozialprodukt zum Ökosozialprodukt. Birkhäuser, Basel Boston Berlin

4.2 Ecosystem and Society: Orientation for Sustainable Development

Hartmut Bossel

Abstract

Human systems are interacting with, and dependent on, the global ecosystem. These complex interconnected systems have to obey natural laws and system laws in order to remain viable in the long-term. If this development is to be sustainable, it has to observe certain constraints and follow certain principles and orientations which derive partially from physical conditions and natural laws, partially from evolutionary adaptation of complex systems, and partially from conditions applying to human systems with conscious actors and normative standards.

Although 'sustainable development' constrains the spectrum of permissible processes, it does not define a final steady state. The development process must be guided by reference to an ethical framework and to the balanced satisfaction of fundamental system interests (= basic orientors) resulting from a system's interaction with its environment. Although in many ways similar, there are fundamental differences with regard to orientations and goal functions between ecosystems and human systems.

4.2.1 Constraints and Orientation of Sustainable Development

Until the advent of the nuclear bomb and the grudging recognition of 'Limits to Growth' (Meadows et al. 1972, 1992), permanent existence and progress of human society was accepted as almost self-evident fact. More than anything else, a few billion years of existence and evolution of the global ecosystem seemed convincing proof of the permanence of life on earth, including human existence. But the consequences of human population and consumption growth (Brown since 1983) are now increasingly threatening the natural system on which humanity depends for its life support. Since the Brundtland commission's report on 'Our Common Future' (WCED 1987), 'sustainable development' has become a growing political and scientific concern, an 'ultimate end' (in the sense of Daly 1973), or an 'eco target' in the sense of this volume.

F. Müller M. Leupelt (Eds.)
Eco Targets, Goal Functions, and Orientors

Recommendations for appropriate responses differ widely. Some neoclassical economists continue to assume the equivalence and substitutability of capital, labor and human-made capital for natural capital, thus interpreting irreversible conversion of natural resources as 'sustainable development' ("weak sustainability" according to Daly 1991). Others acknowledge their complementarity and see the need for maintaining natural capital ("strong sustainability" according to Daly 1991). Still others call for a much more radical 'deep ecology' approach (e.g. Goldsmith 1992) to remodel society in the image, and as part, of ecological systems. The present contribution attempts to summarize conditions and constraints of both human systems[1] and ecosystems[2], and in particular their (implicit or explicit) goal functions and system orientations, in order to identify common concerns as well as differences that must be respected in pursuit of the goal of 'sustainable development'.

The development of human systems is governed by a number of characteristic conditions that are beyond human power to change: physical processes, natural laws, system laws, human psychology, and the interaction of human systems and their dependence on the natural system. These conditions apply to any type of societal development, even unsustainable development. But development of the human system can be sustainable only if these conditions are recognized and respected, and if human action is governed by principles recognizing the dependence of human systems on ecosystems, and rules constraining human interaction and influence to a scale that can be tolerated indefinitely by the total system.

This implies that sustainable development of human societies cannot be discussed by reducing the problem merely to either physical aspects, or those of the natural environment, or those of human society, or even of ideology. The principles governing human society are only part of the story – which is simply not complete without simultaneously considering human goals and values, physical laws, system principles, ecosystem constraints, and 'eco targets'.

To deal with this complex problem, it is necessary to synthesize a perspective by integrating concepts from many different scientific fields. A rough sketch of such a perspective is attempted in this contribution (more detail and extensive bibliography in Bossel 1996). It focuses on the question: "What general principles can be formulated to guide sustainable development?" The emphasis on general principles reflects (i) the complexity of the problem, and (ii) a systems theoretical approach.

In searching for principles to guide sustainable development, it is only natural to have a closer look at the global ecosystem, which has demonstrated sustain-

[1] In this paper, the term 'human system' will be used to denote any system in which humans and human organizations play the major role. Often, but not always, the term will be synonymous with 'society' or 'social system'. 'Social system', 'political system', 'technological system', 'economic system' etc. are all 'human systems' in the sense applied here.

[2] The terms 'ecosystem' or 'natural system' will be used to denote local and regional ecosystems as well as the global ecosystem, and their system components, i.e. systems in which natural processes play the major role.

ability over a few billion years. Ecosystems are working models of sustainable complex systems, and it is reasonable to study them for clues for the sustainable management of the human enterprise. Some typical questions are: What are the essential differences between ecosystems and human systems? Where do their goal functions agree, where do they differ? What are specific properties of human systems? Where do general ecosystem principles apply in human systems? Where not? How are human systems coupled to ecosystems, and how much are they therefore determined by ecosystem principles? What conclusions can be derived from this for sustainable development of human systems, and in particular for matching human objectives to ecosystem functional objectives (eco targets)?

The nature of our physical world, of dynamic and evolving systems in general, and of ecosystems and the peculiarities of humans and human systems in particular, must all be reflected in the search for general principles of sustainable development.

4.2.2 Principles of Societal Development

The issue of sustainable development concerns the *dynamics of the interaction of human systems with the natural environment*. The knowledge and tools for dealing competently with this issue come from different scientific fields and disciplines relating to

- natural laws and constraints of the *physical system* in which ecological and human systems are embedded;
- processes governing the function, dynamics, and evolution of complex systems, as described by *systems science*;
- processes in *ecosystems* and ecosystem dynamics as a function in particular of energetic and evolutionary processes;
- processes determining the operation and behavior of *human systems*;
- processes of *interaction of human systems and ecosystems*.

Obviously, an interdisciplinary, holistic systems approach is required. Anything less can only lead to partial solutions, or worse, produce additional problems because indirect and feedback effects are not considered.

If present trends continue, human society will clearly be on a physically unsustainable path (e.g. Brown et al. since 1983; Brown et al. since 1992; German Bundestag 1995). The effects of resource consumption growth (mainly in the North), and population growth (mainly in the South), severely stress the global ecosystem even now, and have caused local and regional environmental collapses. The risk of local, regional, and global breakdowns increases as some pollution rates exceed environmental absorption rates by far (typical for carbon emissions, German Bundestag 1995). The call for sustainability is a logical consequence. But the long time lags (time constants) inherent in human systems (long delays in problem recognition, value change, research and development, decision-making, investment and implementation) and in ecological systems (adaptation, change in species composition, evolution) make instant switching to another, in particular a sustainable path impossible, even if society could now agree on the path and on the need for a switch.

Even under the best of conditions, sustainability can be reached only after several decades. The present infrastructure, behavioral orientations, and production and consumption patterns of industrialized nations would have to change profoundly. But in contrast to ecosystems, human systems *have* the option to switch to a different (physically possible) development path by making and implementing conscious decisions. Human system development is constrained by the laws of nature, but it is not limited to the slow and 'blind' processes of natural evolution. Conscious planning, choice, innovation, invention, and implementation can guide the direction of development, and accelerate the rate of change significantly. However, if humankind does not act in time to achieve sustainability, nature will eventually enforce sustainability by the forces governing the coevolution of species in a finite world, i.e. by fitness selection in a deteriorating environment to which humanity will be increasingly ill adapted. Nature would then set the conditions on her own terms, and they would not correspond to human goals. If humankind adopts sustainable development as a goal, this also requires learning to match human orientations and actions to the eco targets and goal functions implicit in ecosystem development.

Progress toward sustainability can be made by improving current processes, but the current dynamics of sustainability loss really require significant changes, not just token improvements. When contemplating changes, one should be clear about the differences between ecosystems and human systems – one should know where ecosystems principles can be applied to human systems, and where not (Bossel 1996). Scientific paradigms differ significantly in this respect. Where the paradigm of mainstream economics (Samuelson and Nordhaus 1989) sees few constraints on human systems posed by the natural system, the paradigm of deep ecology (e.g. Goldsmith 1992) would demand that human systems must be carefully matched to the processes and dynamics of the natural system. Where one school of thought finds in the natural system little of relevance to the operation of human systems, the other views the natural system as the only viable model and pattern for sustainable human development. The review in the following two subsections reveals that there are similarities as well as very significant differences.

4.2.2.1 Where Ecosystem Principles Apply to the Development of Human Systems

It is important to realize that human systems are indeed cut from the same cloth as natural systems, and that therefore there are important common aspects, and similar system orientations and goal functions. Some of these are of the nature of natural laws and physical constraints; there is no escape from them by any amount of scientific theory or political ideology.

- Human systems are subject to the same physical forces and processes and their limits as all other systems. Sustainability can only be achieved if material and energy flows stay within natural limits, and if nature's approach to maximize exergy use efficiency (see Jørgensen Sect.s 2.4 and 2.7) and resource use efficiency (see Kutsch et al. Sect. 2.12) is adopted, including recycling of materials (see Patten Sect. 2.8).
- As in ecosystems, balanced attention to, and satisfaction of, the basic system orientors (existence needs, effectiveness, freedom of action, security, adaptability, and coexistence needs; Bossel 1977, 1994, see also Bossel, this volume) will facilitate the path towards sustainability of human systems. It will allow the best possible response to changing conditions, without a teleological prescription of system development. Balanced satisfaction of basic orientors is required at all system levels (subsystems, systems, total system).
- The principle of blind chance evident in evolution is also an essential (but not exclusive) design principle for a sustainable society. In the interest of innovation and successful adaptation to changing conditions, there are advantages in giving new and different solutions a chance to prove themselves.
- In terms of its selective function, the market principle of the human economic system is analogous to the evolutionary principle of natural selection (where unsustainable solutions are penalized and ultimately eliminated). If prices would correctly reflect 'true' costs to viability and sustainability (i.e. if prices would reflect true fitness), the market principle would be a simple and efficient means for furthering sustainability goals. Although a central control apparatus would still have to set and enforce certain minimum standards, it would be restricted to a minimum.
- Diversity can enhance system adaptability in ecosystems as well as in social systems by enlarging the pool from which innovative responses to new challenges can be drawn, and by channeling ruinous competition into mutualistic enterprises and the more effective utilization of available niche opportunities.
- As in ecosystems, emergent properties can be expected, which introduce new qualities not possessed by constituent systems of the human system. Emergence of 'higher' goals such as the basic orientors, or of goal functions like resource use efficiency, or of cooperative processes, is relevant in both natural and human systems. In human systems, the entrainment processes (Haken 1983) connected with emergence dynamics may significantly accelerate transition processes (Bühl 1990).
- As in evolution, prediction of the future development of human systems is not possible. However, major forces and trends shaping future development can be identified. These natural and system forces restrict the spectrum of future

paths to a subset of possible paths. Among these, orientor impact assessment can help to identify desirable paths (Bossel 1997, 1998).

4.2.2.2 Where Ecosystem Principles do not Apply to the Development of Human Systems

It is important to realize that human systems are qualitatively different in important aspects from other biological systems and ecosystems. Conscious goal-setting of humans and their organizations, and the rational and purposive utilization of physical, biological, intellectual and organizational processes can produce effects far surpassing the power and speed of human individuals or natural evolution. Human action can therefore produce impacts on natural systems that cannot be matched or absorbed by them. Matching human systems to the natural system therefore requires control and constraint on the part of humans.

- In natural systems, the processes of evolution under fitness selection prevent any nonhuman species from endangering the survival and sustainability of the *global* ecosystem. Excessive population and demand growth would eventually and effectively be controlled regionally and locally by countervailing forces, which can build up at similar speed as the threat. However, there is no such fail-safe control for human action, since its potential far surpasses that of evolutionary counteraction both with respect to its time constant (too fast), its degree (too massive, and often irreversible), and its spatial reach (global instead of local or regional).
- While ecosystems can only draw on *local* sustainable exergy and material flows, human systems can operate for long periods at the cost of exhausting *global* fossil and mineral resources (by technology, organization, transportation), thus eroding not only the local base for sustainability. Populations of organisms can also show overshoot and collapse as a result of *local* resource depletion, but they are unable to utilize resources from faraway places, and they cannot therefore threaten the global resource base as humans can.
- The dissipative structures of organisms and ecosystems can only grow up to the limits of local sustainability. By contrast, society can build up (and has built up) dissipative structures far exceeding regional and even global sustainable carrying capacities: e.g. urban concentrations, industrial complexes, the dumping of wastes in ocean, atmosphere, and soils.
- In contrast to ecosystems, the amount of dissipative structures built by the human system, and their exergy dissipation rates are not a measure of the degree of sustainable development – on the contrary. In human systems, exergy and material flow rates can be far greater than their sustainable rates, while natural evolution will not normally allow such overshoot in ecosystems.
- Emergent properties of ecosystems are adaptations shaped by evolutionary forces to enhance viability and sustainability of ecosystems. In human systems, emergent properties do not normally appear as a result of processes

controlled by sustainability principles; other forces (e.g. cultural or economic forces) are (still) much more powerful.

- Because of the mental and organizational powers of humans, the impact of human action can be far greater than any countervailing and controlling forces. Human development therefore cannot be left to trial-and-error processes. It requires conscious action by society. As social sustainability implies equity, even for future generations, development towards sustainability must be guided by a Principle of Partnership (see below).
- Unlike ecosystems, human system development is not limited by evolutionary continuity and progress (incremental change for better fitness). Human systems have no definite 'arrow of progress' in the direction of building up more dissipative structure, more effective use of gradients (see Jørgensen Chapt. 2), greater orientor satisfaction, and sustainability. Actors on all levels often tend to move social systems into the opposite direction by destroying dissipative structure, diminishing diversity needed for adaptation, increasing waste rates of exergy and materials, and by impeding progress in controlling population and consumption growth.
- Human systems are characterized by conscious action, made even more powerful by abstract reasoning, goal-directed research, technology and science, institutionalized education, and enormous information storage and processing capabilities and organizational power. Conscious action includes both the danger of the loss of fitness (sustainability) as well as the opportunity for finding a path to sustainability faster than by evolutionary trial-and-error.
- Human beings have the possibility, as no other species does, to reason through the laws of ecological sustainability and understand the necessity of obeying them.
- Human beings have the moral sense, as no other species does, to value other species and their evolutionary potential and protect them, and to value future human and non-human generations and respect their interests.

4.3.3 Ethical Principles and Goal Functions

To sustain something implies valuing it enough to put an effort into maintaining its integrity. The question of what is meant by 'sustainability' is therefore first of all a question of (socially communicated) ethics. If we value humankind and human culture, we must strive for their sustainability. If we agree on this, we face a more difficult question: "*What* is to be sustained?" Sustaining the unsustainable system we have now, will not be possible. So we must develop the vision of a sustainable society which is both physically and organizationally sustainable, which is able to accommodate the ethnic and cultural spectrum of humankind in all its diversity, and which moreover permits change and human development 'indefinitely'.

This society will have to adopt certain goal functions: It must allow development without physical growth (of material and energy flows and population). Its population must remain below a certain limit which is probably less than today's global population. The per capita use of energy and materials must be less than what it is now in the industrialized countries of the North. All energy must be renewable, all materials recyclable. These limited throughputs of resources must support a system which maintains an unlimited potential for nonmaterial cultural, social, and individual growth.

The problem is that we are not free in the choice of the material and energy flows which will support this society indefinitely. These flows are restricted by conditions on earth and by the laws of nature. The flows required by a given demographic, technological, and societal development path may or may not be physically possible. Given this restriction of resource flows, we can still choose how they should be used in a sustainable future, where social sustainability would imply equitable distribution: The spectrum reaches from supporting a small population in luxury to providing a large population with the bare essentials. There are many possible options in between. The ethical implications are obvious. Obviously, a continuation of present trends, where a small minority lives in luxury, partly at the expense of an underprivileged majority, is socially unsustainable and therefore beyond question.

There is therefore no unique state of sustainability. A sustainable world could be a barren wasteland with permanent strife and abject poverty of the remaining human population, or a fertile landscape with an ecologically benign civilization at a high level of individual welfare.

Moving onto a sustainable path which is not dictated by nature's reaction to exploitation requires conscious choices by individuals and society, if only to select and secure the sustainable levels for exergy and material flows.

Even at (constant) sustainable level of exergy and material flows, sustainable society will represent a wide spectrum of human diversity and will continue to develop – if the laws of system evolution are allowed to prevail. Note that the requirement of sustainable material and energy flows on the global scale does not imply constant flows on a regional scale. These may undergo fluctuating or even periodic dynamics, perhaps driven by cycles of exploitation – conservation – release – reorganization which drive much of ecosystem development (e.g. gap dynamics in forests, Holling 1986), and which are also observed in human systems (i.e. Kondratieff cycles; Schumpeter 1939; Sterman 1986). Moreover, the rates of sustainable flows, and of human populations to be supported by them, are matters of choice, partly determined by cultural and political processes.

While goal functions of ecosystems and their components are emergent features of these systems and their interactions with their environments, this is only partially true for goal functions operating in the human system: To a very significant extent, they are determined by human choice.

4.2.4 Blind Chance vs. Partnership

The evolutionary process in nature can be viewed as giving rise to the emergence of an 'evolutionary ethic' (Jantsch 1980), i.e. a set of (implicit) general ethical principles which seem to regulate interactions in the global ecosystem, and which could be summarized as 'Principle of Blind Chance'. Because of the limited reach and potential of (predating or competing) individuals and species, every organism or species has a (sometimes small, but usually finite) chance of succeeding and evolving further in the struggle for survival.

Humankind in all its cultural diversity has not adopted this ethic, however: The inequality and even cruelty it permits at the level of individuals, and the lack of anticipation and responsibility is not morally tolerated by most societies. As conscious beings, humans have the ability to anticipate consequences of developments and actions, and to empathize with those affected. They also have the ability to control developments to some degree, and cannot therefore escape responsibility for their actions. Conscious awareness challenges humankind to adopt an ethic which goes beyond the uncaring, casual, and capricious attitude of evolutionary ethics. If adopted, the goal of sustainability of the human enterprise implies concepts of equity, fairness, and partnership with future generations, and with other components of the global ecosystem (Fox 1996).

As a guiding principle for sustainable human systems, a Principle of Partnership can be formulated: "All systems which are sufficiently unique and irreplaceable have an equal right to future existence and development." (Bossel 1978, p 71). If there exists a sustainable (renewable and recyclable) resource base, this principle implies sustainable development and evolution, since it (i) allows resource use only at sustainable rates (to protect future systems), (ii) ensures fair access to available resources to all, and (iii) preserves fair chances of continuous development for all. Note that this principle differs from evolutionary 'practice' in implying e.g. protection of species, and a fair allocation of life chances to present and future individuals, i.e. it implies restraint in consumption (sufficiency) for the present generation. It protects human individuals as well as the diversity of social and cultural systems. With other words, the Principle of Partnership is not 'blind', as is the evolutionary principle, but is characterized by anticipation and consideration.

In terms of orientation theory (see Bossel, this volume), the Principle of Partnership amounts to introducing all other (present and future) systems that are potentially affected by an actor's action into the decision calculus, by assessing possible impacts on their orientor satisfactions (Bossel 1978, Bossel 1997). In practice, this task can only be dealt with in approximation. However, the important point is simply that the interests of the potentially affected are taken seriously, and responsibility is taken for their welfare. A logical consequence of the Principle of Partnership is that the 'range of action' should be no greater than the 'range of responsibility' – which favors the decentralized organization of small autonomous units, whose 'power' is limited to their own region.

4.2.6 Principles of Sustainable Development

Consideration of the similarities and differences of natural systems and human systems and of their mutual interactions leads to a list of principles operating in both systems. They must be respected in sustainable development. The human component brings in ethical considerations and conscious choice as an inescapable component. There is therefore a qualitative difference between the (implied) goal functions, eco targets and orientations emerging in natural systems, and the goal functions and normative orientations of human systems. Principles and conclusions concerning system development in general, and sustainable development of human society in particular, are collected and reviewed in the following.

4.2.6.1 Integrated View of the Problem

- Laws of nature and the physical constraints of planet earth put physical limits on any type of development of natural or human systems. These limits cannot be eliminated by wishful thinking, economic theory, or ideology.
- All systems on earth, whether living or non-living, conscious or non-conscious, have to obey certain general system laws. It is possible to analyze and understand complex system behavior by analogy or simulation, and to use this understanding for identifying behavioral trends, for planning, and for decision-making.
- The knowledge required for dealing with the issue of sustainable development has to be integrated on a general systems background. It ranges over a wide field of traditional disciplines encompassing the natural sciences, systems science, social sciences, and engineering.
- In human systems, the multitude of interacting conscious actors (individuals and organizations) pursuing different goals and interests complicates things enormously. Holistic analysis of the possible complex behavior must find ways to simplify and aggregate without losing essential features of dynamic processes.
- In response to the emerging problems of the interaction of human systems and natural systems, and the realization of the unsustainability of current developments, there is a growing perception of the need for fundamental change in human institutions, technology, behavioral principles, and ethics (see references in Bossel 1996, p 152).

4.2.6.2 Principles and Properties of Physical Systems

- The First and Second Law of Thermodynamics as well as the (Liebig) Principle of the Minimum put severe constraints on the development of systems in general, and on that of the human system embedded in the global ecosystem in particular.

- Development is rate-limited, not supply-limited. The dominant rate control is effected by the rate at which solar radiation can be received and utilized. Solar radiation drives all important geophysical and biological processes, including exergy capture and recycling of materials, and is therefore crucial.
- Material recycling, as practiced in the global ecosystem, is a prerequisite for sustainable development.
- The productivity of ecosystems is limited by the rate of solar energy input, physical efficiencies, availability of key materials, and waste absorption rate.
- Sustainable development of the human enterprise at a high level of quality is possible in principle, but not possible at present rates of resource destruction and population increase (Brown et al. since 1983; Brown et al. since 1992; Weizsäcker et al. 1995).

4.2.6.3 Principles and Properties of Ecosystem Development

- The coevolution of species, ecosystems, and physical environment brings forth an increase of structural and dynamic complexity in a sustainable overall system. Certain aspects can serve as models for human development, other aspects are general system properties which also apply to human systems.
- The carrying capacity of a given environment is the result of the coevolution of species populations, ecosystem, physical environment, and human activities. It is limited by the energy and material flows which can be mobilized and utilized.
- Ecosystems rely almost exclusively on renewable (solar) energy and on material recycling. This has ensured sustainability of the global ecosystem over billions of years.
- In the course of evolution, individual species tend to maximize exergy efficiency, while ecosystems as a whole tend to maximize the use of the available exergy gradient by maximum build-up of dissipative structure. Evolution therefore implies irreversibility and development.
- The environment not only determines the physical structure and processes of systems evolving in it, it also determines behavioral processes and cognitive structure. In order to be 'fit' a system must give simultaneous and balanced attention to different objectives: basic orientors (see Bossel Sect. 2.2).
- Diversity is essential for optimizing the use of available gradients, and for providing the innovative potential for future progress.

4.2.6.4 Principles and Properties of Human System Development

- As a result of conscious, reasoned, and anticipatory decision-making, and the intellectual, technical, and organizational power of humans, the rate of development of human systems can be much faster than natural system evolution. Hence there is little control of human activity by processes of natural evolution.

- Because of the power of reasoned decision-making and the potential of putting thoughts into practice, the spectrum of possible behavior of human systems is extremely wide.
- Human action therefore requires normative guidance. It can be assumed that human behavioral norms are not completely subjective and arbitrary, but have emerged as value orientations (orientors) and ethical standards (partnership in family and community) in the course of human genetic and cultural evolution.
- The evolutionary process causes the emergence of (implicit) normative standards in the behavior of organisms, which can be interpreted as the emergence of 'evolutionary ethics' (of blind chance of competing species, etc.). This evolutionary ethic is only partially acceptable for human systems. Because of their powers of imagination, anticipation, and reasoning, humans must adopt a wider ethical framework.
- Maintenance of value orientations and ethical standards under changing (systems environmental) conditions may require dynamic change of material and cognitive structure, including their periodic destruction and renewal.

4.2.6.5 Principles for Sustainable Society

- Ecosystem principles should not be applied uncritically to human systems. Some of them are applicable, others are not.
- Valuing humankind and human culture is equivalent to striving for their sustainability. Social sustainability in particular suggests adoption of a Principle of Partnership as ethical principle. It would extend to present and future generations, and to human and nonhuman systems.
- Society's current ethical base, as incorporated in its institutions, is inadequate for the initialization and protection of sustainable development. Changes in the legal system, and in the representation of the interests of human and nonhuman, present and future 'partners', are required.
- Full representation of the interests of affected systems requires assessment of their basic orientor satisfaction in the course of time.
- Population control, efficient resource use, recycling, use of renewable resources, in particular renewable energy resources, and sufficiency in consumption will be required to achieve sustainable development (Weizsäcker et al. 1995).
- Necessary processes of control and self-organization should be self-inducing, self-sustaining, equitable, and effective. The protection and encouragement of diversity on all system levels would provide creativity, innovation, redundancy, niches, resilience and optimum orientor satisfaction for all partner systems.

4.2.6.6 Required Steps for the Transition to Sustainability

- The transition to sustainability hinges on effective control of population growth and consumption growth. In view of the different time constants of these processes, the reduction of per capita consumption in the countries of the North must have absolute priority. There is an enormous potential through (i) improvements in the efficiencies of resource and energy use, and (ii) social acceptance of sufficiency limits.

- A truly holistic and integrated systems approach spanning the global ecosystem and the human system and their interactions is necessary. It requires new approaches in science, research, education, social and political processes, planning and decision-making.
- The normative principles of society should be consistent with an overarching ethical principle. Sustainability seems to require adoption of a Principle of Partnership with other systems and organisms of the global ecosystems, today and in the future. Sustainability cannot be attained under a split ethical framework (as today, with selfish competitiveness in society at large, and partnership in the family).
- In contrast to current trends for globalization, it will be essential for sustainability to reintroduce diversity and regionalization, or even localization, to adapt development to regional carrying capacity, and to decouple local markets from global markets.

4.2.6.7 Overall Conclusions

- Current development of the global human system is unsustainable in more than one sense: physically, ecologically, socially.
- The unsustainability is partly due to processes in human systems proper (population, economics, technologies, etc.), and partly due to destruction and change of the natural environment on which all human systems depend.
- 'Sustainable development' is possible, but it requires a departure from the present path. Many sustainable futures, and paths to achieve them, are possible.
- The choice requires adoption of an ethical principle (Principle of Partnership).
- Society (and environment) cannot afford an evolution of human systems to a sustainable path by trial and error; conscious strategic choices are required.
- Strategic choices must be in agreement with general system principles, and ecosystem principles in particular, in order to be successful in the long run.
- These principles should be identified, in order to provide guidance for structural and strategic changes.
- The emphasis should be on changing, where necessary and possible, the rules of human-determined processes of self-organization to ensure sustainable evolution. Omniscient design of the sustainable society is not possible. The task is rather to determine and apply functional principles which lead to self-organizing evolution of sustainable human systems (of whatever shape) in a sustainable environment.

4.2.7 Orientation in the Natural System and The Human System

Complex adaptive systems such as ecosystems or human systems cannot operate without the normative guidance of an orientation system. While in human systems, much of the orientation content may be explicit, or can be made explicit, it is always implicit in non-conscious and inanimate systems such as ecosystems and their (non-primate) components. Orienting concepts, such as 'eco targets', basic orientors, goal functions etc. can only be inferred from system behavior by an observer in the latter case. Although this process appears highly subjective, there is evidence from experiments with artificial life (Bossel Sect. 2.2) that basic orientors of complex adaptive systems (and hence values like 'security', 'freedom', 'effectiveness' and the like) are not subjective figments of human imagination, but fundamental system requirements that emerge in response to environmental challenges (for 'goal functions', see several authors, this volume).

In the organisms, species, and ecosystems that have developed successfully in millions of years of evolution, balanced attention to the basic orientors has evolved for securing viability and sustainability. In the concrete system/ environment interaction, the basic orientor requirements have been translated into goal functions of practical system behavior and function (for example, 'efficient use of resources' to satisfy the 'effectiveness' orientor, or 'storage' to satisfy the 'security' orientor). In these systems, the relative priority and weight of different goal functions in different situations is a result of evolutionary learning in species evolution, and some adaptive learning in individual development. It is not a matter of individual choice.

Since human systems are complex adaptive systems, their orientation systems are also partially shaped by emergent orientors reflecting the system / environment interaction. But to a significant degree they are also the result of culture, tradition, habit, or personal choice. Moreover, personal choice (or socialization or indoctrination) can override innate orientations. Since they can envision the consequences of their actions (or inaction) human systems – individuals as well as organizations – cannot escape ethical choice, including that of the ethical framework for behavior. Adoption of sustainable development as a societal goal would require conscious adoption of a partnership orientation that is only partially consistent with the 'evolutionary ethic' that can be inferred from the behavior of organisms and ecosystems.

References

Bossel H (1977) Orientors of nonroutine behavior. In: Bossel H (ed) Concepts and Tools of Computer-assisted Policy Analysis. Birkhäuser, Basel, pp 227-265

Bossel H (1978) Bürgerinitiativen entwerfen die Zukunft – Neue Leitbilder, neue Werte, 30 Szenarien. Fischer, Frankfurt/Main

Bossel H (1994) Modeling and Simulation. A K Peters, Wellesley MA and Vieweg, Wiesbaden

Bossel H (1996) Ecosystems and society: Implications for sustainable development. World Futures 47:143-213
Bossel H (1997) Deriving Indicators of Sustainable Development. Environm Mod and Assessment 4:193-218
Bossel H (1998) Paths to a Sustainable Future. Cambridge University Press, Cambridge
Brown LR (ed) (since 1983) State of the World. WW Norton, New York London
Brown LR (ed) (since 1992) Vital Signs. WW Norton, New York London
Bühl WL (1990) Sozialer Wandel im Ungleichgewicht. Enke, Stuttgart
Daly H (ed) (1973) Toward a Steady State Economy. WH Freeman, San Francisco
Daly H (1991) Steady State Economics. 2nd edn. Island Press, Washington DC
Fox W (1996) A critical overview of environmental ethics. World Futures 46:1-24
German Bundestag (ed.) (1995) Protecting Our Green Earth. Economica Verlag, Bonn
Goldsmith E (1992) The Way – An Ecological World-View. Rider, London
Haken H (1983) Synergetics – An Introduction. 3rd edn. Springer, Berlin
Holling CS (1986) The resilience of terrestrial ecosystems: Local surprise and global change. In: Clark WM, Munn RE (eds) Sustainable Development of the Biosphere. Oxford University Press, Oxford UK, pp 292-320
Jantsch E (1980) The Self-Organizing Universe. Pergamon Press, Oxford UK
Meadows DH, Meadows DL, Randers J, Behrens WW III (1972) The Limits to Growth. Potomac Associates, Washington DC
Meadows DH, Meadows DL, Randers J (1992) Beyond the Limits. Chelsea Green, Postmills VT
Samuelson PA, Nordhaus W D (1989) Economics. 13th edn. McGraw-Hill, New York
Schumpeter JA (1939) Business Cycles. New York
Sterman JD (1986) The Economic long wave: Theory and evidence. System Dyn Rev 2:87-125
WCED (Brundtland SH) (1987) Our Common Future. Oxford University Press, Oxford
Weizsäcker EU von, Lovins AG, Lovins LH (1995) Faktor Vier: Doppelter Wohlstand, halbierter Naturverbrauch (Factor Four). Droemer Knaur, München

4.3 Human Orientors: A Systems Approach for Transdisciplinary Communication of Sustainable Development by Using Goal Functions

Ulrich Jüdes

Abstract

Recognizing that the discussion on Sustainable Development is highly diverse and partly contradictory, an analytical approach is presented based on systems theory. Five steps are described:

1. The basic terms and suppositions of the idea of Sustainable Development are defined.
2. The key concepts of the idea of Sustainable Development are described.
3. Sustainable Development is interpreted as a cultural process, which includes an unlimited number of micro processes, leading to a „Culture of Sustainability", which is indicated by its co-evolutionary functionality.
4. Four different levels of a Sustainable Development Paradigm are differentiated (problem analytical-ethical, epistemological-conceptual, organizational, theoretical, realizational) and characterized by specific scientific and practical levels.
5. The goal function concept - e.g. introduced by Jørgensen (1992) and Müller (1996) in systems ecology - is adapted for use in Sustainable Development theory. In this context goal functions are defined as characters (elements, structures, principles), which have a promoting function for Sustainable Development. Unifying goal functions can be used for developing standards and indicators of Sustainable Development. As an application of the Sustainability Paradigm in an interdisciplinary project a list of potential goal functions in ecology, economics, cultural and social sciences is given.

The described model might pave the way for further transdisciplinary discussion and analysis on Sustainable Development and contribute to a deeper understanding and rationale of a Sustainable Development Paradigm.

F. Müller M. Leupelt (Eds.)
Eco Targets, Goal Functions, and Orientors

4.3.1 Introduction

The idea of Sustainable Development, as agreed upon by more than 170 countries at the UNCED 1992, is a reaction to rising global problems in different fields, which have been discussed before, separately (e.g. Adrian and Blatt 1987; Brown 1993). The new step was to aggregate the results of these discussions into one political document (Agenda 21) and to come to agreement on more or less pragmatical suggestions for global, regional and local actions. This offers to the present a vision for the future leading to a new paradigm for science and society.

The idea of Sustainable Development has so far been studied from the very different perspectives of ecology, economics, business and industry, agriculture, forestry, science and technology, energy, transportation, tourism, cities and buildings, health, politics and planning, social science, education, philosophy and ethics (Fritz et al. 1995; Goodland et al. 1992; Haber 1994; Heins 1994; Jüdes 1995, 1996; Kastenholz et al. 1996; Marien 1996; Mathes and Breckling 1995; Pearce and Warford 1993; Pearce et al. 1993; Turner 1993; von Amsberg 1993; Vornholz 1993; Voss 1994 and many others). Criticism of a paradigm of Sustainable Development (e.g. Dovers and Handmer 1993; Lele 1991; Redclift 1987) is based on the very heterogeneous, partially contradictory assumptions and basics and calls for a transdisciplinary comparative approach to analyze different disciplinary perspectives and aggregate their models (Jüdes 1996).

For such purpose a systems theoretical approach seems likely to be helpful. It would allow construction of a theoretical frame for the analysis of the different disciplinary perspectives on Sustainable Development and of their scientific paradigms. Another important step is to find suitable starting points for communication between the disciplines and for mediation between conflicting interests. The need for such a comprehensive framework is urgent to prevent the further degradation of Sustainable Development through misconstrued use in science, politics, and by special interest groups ('sustainabilism' Gligo 1995; Worster 1994).

4.3.2 Analysis of the Idea of Sustainable Development

Communicating the idea of Sustainable Development in society (Fig. 4.3.1) raises four questions of central importance for its understanding:

- Which assumptions underlie the idea of Sustainable Development?
- What is the theoretical reasoning behind Sustainable Development?
- What are the relevant fields for planning of strategies of Sustainable Development?
- Which factors need to be considered for implementing strategies of Sustainable Development into actions?

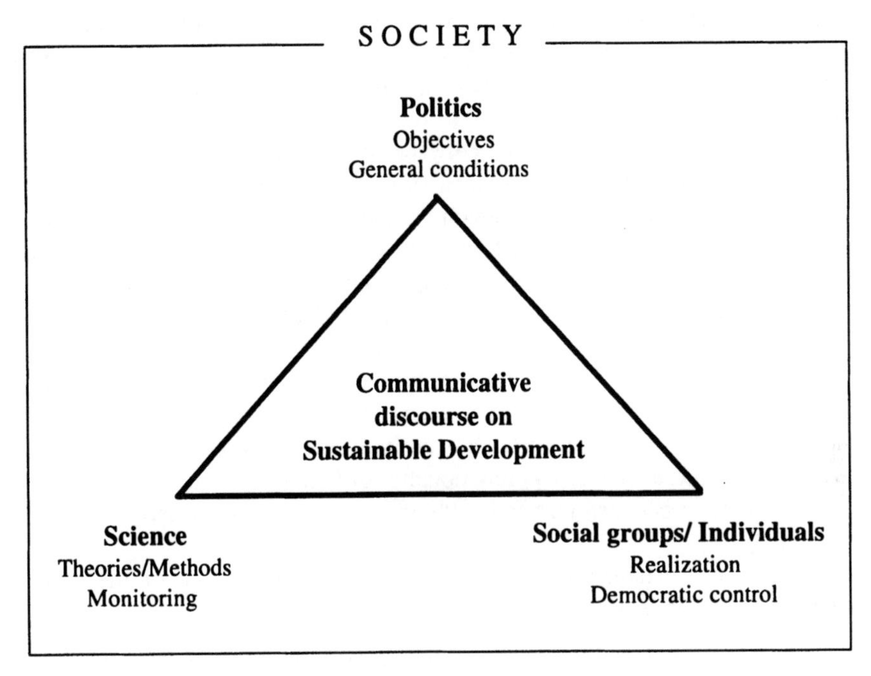

Fig. 4.3.1. Communicating the idea of Sustainable Development

4.3.2.1 Terminology and Basic Suppositions

Development. "Development" is widely understood as a process of gradual qualitative change of a system (progressing from one stage to another), very often including an increase of complexity, differentiation and organization. But it is a cluster-term, which is used in very different contexts (biology, social sciences, politics, technology etc.). The diversity of its use and its dynamic character are responsible for different meanings which resist all efforts to find a single definition fitting all situations.

A second difficulty arises from the fact that development is not related to a static condition but to a goal. Therefore it is a normative term dependent on individual and collective standards of values and varying with time and locality (Nohlen and Nuscheler 1993, 56).

The RIO-Report of the Club of Rome (1977) formulated six principles of human development: equality, freedom, democracy and participation, solidarity, cultural diversity, healthy environment.

Sustainability. "Sustainability" is derived from the Latin verb "sustinere" and describes a characteristic of relations (states or processes) that can be maintained for a very long time or indefinitely (Jüdes 1996). This is a general and very unspecified definition. If applied to complex situations that are based on conflicting structures or competing trends, it definitely leads to misunderstanding. So Tisdell (1993) suggests to specify its meaning within an ecological context.

At present, ecological relations between human and nature seem to include the most prominent problems as compared to others (e.g. social or economic) as they have become highly acute even on the global scale, whereas other problems – although politically first-ranked worldwide – are mostly regional ones. Very often they become global problems through their ecological implications. This is the reason why, in the context of this paper, sustainability is defined as a criterion for evaluation of human relationships to nature and of human behavior towards the environment. Human relationships and behaviors have influenced the non-human nature to a great extent and have constructed their particular cultures and modified their environments. Thus an indicator of sustainability can be described as the (co-evolutionary) functionality of human culture for the endeavor to find a balance between human health and ecosystems health (see Sect. 4.3.3, footnote 3).

The use of "sustainability" as a criterion in the evaluation of the human-nature-relationship is based on five suppositions:

1. The integrity and evolutionary capability of ecosystems are basic conditions to their performance (e.g. carrying capacity).
2. The diversity and performance of natural systems are environmental prerequisites for human life and especially for a "good life".
3. The use of nature for human living and well-being, for cultural and economic purposes results in stress, changes and damages to ecosystems.
4. The performance of ecosystems can only be preserved, if human accepts limits set by natural space, resources and regenerative capability of ecosystems.
5. The destruction of structural diversity and of functions in/of ecosystems by human does not only reduce the ecological capacity, but it results in new uncertainties, in other limitations to human ("good") life and in a loss of vitality of humankind on the whole.

4.3.2.2 Key Concepts of Sustainable Development

A combination of fuzzy terms like "Development" and "Sustainability" cannot create a precise term. So the meaning of "Sustainable Development" is even fuzzier and extremely resistant to clarification. More than 60 definitions are to be found in literature (Marien 1996).

The definition of Sustainable Development cited most is that of the Brundtland-Commission (Hauff 1987), who characterized it as an adjustment of three kinds of relationships:

- relations between human needs and nature's capacity (problems of retinity),
- relations between the needs of the poor and the rich (problems of intragenerational equity) and

- relations between needs of the present and those of the future generations (problems of intergenerational equity).

So far the function of the term "Sustainable Development" has primarily been political. It has become such a prominent political phrase because on a global scale all three relationships mentioned above are unsustainable today. These relations can be classified by two groups:

- relationships between human and nature and
- relationships between human and human.

IUCN/UNEP/WWF (1991) describe Sustainable Development as a process of "improving the quality of human life within the carrying capacity of supporting ecosystems". This definition includes three basic key concepts:

- a co-evolutionary concept of humans and nature,
- the socio-cultural concept of human needs and
- the natural scientific concept of the (limited) ecosystems.

Philosophically Sustainable Development can be regarded as a regulative idea ("regulative Idee" in the sense of Kant) bringing people together which, for many, constitutes a new paradigm. This ethical imperative of Sustainable Development tends to balanced relationships between human needs (today and in the future) and performance of nature. It is safe to add that this balance is not a static one.

As we have learned from history and from ethnology, the quality of both groups of relationships depend on the type of human culture involved. Nevertheless five principles are common to humans in all cultures and can be seen as the basic human characteristics:

- Human reacts to problems when problem pressure is high (principle of reflexive fitness).
- Human creativity allows unexpected solutions (principle of creativity).
- Human awareness reacts to positive guiding principles rather than catastrophical scenarios (principle of hope).
- Human is able to change his behavior, if it is helpful to his group (principle of responsibility).
- Human communicates and cooperates if possible (principle of sociability).

These characteristics have functional qualities and have a long tradition in educational theory and practice. We can assign them as human goal functions[1].

[1] Goal functions are defined as characters of all kind (elements, structures, principles), which have a promoting functional quality in a process for which a goal has been identified and accepted or constituted (e.g. human development or sustainable development). This definition differs from that given by Jörgensen (1992), who introduced the term into ecology, and to that of systems orientors as used by Bossel (1996). The term goal functions emphasizes a primarily qualitative functional perspective to the relations between the elements and results of a process and an ethically preset goal . So it may be helpful for a better understanding of the Sustainable

Human cultures are the result of these functions within specific environments. Glaeser (1992) understands environmental problems as the result of cultural crisis. Following this interpretation we may conclude that it must be possible to overcome this crisis and solve today's problems through cultural change. So Sustainable Development is interpreted here as a cultural developmental process leading to a new paradigm, the "Culture of Sustainability".

4.3.3 The Culture of Sustainability: From Goal Functions to a Sustainable Development Paradigm

What does "Culture of Sustainability" mean? We have suggested above an indication of sustainability through the co-evolutionary functionality of human culture. Coevolutionary functionality is defined here as the degree to which a balance between human health and ecosystems health[2] exists. The intention and endeavor toward a successful contribution to this functionality is the central characteristic of a Culture of Sustainability. A prerequisite for measuring the co-evolutionary functionality is the identification of such characters with a function for the goal of sustainability (see definitions above). For simplification we call the characters themselves goal functions for Sustainable Development as they implicate and constitute the relevant functions.

From this perspective on sustainability the industrialized countries, with their phenomenon of economic saturation and their environmental burdens through overconsumption, as well as the Southern and Eastern countries, where economic growth is necessary to fulfill basic human needs, however it occurs ecologically and socially unsustainable, both are developing countries. These different conditions demand different strategies of Sustainable Development, where economic growth needs and other qualities of human culture are respected and considered. So, contrary to the globalization of the western culture is the necessity to consider cultural heterogeneity in theoretical models of Sustainable Development. The culture of sustainability allows different ways and requires local adaptation.

As described above, discussions on Sustainable Development begin usually from more or less well-established disciplinary perspectives lacking a compre-

---

Development Paradigm. Bossel (1996) described some basic orientors as system-immanent principles, which correspond to fundamental environmental properties, to derive indicators for Sustainable Development. Indicators always intend quantification (e.g. see definitions of social or biological indicators) and so are to be used for verification of sustainability. The difference between both approaches seems a philosophical one based on the origin of the goal (systems-immanent or preset by ethical decision). Both may serve as supplements.

[2] It is clear that the terms "human health" as well as "ecosystem health" are extremely fuzzy (Costanza et al. 1992; Jüdes et al. 1985; WHO 1992), but they are ethical sub-imperatives of Sustainable Development and "are in essence two sides of the same coin" (Di Giulio and Monosson 1996).

hensive and really interdisciplinary view and therefore rarely fulfill theoretical requirements of a paradigmatic concept of Sustainable Development.

From a systems perspective, Sustainable Development can be seen as a macro process (of the global system) consisting of an unlimited number of micro processes (of sub-systems). Both scales differ in their dynamic characteristics: The macro process of Sustainable Development is per definition directed from conditions of unsustainability toward those of sustainability (however, not obligatory for each micro process). The micro processes fit given specific environmental situations and can include highly dynamic and even catastrophic events when seen on their specific scale. This means that although the micro processes may be unsustainable themselves, their results can contribute to sustainability on a higher scale. This idea, well-known in ecosystem research (Holling 1986; Wiegleb 1992; Jax et al. 1996), has been widely neglected in other fields of discussions on Sustainable Development so far.

This is an important reason for serious difficulties with defining adequate measures for Sustainable Development (see e.g. BMU 1996; Cansier and Richter 1995; CSERGE 1996; DGVN 1994, 1996; Statistisches Bundesamt 1994; Trzyna 1995). In reference to Müller (1996) and the idea of unifying goal functions we suggest a discipline related but transversally qualitative approach to define indicators of Sustainable Development, which analyzes multiple levels simultaneously for functional characters promoting coevolutionary processes.

4.3.3.1 An Analytical Scheme of Sustainable Development

Dovers (1989) differentiates two general contexts for viewing Sustainable Development, a goal context and an organizational context. Along the lines of the questions posed in Sect. 4.3.2 my approach to Sustainable Development shall be structured here into four levels (Jüdes 1996):

1. a problem analytical-ethical level,
2. an epistemological-conceptual level,
3. an organizational theoretical level (strategic planning),
4. a realizational (practical) level.

Fields or disciplines relevant to Sustainable Development can be assigned specific roles for specific levels (Fig. 4.3.2). Studies in ecology, economics, business and industry, agriculture, forestry, science and technology, energy, transportation, tourism, architecture and city planning, health, politics and planning, social science, education, philosophy and ethics contain contributions or at least considerations for the process of Sustainable Development. Although disciplinary approaches are helpful for detailed aspects, they cannot deliver a paradigmatic concept of Sustainable Development (see Jüdes 1996).

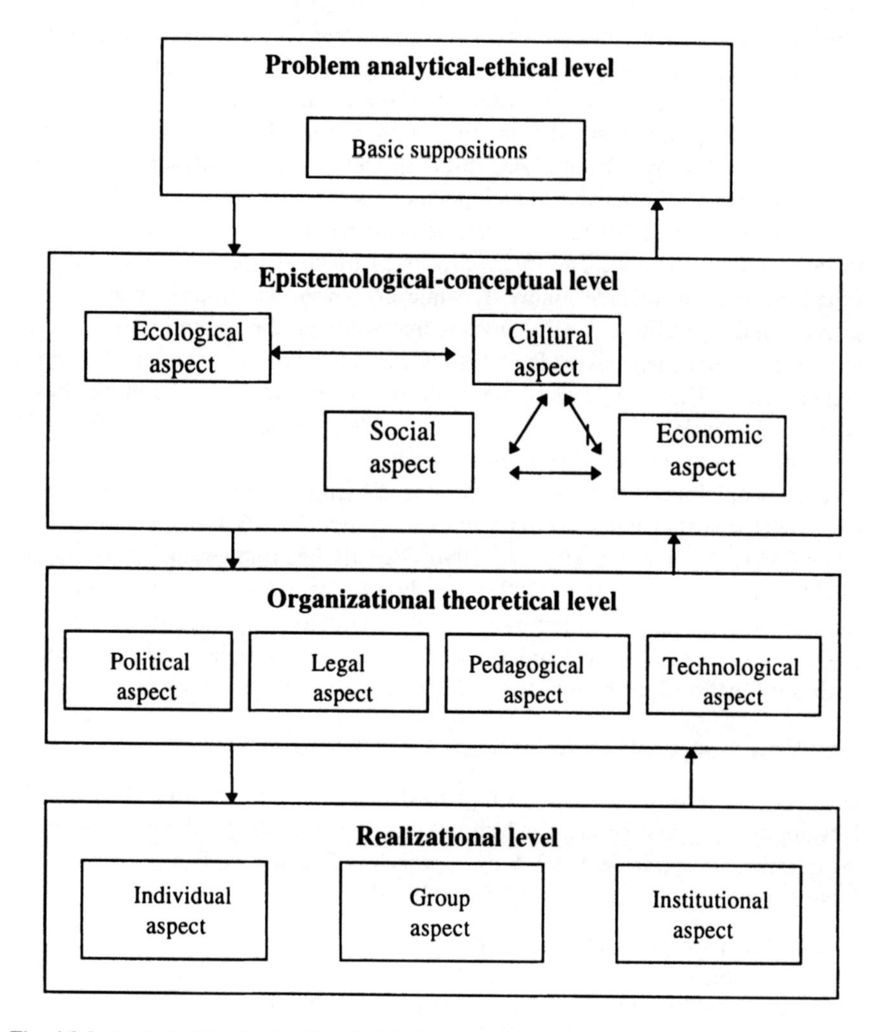

Fig. 4.3.2. Analytical levels of a Sustainable Development paradigm

For example each of these aspects grounds on disciplinary key concepts with specific characters (elements, structure, principles) helpful in understanding subsystem components or micro processes of Sustainable Development (e.g. ecosystem structures, energy flow between trophic levels and productivity in ecology; resource use, allocation and growth in economy; human rights, basic needs and justice in social sciences). Disciplinary concepts and characters are necessary for mediating measures of Sustainable Development in conformity with the accepted scientific standards, but a Sustainable Development paradigm requires a transversal "good sense" ("Vernunft" in the sense of Welsch 1996) unifying controversial ideas and models. Therefore it must be both communicable to individuals and social groups and operationable for pluralistic reasoning and transdisciplinary judgment of Sustainable Development processes.

4.3.3.2 Goal Functions, Orientors, and Indicators of Sustainable Development

Besides the theoretical construction of a conceptual framework of a Sustainable Development paradigm, as explained above, the communicative discourse on Sustainable Development in society (see Fig. 4.3.1) entrusts science with another two tasks: developing of methods for implementation and for monitoring of Sustainable Development.

Table 4.3.1. List of potential goal functions relevant for Sustainable Development

Ecological goal functions	Cultural goal functions	Social goal functions	Economic goal functions
• complexity	• perception	• pluralism	• benefits
• retinity	• rationality/ reasoning	• participation	• economic production
• biodiversity	• sense/meaning	• communication	• market
• stability	• self-formation	• social integration	• innovation
• exergy storage	• religion/myth/ sanctification	• prosocial attitudes/behavior	• costs
• carrying capacity	• morality/value system	• cooperation/partnership	• efficiency
• regeneration	• self conception	• manners	• economic cooperation
• resilience	• common good	• harmony/peace	• sufficiency
• selfregu-lation	• zeitgeist	• collective identity	• circular economy
• fitness	• worldview	• work	• economic risk management
• ascendency	• time perception	• humanization of work	• social compatibility
• cyclicity	• enculturation	• equality of opportunity	• substitution
	• family/cultural group	• self-help	
	• homeland	• solidarity	
	• cultural identity	• community work	
	• tradition	• social security	
	• innovation	• social distance motility	
	• differentiation	• sufficiency	
	• cultural diversity	• structural	
	• adaptive lifestyle	• structural nonviolence	
	• freedom (of rational action)		
	• institutional freedom		
	• social freedom/ liberty		

Three steps lead from theory to practice: First, it is necessary to select characters (elements, structures, principles) promoting Sustainable Development. Characters explicated in disciplinary key concepts are of different types: characters promoting Sustainable Development (goal functions), characters inhibiting Sustainable Development (goal-inhibiting functions) and characters with functions not related to Sustainable Development (goal-differing functions). Some characters are of facultative types which, depending on quantity or a change of co-characters, may either promote or inhibit Sustainable Development (Table 4.3.2). The list of (potential) goal functions given in Table 4.3.1 exemplifies disciplinary terms, which are functional for Sustainable Development on the epistemological-conceptual level (see Fig. 4.3.2). Such lists should be put up for all relevant disciplines and problem fields.

Table 4.3.2. Types of characters (elements, structures, principles) in disciplinary key concepts functionally relevant to Sustainable Development

Characters (elements, structures, principles) with ...	*cumulative characteristics*	*emergent characteristics*
promoting functions for Sustainable Development (goal functions)	Characters may grow or multiply on a continuous scale without changing their typical characteristics. The constituting functions for SD are dose-dependent. Examples are the biodiversity or ecosystem complexity (in ecology), costs of luxury goods (in economy), social security or sufficiency (in social science)	Characters with synergetic effects of components leading to new qualities. They have functions for SD, which cannot be deduced from their parts. Examples are exergy storage or stability (in ecology), market or substitution (in economy), cooperation (in social science)
inhibiting or no functions for Sustainable Development (goal-differing functions)	Characters may grow or multiply on a continuous scale without changing their typical characteristics. They have no function for SD or a negative one leading to detrimental effects. Examples are the biomass or human population density (in ecology), consumption and advertising (in economy), sexual behavior (in social science)	Characters with synergetic effects of components. They have no function for SD or a negative one, which may lead to detrimental effects. Examples are niche competition (in ecology), trade or product competition (in economy), work or sports (in social science)

Second, it is important to identify goal functions associated with more than one discipline or field; they might aggregate different aspects of Sustainable Development (unifying goal functions). This needs interdisciplinary cooperation. Unifying goal functions would help to bring disciplinary theoretical models together and could be used as orientors in interdisciplinary communication on a

Sustainable Development theory. They are starting points for defining standards of Sustainable Development and developing indicators (Fig. 4.3.3). A set of 220 sustainability indicators has been constructed by Bossel (1996) who used systems orientation theory to generate a scheme of basic system orientors.

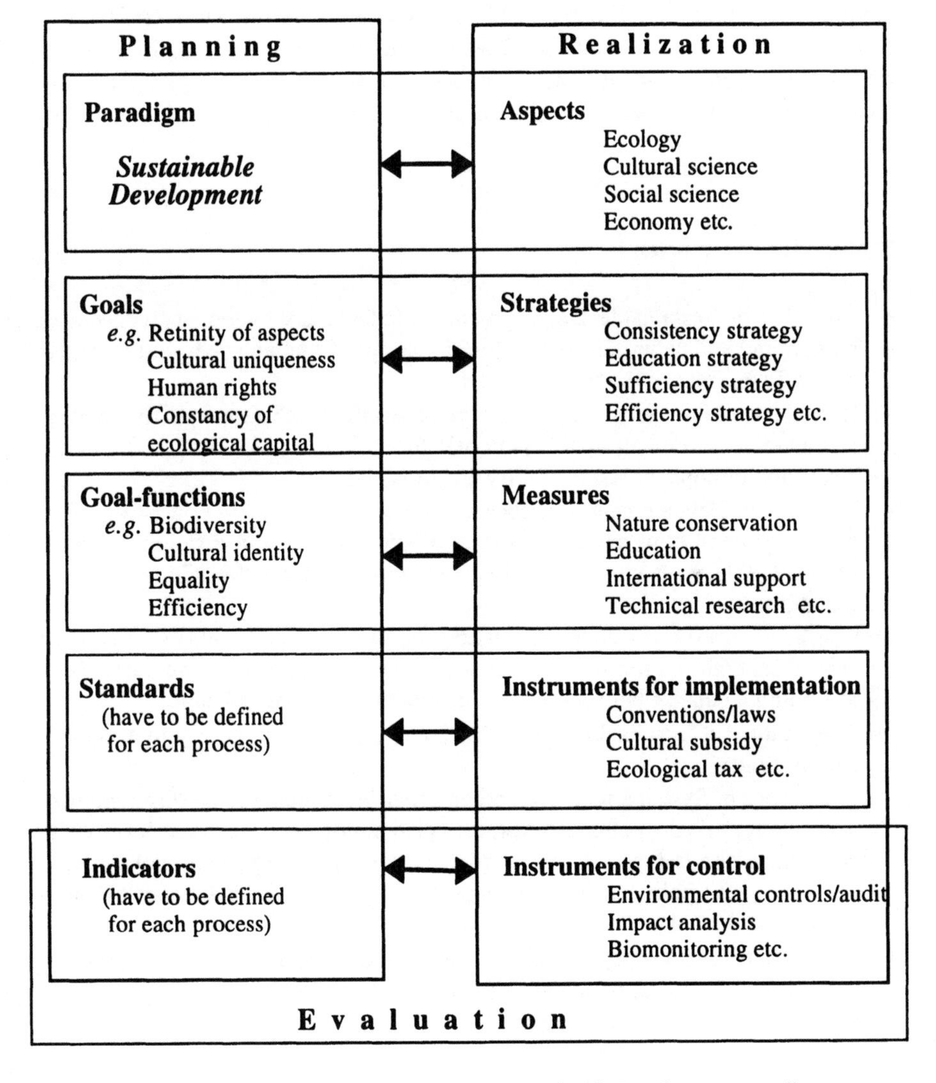

Fig. 4.3.3. Hierarchy of theoretical components of a Sustainable Development paradigm

Third, among these characters another difference has to be put in the focus: some have cumulative and others have emergent characteristics. My impression is that in the context of Sustainable Development studies have been focused on cumulative characters (either with promoting or inhibiting functions) (e.g. environmental pollution, toxicological substances, economic growth) and these characters are predominantly used for modeling so far, whereas emergent characters have attained less attention. This is a serious failure as emergent characters contribute to new qualities and modify the „sense" of a system, subsystem or component (Krohn and Küppers 1992). It seems to be the most critical and difficult point in the Sustainable Development discourse. An empirical pathfinding approach for identification of unifying goal functions with emergent characteristics will be tested by Jüdes et al. (in prep.) using associative networks of scientists.

4.3.4 Conclusions

- The idea of Sustainable Development includes a coevolutionary concept of humankind and nature, a socio-cultural concept of human needs and a natural scientific concept of the limited world.
- It is a „regulative idea" (in the sense of Kant) which refers to five human characteristics (human goal functions): the principles of reflexive fitness, of creativity, of hope, of responsibility and of sociability.
- Sustainable Development describes a cultural macro process consisting of an unlimited number of micro processes (of subsystems) with different, perhaps unsustainable, dynamics and leading to a new paradigm, the "Culture of Sustainability".
- The new paradigm is characterized by the intention and endeavor toward a successful contribution to the (coevolutionary) functionality of human culture in finding a balance between human health and ecosystems health.
- The Culture of Sustainability requires local adaptation (contrary to a globalization trend).
- In a theory of Sustainable Development, four levels ought to be differentiated: a problem analytical-ethical level, an epistemological-conceptual level (with ecological, cultural, social and economic aspects), an organizational theoretical level (with political, legal, pedagogical and technological aspects) and a realizational level (with individual, group and institutional aspects).
- In the context of this paper goal functions are defined as characters (elements, structures, principles) which have a functional quality for promoting the process of Sustainable Development. They may have collective or emergent characteristics.
- Unifying goal functions might be used as orientors in interdisciplinary communication on Sustainable Development and for defining standards and indicators of Sustainable Development.

Acknowledgment: The author is very grateful to all colleagues who participated in discussing aspects of this paper. He would like to thank his colleague Rachael Dempsey and his wife Renate Jüdes for kindly checking and correcting the English translation.

References

Adrian W, Blatt H (1987) Globale Entwicklungsanalysen im Nord-Süd-Kontext: Die nicht-ökonomischen Weltberichte. Themendienst 6. Deutsche Stiftung für internationale Entwicklung, Bonn

Bossel H (1996) Deriving Indicators for Sustainable Development. Environmental Modelling and Assessment 1 (4):193-218

Brown LR et al. (1993) Zur Lage der Welt – 1993. Daten für das Überleben unseres Planeten. Fischer, Frankfurt/M

BMU (Bundesministerium für Umwelt, Naturschutz und Reaktorsicherheit) (ed) (1996) Umweltökonomische Gesamtrechnung. Zweite Stellungnahme des Beirats Umweltökonomische Gesamtrechnung. Bonn

Cansier D, Richter W (1995) Nicht-Monetäre Aggregationsmethode für Indikatoren der nachhaltigen Umweltnutzung. Z f angew Umweltforsch 8 (3):326-337

Club of Rome (1977) Reshaping the International Order (RIO-Report). Hutchinson, London

CSERGE (Centre for Social and Economic Research on the Global Environment) (eds) (1996) The Measurement and Achievement of Sustainable Development. Economic and Social Research Council, Norwich

Costanza R et al. (eds) (1992) Ecosystem Health. New goals for environmental management. Island Press, Washington

DGVN (Deutsche Gesellschaft für die Vereinten Nationen) (ed) (1994) Bericht über die menschliche Entwicklung 1994. UNO-Verlag, Bonn

DGVN (Deutsche Gesellschaft für die Vereinten Nationen) (ed) (1996) Bericht über die menschliche Entwicklung 1996. UNO-Verlag, Bonn

Di Giulio RT, Monosson E, (eds) (1996) Interconnections between human and ecosystem health. Chapman & Hall, London

Dovers S (1989) Sustainability: Definitions, clarifications and contexts. Development 2/3:33-36

Dovers SR, Handmer JW (1993) Contradictions in sustainability. Environ Conserv 20(3):217-222

Fritz P et al. (1995) Nachhaltigkeit in naturwissenschaftlicher und sozialwissenschaftlicher Perspektive. Hirzel, Stuttgart

Glaeser B (1992) Natur in der Krise? Ein kulturelles Mißverständnis. Gaia 1 (4):195-203

Goodland R et al. (1992) Population, technology, and lifestyle. The Transition to Sustainability. Island Press, Washington

Haber W (1994) Ist „Nachhaltigkeit“ (Sustainability) ein tragfähiges ökologisches Konzept? Verh Ges f Ökol 23:7-17

Harborth H-J (1993) Dauerhafte Entwicklung statt globaler Selbstzerstörung. Eine Einführung in das Konzept des "Sustainable Development". 2nd ed., Ed. Sigma, Berlin

Hauff V (ed) (1987) Unsere gemeinsame Zukunft. Der Brundtland-Bericht der Weltkommission für Umwelt und Entwicklung. Eggenkamp, Greven

Holling CS (1986) The Resilience of Terrestrial Ecosystems: Local Surprise and Global Change. In: Clark WM, Munn RE (eds) Sustainable development of the biosphere. Cambridge Univ Press, Cambridge

Heins B (1994) Nachhaltige Entwicklung - aus sozialer Sicht. Z f angew Umweltforsch 7:19-25

IUCN/UNEP/WWF (1991) Caring for the Earth: a Strategy for Sustainable Living. Gland, Schweiz

Jax K et al. (1996) Skalierung und Prognoseunsicherheit bei ökologischen Systemen. Verh Ges Ökol 26:527-535

Jörgensen SE (1992) Integration of Ecosystem Theories. A pattern. Kluwer, Dordrecht

Jüdes U (1995) Die Bedeutung von "Sustainable Development" (SD) für die Umweltpädagogik. In: GfÖ-Arbeitskreis Theorie in der Ökologie. Nachhaltige Entwicklung - Aufgabenfelder für die ökologische Forschung. EcoSys - Beiträge zur Ökosystemforschung 3:30-34

Jüdes U (1996) Hat Umweltbildung eine Zukunft? Zur Bedeutung von "Sustainable Development" und "Öko.Konto" für die Umweltpädagogik. In: Bitz A et al. (eds) Runder Tisch Umweltbildung. AG Umweltbildung Rheinland-Pfalz, Mainz, pp 14-30

Jüdes U (1996) Das Paradigma „Sustainable Development". Nachhaltige Entwicklung im Hinblick auf ökologische, kulturelle, soziale und ökonomische Dimensionen. Gutachten i. A. des Bundesministeriums für Bildung, Wissenschaft, Forschung und Technologie, unpublished

Kastenholz HG et al. (eds) (1996) Nachhaltige Entwicklung. Zukunftschancen für Mensch und Umwelt. Springer-Verlag, Berlin

Kopfmüller J (1995) Die Idee einer zukunftsfähigen Entwicklung ("Sustainable Development") – Hintergründe, Probleme, Handlungsbedarf. In: Bechmann G (ed) Gesellschaftliche Probleme und Technikfolgenabschätzung. Campus, Frankfurt/M

Krohn W, Küppers G, Hrsg (1992) Emergenz: Die Entstehung von Ordnung, Organisation und Bedeutung. Suhrkamp, Frankfurt/Main

Lele SM (1991) Sustainable Development: a Critical Review. World Developm 19 (6):607-621

Marien M, Ed (1996) Environmental Issues and Sustainable Futures. World Future Society, Bethesda

Mathes K, Breckling B (1995) Nachhaltige Entwicklung: Aufgabenfelder für die ökologische Forschung. EcoSys 3:71-73

Mathes K et al. (1996) Systemtheorie in der Ökologie. Ecomed, Landsberg

Müller F (1996) Ableitung von integrativen Indikatoren zur Bewertung von Ökosystem-Zuständen für die Umweltökonomische Gesamtrechnung. Unveröff. Projektstudie. Projektzentrum Ökosystemforschung, Kiel

Nohlen D, Nuscheler F, Hrsg (1993) Handbuch der Dritten Welt. Bd. 1: Grundprobleme, Theorien, Strategien. Dietz, Bonn

Pearce D (1993) Blueprint 3. Measuring Sustainable Development. Earthscan Publ, London

Pearce DW, Warford JJ (1993) World Without End. Economics, Environment, and Sustainable Development. Oxford Univ. Press, New York.

Redclift M (1987) Sustainable Development: Exploring the Contradictions. Methuen, London

Redclift M (1994) Reflections on the 'Sustainable Development' Debate. Int J Sustain Dev World Ecol 1 (1):3-21

SRU (Der Rat von Sachverständigen für Umweltfragen) (1994) Umweltgutachten 1994 für eine dauerhaft-umweltgerechte Entwicklung. Metzler-Poeschel, Stuttgart

SRU (Der Rat von Sachverständigen für Umweltfragen) (1996) Umweltgutachten 1996 zur Umsetzung einer dauerhaft-umweltgerechten Entwicklung. Metzler-Poeschel, Stuttgart

Statistisches Bundesamt (ed) (1994) Umweltökonomische Gesamtrechnungen - Basisdaten und ausgewählte Ergebnisse 1994. Reihe 4, Fachserie 19 Umwelt, Metzler-Poeschel, Stuttgart

Tisdell CA (1993) Ecologically Sustainable Development (ESD) and industry: policies, economic attitudes and Australian proposals. Univ Queensland Dept Econ Disc Pap 109:1-14

Trzyna TC, Ed (1995) A Sustainable World. Defining and Measuring Sustainable Development. Int. Center for the Environment and Public Policy, Sacramento Claremont

Turner RK, Ed (1993) Sustainable Environmental Economics and Management. Principles and Practice. Belhaven Press, London New York

von Amsberg J (1993) Project evaluation and the Depletion of Natural Capital: an Application of the Sustainability Principle. Environment Working Paper No. 56, World Bank, Washington

Voss G, Hrsg (1994) Sustainable Development: Leitziel auf dem Weg in das 21. Jahrhundert. Deutscher Instituts-Verlag, Köln

Welsch W (1996) Vernunft. Die zeitgenössische Vernunftkritik und das Konzept der transversalen Vernunft. Suhrkamp, Frankfurt/M

WHO (1992) Our Planet, our Health. Report of the WHO Commission on Health and the Environment, Geneva

Wiegleb G (1992) Explorative Datenanalyse und räumliche Skalierung - eine kritische Evaluation. Verh Ges Ökol 21:327-338

Worster D (1994) Auf schwankendem Boden. Zum Begriffswirrwarr um "nachhaltige Entwicklung". In: Sachs W (Hrsg.) Der Planet als Patient. Über die Widersprüche globaler Umweltpolitik. Birkhäuser, Berlin, pp 93-112

4.4 Environmental Law and the Science of Ecology

Hans-Ulrich Marticke

Abstract

The paper first discusses whether the term 'goal function' might be useful as a new term in environmental law. In the second place it focuses on general aims and values in environmental law like intergenerational equity, the intrinsic value of biodiversity and other values. It states that there is no general hierarchy of values and that the aims and values of environmental protection have to be balanced against other constitutional values. The third part deals with the legal implementation of the concept of sustainable development. The forth part treats the problem of breaking down general aims and values to specific eco targets in standard setting procedures, zoning law and landscape planning, the Environmental Impact assessment and the Integrated Pollution Control. The underlying intention of the paper is to demonstrate the interaction between ecological research and environmental law and to point out possible fields of interdisciplinary cooperation.

> 'When *I* use a word,' Humpty Dumpty said in rather a scornful tone, 'it means just what I choose it to mean – neither more nor less.'
> 'The question is,' said Alice, 'whether you *can* make words mean so many different things.'
>
> LEWIS CARROL

4.4.1 Introduction

Originally, I have prepared this paper for the interdisciplinary workshop on 'unifying goal functions' which was held in September 1996 in Salzau. That's why it starts with a discussion on whether 'goal function' might be useful as a new term in environmental law. As many other participants I have questioned the usefulness of this term for humanities, and subsequently the title of this book has been extended to 'eco targets'. The second part deals with aims and values in environmental law one might call 'eco targets'. This is also true for the concept

F. Müller M. Leupelt (Eds.)
Eco Targets, Goal Functions, and Orientors

of sustainable development (Section 4.4.4). In this context I will try to describe the interrelationship between environmental law and the science of ecology: how legal systems took up ideas from ecology and how technology forcing concepts in environmental law create a demand for information from ecology. Legislation, standard setting procedures, landscape planning and the Environmental Impact Assessment are instruments, where knowledge from ecology is needed and may contribute to develop and implement binding norms for the protection of the environment (Section 4.4.5).

4.4.2 Weasel Words, Treacherous Twins and the Naturalistic Fallacy

The usage of language obviously is a major concern for lawyers, as it is their most prominent tool. We have to be careful when we meet weasel words and treacherous twins. *Weasel words* are terms that lack force or a precise meaning. For example, 'eco-' and 'ecological' have became such words. *Treacherous twins, faux amis* in French and *falsos gemelos* in Spanish characterize an experience in translation and in comparative law: often you find two words, ideas or concepts, that seem to be very similar. But if you take a closer look, they have a completely different meaning. When lawyers and economists, for example, discuss 'property rights', they will soon realize that they are talking about two different subjects.

And what about 'goal functions' or 'Zielfunktionen'? I did not know the term in the context of environmental protection and ecology before I came to the workshop. I could not find it neither in English nor in German manuals on ecology. I was told, that goal function is a technical term in mathematics and mathematical modelling. It was the intention of the organizors of the workshop to extend this term to the description of ecosystems and societies. During the workshop, goal function was used in at least four different meanings:

1. a mathematical function to describe certain dynamics of system models,
2. a trend or tendency in real-live ecosystems,
3. a consciously pursued aim of a person, organisation or society,
4. a hidden goal of human societies and history.

In the everyday meaning of the word, goals are connected to an intention. Human beings, organizations and societies may pursue goals (meaning no. 3). But to say, that ecosystems have inherent goals (meaning no. 2), is rather a metaphor.[1] The question is, "whether a biological system as a whole can have a goal that is in some way similar to a human goal - *i.e.* whether it is programmed with an ultimate purpose."[2] The most problematic meaning is no. 4, which seems to

[1] For the different metaphors underlying ecological research see *F.B. Golley,* The History of the Ecosystem Concept in Ecology, 1993.

[2] Philosophy of Biology, The New Encyclopaedia Britannica, 15th ed., Vol. 25, 677 (1991).

be ideologically biased. It may presuppose that human beings, organisations and societies as well as the history of mankind are driven by a hidden force called 'unified goal function' which scientists can reveal with the help of systems theory and mathematical modelling.[3] In any case, goal function is a treacherous twin. It will lead to confusion between model (no. 1) and reality (no. 2 and 3) and may favor simplicistic analogies from biology to humanities and *vice versa* (no. 2 and 4).

Another general remark refers to the difference between *pre*scriptive and *de*scriptive statements and their interrelation. Lawyers and judges apply legally binding, normative rules to specific cases. In the philosophy of law, there has been a long debate on whether one can derive value judgements from the 'essence' or 'true nature' of a situation.[4] Those, who oppose this argumentation, call it the *naturalistic fallacy* (G.E. Moore): You cannot conclude what *should be* from what *is* or what *could be*. Lawyers sometimes confuse questions of fact with questions of law, description with value judgement. Some lawyers, for example, believe that the term 'ecological damage' is descriptive and directly derived from ecology,[5] though 'damage' necessarily implies a value jugdement. On the other hand ecology sometimes applies concepts that seem to be normative. I am not sure, if you can define 'disturbance' which has a strong negative connotation on a purely descriptive level. During the workshop, the statement was made, that the way Mother Nature does things is good. It was the only reason given for the conclusion that maturity should be used as a criterion in environmental evalution. This is a typical example of the naturalistic fallacy.[6] And as the term 'goal function' implies intentions, it may also lead to confusion between what *is* or *could be* and what *should be.*

Therefore, in my opinion, the term 'goal function' is not helpful in the context of environmental law. It will not facilitate communication between scientific ecologists and others who try to implement environmental protection.

4.4.3 Aims and Values in Environmental Law

Aims and values in legal systems often are expressed in written constitutions, in the first articles of statutes stating their purpose or in the preamble of interna-

---

[3] See *K. Popper,* The Poverty of Historicism, 2nd ed. 1960 and *item,* The Open Society and Its Enemies, Vol. 1: The Spell of Plato, 1945.

[4] *A. Scheuerle,* Das Wesen des Wesens, Archiv für civilistische Praxis 163 (1964), 429 and *R. Dreier,* Zum Begriff der „Natur der Sache", 1965.

[5] *N. Wenk,* Naturalrestitution und Kompensation bei Umweltschäden, 1994, p. 43; *C. H. Seibt,* Zivilrechtlicher Ausgleich ökologischer Schäden, 1994, p. 6 *et.seq.*; both hide their position behind the vague term 'phenomenological definition'.

[6] If maturity should be used as a criterion in environmental assessment, this needs another, more convincing argumentation: What maturity indicates and if it reflects societal values which are protected in legislation.

tional conventions. Mostly, they do not provide directly applicable rules but may support a teleological interpretation of more specific provisions.

4.4.3.1 Intergenerational Equity

The Preamble of the 1979 Bonn Convention on the Conservation of Migratory Species of Wild Animals clearly states the concept of intergenerational equity: "each generation *holds* the resources of the earth for future generations and has an *obligation* to ensure that this *legacy* is conserved and, when utilized, is used wisely". In international law, the concern for future generations was recognized as early as 1946 in the Preamble of the Whaling Convention. This generational perspective also underlies the 1972 Stockholm Declaration, the 1982 World Charter for Nature and the 1992 Rio Declaration. Whereas these declarations are 'soft law', the 1992 Convention on Biological Diversity, which has been ratified by more than 110 Nations, is legally binding. Its Preamble declares the determination "to *conserve* and *sustainably use* biological diversity for the benefit of present and *future* generations."

On the national level, the US-National Environmental Policy Act of 1969 mentions the "social, economic, and other requirements of present and future generations of Americans". In Germany, environmental protection is a constitutional aim for governmental action (Staatsziel). According to Article 20a Grundgesetz, as amended in 1994, the State protects the "natural bases for the existence of life" (natürliche Lebensgrundlagen) "also in responsibility for future generations".

The ethical concept of intergenerational equity thus has entered international law and many national legal systems.[7] However, it is more difficult to find out whether this goal already has an influence on the practice of environmental law. We have to go as far as the Philippines to find a *leading case*, in which intergenerational equity has played a decisive role: In *Minors Oposa v. Secratary of the Department of Environment and Natural Resources* the Supreme Court granted the plaintiffs standing to represent their "yet unborn posterity".[8]

4.4.3.2 Intrinsic Value of Biodiversity

The mainstream in German environmental law dislikes the debate on anthropocentrism and ecocentrism. Lawyers argue, that there is hardly a difference in

---

[7] See *E. Brown-Weiss*, In Fairness to Future Generations, 1989.

[8] 33 International Legal Materials 173 (1994); See *A. Rest*, The Oposa Decision: Implementation of the Right to a Sound Environment and the Principle of Intergenerational Responsibility, The Layers Review (Manila), August 1995, 5.

results between an ecocentric and an 'enlighted' antropocentric view.[9] Both are able to acknowledge that nature, or parts of it, may have an intrinsic value.

It is little known that the 1992 Convention on Biological Diversity explicitly recognizes the instrinsic value of biodiversity at the very beginning of its Preamble. As early as 1979, the term 'intrinsic value' is found in the Council of Europe's Berne Convention on the Conservation of European Wildlife and Natural Habitats. And the 1982 World Charter for Nature states, that "every form of life is unique, warranting respect regardless of its worth to man."

In national legislations, New Zealand's Resource Management Act of 1991 provides a definiton which extends the intrinsic value to characteristics of ecosystems:

"'intrinsic values', in relation to ecosystems, means those aspects of ecosystems and their constituent parts which have *value in their own right*, including –

(a) their biological and genetic diversity; and
(b) the essential characteristics that determine an ecosystem's integrity, form, functioning, and resilience." (Article 2 (1))

In Germany, the legislative history of Article 20a Grundgesetz clearly reveals a disagreement on whether an anthropocentric or an ecocentric view should be expressed in the Constitution, and that the question has been left open.[10] Article 39 III of the Constitution of Brandenburg (1992) goes further: "Animals and plants are respected as living beings." On the other hand, the Federal Nature Conservation Act of 1976 is still strictly anthropocentric: its purpose is to maintain "the potential for services of the household of nature (Leistungsfähigkeit des Naturhaushalts) [...] as the bases for the existence of man (Lebens-grundlage des Menschen)".[11]

It is difficult to say what the consequences of the recognition of an intrinsic value of biodiversity or of aspects of ecosystems will be. A priority in the hierarchy of values? A shift in the burden of proof? A priority of preservation over mitigation and compensation? Probably one of the most important consequences is that an endangered species is to be preserved as an irreplaceable resource, even if its worth for human use is not yet known or may be substituted.

Recognizing intergenerational equity and the intrinsic value of biodiversity will, in my opinion, lead to a shift in paradigms. The dominant paradigm of the liberal and democratic societies has been equal rights and the reciprocity between rights and duties. It is the intention of those who advocate 'rights' of na-

[9] *M. Kloepfer*, Umweltrecht, 1989, p. 13 *et.seq.*; *Hoppe W, Beckmann M (1989)* Umweltrecht. p. 20 ss.

[10] See *K. Waechter*, Umweltschutz als Staatsziel, Natur und Recht 1996, 321, at 324.

[11] The proposal for a revision of the Federal Nature Conservation Act of 1996 is as ambigious as Art. 20a Grundgesetz ("Verantwortung des Menschen für die natürlichen Lebensgrundlagen").

ture and of future generations to extend this paradigm.[12] But whereas women and former slaves can speak for themselves, future generations, animals, plants and nature as a whole cannot. They would have to be represented. This is the crucial point in the debate on 'rights' of future generations and nature: who shall have the right to speak 'in their name'? Our responsibility for future generations and the biosphere is necessarily *paternalistic* or *parental*. Therefore we should discuss duties rather than rights. This is another paradigm: responsibility and fiduciary obligations in a non-reciprocal context. If we look for models in contemporary legal systems, we may end up in family law (to which the term 'parental' refers) and the law of succession (where the term 'legacy' stems from). In constitutional law, we have to go back to feudal times and to ideas which were originally religious. Another way to develop the fiduciary obligation of the present generation is the concept of public trust. There are early hints in US-legislation: According to § 107 Comprehensive Environmental Response, Compensation, and Liability Act (CERCLA) of 1980 and § 1006 of the Oil Pollution Act (OPA) of 1990 governmental entities as *trustees* may claim or even have to claim compensation for natural resource damages. The monies collected must be used to restore or to acquire the equivalent of the injured resources.

In the outcome, the goals which the 'rights'-of-nature-movement wants to achieve in legislation are already met by the recognition of the instrinsic value of parts of nature, by the concept of public trust and by standing of environmental groups in decision making processes and in court.[13] From a laywer's point of view, the idea of 'rights' of nature therefore is *hypertrophic*.

4.4.3.3 Other Values

Economic values are not the only and not the most important values that environmental law protects. Human health, for instance, should come first. Many conventions in international law, but also § 2 (a) (3) of the US-Endangered Species Act of 1973 name several values. One of the longest lists is found in the Preamble of the Convention on Biological Diversity: "Conscious of the intrinsic value of biological diversity and the ecological, genetic, social, economic, scientific, educational, cultural, recreational and aesthetic values of biological diversity and its components, [...]"

Ecology as a descriptive natural science cannot directly contribute to operationalize these values in the decision making process. But ecologists can leave the safe ground of hard sciences. 'Applied ecology' or 'conservation biology'

---

[12] *C. Stone,* Should trees have a Standing, 45 S. Cal. L. Rev. 450 (1972) = Umwelt vor Gericht, 2nd ed. 1992; *R.F. Nash,* The rights of Nature, 1989, *K. Bosselmann,* Im Namen der Natur, 1992.

[13] In this respect, *C. Stone* is much more pragmatic than many of his followers. See the postscript to the German edition, Umwelt vor Gericht, 2nd ed. 1992, p. 107 *et.seq.* and *J.-F. Blume,* Zusammenfassung der Diskussion mit Prof. Stone, in: *H. Burmeister (Hrsg.),* Wege zum ökologischen Rechtsstaat, 1994, 39.

('naturschutzfachlich', 'wissenschaftlicher Naturschutz') is a practical, value-orientied science like medicine and can help to operationalize the ecological and genetic values of ecosystems: to set priorities in red data books, to select areas for the designation of protected sites, to put landscape planning into practice, and to assess the impact of projects in the Environmental Impact Statements. In applied ecology, the scientific work of ecologists is qualified by societal aims as laid down in legislation.

There are some traditional values of nature conservation, however, that the ecosystem-approach cannot reflect as easily. One of them is beauty. The German Federal Nature Conservation Act of 1976, for example, plainly speaks of the 'beauty' of nature and landscapes. And the CITES-Convention of 1973 protects wild fauna and flora "in their many *beautiful* and varied forms". Other legal texts use the less offensive term 'aesthetic value'. Beauty does not really fit into our modern, matter-of-fact way of thinking, nor into the functional, scientific concept of ecosystems. Of course, one can speak of 'aesthetic services' or even of 'aesthetic productivity'[14], but this means to express a non-instrumental experience in intrumental terms. Beauty, hopefully, will resist quantification.

Another idea is particularity ('Eigenart', as the German Federal Nature Conservation Act calls it), as well as uniqueness. Both are qualitative criteria which are not completely reflected in the quantative criterion of rarity.[15] They refer to the specific, individual characteristics of a given ecosystem in time and space, which are a subject for natural history[16] rather than for the generalizing, causal ecosystem-approach. The functional approach is closely connected to the idea of 'functional substitutability'.[17] Uniqueness, on the other hand, refers to objects that, from a certain point of view, resist generalization and substitut-ability. And if you really try to quantify uniqueness, you may soon end up in infinity...[18] In the discussion, singular circumstances of an ecosystem (particularity) and uniqueness often are not clearly distinguished.[19] Sometimes, each and every habitat is called unique. This leads to an inflation of uniqueness depriving this value judgement of its sense.

---

[14] *W. Haber*, Ökologische Grundlagen des Umweltschutzes, 1993, 48.

[15] It should be noted that manuels on nature conservation generally do not mention particularity and uniqueness as criteria for assessment, see *G. Kaule*, Arten- und Biotopschutz, 1986, p. 248 *et.seq.*, *H. Plachter*, Naturschutz, 1991, p. 180 *et.seq.*, *M.B. Usher/W. Erz (Hrsg.)*, Erfassen und Bewerten im Naturschutz, 1994, p. 17 *et.seq.*

[16] See *Trepl*, Geschichte der Ökologie, 2nd ed. 1994, p. 44 *et.seq.*, p. 231 *et.seq.*

[17] Most clearly stated by an economist: *C. Perrings*, Ecological and economic values, in: *K.G. Willis/J.T. Corkindale*, Environmental Valuation, 1995, 56, at 59.

[18] See *Hampicke*, Naturschutz-Ökonomie, 1991, 104.

[19] From a historical perspective see *G. Canguilhem*, Du singulier et de la singularité en épistémologie biologique, 1962, German translation in: *item*, Wissenschaftsgeschichte und Epistemologie, 1979, 59.

4.4.3.4 Hierarchy of Values and Balancing of Interests

The aims and values of environmental protection and nature conservation may conflict with one another as well as with other aims, values and rights. Legal systems usually do not state a general hierarchy of values or a fixed program for the balancing of interests. In German constitutional law human health, for example, has a certain priority, whereas environmental protection is one aim among others. Law provides procedures and general rules for the solution of conflicts like the principle of proportionality, which plays a dominant role in German constitutional and administrative law.

In the field of pollution control, a more or less clear priority for the protection of human health can be found.[20] In German environmental law, there is a basic distinction between hazards or dangers (Gefahren) which have to be avoided at all costs and mere risks which should be minimised according to the best available technology and subject to the principle of proportionality.[21] The priority for human health is not absolute: the threshold of danger is usually defined for an average, healthy person, so that young, elderly or sick people are not necessarily protected.[22]

In the field of nature conservation, many statutes contain substantive rules that give a priority to the protection of habitats of special ecological value. In the United States, the Endangered Species Act of 1973 affords endangered species the "highest of priorities".[23] Still, exemptions may be granted by a special committee (§ 7 (g), (h)) which environmentalists call the "God Squad". In the European Union, the 1979 Directive on the conservation of wild birds and the 1992 Directive on the conservation of natural habitats and of wild fauna and flora provide special protection. The member states are obliged to designate protected areas, following ornithological and ecological criteria.[24] Any plan or project that might have an impact on designated areas is subject to an impact assessment. If the assessment is negative, the plan or project may only be carried out "in the absence of an alternative solution" and "for imperative reasons of overriding public interest".[25] Compensatory measures have to be taken. In Ger-

---

[20] For the protection of human health as a constitutional obligation of the State see Federal Constitutional Court, BVerfGE 49, 89, at p. 140 *et.seq.*; BVerfGE 53, 30, at 57 ss.; BVerfGE 56, 54, at p. 73 *et.seq.*

[21] See § 5 I Nr. 1, 2 of the Federal Immission Control Act; *A. Reich,* Gefahr - Risiko - Restrisiko, 1989; *F. Petersen,* Schutz und Vorsorge, 1993.

[22] This statement only refers to the abatement of dangers where health has a stated priority over economic interests. See also *R. Wulfhorst,* Der Schutz „überdurchschnittlich empfindlicher" Personen im immissionsschutzrechtlichen Genehmigungsverfahren, Natur und Recht 1995, 221.

[23] See *TVA v. Hill,* 437 U.S. 153, 98 S.Ct. 2279.

[24] See European Court of Justice, Case C-355/90, *Commission v Spain,* Zeitschrift für Umweltrecht 1994, 305 mit Anmerkung G. *Winter; K. Iven,* Schutz natürlicher Lebensräume und Gemeinschaftsrecht, Natur und Recht 1996, 373.

[25] Art. 6 III, IV of the Habitat Directive. See also European Court of Justice, Case C-57/89, *Commission v Germany* (1991) ECR I 883, *G. Baldock,* The status of special protection areas

many, some types of natural habitats are strictly protected without prior designation (§ 20c Federal Nature Conservation Act).

In most other cases, environmental law does not give a general priority to environmental protection. Each case is treated on its own merits. Conflicting interests are balanced according to the principle of proportionality. Whereas the Environmental Impact Assessment provides for a procedure, the provisions regarding "interventions in the natural surroundings and landscape" (Eingriffe in Natur und Landschaft, § 8 Federal Nature Conservation Act) constitute a substantive obligation to avoid and, if that is not possible, to compensate or mitigate the impact of projects in all areas of the country.[26]

4.4.4 The Concept of Sustainable Development

The concept of sustainable development has been acknowledged by the international community in the Rio Declaration of 1992. In international law, "it is a notion around which legally significant expectations regarding environmental conduct have begun to crystallize."[27] It is also beginning to appear in national legislations: for example in New Zealand's Resource Management Act 1991, in the Proposed Swedish Environmental Code (1996)[28], and in an experts' proposal for a General Part of a Federal Environmental Code in Germany (1989).[29]

The concept tries to balance the needs of present and future generations and to integrate conflicting social, economic and environmental goals. This is why 'sustainable development' is a *weasel word.* To exemplify its ambiguity, one only has to look at the different German translations. The technically correct, traditional term is 'nachhaltig' and can be found in the Prussian Mining Competences Order of 1861, in § 11 of the Federal Forest Act of 1975 and § 1 of the Federal Nature Conservation Act of 1976. Still, this term is not very popular, as it also connotes 'effectively' and 'up to the limit'. Others translate 'sustainable'

---

for the protection of wild birds, Journal of Environmental Law 1992, 139; *G. Winter,* Der Säbelschnäbler als Teil für's Ganze, Natur und Recht 1992, 21; *W.T.J. Wils,* The bird's directives 15 years later: survey of the case law and a comparison with the habitat directive, Journal of Environmental Law 1994, 219.

[26] For an overview in English see *G. Winter,* Substantive Criteria of Environmental Protection, in: *G. Winter (ed.),* European Environmental Law, 1996, 37, at p. 44 *et.seq.* An English translation of the Federal Nature Conservation Act is published in *G. Winter (ed.),* German Environmental Law, Basic Texts and Introduction, 1994, 81.

[27] *G. Handl,* Environmental Security and Global Change: The Challenge to International Law, Yearbook of International Environmental Law 1 (1990), 3, at 25.

[28] The Environmental Code, Focused and Co-ordinated Environmental Legislation for Sustainable Development, Stockholm 1996, p. 19 *et.seq.*

[29] §3 (2) Professorenentwurf, in: *M. Kloepfer et al.,* Umweltgesetzbuch - Allgemeiner Teil -, 1990. For a rather peculiar English translation see *Federal Environmental Agency,* Environmental code - General Part -, A Proposal for a German Federal Environmental Code, 1993.

with 'dauerhaft-umweltgerecht' (environmentally durable).[30] Some economists even invented the new word 'zukunftsfähig'[31] (which is something like 'futureable' or 'futureability'). It sounds good, but may mean anything.

However, the concept of sustainable development may be helpful to reach consensus and at least to clarify which practices of resource management, consumption and pollution are definitely not sustainable. The concept focuses on practical problems and expresses a minimal consensus, that environmental, economic and social interests of present and future generations have to be brought into balance. Like many basic ethical, political and legal ideas it is a regulatory idea, which cannot be made more precise by a good definition, but can only be implemented by a continuous discussion in any new situation.

The basic economic idea of sustainability is that the capital stock should stay constant over time.[32] According to 'weak sustainability', it is not relevant in which form the capital stock is preserved. The loss of natural capital may be offset by more roads, machinery or other man-made-capital. 'Strong sustainability' states that not all forms of capital are substitutable and that at least the critical natural capital has to be preserved, applying 'safe minimum standards'. A broader interpretation is that the overall stock of natural capital should not be allowed to decline. This is indeed quite a wide range of options! Only strong sustainability seems to be compatible with the legal concept of intergenerational equity and the precautionary principle.[33]

In any case, to put the concept of sustainable development into practice, an enormous amount of information is needed: physical accounting, indicators for the state of the environment, knowledge about the regenerative capacity of ecosystems, their carrying capacity, and their resilience to pollution, and so on. To operationalize the concept, usually four sub-goals are distinguished:

- a cautious use of non-renewable resources
- a use of renewable resources which does not exceed the natural regenerative capacity of ecosystems
- an amount of pollution and physical change which does not exceed the assimilation and carrying capacity of ecosystems
- the avoidance of risks to human health.[34]

---

[30] Rat von Sachverständigen für Umweltfragen, Umweltgutachten 1994, Für eine dauerhaft-umweltgerechte Entwicklung.

[31] *Schmidt-Bleek,* Ohne De-Materialisierung kein ökologischer Strukturwandel, in: Jahrbuch Ökologie 1994, 1993, 94; Enquete-Kommission „Schutz des Menschen und der Umwelt" des Deutschen Bundestages (Hrsg.), Die Industriegesellschaft gestalten, 1994, p. 26 *et.seq.* („nachhaltig zukunftsverträgliche Entwicklung").

[32] For an overview see *D. Pierce,* Blueprint 3, Measuring sustainable development, 1993, p. 15 *et.seq.*

[33] See Principle 15 of the Rio Declaration and Article 130 r (2) EC Treaty. The precautionary principle originates from German environmental law, see *E. Rehbinder,* Das Vorsorgeprinzip im internationalen Vergleich, 1991. It has been introduced in France by the Law no. 95-101 from February 2nd 1995.

[34] See note 30, at Nr. 11 *et.seq.*

Scientific knowledge of ecology is demanded, especially for the sustainable use of renewable resources and the acceptable rate of pollution. The legal instruments to set targets in most cases already exist.

4.4.4.1 Sustainable Use of Renewable Ressources

The use of renewable resources covers a wide body of law, from fishery and hunting to agriculture, forestry, and the use of fertilizers and pesticides.

Concerning the sustainable use and conservation of living marine ressources, there are numerous conventions and commissions which are entitled to prevent over-exploitation.[35] The 1958 Geneva Convention on Fishing and the Conservation of Living Resources of the High Seas, states that "'conservation' of the living resources of the high seas means the aggregate of the measures rendering possible the *optimum sustainable yield* for the resources to secure a *maximum supply* of food and other marine products." The 1982 UN Convention on the Law of the Sea lays down certain objectives of conservation and management for coastal states. "Taking into account the best scientific evidence available", the "measures shall be designed to maintain or restore populations of harvested species at levels which can produce *the maximum sustainable yield,* as qualified by relevant environmental and economic factors..." The concept of 'optimum' or 'maximum sustainable yield' has been critized for not taking into accound the interdependence of ecosystems. A broader ecosystem approach is found in the 1980 Convention on Conservation of Antarctic Marine Living Resources, which protects the "integrity of the ecosystem of seas surrounding Antarctica." Conservation includes 'rational use', which has to assure the stable recruitment of the harvested population, the "maintenance of the ecological relationships between harvested, dependent and related populations", and the "prevention of changes or minimizing the risk of changes in the marine environment which are not potentially reversible over two or three decades" (Article II).

As to the sustainable management of forests, States until now could only agree on a "non-legally binding authoritative *(sic)* statement of principles for a global consensus on the management, conservation and sustainable development of all types of forest" (1992). Whereas sovereign rights to exploit and clearcut forests are clearly stated, substantive restrictions are not. Instead, a broad holistic approach is advocated: "Forestry issues and opportunities should be examined in a holistic and balanced manner within the overall context of environment and development."

Coming to agriculture, I only want to comment on the privilege of agriculture and forestry in the German Federal Nature Conservation Act. § 1 III states as a general rule that "orderly pursued" (ordnungsgemäße) agriculture and forestry serve the purposes of the Act. They are exempted from the provisions on interventions into natural surroundings and landscapes (§ 8 VII). These paragraphs

[35] See *Birnie/Boyle,* International Law and the Environment, 1992, p. 435 *et.seq.*, 490 *et.seq.*

constitute a legal fiction which provoced strong criticism. The courts have interpreted the exemption in a narrow sense[36] and have stated, that 'orderly' management cannot be based on a purely micro-economic view.[37] The proposal for a revision of the Federal Nature Conservation Code of 1996 now speaks of the 'good practice' of agriculture and forestry which takes into account the purposes of nature conservation (§ 17 II).

4.4.4.2 Limits to Pollution

The concept of sustainable development demands for the elaboration of environmental quality standards and indicators, and of standards for critical loads and critical structural changes in ecosystems. From this, in theory, the tolerable rate of emissions could be derived. In this context, I only want to explain why German environmental lawyers are sometimes sceptical about this concept.

There are two approaches to control pollution – control of emissions and immissions. The term 'immission' refers to the end point where an 'emission' arrives.[38] In the European Union, Germany advocates emission control using the best available technology (Stand der Technik), whereas the United Kingdom favors environmental quality objectives[39] and pollutant input limit values. In Germany, many people think that this has something to do with favorable climatic and geographic conditions on the British Isles.

Emission control according to the best available technology is based on the precautionary principle. As long as we do not know exactly how much pollution is tolerable and acceptable, we should try to avoid it, as far as possible. The need for clear scientific evidence is much smaller than in a system of environmental quality objectives. The problem is, however, that even emissions at the lowest possible rate may exceed the tolerable level of pollutant inputs. To define environ-mental quality objectives, you have to know what the safe thresholds and critical loads are needed to protect human health and to preserve the integrity of ecosystems and their resiliance. Apart from the many uncertainties involved, the acceptable rate of pollution correlates to the quality of the environment we are willing to pay for and the change in ecosystems we are willing to accept.

In most countries, environmental law provides procedures to set both emission limit or threshold values and environmental quality objectives. Usually both concepts are applied together (cumulatively): polluters have to apply the best available technology of emission control *and* must not exceed pollutant input

---

[36] Bundesverwaltungsgericht, Natur und Recht 1983, 272; Oberverwaltungsgericht Koblenz, Natur und Recht 1987, 275.

[37] Bundesverwaltungsgericht, Natur und Recht 1989, 257, at 259.

[38] The term stems from the language of German environmental law and is used in the English version of EC Directives, but is hardly known in England or the United States. See also *G. Winter*, Standard-setting in Environmental Law, in *G. Winter (ed.)*, European Environmental Law, 1995, 109.

[39] See Sec. 82 to 84 Water Resources Act 1991.

limit values. When discussing the proposed directive on Integrated Pollution Control in the European Union, a major issue concerned the question, whether exemp-tions from the best available technology should be allowed in relatively unpolluted areas.[40] One could argue, that a higher tolarable rate of emissions in relatively unpolluted areas is the consequence of natural geographic advantages, and that applying the best available technology may be too expensive.[41] But the experience in the last decades has shown that the problems of long-range airborne pollution and of the pollution of the seas have been severely underestimated. Higher chimneys and the emission of sewage into the sea may lead to a policy of 'beggar my neighbourhood'.

From a legal point of view, standard setting is to a high degree a decision under conditions of unquantifiable uncertainty.[42] Therefore, the precautionary principle should be applied, and a steady monitoring of the impacts of pollution is vital.

4.4.5 Instruments

Until now, I have only discussed very general aims on the top of legal systems and legislation. The problem is how to put these aims into practice. Some lawyers strongly dislike general declarations of aims, values and principles calling them 'symbolic legislation'. It is true, that legislation will stay mere law on the books, as long as it is not implemented and enforced. The former Soviet Union, for example, had a far-reaching article on environmental protection in its constitiution and had a severe environmental legislation which nevertheless had little or no influence on the actual behavior of government, enterprises and citizens. In all countries, there is still a huge gap between declared aims concerning environmental protection and reality.

As the English saying goes, God dwells in the details. The corresponding German saying is more sceptical: It is the devil who is hidden in the details. Environmental legislation in most cases does not break down its aims into spe-

---

[40] *I. Appel,* Emissionsbegrenzung und Umweltqualität, Zu zwei Grundkonzepten der Vorsorge am Beispiel des IPPC-Richtlinienvorschlags der EG, Deutsches Verwaltungsblatt 1995, 399. Art. 8 (2) of the latest proposal obliges to use the best available technology in any case, but allows to take into account the geographic and local conditions of an installation.

[41] It should be noted that the German standard of Best Available Technology is qualified by the principle of proportionality which allows to take into account economic considerations. Therefore the standard is more or less identical to the "best available technology, provided that the application [...] does not entail excessive costs" (Article 4 Nr. 1 of the EC-Directive 84/360/EEC on the combating of air pollution from industrial plants). When there are in practice great differences to the British standard of Best Available Technique Not Entailing Excessive Costs, this is not due to the different definitions.

[42] See *U. de Fabio,* Risikoentscheidungen im Rechtsstaat, 1994 and *K.H. Ladeur,* Das Umweltrecht der Wissensgesellschaft, Von der Gefahrenabwehr zum Risikomanagement, 1995.

cific, operational targets. This is left to regulations, planning procedures and the permit process.

4.4.5.1 Standard Setting Procedures

Some standards are directly set in legislation, for example in EC-directives concerning water pollution and air pollution. In German environmental law, it is rather unusual that § 40a Federal Pollutant Input Control Act, as amended in 1995, sets a specific pollutant input limit value for the ozone concentration. Normally this is left to regulations. In the rule making process, people from government, science, industry as well as environmentalist have to be heard (§ 51 Federal Pollutant Input Control Act). But there is not such an elaborated procedure with a broad public participation as in the USA (according to the Administrative Procedure Act, 5 U.S.C. § 551). Many technical standards which operationalize the 'best available technology' are based on rules elaborated by private institutions - on a worldwide level by ISO (International Standard Organisation), on the European level by CEN (Comité Européen de Normalisation), and in Germany by DIN (Deutsches Institut für Normung). According to a covenant between the Federal Government and DIN, in 1993 a body for environmental standard setting was founded (Normausschuß Grundlagen des Umweltschutzes). It provides the participation of government, industry, environmentalists, trade unions, and scientists. Still, standard setting is to a wide extent an uncontrolled bargaining process, in which there is no clear separation between scientific advice and political and social participation.

4.4.5.2 Zoning Laws and Landscape Planning

In densly populated countries like the Netherlands, Germany and England, the land-use planning has a much stronger tradition than in the USA.[43] In Germany, landowners have no general proprietarian right to develop their land. Therefore, concerning the 'taking issue', the borderline between police powers and eminent domain is set more in favor of environmental protection than in many other countries. In zoning regulations, the objectives of environmental protection and nature conservation have to be taken into account. Compensation and mitigation measures are not completely left to the discretionary powers of government, but are to a certain degree mandatory (§ 1 V Construction Act, § 8a Federal Nature Conservation Act).

In § 5 *et.seq.* Federal Nature Conservation Act, there are provisions for landscape planning which only draw very general lines. The landscape planning

---

[43] See *D. Vogel,* National Styles of Regulation, Environmental Policy in Great Britain and the United States, 1986, 31, who states that Britain is supposed to have "the world's most extensive system of land-use planning".

might have the potential for an integrated, ecological planning process.[44] The practice, however, differs from one German State (Land) to the other, and does not seem to be very promising.[45] There are other planning instruments for air pollution control, noise abatement, water resources and waste management. But the euphorical expectations of the 60s and 70s have vanished. Governmental agencies simply do not have the resources for comprehensive and up-to-date environmental planning. Nevertheless, thorough taxonomy and a complete mapping of critical habitats may already serve as a useful input to planning and to the permit process.

4.4.5.3 Environmental Impact Assessment

The introduction of the Environmental Impact Assessment (EIA) in the US-National Environmental Policy Act of 1969 led to a worldwide success (at least, as far as law on the books is concerned). France introduced it in 1976, the EC-directive was issued in 1985, and more than 70 developing countries have followed.[46] Whereas NEPA covers governmental programmes, plans and projects, the EC-Directive applies to public and private projects. Germany had problems integrating this new procedure into the existing body of environmental legislation.

The aim of EIA is to assure that all direct or indirect effects of a project or plan on the environment will be identified, described and assessed. The information gathered must be taken into consideration in the decision making process. The integrated approach of EIA is most clearly stated in §100 (c) (A), (B) NEPA: The Federal government shall "utilize a systematic, interdisciplinary approach which will insure the integrated use of the natural and social sciences and the environmental design arts in planning and in decisionmaking", and "indentify and develop methods and procedures [...] which will insure that presently unquantified environmental amenities and values may be given appropriate consideration in decisionmaking."

In German environmental law, there is a debate on the criteria for assessment. Those who advocate the so-called 'legal assessment', say that all criteria have to be derived from legal norms.[47] This is also laid down in the administrative rule

---

[44] *W. Erbguth,* Ausschöpfung der Möglichkeiten der Landschaftsplanung im Sinne einer Umweltplanung, 1992.

[45] See *Rat von Sachverständigen für Umweltfragen,* Konzepte einer dauerhaft-umweltgerechten Nutzung ländlicher Räume, 1996, Nr. 20 *et.seq.*; *E. Gassner,* Möglichkeiten und Grenzen einer rechtlichen Stärkung der Landschaftsplanung, Natur und Recht 1996, 380.

[46] *M. Yeater/L. Kurukulasuriya,* Environmental Impact Assessment Legislation in Developing Countries, in: UNEP's New Way Forward: Environmental Law and Sustainable Development, 1995, 257; *R. Coenen/J. Jörissen,* Umweltverträglichkeitsprüfung in der Europäischen Gemeinschaft, 1989.

[47] See *E. Bohne,* Grundprobleme des UVP-Gesetzes und seiner Ausführungsvorschriften, in: *K.H. Hübler/M. Zimmermann (Hrsg.),* UVP am Wendepunkt, 1992, p. 3, at p. 12 *et.seq.; A.*

for EIA.[48] But this administrative rule which has been prepared by lawyers is so vague that it is hardly operational.[49] Therefore, as others argue, the expertise of applied ecology and other sciences should not be ignored.[50] This debate somehow is centered around the question whether the hen or the egg came first. Some believe, for example, that the criteria for ecological evaluation used in nature conservation are derived from the Federal Nature Conservation Act.[51] It is true, that they are compatible with that law, as the act took up ideas from ecology (like the 'household of nature'). But the same criteria are applied all around the world in completely different legal systems.

Another issue is the quantification and aggregation in evalution methodologies. In Germany, the administrative rule for EIAs is very sceptical towards aggregation.[52] The same is true for the Supreme Administrative Court.[53] It states, that these methodologies may be helpful to stress the value of some environmental goods as well as their quantity and quality. But a common denominator for a calculus, for example, between polluted water and fresh air, or the integrity of habitats and noise abatement, might pretend precision and rationality. Lawyers are used to solving conflicts of interests and values in a process of deliberation. This procedure would only be hampered by a high aggregation which hides value judgements beneath equitations.

The greatest challenge for ecological evaluation methods is to lay down the kind and amount of necessary compensation and mitigation measures. In Germany, there are more than fifty methodologies in use which lead to highly diverging results. From an investor's point of view, the outcome often is unpredictable and arbitrary. Attemps have been made to unify the methodologies, but, until now, with little success.[54]

Environmental evaluation is one of the main fields for interdisciplinary cooperation especially between applied ecology, economics and jurisprudence. Together,

---

Vorwerk, Die Bewertung von Umweltauswirkungen im Rahmen der Umweltverträglichkeitsprüfung nach § 12 UVPG, Die Verwaltung 29 (1996), 241.

[48] Allgemeine Verwaltungsvorschrift zur Ausführung des Gesetzes über die Umweltverträglichkeitsprüfung, 18 September 1995, Nr. 0.6.1.1.

[49] To a high degree, this is due to the bargaining process between diverging interests represented by the Ministry for environmental protection, economy, traffic, housing, agriculture and so on. The Ministry for environmental protection is not the strongest one, and the Department for nature conservation is not a strong one within that Ministry.

[50] *E. Gassner,* Umweltverträglichkeitsprüfung in der planerischen Abwägung, Umwelt- und Planungsrecht 1993, 241, at 244.

[51] *A. Vorwerk,* note 47, at 258.

[52] See note 48, Nr. 0.6.2.1.

[53] Bundesverwaltungsgericht, Natur und Recht 1995, 537, at p. 542 *et.seq.*

[54] See *H. Kiemstedt et al.,* Methodik der Eingriffsregelung, Teil III, Vorschläge zur bundeseinheitlichen Anwendung der Eingriffsregelung nach § 8 Bundesnaturschutzgesetz, 1996; *Planungsgruppe Ökologie + Umwelt,* Richtwerte für Kompensationsmaßnahmen beim Bundesfernstraßenbau, Untersuchung zu den rechtlichen und naturschutzfachlichen Grenzen und Möglichkeiten, 1995; for the methodological problems of a mitigation fee see *H.-U. Marticke,* Zur Methodik einer naturschutzrechtlichen Ausgleichsabgabe, Natur und Recht 1996, 387.

they have to fill the gap between *de*scription and *pre*sciption. Environmental assessments and evaluations are based on value judgements, which should take into account both knowledge from ecology and societal aims for ecosystem management. Often, politicians as well as lawyers leave these questions to scientific experts. They try to avoid decisions or try to use the authority of science to back up their position. In many instances, society demands clear answers from ecology which ecology as a science cannot give. This is a great temptation for scientists, even for those with the highest reputation. In his recent book on environmental accounting, *Howard T. Odum* writes: "Whereas environmental issues are now characterized by adversarial decision making, rancor, and confusion, theses conflicts may not be necessary in the future. A science-based evaluation system is now available to represent both the environmental values and the economic values with a common measure. Emergy [...]"[55] I must admit that I do not quite understand what emergy as well as exergy mean and may indicate. But I am positive that a single common denominator will not be able to reflect all the values and conflicting interests that have to be taken into account in environmental evalution and that have to be balanced in decision making processes.

There are other very practical problems in the cooperation between lawyers and ecologists: Trying to avoid litigation investors often hire consultants who offer all kinds of time-consuming and expensive research. Scientists collect a lot of data for the EIAs which are of little or no relevance for the decision. The result may be characterized by a short dialogue, I want to call the 'escalation of scientific expertise':

> *Investor:* "We have done much more than we are legally obliged to. Our experts have studied all relevant environmental impacts in 27 volumes, including the effects on butterflies, beetles and ants."
> *Environmentalists:* "Your studies are only additive and not truly holistic. You have not taken into account all possible cumulative and synergetic effects. We need at least ten more studies and three more vegetation periods."

From a legal point of view, the EIA will only become effective if its application is controlled by the courts. The German Supreme Administrative Court is rather restrictive in this respect: deficiencies in the EIA are only relevant, if a correct procedure could have conceivably led to another result.[56] It is difficult to prove this causal link between procedural deficiencies and the substantive outcome of the decision. To effectuate judicial control, the Citizen Group Action (Verbandsklage), which most of the German States (Länder) have introduced in their nature conservation acts, should be incorporated into the Federal Nature Conservation Act.

---

[55] *Howard T. Odum*, Environmental Accounting, Emergy and Environmental Decision Making, 1996, at p. 1.

[56] Bundesverwaltungsgericht, Deutsches Verwaltungsblatt 1994, 763; 1995, 1012, at 1017; critical *R. Steinberg*, Chancen zur Effektuierung der Umweltverträglichkeitsprüfung durch die Gerichte?, Deutsches Verwaltungsblatt 1996, 221, at p. 226 *et.seq.*

4.4.5.4 Integrated Pollution Control and the Best Practicable Environmental Option

In the United Kingdom and in the Netherlands, the different permits concerning air pollution, water pollution, waste management, nature conservation and others have been substituted by a 'one-stop' permit. In the European Union, a Directive on integrated pollution prevention and control has been passed in September 1996. It will not only lead to a unified procedure, but also to an integrated approach to standard-setting. It is the goal to prevent pollution abatement in one medium from causing negative impacts on another.[57] The definition of 'best available technology', which in the German Federal Pollutant Input Control Act focuses on the control of air pollution, will also have to take into account the use of hazardous substances, waste production, emissions into water and soil, and the use of ressources and energy.

The model for the EC-Directive is the Integrated Pollution Control which has been introduced in the United Kingdom in 1990. Sec 7 (7) of the Environmental Protection Act 1990 adopts a new, integrated standard, called the 'Best Practicable Environmental Option':

"The objectives referred to in subsection (2) above shall, where the process -

(a) is one designated to central control; and
(b) is likely to involve the release of substances into more than one environmental medium;

include the objective of ensuring that the best available technique not entailing excessive cost will be used for minimizing the pollution which may be caused to the *environment taken as a whole* by releases having regard to the best practicable environmental option available as respects the substances which may be released."

This technology forcing standard demands for a comparison between the impacts on different environmental media and a cost-benefit-analysis. A consultant document, released by Her Majesty's Inspectorate of Pollution in 1994, proposes an 'Integrated Environmental Index' for the aggregation.[58] Though the basic idea of Integrated Pollution Control is very appealing and convincing, I am not quite sure, if this integrated, 'holistic' approach in practice will lead to more environmental protection than the old-fashioned, segregated emission limit values for each single environmental medium.[59]

---

[57] For example, filters to minimize air pollution produce a lot of hazardous wastes.

[58] Environmental, Economic and BPEO Assessment Principles for Integrated Pollution Control, 1994.

[59] See *S. Owens*, The Unified Pollution Inspectorate and Best Practicable Environmental Option in the United Kingdom, in: *N. Haigh/F. Irwin*, Integrated Pollution Control in Europe and North America, 1990, 169; *O.A. Wagner*, Die umweltrechtliche Anlagenaufsicht in England und Wales (Integrated Pollution Control), 1996.

Coming to the end, the reader may wonder why I have not mentioned all the economic instruments which have been discussed so intensively during the last few years: eco-taxes, emission permit trading, liability, eco-auditing, other self-regulatory instruments and so on. The answer is very simple: At least in Germany, there is a remarkable disproportion between the amount of scholarly writings and the amount of practical experience as well as between theoretical and empirical studies.[60] It is too early to state whether economic instruments really are more effective and efficient than the traditional instruments of administrative law.

4.4.6 Conclusions

The workshop held in 1996 and this book are closely related to the research in the Bornhövel Lakes Region, done by the Project Centre for Ecosystems Reserch in Kiel. This project has reached the evaluation and synthesis phase. In this final phase, conclusions for ecosystem management shall be drawn. This is the point where expertise from social sciences and the practical skill of jurisprudence in conflict solving, rule making and authoritative value judgements may be helpful. When scientific ecologists contribute with their knowledge to decision making processes, they cease to be observers of systems, but act within 'the system'. They have to convince others who do not share their knowledge and may not share their personal convictions. Legislation, standard setting procedures, landscape planning and the Environmental Impact Assessment are extremely time-consuming and may be very frustrating. You have to look for common points of interest and consensus, you have to compromize and sometimes you even have to accept decisions which in your opinion are completely wrong, biased or irrational. Yet, this is the only way democracy and the rule of law provide for.

---

[60] See for example a study for the European Commission: *ERM Economics*, Economic Aspects of Liability and Joint Compensation Systems for Remedying Environmental Damage, 1996; for the failure of the 'bubble concept' in German air pollution control see *D. Ewringmann/E. Gawel*, Kompensationen im Immissionsschutzrecht, Erfahrungen im Kannenbäckerland, 1994.

4.5 Coupled Economic and Environmental Growth and Development

Matthias Ruth

Abstract

This section investigates constraints on growth and development in the economic and environmental system if both systems are coupled with each other through exchanges of material and energy flows. The concept of exergy is used to aggregate and assess these flows and their impacts on system change. Dynamic modeling is used to investigate the time domain of system change. The results of the model indicate that economy-environment interactions may cause economic activity to significantly fluctuate even if the environment on which the economy depends for materials and waste absorption is modeled without exhibiting seasonality, randomness or discontinuities.

4.5.1 Introduction

It is a truism that the economy is part and parcel of an ecosystem from which it extracts materials and fuels and into which it releases its waste products. Despite the importance of these linkages for the growth and development of each system, the economy and its environment have traditionally been separated for conceptual and analytical purposes and the trajectories of both systems have been portrait as decoupled. The assumption of decoupled system behavior may have been appropriate for a world in which the economy is small when compared to the rest of the ecosystem. In such a world, extraction of materials and fuels may not significantly impair the economy's ability to extract them in the future. Similarly, the release of waste products is likely to have only small impacts on the ecosystem's waste absorption capacity. As a consequence, the goal of perpetual economic growth and development may seem attainable through continued use of natural resources, substitution of resources for one another in the case of increasing relative scarcity, and sustained technological change.

F. Müller M. Leupelt (Eds.)
Eco Targets, Goal Functions, and Orientors

The notion of continued growth and development of the economy has, to some extent, been paralleled by the discussion of ecosystem goals. For example, the concept of succession reflects an understanding that ecosystems tend to resume movement along their trajectories after disturbances have been temporarily imposed on them. The concept of adaptation mirrors the economists' notion of technical change in a world in which constraints are imposed on a particular trajectory. Although caution must be taken when using such analogies (Ruth 1996), the many parallels that exist in the way in which we perceive economic and ecosystem goals prompts us to explore the ability to sustain growth and development in a world that is characterized by an increased appropriation of materials, fuels and waste absorption capacity by the economy at the expense of future generations and other species (Houghton et al. 1996; Heywood 1995; Postel et al. 1996; Vitousek 1994). Even casual observation suggests that growth cannot persist indefinitely in a finite world, and is not even a necessary condition for development.

Recent insights from the mathematics of nonlinear dynamics have come to challenge traditional concepts of growth and development in both economics and ecology (Rosser 1991, 1992), and the resurgence of systems theory has highlighted the need for a treatment of system goals and trajectories that explicitly acknowledges system-environment interactions (Costanza 1996; Costanza and Ruth 1997). Particular emphasis is now being given to the understanding of the effects that nonlinear, time-lagged interrelationships among systems have for their behavior, and how these effects permeate from one system to another, and across various hierarchies of system organization (Allen 1994).

Modeling is an essential part in this efforts to assess the effects of feedback mechanisms on system behavior. Particularly, recent advances in dynamic modeling (Beltrami 1987; Bossel 1994; Hannon and Ruth 1994, 1997; Ruth and Hannon 1997) have made it increasingly possible to investigate the time domain of economy-environment interactions and have helped scrutinize the assumptions that are traditionally made in economic models of material and energy use, production and emissions. Of particular interest to the discussion that follows below is the question how the behavior of a traditional model of the economy changes if economy-environment linkages are taken into account. Under which scenarios is the emergent behavior consistent with conventional goals of growth and development? The answers to these questions make it necessary to discuss how flows across the economy-environment boundary are measured and aggregated, and how the extraction of materials and fuels, production of desired products and emissions of wastes are modeled. Each of these issues are briefly addressed in the following five sections. Then a simple model is presented to illustrate various points of the discussion. The paper closes with a set of conclusions regarding the role of economy-environment interactions for future growth and development of the economy and its environment.

4.5.2 Modeling Material and Energy Flows Across the Economy-Environment Boundary

It is a well-recognized fact that all systems require materials and energy to maintain their structures, grow and develop. To identify the limits to maintenance, growth and development and to influence the behavior of a system requires that we understand how materials and energy can be supplied to the system, how materials and energy are transformed from one state to another, and how the system reacts in response to changes in its environment following the release of undesired waste products. One type of systems of particular interest to human decision making are the production systems that supply us with desired goods and services for final consumption. Analyses of these systems have frequently concentrated on the flow of goods and services through the economy, focusing on the extraction stage, production stage or the economic repercussions of pollution emissions (Fig. 4.5.1). Changes in each part of that system have been evaluated with respect to a set of economic goal functions, such as profits, consumer utility or, most generally, welfare.

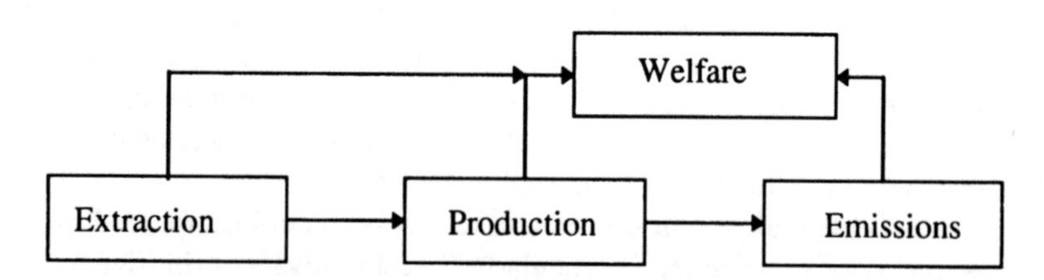

Fig. 4.5.1. Linear Approach to Modeling Economy-Environment Interactions

Examples of early attempts to deal with the effects of resource depletion on profits and welfare include the models by Gray (Gray 1913, 1914) and Hotelling (Hotelling 1931) which spurred much of the research that has been done in natural resource economics since then (Conrad and Clark 1987; Dasgupta and Heal 1979). With respect to emissions, the works by Coase (Coase 1960) and Pigou (Pigou 1932) significantly contributed to the emerging field of environmental economics which attempts to inform policy how to incorporate into economic decision making the externalities caused by the unwanted by-products of production (Baumol and Oates 1988).

The "middle piece" that links extraction and emissions – the transformation of goods and services with low economic value into goods and services of high economic value – has traditionally received most of the attention. In the context of system optimization, the goal has often been stated as one of profit maximization in the light of a given technology and inputs that need to be purchased at positive prices. With the concentration on priced inputs and outputs, a host of nonpriced material and energy flows into and out of the production process had been neglected, and many of the physical realities of material and energy transformations have been removed into the background or even overlooked (Christensen 1989).

A key concept for the description of the production process is the production function which has been invoked by Wicksteed (Wicksteed 1894) and, together with the cost function, now stands at the heart of traditional production theory (Varian 1992). It has been criticized for its lack to ensure that the physical aspects of the production process link up with the physical aspects of resource extraction and pollution emission (Ayres 1978; Daly 1968; Georgescu-Roegen 1971). Recent efforts attempt to modify traditional economic analyses to account for the physical reality within which production takes place (Månsson 1986). The laws of thermodynamics provide the context for these modifications. These laws are used to constrain production functions to the physically feasible range (Berry et al. 1978; Islam 1985; Lesourd 1985; Ruth 1993; Kümmel et al. 1996).

The model outlined below only uses a simple modification of the traditional production function and concentrates more on the performance of the economy in interaction with its environment. To modify the traditional production function and to couple the economy to its environment, concepts from thermodynamics are used. These concepts help us to assess changes in environmental quality in response to resource extraction, materials and energy use in production, emissions of waste products, and changes in environmental waste absorption capacities. The linear view of Fig. 4.5.1 is replaced with a perspective of the economic system that admits feedback processes through which, for example, extraction influences the resource base, and, vice versa, through which changes in environmental quality influence the productivity of the economic system (Fig. 4.5.2). To establish these feedback processes requires that those material and energy flows are captured whose appropriation is "incidental" and typically unaccounted in economic models (Allen 1994; van der Voet et al. 1995). Examples are flows of oxygen used in combustion processes and solar energy incident upon agricultural plants.

4.5.3 A Physical Perspective of Economy-Environment Interactions

Economic processes temporarily transform materials and energy from less desired into more desired forms. If the initial and final states are evaluated in monetary terms, the net effect of production with regard to a change in economic value can easily be computed for a set of diverse products by multiplying quantities by their respective market prices and then subtracting the value of inputs from those of the outputs.

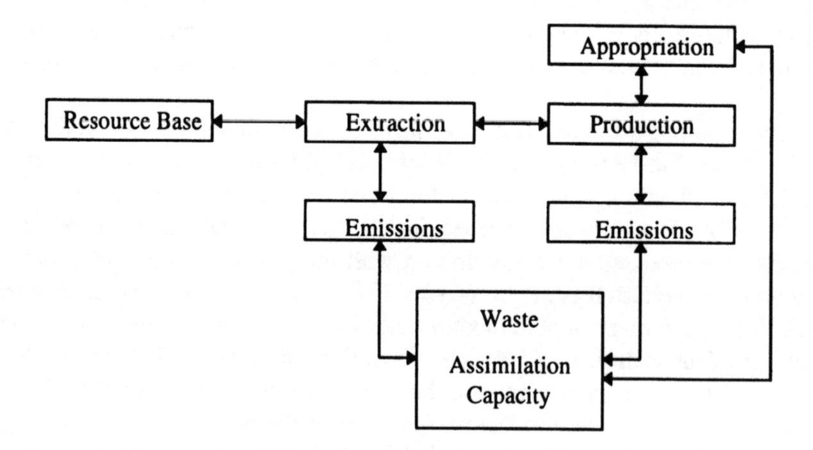

Fig. 4.5.2. Key Economy-Environment Interrelations.

However, prices do not exist for all inputs, making it difficult to judge the implications that the changes in production may have for the future availability of a resource or for environmental quality. Furthermore, an economic description of a production process in terms of the economic value that it generates is typically highly aggregate and may not necessarily retain enough information on the physical limits of substitution among inputs and of the physical input-output relationships.

In contrast to the economics approach to economy-environment interactions, engineering studies provide detailed lists of material and energy requirements and emissions that often cannot be aggregated in a meaningful way when measured in mass and energy units. Yet, these mass and energy balances are a prerequisite for an assessment of economy-environment interactions. Given the detail of engineering information, it is difficult to maintain the elegance and powerful simplicity of an economics approach and at the same time retain the information that is necessary to make useful judgments on these interactions. Thus, what is needed are means to aggregate physical measures such that these measures retain important information on material and energy use and can be interpreted more broadly in the economic context.

Mass and energy balances provide a complete description not only of the quantities used and released by a process but are also the starting point for an analysis of the change in the thermodynamic state of materials and the quality of energy as a process occurs. The concept of exergy captures these qualitative changes and can provide a basis for aggregation across different types of materials and energy.

To calculate the exergy of a system, a reference system must be defined. The choice of the reference system can only be done on the basis of first principles in the context of statistical and quantum mechanics (Månsson 1991). For many practical applications, the reference system is an idealized system with material composition, pressure, and temperature equal to some average of the earth's crust, atmosphere or oceans (Ahrendts 1980). Given the reference system, the exergy A of an input into or output from the system of interest is defined as

$$A = (E - T_0 + P_0 V - \sum_{ii=1}^{n} N N_i \mu_{0i}) \tag{4.5.1}$$

with E as internal energy, T_0 as the temperature of the reference environment, S as entropy, P_0 as pressure of the reference environment, and N_i as the number of molecules which have in the reference environment chemical potentials μ_{0i} (Gagglioli 1980; Howell and Buckius 1992). The exergy is thus the maximum useful work that can be done as the conditions of the reference environment are approached.

Exergy analyses can be performed for any process inside or outside the system boundaries that delineate the economy from its environment. Several detailed illustrations of their application to extraction and manufacturing industries can be found in Szargut et al. (1988). These applications typically are concerned with the optimization of processes or technologies with regard to either material or energy efficiencies. When both are optimized simultaneously, special care has to be taken in the interpretation of the results as the exergy of fuels can be very large in comparison to the exergy of materials.

On a fundamental physical level second law analyses offer insight into the advancement of a process itself. Exergy is used to maintain or generate arrangements of materials in unlikely thermodynamic states, such as high concentrations of copper, iron, and plastics in an electric motor, distinct arrangements of inks on cellulose in books, or specific genetic information in the cells of an organism. The ability to generate those desired arrangements of materials at given expenditures of exergy may be used as a measure of the advancement of a process and the potential for its improvements. Changes in the ability to cause desired arrangements of materials, in turn, require exergy expenditures to generate the knowledge necessary for system change.

The relationships between exergy, order, information and knowledge have been developed theoretically by Szilard (1929), Evans (1969), Tribus and Mc Irvine (1971), and others. They now begin to inform the analysis of production and consumption processes (Ayres and Miller 1980; Berg 1980; Chen 1992; Chen 1994; Spreng 1993; Ruth 1995a) and of economy-environment interactions (Ayres et al. 1996; Ruth 1995b). Extending these studies, the following three sections provide a basis to assess resource extraction, production, and changes in waste absorption capacity from a physical perspective and to link physical changes to the growth and development of the economic system. Extensions and applications of the exergy concept to ecosystem growth and development can be found, for example, in Jørgensen (1992), Schneider (1988), and Schneider and

Kay (1990, 1993, 1994a, 1994b). Common to both areas of application of the exergy concept is a struggle to properly reflect the fact that exergy is essential for the growth and development of structures, yet these struc-tures have values that are not fully reflected in their exergy values.

4.5.4 Resource Extraction

Physical scientists note that economic indicators such as cost and price are determined partially by market structure, human preferences, social institutions, regulations, and other forces that are not related directly to the quality and quantity of the resource base. They argue for an approach that captures the physical and technological aspects of resource availability and transformation, and which minimizes the influences of factors such as market structure and preferences. Thermodynamics provides such an approach because it tells us where we are with respect to a fixed reference system (Ruth and Bullard 1993; Cleveland and Ruth 1997). Starting with this perspective, physical scientists emphasize an important consequence of mineral depletion – the inverse relation between ore grade and the energy cost of extracting a unit of the mineral. This inverse relation is well-defined for many metals (Page and Creasey 1975). At a given technology, energy costs increase as ore grade declines for a variety of reasons. More energy per unit of metal is required to mine, break, transport, crush, grind, and beneficiate a larger amount of rock. At very low grades, a metal exists in the rock as either finely divided grains, or as replacement ions in the crystal lattice of the waste rock (Chapman and Roberts 1983; Skinner 1976). The low concentration requires that the rock be ground to a finer size, greatly increasing energy cost. The most abundant forms of minerals in common rock are silicates, which require much more energy to process than the oxides and sulfides we now mine. For minerals that are extracted with surface mining, depletion often leads to larger stripping ratios - the quantity of material handled per unit of ore mined. An increase in the stripping ratio generally increases energy cost (Gelb 1984).

The inverse relationship between ore grade and energy cost suggests that energy cost can be used to establish the economic limit or cut-off grade for a particular mineral. For example, at conventional extraction methods a copper ore grade between about 0.2 to 0.5 percent may cause energy costs to increase sharply, triggering substitution of other resources or drastically rising prices (Govett and Govett 1978). Not explicitly taken into account in these studies are the changes in the quality of the energy sources that are used in mining and refining processes. In the history of the US extractive industries a shift occurred from human and animal labor as the prime movers to high quality fuels such as oil, natural gas and electricity (Hall et al. 1986). This shift towards an increased use of high exergy paralleled the movement towards lower grade ores. As a result, the ratio of gains in the order inside the economy that results from concentrating minerals and metals to the loss of order by burning fossil fuels and depleting high quality ores is smaller than the ratio of the quantity of materials extracted to energy use.

Table 4.5.1. Total US Energy Use and Estimated Increase in Order in Selected Mineral Industries in the 1980s.

Mineral	Energy Use in Mining[1)] (MJ)	Energy Use in Manufacturing[2)] (MJ)	Estimated Increase in Order in Mineral[3)] (MJ/K)
Bauxite/Aluminium	$0.19 \cdot 10^9$	$858.40 \cdot 10^9$	$2.09 \cdot 10^3$
Copper	$43.25 \cdot 10^9$	$64.67 \cdot 10^9$	$0.91 \cdot 10^3$
Iron	$50.96 \cdot 10^9$	$1477.00 \cdot 10^9$	$12.60 \cdot 10^3$
Lead and Zinc	$2.32 \cdot 10^9$	$18.56 \cdot 10^9$	$0.05 \cdot 10^3$

1) Values for Copper, Iron, Lead and Zinc are for 1987; calculated from Bureau of the Census, Census of Mineral Industries (1987) and Bureau of Mines, Minerals Yearbook (1987). Value for Bauxite Mining is for 1989; calculated from Statistical Department, The Aluminum Association, Inc., personal communications.
2) Values for 1981. More recent data is not available, yet. Calculated from Bureau of the Census, Census of Manufacturers (1981).
3) Based on 1987 production data.

The net effect of resource extracting activities is the achievement of a local increase in order in the economic system by upgrading the thermodynamic state of the "raw material" from its natural state to a more desired state (Table 4.5.1). This local increase in order requires information on the location of materials – expressed, for example as differences in concentrations in comparison to reference states such as the average crustal abundance – and knowledge – embodied in the technologies used for extraction and refining and embodied in the socioeconomic system as a whole that makes use of the finished product.

Increases in local order are accompanied by decreases in order elsewhere as a result of the degradation of exergy in the fuels used for extractive processes, and the dispersal of overburden and waste materials. To reduce, for example, the degradation and dispersal of materials requires increasingly sophisticated technologies to locate raw materials, change their thermodynamic state and trace waste products. Knowledge about the state and fate of waste products is an essential prerequisite for closing material cycles. However, without proper knowledge embodied in technologies and socioeconomic institutions an industrial ecology (Jelinski et al. 1992; Graedel and Allenby 1995) based on closed material cycles cannot be achieved.

4.5.5 Economic Production

To be able to link economic production to changes in the resource base and environmental quality requires that production processes are described in accordance with the physical laws that govern all transformations of materials and energy. Towards this end, three categories of inputs into production are often distinguished: a set of agents such as capital goods and labor, materials, and energy (Georgescu-Roegen 1971; Anderson 1987). Since the ability of the agents to provide productive services may decline as production takes place, the list of material and energy inputs must include those that are used to maintain the agent's ability to provide productive services. In the absence of maintenance flows that keep the agents of production in a state that enable them to perpetually provide services, accounts are required that capture the loss of productive services. These accounts must reflect the physical deterioration of the agents rather than the depreciation that is frequently calculated for tax and other purposes. Outputs from the production process include the desired goods and services, waste materials, and waste heat.

In order to describe the transformation of materials and energy with the influence of various agents, four interrelated aspects need to be captured. First, mass and energy balances must be satisfied. Second, as materials and energy are transformed in the production process, their thermodynamic states change. Second law analysis enables us to capture the changes in the quality of energy and the organization of materials. Third, within a production process it may be possible to substitute various inputs for each other. For example, additional labor inputs may make it possible to more carefully handle material inputs, and thus reduce material input requirements. The substitutability of inputs, however, is constrained by the thermodynamic laws. For example, it is not possible to continue to replace materials by labor inputs indefinitely. A minimum amount of materials will always be required for a unit of finished (material) output. Production functions relate inputs to outputs and capture, through their functional form, the extent to which inputs can be substituted for each other.

A fourth aspect of the production process that needs to be considered in its description is the time over which production takes place. Obviously, the issue of timing is relevant for the proper choice of inputs and the rate at which outputs are generated (Winston 1982). It is also important in any attempt to link economic production to changes in environmental quality. Cyclical fluctuations in inputs and outputs are notable, for example, in electricity generation, as daily peak loads and seasonal variations in demand must be met. These fluctuations not necessarily coincide with the daily and seasonal changes in the environment's ability to assimilate the waste heat that is generated in electricity generation.

The description of production processes in terms of material and energy balances, exergy degradation, production functions and the time domain of production extends the standard description of substitution possibilities and cost of production into the realm of physically relevant concepts. However, it requires that more attention is given to the context within which production takes place. Some inputs into one production process may be in the form of funds. The same type of agents may enter another production process as a material flow. For example, a plastic container in one case provides a service of con-taining a liquid – it is a fund – and in another case, without having altered its physical or chemical appearance, enters a recycling process as a material flow. These distinction are essential to properly reflect at a process level limits to substitutability.

An approach that combines thermodynamically-based production functions with first and second law analysis has been used to compare production processes in minerals and metals industries with regard to the savings in exergy that can be achieved over time as resource deposits are depleted and as technological change occurs (Ayres 1988; Ruth 1995b). Since these studies concentrate on the changes in resource endowments as a result of extractive processes in an economy they typically treat waste flows as if they occurred at ambient temperature and concentration, thus neglecting the processes these flows may trigger in the environment. To provide a complete description of the impli-cations of economic production for environmental change – and thus a valuation of the contribution of materials and energy flows to maintenance, growth and development of the economy – requires that impacts of emissions on the environment's waste absorption capacity are taken into account. The simple model developed below illustrates this issue at an aggregate level.

4.5.6 Waste Generation

First law analyses of production processes provide a quantification of material and energy waste streams from a production process. Second law analyses capture the qualitative change that occurs as materials change their thermodynamic state and as the quality of energy decreases. A characterization of the waste stream in terms of its exergy may be used as a measure of its ability to affect physical, chemical and biological processes in the environment (Ayres et al. 1996). High-exergy waste flows have a high ability to trigger environmental change; waste flows of zero exergy have a material composition, temperature and pressure that is indistinguishable from the environment, and as a result do not affect environmental processes. However, even though exergy flows have the ability to cause system change, a measure of these flows not necessarily indicates whether and how much a system actually changes when it receives those flows. Some environmental systems, for example, have evolved strategies to cope with specific materials even in high concentrations while others have not.

To make judgments about the impact of exergy flows on biological structure and function requires that we know the history and current state of the systems that receive those flows, their ability to assimilate exergy flows and the distance these systems are away from thresholds. The latter is frequently impossible to know a priori. As a consequence, significant efforts in field research and modeling are required to anticipate the impacts of exergy flows and to guide technology choice towards a reduction in resource (exergy) depletion and environmental impact (exergy loading). Yet, even if all the data were available to describe the systems' past and current states, the nonlinearities that underlie feedback mechanisms in the system will make it inherently impossible to predict their future behavior. Even small variation in initial conditions or parameter values may lead to fundamentally different trajectories for the system.

Given the lack of empirical information on the ability of ecosystems to degrade the exergy flows that they receive, the spatial and temporal variation in waste assimilation capacity, and the role of nonlinear feedback mechanisms in shaping growth and development, theoretical considerations and numerical examples based on the concepts outlined above can help us elucidate potential impacts of material and energy flows across the economy-environment boundary on growth and development of each system. The following section provides a simple illustration of the behavior of the combined economy-environment system. In lieu of empirical process-level information the model takes on an aggregate view. The purpose of this model is to highlight the implications of potential environmental repercussions for the time paths of economic activities, not to provide any real analysis of an actual system.

4.5.7 A Simple Dynamic Model of Extraction, Production, and Waste Assimilation

4.5.7.1 Model Features

This model distinguishes two types of agents of production: economic funds, such as capital goods or knowledge embodied in the minds of the work force, and environmental funds such as forest ecosystems. To produce economic funds, the economy extracts materials and fuels from stocks that are at any point in time finite in size but that can be enlarged through discoveries and development. The production of economic funds itself requires economic funds – for example, capital goods are used to produce capital goods and knowledge is used to produce capital goods and knowledge. Furthermore, the economic production process requires environmental funds, for example to supply waste assimilation services or biomass fuels. In the process of economic production, waste is generated from the materials and fuels that are used and by deteriorating economic funds – capital goods wear out and obliterate the landscape. By the same token, experiences and skills of a workforce can be lost over time from society with the death of the members that carry those experiences and skills.

Environmental funds capture an exogenously given influx of exergy to maintain themselves, grow and develop. They provide waste assimilation ser-vices to the economy and can be consumed in economic production processes to provide a stock of exergy – for example in the form of a biomass fuel – for the economy. In the model, the maintenance, growth and development of environ-mental funds is allowed to be, in principle, exponential. However, increased economic production leads to a decline in the growth rate of environmental funds. For example, an expansion of economic activity decreases the ability of the fund to grow because exergy flows that are necessary for growth and development of the environmental fund are appropriated by the economy or because the economy releases waste flows that degrade environmental funds.

In the model, neither the economy nor for the environment are assumed to *optimize* their behavior with regard to some given goal function. Rather, each of the systems are set up to be able to simply continue to grow and develop indefinitely if they are isolated from each other.

The model is based on a number of assumptions to maintain as simple a representation of economy-environment interactions as possible. Among the most restrictive assumptions are the absence of technological change in the economy, the ability to keep over time exergy losses in the extraction of materials and fuels proportional to extraction quantities, and the absence of seasonality, randomness, time lags, and threshold effects in the ecosystem. The discussion of the model results, however, presents the implications if some of these assumptions are relaxed.

4.5.7.2 Notation

In the model, exergy is denoted A. The subscripts m and f refer to materials and fuels, respectively. Albeit this distinction is in principle not necessary, it enhances the model's transparency and keeps it more comparable to economic models of materials and energy use. The subscripts y and e stand for economic and environmental funds, respectively. The subscript 0 denotes the value of a material or energy flow in some base period. These base period values are used to normalize input and output quantities of material and energy transformation processes. The superscript l refers to a loss of exergy, such as it takes place during metabolism or when a machine deteriorates. The superscript D refers to discoveries (in the case of resource endowments) or growth and development (in the case of environmental funds). The superscript T denotes totals. The superscript * refers to the theoretical minimum of an input that is required in a production process to achieve the desired output. That minimum may be defined by thermodynamic laws.

4.5.7.3 Resource Extraction

The economy extracts from a stock of materials A_m an amount ΔA_m per time period. The more of the material endowment A_m is available (e.g. the higher the ore grades, the less diluted a mineral) and the more of the economic fund is available (e.g. the more sophisticated the capital equipment A_y or the more skilled the labor force), the more ΔA_m can be generated. This assumption about the relationship between the quality of the resource endowments and the use of economic funds in resource extraction is consistent with the literature discussed above. The relationship between material extraction and availability of materials and economic goods is given by:

$$\Delta Am = C_1 \cdot \left(\frac{Am - A_m^*}{Am0 - A_{mo}^*} \right)^{\beta_{1m}} \left(\frac{Am - A_m^*}{Am0 - A_{mo}^*} \right)^{\beta_{2m}} \quad (4.5.2)$$

The extraction process is assumed to be a simple mechanical sorting in which none of the exergy of the inputs of fuels enters the product. The production function (4.5.2) is of a modified Cob-Douglas type with isoquants that asymptotically approach the minimum input requirements

$$Am \longrightarrow \infty \ \textit{for} \ A_y \longrightarrow A^* \ \textit{and} \ A_y \longrightarrow \infty \ \textit{for} \ A_m \longrightarrow A_m^* \quad (4.5.3)$$

These minimum input requirements constrain the process within technically or physically feasible limits (Ruth 1993, 1995b).

Exergy consumption in the extractive process is assumed to be proportional to the total exergy of the materials being extracted:

$$\Delta A^1{}_m = \mu_{A_m} \cdot \Delta A_m, \ \textit{with} \ \mu_{A_m} > 0 \quad (4.5.4)$$

Additions to the stock of material endowments are directly proportional to the size of the fund A_y:

$$A_m^D = G_{Am} \cdot A_y, \ \ \textit{with} \ \ G_{Am} > 0 \quad (4.5.5)$$

where G_{Am} is a fixed rate of proportionality. For example, the larger the knowledge in the economic system, the larger the discovery of new material endowments. For simplicity, there are no decreasing returns to the discovery and development of new sources, and a potentially infinite source of substitutes. The exergy loss associated with the use of fuels or materials in the development of new material sources is assumed to be proportional to the exergy generated in the development process.

With the proportionality factor $\mu_{A_m^D}$, the exergy loss is calculated as

$$A_{m1}^{D1} = \mu_{A_m^D} \cdot A_m^D, \quad with \quad \mu_{A_m^D} > 0. \tag{4.5.6}$$

Extraction from and additions to the exergy stock of fuels, A_f, are modeled analogously to materials:

$$\Delta A_f = C_2 * \left(\frac{A_f - A_f^*}{A_{fo} - A_{fo}^*} \right)^{\beta_{1f}} \left(\frac{A_y - A_y^*}{A_{yo} - A_{yo}^*} \right)^{\beta_{2f}} \tag{4.5.7}$$

$$with \; C_2, \beta_{1f}, \beta_{2f}, A_f^*, A_f^* > 0, \beta_{1f} + \beta_{2f} \leq 1.$$

$$A_f^D = G_{A_f} \cdot A_y, \quad with \quad G_{A_f} > 0. \tag{4.5.8}$$

The exergy losses associated with extraction and discovery and development of new fuel sources are, respectively,

$$\Delta A_f^1 = \mu_{A_f} \cdot \Delta A_f, \quad with \quad \mu_{A_f} > 0. \tag{4.5.9}$$

$$\Delta A_{f1}^{D1} = \mu_{A_f^D} \cdot \Delta A_f^D, \quad with \quad \mu_{A_f^D} > 0. \tag{4.5.10}$$

where μ_{A_f} and $\mu_{A_f^D}$ are fixed proportionality factors that relate the exergy loss from burning fuels and using materials in the extractive process to the exergy gains from those processes.

4.5.7.4 Environment of Model

The environment of this model consists of two parts. One of these parts contains the stock of materials and fuels from which the economy extracts exergy flows for productive purposes. The other part of the environment contains the biotic components that capture solar radiation and materials for their maintenance, growth and development. The linkages between these two parts, as they are present, e.g., in the form of biogeochemical cycles, are not modeled here.

The environmental fund (A_e) is assumed to be able to grow in proportion to its size at a rate $G_{A_e}^D$. The incremental rate of increase in each time period is

$$A_e^D = G_{A_e}^D \cdot A_e, \quad with \quad G_{A_e}^D > 0. \tag{4.5.11}$$

$G_{A_e}^D$ is inversely proportional to the total amount of waste products generated by the economy, i.e. the total exergy loss A_e^1.

$$G_{A_e}^{D} = \frac{B}{A^{TI}}, \quad with \quad B > 0. \tag{4.5.12}$$

where B is a fixed constant. If A^{II} is small, environmental funds grow rapidly. If A^{TI} is large, little regeneration takes place.

A decline of environmental funds A_e^1 is assumed to be proportional to the production of economic funds ΔA_y:

$$A_e^1 = \mu_{A_e}^1 \cdot \Delta A_y, \quad with \quad \mu_{A_e}^1 > 0, \tag{4.5.13}$$

i.e. the higher the rate of economic production ΔA_y, the higher the rate of destruction of environmental funds.

4.5.7.5 The Economy

The "production process" in the economy requires exergy of materials and fuels that have been extracted from the environment, i.e. ΔA_m and ΔA_f, the environmental funds A_e, and the funds of the economy A_y. The relationship between inputs into production and output is given by the following production function:

$$\Delta A_y = R \cdot \left(\frac{\Delta A_m - \Delta A_m^*}{\Delta A_{m0} - \Delta A_{m0}^*} \right)^{\alpha_1} \cdot \left(\frac{\Delta A_f - \Delta A_f^*}{\Delta A_{f0} - \Delta A_{f0}^*} \right)^{\alpha_2} \cdot \left(\frac{\Delta A_e - \Delta A_e^*}{\Delta A_{e0} - \Delta A_{e0}^*} \right)^{\alpha_3} \cdot \left(\frac{\Delta A_y - \Delta A_y^*}{\Delta A_{y0} - \Delta A_{y0}^*} \right)^{\alpha_4} \tag{4.5.14}$$

$$with \ R, \alpha_1, \alpha_2, \alpha_3, \alpha_4, \Delta A_m^*, \Delta A_f^*, A_e^*, A_y^* > 0, and \sum_{i=1}^{4} \alpha_i \le 1.$$

Since the production process is not perfectly efficient, not all exergy of materials and fuels ultimately enters into the product A_y. The loss of exergy of materials and fuels is assumed proportional to the quantities being used, with $\mu_{A_m}^1$ and $\mu_{A_f}^1$ as fixed proportionality factors for materials and fuels, respectively:

$$A_m^1 = \mu_{A_m}^1 \cdot \Delta A_m, \quad with \quad \mu_{A_m}^1 > 0 \tag{4.5.15}$$

$$A_f^1 = \mu_{A_f}^1 \cdot \Delta A_f, \quad with \quad \mu_{A_f}^1 > 0 \tag{4.5.16}$$

Deterioration of the fund A_y occurs in proportion to its size:

$$A_y^1 = \mu_{A_y} \cdot Ay, \quad with \quad \mu_{A_y} > 0 \tag{4.5.17}$$

where μ_{A_y} is a proportionality factor.

The total loss of exergy, A^{Tl}, is the loss of exergy that is generated in the discovery or development of new sources of materials and fuels (A_m^{D1} and A_f^{D1}), in the extraction of materials (A_m^1) and fuels (A_f^1), the production of economic funds (ΔA_m^1 and ΔA_f^1) and the deterioration of the endowments in the economy (A_y^1):

$$A^{Tl} = A_m^{D1} + A_f^{D1} + A_m^1 + A_f^1 + \Delta A_m^1 + \Delta A_f^1 + A_y^1 \tag{4.5.18}$$

A^{Tl} is assumed, via equation (4.5.11), to influence the incremental rate of increase in environmental funds.

4.5.7.6 Results

In the model, two linkages between the economy and its environment are present that lead to production-induced changes in environmental funds. These linkages manifest themselves via the impact that economic activity has on the rate of growth and the rate of deterioration of environmental funds. Obviously, if neither link is considered, and given the ability to discover new exergy stocks of materials and fuels, the economy can grow exponentially. The more interesting cases are those in which both linkages are considered. For the following discussion only changes in the impacts of the economy on the deterioration of environmental funds are considered. The parameter $\mu_{A_e}^1$ captures the extent of this impact. Analogous results can be found by varying the impacts of the economy on the growth rate for given deterioration rates $0 \leq \mu_{A_e}^1 \leq 1$.

In the case of $\mu_{A_e}^1 = 0$, the only effect the economy has on its environment is through reductions of the growth rate G_{A_e}. A growing economy leads environmental funds to increase not at an exponential but a decreasing rate (Fig. 4.5.3). The increase in the exergy stock A_y enables the economy to continuously discover and develop new exergy stocks of materials and fuels.

Changes in the impact of economic activity on the deterioration of the environmental fund alone can lead to, alternatively, steady state, a fluctuating economic output, and rapid economic collapse. The results of Fig. 4.5.4 are derived solely by changing the parameter $\mu_{A_e}^1$ that captures the degradation of environmental funds in response to economic production. Curve 1 in the graph refers to a low value of $\mu_{A_e}^1$, Curve 2 to an intermediate value of $\mu_{A_e}^1$ and Curve

3 to a high value of $\mu_{A_e}^1$. When $\mu_{A_e}^1$ is low, economic production has little effect on the environment's waste absorption capacity. In that scenario, economic output can grow to the highest level of the three scenarios and can be maintained in the long run at a positive steady-state. In that case, regeneration of environmental funds balances losses and discoveries of materials and fuels.

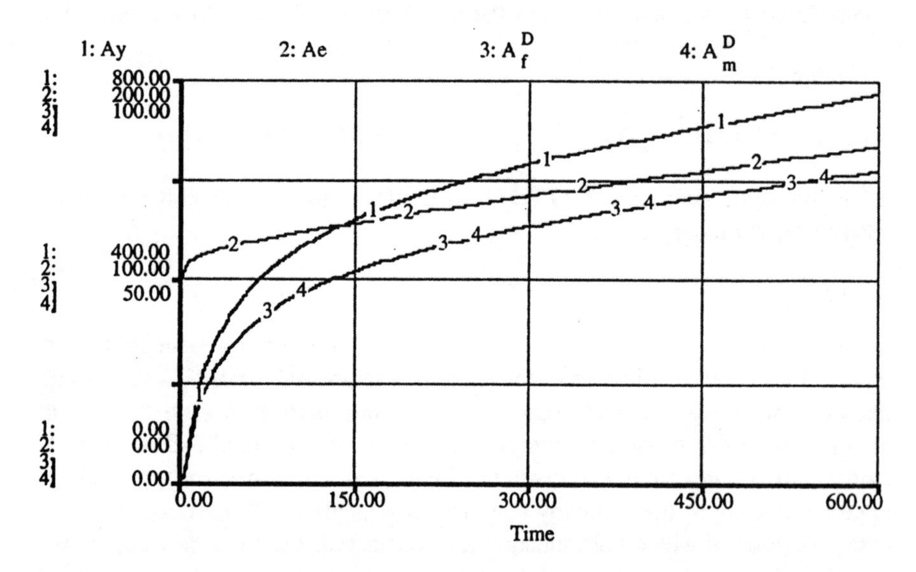

Fig. 4.5.3. Impacts of Economic Activity on the Rate of Growth of Environmental Funds ($\mu_{A_e}^1 = 0$).

Analogous results are achieved if economic production has large effects on the regeneration of environmental funds, with the exception that the ultimate steady state is one of zero economic production. In the case of intermediate effects of economic production on environmental funds, economic production first increases to a point at which the exergy flows of the waste reduce environmental funds and thus economic output, but these reductions in output occur neither too rapidly nor are they too severe to entirely halt economic production, allowing it to recover again but to a steady state that lies below the one that is achieved in the first scenario.

The model does not deal explicitly with issues of technical change. However, in one sense the model can be interpreted to include autonomous technical change: The assumption that a depletion of the stocks of exergy of materials and fuels does not require increasing expenditures of exergy in essence presupposes that technical change just counteracts the depletion of these stock. Including exogenous technical change into the model does not alter the fundamental insights that it generates. As long as there are adverse effects of economic production on environmental goods and services, the results above hold in the presence of technical change, albeit the economy may be increasingly decoupled from its environment, i.e. steady state may occur at higher levels or collapse may be postponed for several time periods. The extent to which this decoupling can take place, in turn, depends on the exergy consumption that is required to bring about technical change.

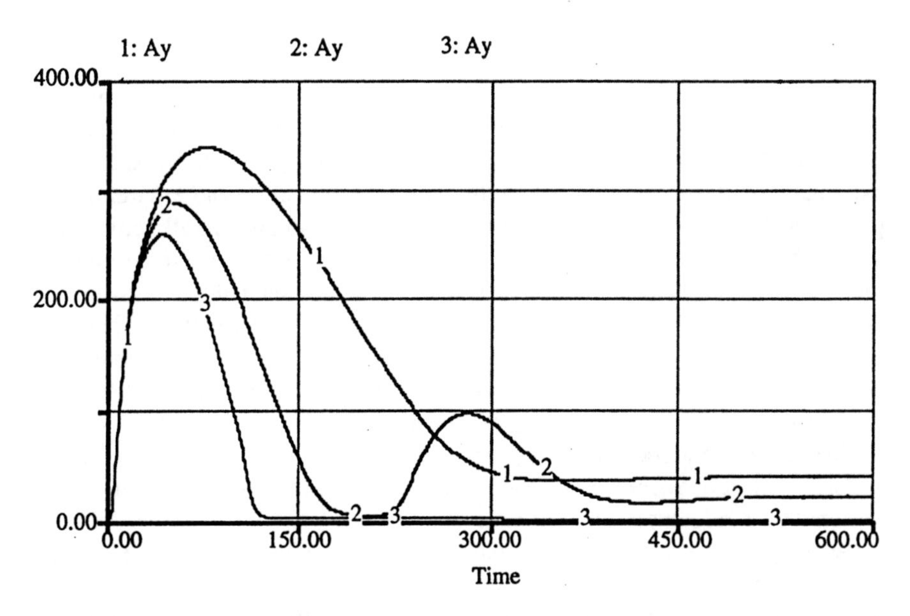

Fig. 4.5.4. Impacts of Economic Activity on the Rate of Growth and Deterioration of Environmental Funds ($\mu_{A_e}^1 > 0$).

The model also does not assume randomness, time lags in ecosystem responses, and threshold effects in the environment. Nevertheless, the long-run implications of economic production are far from obvious, containing the potential for significant fluctuations solely due to the strength of the economy-environment interactions. Introduction of these features would tend to add to the possibility for the system to undergo reversals in its time paths.

4.5.8 Summary and Conclusions

Exergy is the fundamental input that fuels all natural and economic processes. The exergy concept contains the first and second law of thermodynamics and enables us to quantify the maximum useful work that can be done in the economy and its environment. It provides a powerful means for aggregating funds and flows in the combined economy-environment system, as it is done for the simple dynamic model of economy-environment interactions developed above. The disadvantage of using the exergy concept in this way lies in glossing over the differences that exist between exergy of materials and fuels and the product form of the funds that are generated from them. Hence, change in size of a system is difficult to distinguish from a change in structure and organization. As a consequence, growth and development are not easily separable. However, concepts that are closely related to exergy are being developed in information theory and may ultimately be used to capture these differences (Eriksson et al. 1987; Ayres 1994).

The dynamic model developed here illustrates interactions between an exergy-extracting and releasing economy and the environment that provides the exergy flows and assimilates the exergy in the economy's waste stream. Albeit extremely simplistic with regard to the nature of economic and environmental processes, the model explicitly obeys second law constraints on all activities in the combined economy-environment system. It illustrates the need to account for the feedback mechanism by which economic expansion reduces growth and development of the environmental sector or increases its deterioration. That linkage can have repercussions for the economy that determine not just its long-term potential for maintenance and growth but that can also significantly influence its medium-term trajectory – including rapid collapse or temporary fluctuations.

The fact alone that the economy can undergo some form of oscillation if it is coupled with its environment – even though that environment does not exhibit seasonality, randomness or discontinuities – should highlight the need for an expansion of the traditional economic analysis of economy-environment interactions. One concept that can link both systems is exergy. One means by which the feedback mechanisms between the systems and among their subsystems can be investigated are dynamic models that capture, at least in part, the appro-priation and release of exergy by each system and the corresponding change in the funds in those systems. The system behaviors that emerge may elucidate the extent to which each system is able to pursue growth and development.

References

Ahrendts J (1980) Reference states. Energy 5: 667-677

Allen P (1994) Evolution, sustainability, and industrial metabolism. In: Ayres RU and Simonis UE (eds) Industrial Metabolism: Restructuring for Sustainable Development, United Nations University Press, Tokyo, pp 78-100

Anderson, C.L. (1987) The production process: inputs and waste. Journal of Environmental Economics and Management 14:1-12

Ayres RU, Ayres LW, Martiñas K (1996) Eco-thermodynamics: Exergy and life cycle analysis. INSEAD, Center for the Management of Environmental Resources

Ayres RU (1994) Information, Entropy, and Process: A New Evolutionary Paradigm. AIP Press, New York
Ayres RU (1988) Optimal Investment Policies with Exhaustible Resources: An Information Based Model. Journal of Environmental Economics and Management 15:439-461
Ayres RU (1978) Resources, Environment, and Economics: Applications of the Materials/ Energy Balance Principle. John Wiley and Sons, New York
Ayres RU, Miller SM (1980) The role of technical change. Journal of Environmental Economics and Management 7: 353-371
Baumol WJ, Oates E (1988) The theory of environmental policy. Cambridge University Press, Cambridge
Beltrami EJ (1987) Mathematics for Dynamic Modeling. Academic Press, Boston, Massachusetts
Berg CA (1980) Process Integration and the Second Law of Thermodynamics: Future Possibilities. Energy 5:733-742
Berry RS, Salamon P, Heal GM (1978) On a Relation between Economic and Thermodynamic Optima. Resources and Energy 1:125-137
Bossel H (1994) Modeling and Simulation. A.K. Peters, Ltd., Wellesley, Massachusetts
Chapman PF, Roberts F (1983) Metal Resources and Energy. Butterworths, London, 238 pp
Chen X (1992) Substitution of Information for Energy in the System of Production. ENER Bulletin 12: 45-59
Chen X (1994) Substitution of Information for Energy. Energy Policy 22:15-27
Christensen P (1989) Historical Roots for Ecological Economics - Biophysical Versus Allocative Approaches. Ecological Economics 1:17-36
Cleveland CJ, Ruth M (1997) When, Where and By How Much Does Thermodynamics Constrain Economic Processes? Ecological Economics, in press
Coase RH (1960) The Problem of Social Cost. Journal of Law and Economics 3:1-44
Conrad JM, Clark CW (1987) Natural Resource Economics. Cambridge University Press, Cambridge
Costanza R (1996) Ecological Economics: Reintegrating the Study of Humans and Nature. Ecological Applications 6:978 - 990
Costanza R, Ruth M (1997) Dynamic Systems Modeling for Scoping and Consensus Building. In Dragun A, Jakobsson K (eds.) New Horizons in Environmental Policy, Edward Elgar, pp 281-308, in press
Daly HE (1968) On economics as a Life Science. Journal of Political Economy 76:392-406
Dasgupta PS, Heal GM (1979) Economic Theory and Exhaustible Resources. Cambridge University Press, Oxford
Eriksson K-E, Lindgren K, Månsson BÅ (1987) Structure, Context, Complexity, Organization: Physical Aspects of Information and Value. World Scientific, Singapore
Evans RB (1969) A proof that essergy is the Only Consistent Measure of Potential Work. Dartmouth College, Hanover, New Hampshire
Gagglioli RA (ed) (1980) Thermodynamics: Second Law Analysis. American Chemical Society, Washington, DC
Gelb B (1984) A Look at Energy Use in Mining: It Deserves it. International Association of Energy Economists, San Francisco, pp 947-959
Georgescu-Roegen N (1971) The Entropy Law and the Economic Process. Harvard University Press, Cambridge, Massachusetts, London, England
Govett MH, Govett GJS (1978) Geological Supply and Economic Demand: The Unresolved Equation. Resources Policy 4:107-114
Graedel TE , Allenby BR, (1995) Industrial Ecology. Prentice Hall, Englewood Cliffs
Gray LC (1913) The Economic Possibilities of Conservation. Quaterly Journal of Economics 27:497-519

Gray LC (1914) Rent under the Presumption of Exhaustibility. Quaterly Journal of Economics 28:466-489

Hall CAS, Cleveland C J, Kaufmann R (1986) Energy and Resource Quality: The Ecology of the Economic Process. Wiley-Interscience, New York

Hannon B, Ruth M (1994) Dynamic Modeling, Springer-Verlag, New York

Hannon B, Ruth M (1997) Modeling Dynamic Biological Systems, Springer-Verlag, New York

Heywood VH (ed) (1995) Global Biodiversity Assessment. Cambridge University Press, New York, 1140 pp

Hotelling H (1931) The economics of exhaustible resources. Journal of Political Economy 39:137-175

Houghton JT, Filho LGM, Callander BA, Harris N, Kattenberg A, Maskell K (eds) (1996) Climate Change (1995) Cambridge University Press, New York, 570 pp

Howell JR, Buckius RO (1992) Fundamentals of Engineering Thermodynamics. McGraw-Hill Book Company, New York

Islam S (1985) Effects of an Essential Input on Isoquants and Substitution Elasticities. Energy Economics 7:194-196

Jelinski LW, Graedel TE, Laudise RA, McCall DW, Patel CKN (1992) Industrial Ecology: Concets and Approaches. Proceedings of the National Academy of Science 89:793-797

Jørgensen SE (1992) Integratiion of Ecosystem Theories: A Pattern. Kluwer Academic Publishers, Dortrecht, The Netherlands

Lesourd J-B (1985) Energy and Resources as Production Factors in Process Industries. Energy Economics 7:138-144

Kümmel R, Kunkel A, Lindenberger D (1996) Energy-Dependent Production Functions, Technological Change and Industrial Evolution. In: Baranzini A, Carlevaro F (eds) Econometrie de L'Environment et Transdisciplinarite. Proceedings, Applied Econometrics Association, International Conference, Lisbon, Portugal, pp 593-611

Månsson BÅG (1986) Optimal development with flow-based production. Resources and Energy 8:109-131

Månsson BÅG (1991) Fundemental Problems with Energy Theories of Value. In: Hansson LO, Jungen B (eds) Human Responsibility and Global Change. Proceedings from the International Conference in Göteborg, June 9-14, 1991, University of Göteborg, Section of Human Ecology, Göteborg, Sweden, pp 197-206

Page NJ, Creasey SC (1975) Ore Grade, Metal Production, and Energy. Journal Research U.S. Geological Survey 3:9-13

Pigou A C (1932) The Economics of Welfare. Macmillan, London

Postel S, Daily GC, Erlich PR (1996) Human Appropriation of Renewable Fresh Water. Science 71:785-788

Rosser JB (1991) From Catastrophe to Chaos: A General Theory of Economic Discontinuities. Kluwer Academic Publishers, Dortrecht, The Netherlands

Rosser JB (1992) The Dialog Between the Economic and Ecologic Theories of Evolution. Journal of Economic Behavior and Organization 17:195-215

Ruth M (1993) Integrating Economics, Ecology, and Thermodynamics. Kluwer Academic, Dordecht, 251 pp

Ruth M (1995a) Information, Order and Knowledge in Economic and Ecological Systems: Implications for Material and Energy Use. Ecological Economics 13:99-114

Ruth M (1995b) Thermodynamic Implications for Natural Resource Extraction and Technical Change in U.S. copper mining. Environmental and Resource Economics 6:187-206

Ruth M (1996) Evolutionary Economics at the Crossroads of Biology and Physics. Journal of Social and Evolutionary Systems 19:125-144

Ruth M, Bullard CW (1993) Information, Production and Utility. Energy Policy 21:1059-1067

Ruth M, Hannon B (1997) Modeling Dynamic Economic Systems. Springer-Verlag, New York

Schneider ED (1988) Thermodynamics, Ecological Succession and Natural Selection. In: Weber B, Depew D (eds) Entropy, Information and Evolution. MIT Press, Cambridge, Massachusetts

Schneider ED, Kay JJ (1990) Life as a Phenomenological Manifestation of the Second Law of Thermodynamics. Mimeo, Environment and Resource Studies, University of Waterloo, Waterloo, Ontario

Schneider ED, Kay JJ (1993) Energy Degradation, Thermodynamics, and the Development of Ecosystems. Proceedings of the International Conference on Energy Systems and Ecology, The American Society of Mechanical Engineers, Advanced Energy Systems, Cracow, Poland

Schneider ED, Kay JJ (1994a) Life as a Manifestation of the Second Law of Thermodynamics. Mathematical and Computer Modelling 19:25-48

Schneider ED, Kay JJ (1994b) Complexity and thermodynamics: towards a new ecology. Futures 26:626-647

Skinner B (1976) Earth Resources. Prentice-Hall, Englewood Cliffs, New Jersey

Spreng DT (1993) Possibilities for subsitution between energy, time and information. January 1993, pp 13-23

Szargut J, Morris DR, Steward FR (1988) Exergy Analysis of Thermal, Chemical, and Metallurgical Processes. Hemisphere Publishing Corporation, New York

Szilard L (1929) Über die Entropieverminderung in einem thermodynamischen System bei Eingriffe intelligenter Wesen. Zeitschrift für Physik 53:840-960

Tribus M, McIrvine EC (1971) Energy and information. Scientific American 225:179-188

Van der Voet E, Kleijn R, Huppes G (1995) Economic characteristics of chemicals as a basis for pollution policy. Ecological Economics 11- 26

Varian HR (1992) Microeconomic Analysis. WM Norton, New York

Vitousek PM (1994) Beyond global warming: ecology and global change. Ecology 75:1861-1876

Wicksteed PH (1894) The co-ordination of the law of distribution. Macmillan, London

Winston GC (1982) The timing of economic activities. Cambridge University Press, Cambridge.

4.6 Societies' Maneuver Towards Sustainable Development: Information and the Setting of Target Values

Walter Radermacher

Abstract

Sustainable development is by definition a multidimensional concept linking environmental, economic and social aspects. The central element of a shift to a sustainable development is the definition of target values defining society's preferences for the different aspects of sustainability, in particular the social preferences for natural elements. The definition of these target values can by no means be restricted to pure scientific work. In contrary, a co-operation of science and political decision processes is needed to achieve "procedural rationality". Complex systems, uncertainties, etc., delimit the role of science and statistics to the provision of indicators which can and should be used as an input to the political bargaining processes. The quantitative results of scientific models and statistical measurements must be interpreted as representations of possible margins for the manoeuvre to sustainable development. Bringing together scientific, social, and economic considerations in real time, as a sort of interdisciplinary dialogue, social learning and conflict resolution process, provides a basis for prioritizing and revising actions in the environmental domain. Goal functions, target values, scientific models and statistical indicators are closely connected elements in the processes leading to an evolution of societies. Based on these theoretical fundaments the German Federal Statistical Office has developed the framework for an Environmental Economic Accounting System. The objective is to add meaningful modules to the traditional System of National Accounts which are designed to quantify the external (environmental) effects of economic activities.

F. Müller M. Leupelt (Eds.)
Eco Targets, Goal Functions, and Orientors

4.6.1 Defining Target Values - Scientific Task or Bargaining Process

Before it comes to the substantial work on the definition of target values we should lean back and reflect seriously the following alternatives that are essential for the expectations we have and for the way we tackle problems:

1. Are we convinced that (as a matter of principle and as an eternal truth) goal functions exist from which optimal solutions for our decision problems can be deducted? Is it, hence, our main scientific task to detect or discover these goal functions by means of analytical models? How essential is the need to check theoretical models by empirical measurements? And if we do not succeed in providing the required statistical figures, can we substitute them by theoretical axioms or assumptions?
2. Or do we suppose that goal functions (and unique solutions of decision problems) only exist as long and as far we are actually able to define and measure them? Is truth depending on knowledge, awareness, scientific skills, and other factors which are results of an evolution of complex socio-economic systems? Do we have to adjust the theoretical models correspondingly even if they are not further able to give a clear-cut advise but leave some space for actual decision-making?

Of course, this presentation of the alternatives seems to be much more radical than the practical differences really are. Nevertheless, there are some consequences which have to be drawn if the second position is taken. Compared with the first it requires a change of the paradigm concerning the relation of scientific theory, empirical measurements and political processes. The debate with regard to these opposite points of view is absolutely not new. In socio-economic sciences the historical roots of this debate reach (at least) back to the dispute between the "Historical School" and the "Neo-Classical School" in the end of the last century ("Methodenstreit"; see for example Reuter 1996). A recent version of that debate is incorporated in the current discussion between "Environmental Economics", which is oriented towards the application of the deductive methods of neo-classical economic theory, and "Ecological Economics", which combines a widespread variety of approaches in a more loosely linked and inductive conceptual framework.

A question of major political importance is whether we can change the direction of economic and social development to 'sustainability' or (better) to 'sustainable pathways'. Sustainable development is by definition a multidimensional concept linking environmental, economic and social aspects. A recently finished research study for the European Commission has investigated the requirements of the sustainability concept for the elaboration of 'Green National Accounting' (see Brouwer, O'Connor and Radermacher 1996). Within this study the most important finding was that the second of the above mentioned perspectives is needed as a starting point and theoretical background for a theoretically sound and empirically realistic framework. The central element of a shift to a sustainable development is the definition of target values defining society's preferences for the different aspects of sustainability, in particular the social preferences for natural elements. The definition of these target values can by no means be restricted to pure scientific work. In contrary, a co-operation of science and political decision processes is needed to achieve 'procedural rationality' (see Fig. 4.6.1).

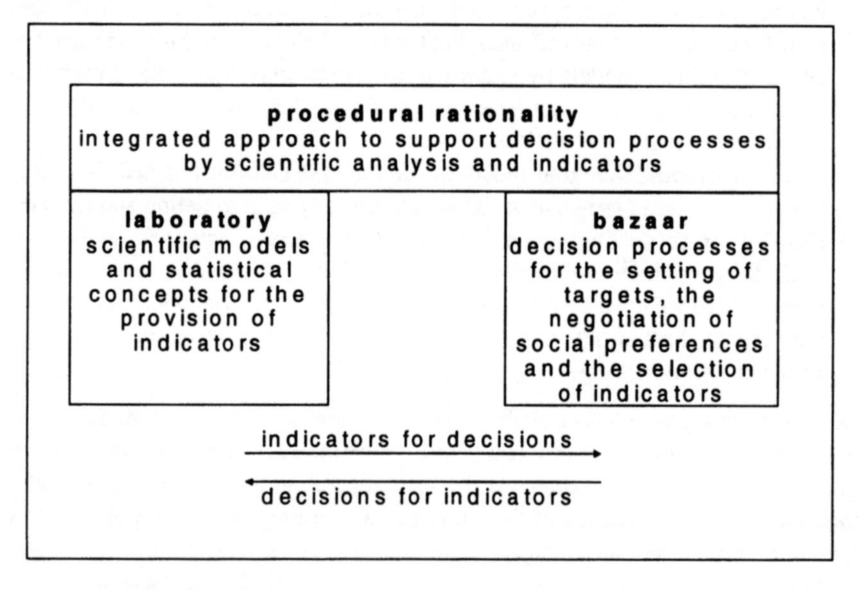

Fig. 4.6.1. Procedural Rationality

Complex systems, uncertainties, etc., delimit the role of science and statistics to the provision of indicators which can and should be used as an input to the political bargaining processes.

4.6.2 Objectives of Indicators for Sustainability: the Need for an Accounting System

Indicators for sustainable development must be interpreted as representations of possible margins for the manoeuvre of society. This reflects the fact that the validation of analysis for policy support depends not only on standards of theoretical rigor and internal coherence, but also on external societal considerations (see O'Neill 1993; Holland, O'Connor and O'Neill 1996; Brouwer, O'Connor and Radermacher 1996). It is convenient to distinguish five broad sets of considerations which are interlocking:

1. scientific adequacy: do the description and evaluation methods deal well with the important features of the natural world and its characteristic processes of change?
2. social adequacy: do the methods provide information in ways that respond to stake-holders' needs and that support social processes of decisionmaking?
3. economic rationality: do the suggested courses of action that emerge from the valuation process respect economic efficiency, in the sense of appearing to be a reasonably cost-effective way of arriving at the envisaged outcome?
4. statistical adequacy: are the empirical measurement and subsequent aggregation procedures consistent with the guiding theoretical precepts, and do they conform to norms of reliability, coverage or representativeness?
5. budgetary realism: can the proposed approaches be expected to yield reliable and useful information (judged in terms of the four sets of considerations above) within the limits on resources that can be committed to the research?

Bringing together scientific, social, and economic considerations in real time, as a sort of interdisciplinary dialogue, social learning and conflict resolution process, provides a basis for prioritizing and revising actions in the environmental domain. These requirements are, evidently, not easy to meet. The solution can only be a trade-off between the different objectives. The appropriate answer seems to be the development of an environmental economic accounting system. On the one hand an accounting system is more committed to consistency than pure lists of indicators are. On the other it favors a 'holistic' perspective which gives higher priority to comprehensive information and less to the precision of specific and detailed data.

4.6.3 The Quality of Statistical Data

Like every other economic good or service, statistical figures can be produced and delivered in very different quality and with very different costs. Hence, it is a task for negotiations between data producers and data users to find an acceptable quality. The quality of data depends on several different components. In general, data have to be reliable, up-to date, and so on. More particularly, data must meet the requirements of a specific user, and this will have a strong influence on the quality priorities. Due to limited capacities (personal or institutional budgets, personnel, data processing know-how and equipment, etc.), a trade-off between the different components of quality - and hence statistical validity - is unavoidable.

Fig. 4.6.2 depicts the various elements of a situation which is characteristic for the design of statistical systems. Quality depends on a complex composition of various aspects. In practice, these aspects are not independent since there exist financial, personnel and time constraints which do not allow the maximum of all quality aspects to be achieved at the same time.

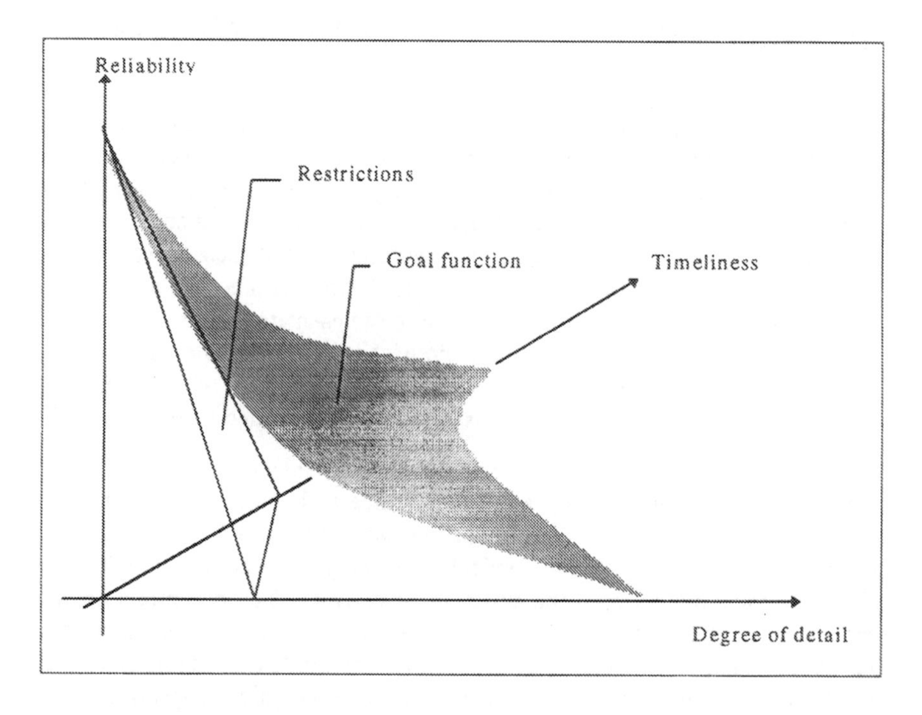

Fig. 4.6.2. The Quality of Statistical Information: Finding a Good Compromise (Source: Radermacher et al. 1995)

The questions of data quality and research ambition have to be resolved in such a way that all these quality determining elements and the negotiation between data producer and data user are taken into account. Some of the aspects of data quality are quantitatively measurable, some are not. In addition to the scientific considerations, the quality of data and scenario modelling results can also be examined critically from the perspective of the data user and his or her data requirements. The quality of information is satisfactory when it gives a good basis for finding a solution to the user's (or users') decision-making problem.

Together the various components influencing the quality of data determine the validity of statistical figures (Radermacher et al. 1995). The validity of statistical figures refers to the extent to which it measures what it is expected to measure. The correspondence of a theoretical concept (for example, income in economic theory) and a statistical definition (income in a specific survey) is important. Other factors that determine the suitability of statistical figures, for example: Are the data up-to-date? Does the classification provide enough detail? Are the results reliable? The amalgam of these interrelated factors determines the validity of statistical figures (Fig. 4.6.3). The validity of a statistical working system can only be examined critically with reference to a specific utilization and the relevant restrictions (financial capacities, personnel, time). Theoretical questions have to be translated into adequate statistical methods. These methods provide sufficient data quality of surveys which results can be interpreted accordingly.

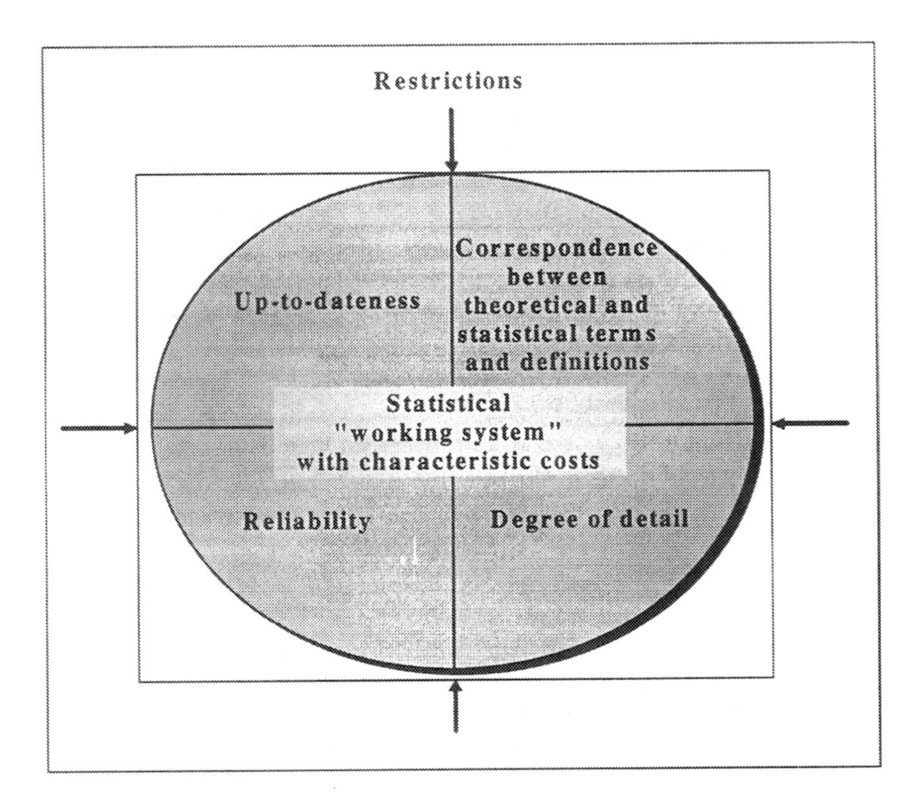

Fig. 4.6.3. The Components of a Statistical "Working System"

4.6.4 Statistical Adequation: an Iterative Process

The development of a statistical system is not a linear process, but should be understood as a feedback loop in which a balance is sought between theory and empirical practice. An extreme situation in which a society has no awareness for environmental problems at all can exemplify this: In this (hypothetical) situation there would evidently be no willingness to pay for environmental research and environmental statistics either. Poor quality of data (and small potential to create awareness) would be the consequence. Hence, the tuning is an iterative adjustment of what should be measured and what is actually measured. For example, finding out where the financial possibilities of data users can be matched to a feasible mixture of quality components belongs to the consultant's and the official statistician's everyday tasks. In the statistical adequation process, concrete requirements from the specific groups of data users are necessary. For the elaboration of a 'green' accounting system, this means specification of:

- the range of application (e.g. a periodic accounting system);
- the environmental scale (local, regional, national, continental, global);
- the economic scale (micro, meso, macro).

Correspondingly, also the factors degree of detail or up-to-dateness will differ. Therefore, statistical description and scientific analysis are seen as complementary activities with a fuzzy borderline. Important interfaces in this process are the definition of statistical methods and the correct interpretation of statistical results and their status, relative to the uses desired of them. It is necessary to delimit clearly the different spaces of work. For example, in order to respect the expectations of policymakers and the public regarding validity and reliability of official statistical figures, the statistical working area which collects and transforms historical data must be distinguished from the domain of scenario modelling which, in addition to its dependency on historical data, introduces speculative hypotheses of quite different orders.

4.6.5 Cost-Effectiveness: an Iterative Target Setting Process

The previous considerations lead to a cost-effectiveness approach used as a guiding concept in the integration of national economic accounts with information on the state and change of the biophysical environment. The concept helps to define the starting points for a scientific investigation. In this case, the starting points are environmental quality standards, or norms, which define limits of environmental pressures caused by economic activities. From the perspective of the recommended economic evaluation procedure standards are taken as exogenous parameters. The definition of standards is interpreted as the core of the political decision process under consideration. It can be supported by multiple criteria from the 'laboratory'. In the end, however, it will be characterized by negotiations and bargaining between the different interest groups of the society. Advantages of this approach and reasons for its selection are discussed extensively in the work carried out in the European Commission project 'Methodological problems in the calculation of environmentally adjusted national income figures' (Brouwer and O'Connor 1996; Brouwer, O'Connor and Radermacher 1996).

The idea of this concept can be summarized in the following way: Achievement of economic and ecological sustainability objectives requires resource management to assure the maintenance of essential environmental functions as well as economic capital stocks. In this way, the legacy of the past - natural and economic capital - is transformed into the welfare base for the future, laying the foundation for a high and sustainable national income (Brouwer, O'Connor and Radermacher 1996; Faucheux and O'Connor 1997). Economic resource management must then fulfill two complementary functions:

1. the delivery of an ecological welfare base through assuring maintenance of key environmental functions; and
2. the delivery of an economic welfare base through production of economic goods and services.

Looking at nature in anthropomorphic terms, the 'function' of the environment is to provide a range of ecological goods and services (energy and material resources, amenities, waste reception, environmental life support functions). These ecological goods and services are broadly complementary with economic goods and services. Since some of the services which are provided by nature are essential and cannot be substituted by economic goods and services, a direct monetarization of environmental goods and services cannot be carried out in a meaningful way. But in the cost-effectiveness approach to environmental management, money valuations are not applied to the environmental functions as such. Environmental functions of various sorts are primary or 'basic' requirements for human welfare, and hence for sustainable economic activity. Economic resources must be committed (directly or indirectly, in ways that we will discuss) in order to maintain the desirable level of environmental functions. This corresponds to what can be called the 'social demand' for maintenance of environmental functions. This 'demand', which includes making provision for future generations and protection from present and future environmental harms, cannot be expressed in a marketlike institution. The concerned future parties cannot be present, and many of the benefits in question are 'public' (and largely indivisible) in character. Therefore, the best operational specification will be in non-monetary terms, by defining environmental standards, or norms, which represent a society's objectives for the delivery of the ecological welfare base to present and future generations.

Establishing priorities in environmental norm-setting is a political process which involves an integration of scientific, economic and social considerations (O'Neill 1993). Because of uncertainties and differences of opinion within a society, it is more appropriate to adopt a multiple-criteria decision support framework and to speak of 'satisfactory' (contentment) rather than optimizing (better and best). Furthermore, environmental priorities and policies will undergo continual revision as a social adaptation and negotiation process (see Radermacher 1994; Holland, O'Connor and O'Neill 1996; Brouwer, O'Connor and Radermacher 1996). Within an economic system, trade-offs will exist between economic output and environmental maintenance and enhancement goals. In this respect, it is possible to apply the principle of cost-effectiveness or 'least economic cost' for achievement of environmental objectives (see Baumol and Oates 1971).

4.6.6 The Framework for Environmental Economic Accounting

Based on these theoretical fundaments the German Federal Statistical Office has developed the framework for a 'green' accounting system (see for example Radermacher, Schäfer and Seibel 1995; Radermacher 1996; Radermacher and Stahmer 1996). The objective is to add meaningful modules to the traditional System of National Accounts which are designed to quantify the external (environmental) effects of economic activities. As has been laid out, the design of such an accounting system itself reflects the requirements and data needs of a society. For instance, the German work to elaborate that system has been started after a debate in the German Parliament. Within this debate there had been a broad consensus that the traditional way to compute the Gross Domestic Product leads to an indicator which must not be interpreted as a welfare measure. Nevertheless, it is often taken as a general signal for the success of an economy or – even worse – for the well-being of a society. Wrong signals cause wrong political decisions. Hence, the first idea was to adjust the calculation method and to create a proper measure of the National Income via a deduction of the depreciation on natural capital. This starting initiative and the corresponding task for the Statistical Office was in 1990. At that time the expectations for the provision of an adjusted 'Eco-Domestic Product' (which is an ecologically adjusted Net Domestic Product) could be quite optimistic: The first experiences from economic evaluation of environmental problems and for resource depletion had been rather successful, at that time.

Predominantly, those economic valuation approaches took into account the environmental problems of the seventies and the eighties, i.e. air and water pollution, the depletion of mineral resources, of fish, etc. Furthermore, they were founded in micro-economic theory. Facing these approaches with the more complex problems of sustainability and trying to apply them to a macro-economic statistical system caused, however, a series of major troubles. Not all of these problems could be solved in a common international framework and there is still a discussion of experts concerning essential methodological elements. United Nations' "System for Integrated Environmental and Economic Accounting (SEEA)" (United Nations 1993) is called an "interim version" accordingly. Meanwhile, it can at least be stated that the leading nations in the development of 'green' accounting (besides Germany e.g. the Netherlands, Canada, Sweden) prefer modular approaches similar to the German concept. The variety of indicators that are needed for actual decision making depends heavily on the special mix of issues in a country. Nevertheless, in principle always the same type of indicators, matrices or accounts is needed: They have to describe the driving forces that are responsible for the most relevant environmental pressures, the damages caused by those pressures in the natural assets, and the responses (i.e. environmental protection measures) which are actually taken or hypothetically needed by the society. Following this common structure Fig. 4.6.4 describes the different modules and their relation to decision making and target setting.

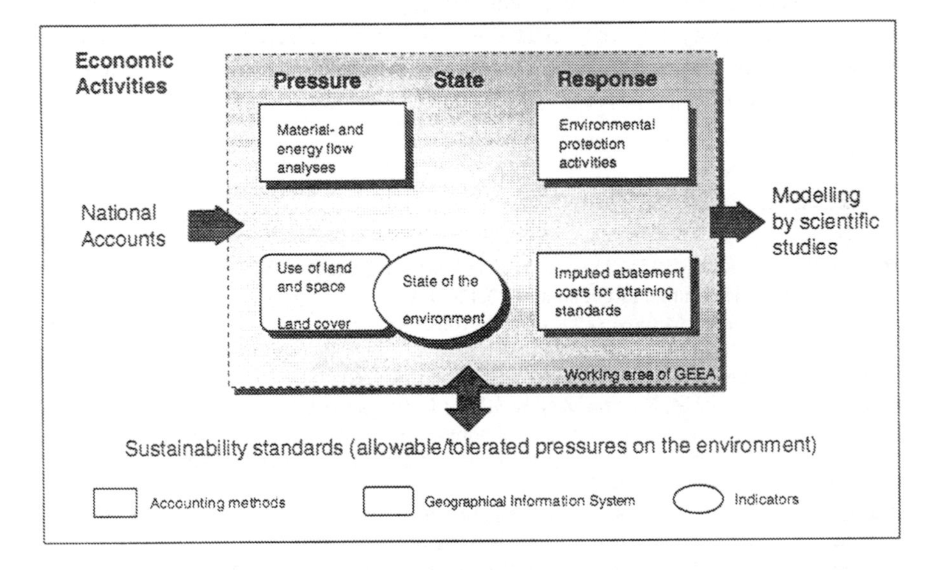

Fig. 4.6.4. Environmental Economic Accounting in Germany (GEEA)

The most time-consuming procedure in the construction of a new statistical system is the search for an appropriate compromise between the user's needs, the consistent general framework and the data availability. Since Environmental Accounting and sustainable development are relatively new items it is a rather normal situation that the data which are available from existing measurement programs and statistical surveys do not fit easily into the theoretically preferable concepts and do not fulfill the user needs. Hence, an iterative procedure is required which can provide some kind of an 'early harvest' with the option of methodological improvements during the next years.

The framework of the GEEA could already be realized and published to an extend that is relevant for actual policy making in Germany. However, the questions of aggregation and evaluation are in a stage that is still characterized by research and development. One aspect of major importance within the current research projects is the integration of the 'bazaar' in order to link normative settings with descriptive indicators.

4.6.7 Conclusions

The existence of an Environmental Economic Accounting system is seen as one of the essential preconditions for a rational procedure towards the realization of sustainable development of a society. The definition of goal functions and the breakdown to concrete target values or sustainability standards is a process in parallel being closely connected to the generation of meaningful information in a mutual sense. A step by step approach keeping the balance between the ambition of consistent theory and the restrictions of empirical reality seems to be the prescription for a successful contribution of information providers in societies' decision processes.

References

Baumol WJ, Oates WE (1971) The Use of Standards and Prices for the Protection of the Environment. Swedish Journal of Economics 73:42-54

Brouwer R, O'Connor M (eds) (1996) Final Project Report: Methodological Problems in the Calculation of Environmentally Adjusted National Income Figures, Research Report for the European Commission DG-XII, Contract EV5V-CT94-0363

Brouwer R, O'Connor M, Radermacher W (1996) Defining Cost Effective Responses to Environmental Deterioration in a Periodic Accounting System. In: Statistics Sweden (ed) Proceedings of the Third Meeting of the London Group on Natural Resource and Environmental Accounting. Stockholm, pp 397-421

Faucheux S, O'Connor M (eds) (1997) Valuation for Sustainable Development, Methods and Policy Indicators. Edward Elgar, Cheltenham (forthcoming 1997)

Holland A, O'Connor M, O'Neill J (1996) Costing Environmental Damage: A Critical Survey of Current Theory and Practice and Recommendations for Policy Implementations, Report for the Directorate General for Research, STOA programme, European Parliament

O'Neill J (1993) Ecology, Policy and Politics: Human Well-Being and the Natural World. Routledge, London

Radermacher W (1994) Sustainable Income: Reflections on the Valuation of Nature in Environmental-Economic Accounting. Statistical Journal of the United Nations ECE 11:35-51

Radermacher W (1996) Land use accounting - pressure indicators for economic activities. In: Bartelmus P, Uno K (eds) Environment Accounting in Theory and Practice. Kluwer Academic Publishers, Dordrecht (forthcoming 1997)

Radermacher W, Schäfer D, Seibel S (1995) Remote sensing for physical accounting and measuring changes in land use. In: Eurostat (ed) Proceedings of the Esquilino Seminar on the Impact of Remote Sensing on the European Statistical Information System. Luxembourg (forthcoming 1997)

Radermacher W, Stahmer C (1996) Material and Energy Flow Analysis in Germany —Accounting Framework, Information System, Applications. In: Bartelmus P, Uno K (ed.) Environment Accounting in Theory and Practice. Kluwer Academic Publishers, Dordrecht (forthcoming 1997)

Reuter N (1996) Der Institutionalismus. Metropolis, Marburg

United Nations (1993) Integrated Environmental and Economic Accounting. Studies in Methods, Series F, No. 61, New York

4.7 Targets of Nature Conservation – Consequences for Ecological and Economic Goal Functions

Klaus Dierßen

Abstract

The protected goods of classical wildlife conservation are populations of animal and plant species, their guilds and habitats, the pattern of selected landscapes, limited abiotic resources, and landscape functions. Some of the difficulties considering all these aspects in landscape planning are discussed, and a gap between knowledge transfer and practical application concerning functional correlations in, and between ecosystems is pointed out. Involving ecological orientors in the description and modeling of ecosystem functions may be a promising way for future approaches in landscape evaluation, management and monitoring, but the empirical base must be tested and expanded. For a more satisfactory acceptance of landscape management, planners have to consider the economic background. However, there is no real market value for the benefits of sustainable land use at date.

4.7.1 Introduction

The interaction between nature conservation and the ecological goal function concept are documented in a series of papers in this volume. In the following paper some aspects are discussed from the viewpoint of a conservation biologist, who is to some extent involved in landscape planning and who intends to use ecosystem research as a means to optimize conservation strategies. In this paper, the current conservation targets are discussed, including the problems of finding priorities. From a holistic point of view, it will be useful to find answers to the question of how to involve aspects of theoretical ecosystem research and economy in further—more satisfying—concepts of landscape planning and wildlife management.

F. Müller M. Leupelt (Eds.)
Eco Targets, Goal Functions, and Orientors

4.7.2 Targets of Nature Conservation

Up-to-date approaches to solve nature conservation problems take place at different levels of integration:

- Conservation and/or restoration of animal and plant species on the population level;
- conservation and/or restoration of animal guilds and plant communities at defined habitats and sites;
- conservation and/or restoration of habitats and sites of a given landscape section;
- conservation and/or restoration of a specific pattern of landscape features, both, in native and man-made landscapes;
- conservation and/or restoration of so-called abiotic resources of defined landscape sections, and
- conservation and/or restoration of selected ecosystem functions of regionally different landscape patterns.

All these partial aims are legally defined by national and international laws or regulations. In the following paper, it seems useful to distinguish between measures concerning a whole landscape's 'sustainable development' and measures, which refer to areas of special interest for wildlife preservation, i. e. those sites, where the demands of nature conservation take a high priority over any other land use form. Comparing the integration levels from populations to ecosystems from a purely theoretical as well as from a practical point of view, an increasing complexity of the targets is obvious. Some difficulties might be stressed:

4.7.2.1 Conservation of Rare Species

Conservation and efficiency monitoring of selected populations, especially for rare and threatened taxa is probably the most commonly used practice of conservation biology, predominantly in selected areas. The following limitation is striking: Nearly each population shows an individual optimum of development in time and space concerning fluctuating abiotic site conditions as well as biotic interactions. A total fundamental niche overlapping between different populations does not exist in one and the same place. If we accept this hypothesis, we may have no real chance to preserve and develop more than one selected population on a given site for a given span of time. Of course, this extreme but consistent opinion is troublesome for conservation practices.

4.7.2.2 Spatial and Temporal Aspects of Conservation

Conservation has a spatial dimension and a time dimension: the patchiness of species distribution and the period, for which a particular project, e. g. a nature reserve, is expected to remain operative. Therefore conservation planning inevitably has implications for site, size, design and management and it depends on the prevailing natural and social conditions in the future. It can extend from a year or a generation to an indefinite future, at least in an evolutionary perspective.

4.7.2.3 Biodiversity

Trying to involve both ecological as well as evolutionary aspects of wildlife conservation, conservationists try to highlight the concept of biological diversity. This term refers to the fact that heterogeneity at different site levels is a fundamental property of natural systems. Biodiversity management intends to maintain or to establish the structural design of separate micro-sites in space and time and gradients or corridors (ecotones, ecoclines) in order to allow various populations of species with different life strategies to coexist even on a small scale.

There are at least three different research traditions dealing with biological diversity: species diversity, genetic diversity within populations, and environmental heterogeneity. Since the beginning of this century, ecologists tended to assume instinctively that morphological and genetical diversity at least of key species such as trees enhances community stability (Goodman 1975; Gregorius 1989; Müller-Starck 1989; Ziegenhagen et al. 1997). On the other hand, populations near their distribution borders may have a higher degree of morphological and genetic differentiation than in their distribution centers, but nevertheless a low adaptive capacity (Tigerstedt 1997). Currently it seems impossible to assess whether species diversity is important for the stabilization or the destabilization of community dynamics and functioning. The knowledge of genetic consequences of small population sizes is limited (i. g. Falk and Holsinger 1991). Because the majority of rare species occur in small populations, often with decreasing numbers, it is important to assess whether significant genetic deterioration can arise through sudden or gradual decreases in number. Stochastic forces such as bottlenecks (large fluctuations in reproductive capacity from year to year), founder events and genetic drift may influence genetic variation. These processes must be studied in detail in order to enhance the chances of survival of an endangered species before starting expensive management manipulations.

Because the frame conditions changes from site to site and the situation fluctuates in time, evaluation of theoretical deductions should be performed with care. Concerning phanerogames, the geneflow depends on the mode of pollination, increasing from autogames through entomogames to anemogames (Govindaraju 1988). Nevertheless, facultative autogames, apomicts and exclusively vegetative propagating species often are particularly successful in colonizing new and harsh environments or are favoured by human impact (Asker and Jerling 1992; Matzke-Hajek 1997). Today it seems unwise or impossible to define, how much genetic diversity is necessary or even useful to establish or to maintain in a given population, if this should be technically possible at all. One further difficulty in the research of genetic diversity, is the lack of adequate methods to summarize the vast natural variability in genomes (Hodgkin and Guarino 1997). One may assume that genetic factors appear to have been overemphasized during the last decade (Hansson and Larsson 1997).

4.7.2.4 Measurements of Biodiversity

The determination of biodiversity involves serious, unsolved problems concerning the scales of the areas involved (population, community, ecosystem, reservate, country, biome), the parameters recorded, (the allelic richness or gene diversity within populations, the richness of species within one site, the richness of site properties in a reservate, or the equality or evenness in frequency of different types from the population to the ecosystem level), and, last but not least, the organism groups investigated, for example the woodpecker assemblages in natural and managed boreal forests (Angelstam and Mikusinski 1994). Generally, biodiversity increases with the area included (summarized by Rosenzweig 1997). Nearly all biodiversity measurements (richness or evenness per area) include only selected species groups or abiotic parameters. Generally speaking: The concept of biodiversity is of low predictive value without any further characterization of the environmental limitations.

4.7.2.5 The Different Degree of Human Impact

The key conditions for sites with a high biodiversity, vary and deviate from each other in different landscapes. In ahemerobic and oligohemerobic areas with a negligible human impact, the continuous development of various ecosystem types normally may lead to an increasing biodiversity. This might be interpreted as an autogenic succession leading to an increasing niche differentiation. In meso- and euhemerobic cultural landscapes, however, the diversity of phanerogames mainly depends on a balance between environmental stress, disturbance and productivity, the highest diversity occurring on sites, where periodic reductions of phytomass take place (Huston 1979; Grime 1979).

4.7.2.6 Evaluating the 'Intrinsic Appeal' of Landscape Sections

The design of landscape features, their 'intrinsic appeal', is often discussed from socioeconomic and politic points of view, and the discussion is far less connected with scientific investigations. The main problem for ecologists involved in landscape planning is how to evaluate 'intrinsic appeal'. The planning of hedgerows in coastal lowland areas of western Europe is fixed by legal regulations without considering the aspects of biodiversity or ecosystem functions. Landscape structures, instead of ecosystem functions, to date are probably the most prominent indices of some questionable degree of quality, which is accessible for politicians, and effectively determines their decisions.

4.7.2.7 Managing Resource Preservation and Restoration

Unlike biological conservation, resource restoration and management follow socio-economic and ecological necessities. When evaluating the sustainability of land use practices, i. e. knowing functional interactions within and between ecosystems in landscape sections, is essential. On the other hand, sustainability has quite different meanings, depending on whether priorities are economic, ecological or even political.

4.7.2.8 Involving Processes in Landscape Planning

At least in central Europe, the priorities of nature conservation and landscape planning are the preservation and restoration of threatened species, communities and selected structural landscape features. The management of populations is primarily based on the structure of landscapes and their elements, far less on their functional interactions. Detailed knowledge even of key processes concerning energy and nutrient budgets and flows, hydrology etc. is usually limited and in most cases includes generalizations instead of detailed case studies from the areas involved. Because it is not really possible to protect and manage all species as separate units (see, e. g. Tear et al. 1995), there has been an increase of attention for the physical environment as a prerequisite for perpetuation of phytocoenoses as parts of ecosystems.

4.7.3 Approaches of Theoretical Ecology for Wildlife Management and Landscape Planning

Most individuals involved in wildlife conservation and environmental impact studies are practical-minded and find it unnecessary to reflect much about the theoretical, ethical and even economic aspects of their work. However, with time they have learned to market their efforts, but often in a simplified and unsophisticated way. Like politicians they believe, that differentiated arguments are likely to be misunderstood and therefore neglected by 'the man in the street'.

Nowadays, some fashionable ecological terms are becoming popular and commonly used by conservationists: the habitat corridor concept derived from the island theory, the metapopulation concept, the call for genetic diversity etc.. The reason for this is perhaps political: scientists involved in such projects often expect to obtain more money, easier, by introducing new concepts and ideas, even if their general applicability and their limitations still need to be proved. Their risk seems to be a loss of reputation.

It could be that too much emphasis has been placed on this situation in order to demonstrate, that conservationists and planners need serious support from theoretically-working ecologists. I am sure that conservationists and planners will accept this assistance or cooperation, as soon as they see and understand the chances of practical application.

Theoretical ecosystem research attempts to define universally valid natural laws. One probable and useful contribution may be to find and to prove general concepts for characterizing buffer capacities and defining paramount parameters for system functions such as 'ecosystem integrity', 'exergy' or 'ascendancy' (Costanza 1992; Jørgensen 1997; Woodley et al. 1993) as well as interactions between different ecosystems in selected landscapes.

Modeling the main pathways of energy, water and nutrient flows and pools may contribute to a better understanding of the (eco)systems and for instance may allow to develop various scenarios characterizing the probable system development, for instance, land use changes. A search for decisive key factors may help to generate new and more precise questions in order to obtain clear answers with a high prognostic value for landscape planning.

These general rules should be applied in order to structurize and characterize phenomena of the real world object level for planning and management processes. This way, theoretical approaches and modeling force us to formulate and pursue distinct questions. However, general theoretical ideas for conservation have to be evaluated in regional and local contexts, e. g. for separate species or nature reserves (Hansson and Larsson 1997).

One problem is to find the appropriate time scale for extrapolations. Many ecologists believe that productivity may be one of the most pervasive influences on diversity. Tilman (1987) reports on the effects of nutrient enrichment in various successional stages from new fields to forests and points out, that increasing the productivity decreases the diversity. Rosenzweig (1997) on the other hand, advocating the empiricists scope, emphasized the risks of extrapolating the results of smaller time scales to larger ones and questions the results of Tilman from an evolution biologists time scale.

Future conservation research will not take place in isolated reservates only, but instead include the pattern of different sites and the interactions between them in whole landscapes. Beside the examination of key species assemblages it will be necessary to consider structures, processes and disturbance dynamics of ecosystems or landscape sections. Selecting 'representative' species and communities on the population and community ecology level, will perpetuate the key problem for indication and monitoring of important structures and processes of the habitats, and the search for appropriate orientors of emergent properties of ecosystems and their self-organizing processes, i. e. described in terms of energy and information content (Müller 1996; Jørgensen and Nielsen, this volume). A synthesis of these ecotargets on different system levels may allow a rational approach in order to obtain useful ecological quality objectives.

4.7.4 Economic Aspects of Wildlife Conservation and Landscape Planning

The conditions for nature conservation in an economic environment have been pointed out, among others, by Hampicke (1991).

The further development of important areas from a nature conservation point of view depends on the input of nutrients, the hydrodynamics and in many regions edge effects of 'conserved' islands surrounded by intensively cultivated areas. The development of useful nature conservation measures in cultural landscapes decisively depends on the sustainability of the agricultural and forest land use practiced. Van Latesteijn and Rabbinge (1994) attempted to model land use scenarios based on different agricultural policy decisions in the EC. Such scenarios may serve to illuminate the consequences of possible further development. Without discussing details of the underlying model, the preconditions imply, that the efficiency of input in the agricultural system will be maximized, leading to minimal losses of necessary inputs per unit of output. This may lead to a highly 'leakproof' agricultural system, which is self-sustainable. Alternative scenarios were used to explore options that materialize when policy aims are differently prioritized. Two scenarios are compared in Fig. 4.7.1:

- a free market scenario (agricultural cost minimized, free trade and new market balance, most cost-efficient types of land use), and an
- environmental protection scenario (like the former, but strict limitations are given regarding the use of fertilizers and pesticides; the agricultural production is fitted to self-sufficiency).

Comparisons like these open possibilities for strategic policy planning, the consequences for the land use and the landscape development on the one hand and conflicts to be expected concerning regional development and its realization on the other.

With this example in mind, it can be discussed, whether the sustainable use of biodiversity may be attractive from an economical point of view. The answer depends on the following factors:

Land use based on environmental protection (sustainable land use) may vary according to climate, soil conditions, topography, hydrology, infrastructure, nearness to market etc. Sustainable land use today fails to be able to compete, because conventional land use is subject to pervasive privilege, special fiscal treatment and distorted property rights.

There is no real market value for the benefits of sustainable use; prices for rare species, a high biodiversity or even ecosystem health do not exist today (e.g. Pearce and Moran 1994).

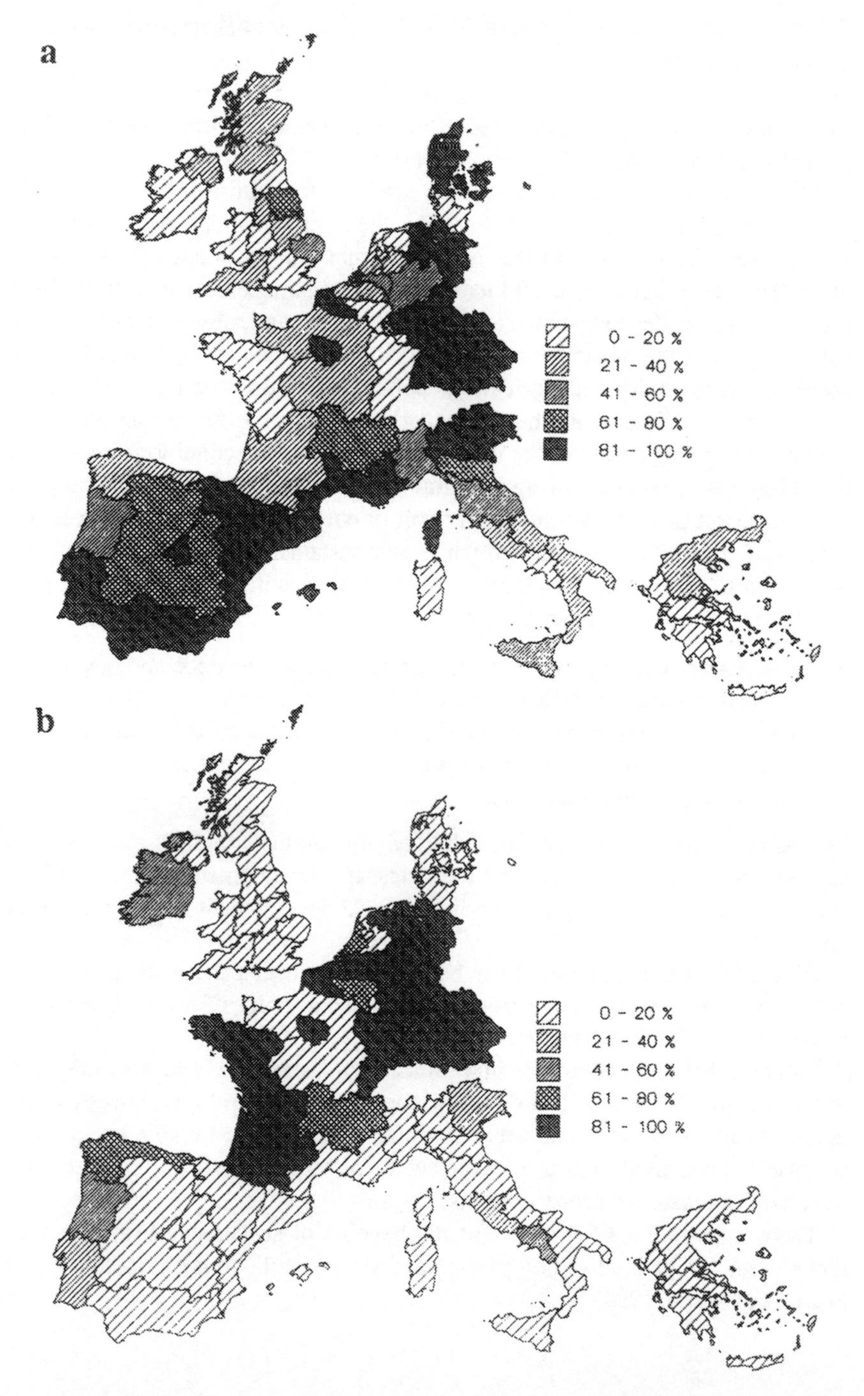

Fig. 4.7.1 Percentage of utilized agricultural area per region in two different scenarios (from Latesteijn and Rabbinge 1994); **a)** environmental protection scenario, **b)** free market scenario

4.7.5 Conclusions

All decisions concerning wildlife management, the conservation of biodiversity, sustainable land use and landscape planning are inevitably shaped by a mixture of ecological, economic, social and political circumstances (i. e. Frankel 1974, Frankel et al. 1995).

Conservation biologists will not accept those aspects of ecosystem management, which totally neglect the essentials of biological conservation. They are mainly interested in detailed information about individual site and community structures and conditions as well as their possible development.

Biology and ecology are highly diverse territories of science. Theorists risk to get lost in areas, which are without interest in ecosystem research application, while conservationists risk to fall in love with their objects or get entangled in unnecessary details and by this way lose their initial working concepts. Theoretical background and application in landscape planning should gradually be joined together in an iterative process in order to find a convincing synthesis between generalistic (reductionist) and specialist approaches.

Neither the consideration of population and community ecology nor the consideration of ecosystem functioning are sufficient to give predictions and recommendations for wildlife management and landscape planning. In order to develop ecological quality objectives for ecosystem types, a combination of biotic, abiotic and ecosystem orientors seems appropriate.

Scientists (involved in ecosystem research, nature conservation, economy) are specialized supervisors; politicians are moderators with high responsibility for mediating between divergent social groups (i.e. Ravetz 1986). The paramount decisions for landscape planning and development – at least on a regional scale – are neither made by scientists, nor by planners or politicians, but by the population in the area affected.

References

Angelstam P, Mikusinski G (1994) Woodpecker Assemblages in Natural and Managed Boreal and Hemiboreal Forests - A Review. Ann zool Fenn 31:157-172

Asker SE, Jerling N (1992) Apomixis in Plants. CRP-Press, Boca Raton, 298p

Costanza RT (1992) Toward an Operational Definition of Ecosystem Health. In: Costanza R, Norton BG and Haskell BD (eds) Ecosystem Health: New Goals for Environmental Management. Island Press, Washington DC Covelo California, 239p

Falk DA, Holsinger KE (eds) (1991) Genetics and Conservation of Rare Plants. Oxford University Press, 283 p

Frankel OH (1974) Genetic Conservation: Our Evolutionary Responsibility. Genetics 78:54 -65

Frankel OH, Brown AHD and Burdon JJ (1995) The Conservation of Plant Biodiversity. Cambridge Univ Press, 299 p

Goodman D (1975) The Theory of Diversity-Stability Relationships in Ecology. Quart Rev Biol 50:237-266

Govindaraju D (1988): Relationship Between Dispersal Ability and Levels of Gene Flow in Plants. Oikos 52:31-35

Gregorius H-R (1989) The Importance of Genetic Multiplicity for tolerance of Atmospheric Pollution. In: Scholz F, Gregorius H-R and Rudin D (eds) Genetic Effects of Air Pollutants in Forest Tree Populations. Springer, Berlin Heidelberg, pp 163-172

Grime JP (1979) Plant Strategies and Vegetation Processes. John Wiley & Sons, Chichester

Hampicke U (1991) Naturschutz-Ökonomie. UTB 1650, Ulmer, Stuttgart

Hansson L, Larsson TB (1997) Conservation of Boreal Environments: a completed research program and a new paradigm. Ecol Bull 46:9-15

Hodgkin T, Guarino L (1997) Ecogeographical Surveys: a Review. Bocconea 7:21-26

Huston M (1979) A General Hypothesis of Species Diversity. Americ Natural 113:81-101

Jørgensen SE, Nielsen SN (1997) Thermodynamic Orientors: Goal Functions and Environmental Indicators, this volume

Latesteijn van HC and Rabbinge R (1994) Sustainable Land Use in the EC: an Index of Possibilities. In: Van Lier HN, Jaarsma CF, Jurgens CR and de Buck AJ (eds) Sustainable land use planning,Elsevier, Amsterdam, pp 31-45

Matzke-Hajek G (1997) Zur Evolution und Ausbreitung apomiktischer Rubus-Arten (Rosaceae) in Offenland-Ökosystemen. Bull Geobot Inst ETH Zürich 63:33-44

Müller F (1996) Emergent Properties of Ecosystems - Consequences of Self-Organizing Processes? Senckenbergiana maritima 27:151-168

Müller-Starck G (1989) Untersuchungen über die Wirkungen von Immissionsbelastungen auf die genetischen Strukturen von Buchenpopulationen. Abschlußber Waldschäden/Luftverunreinigungen UBA - F+E Vorh. 10607046/23

Pearce D, Moran D (1994) The Economic Value of Biodiversity. IUCN, Earthscan Publ. Ltd, London, 172 p

Ratcliffe D (1977) A Nature Conservation Review I. Cambridge Univ press, Cambridge, 401p

Ravetz JR (1986) Usable Knowledge, Usable Ignorance: Incomplete Science with Policy Implications. In: Clark WC and Munn RE (eds) Sustainable Development of the Biosphere. Cambridge Univ Press, Cambridge, pp415-432

Rosenzweig ML (1997) Species Diversity in Space And Time. Cambridge Univ Press, Cambridge, p436

Tear TH, Scott JM, Hayward PH and Griffith B (1995) Recovery Plans and the Endangement Species Act: Are Critisms Supported by Data? Conserv Biol 9:182-195

Tigerstedt PMA (1997) Marginal Plant Populations – Species Ecogeographic Adaptations that need In Situ Conservation. Bocconea 7:121-123

Tilman D (1987) Secondary Succession and the Pattern of Plant Dominance Along Nitrogen Gradients. Ecol Monogr 57:189-214

Woodley S, Kay JJ and Francis G (eds) (1993) Ecological Integrity and the Management of Ecosystems. St. Lucie Press

WRR [Scientific Council for Government Policy] (1992) Grond voor keuzen; vier perspectieven voor de landelijke gebieden in de Europese Gemeenschapp. [Ground for choices; Four Scenarios for Rural Areas in the European Community]. Rapp Regering 42, Sdu Uitgeverij, 's-Gravenhage, p49

Ziegenhagen B, Llamas G¢mez L, Bergmann F, Braun H and Scholz F (1997) Protection of Genetic Variability in Polluted Stands. A Case Study with Silver Fir (*Abies alba*, Mill.) Bocconea 7:357-365

4.8 Conclusion: Sustainability as a Level of Integration for Diverging Targets?

Ernst-Walter Reiche and Maren Leupelt

Using the terms goal functions, targets or orientors we have to clarify their underlying concepts. The scientist means a universal trend of ecosystem development. The social sciences use the term to stress the intentional character of human actions, e.g. conscious aiming. The first Chapter of this volume as well as the third Chapter focus on this differentiation.

Marticke (Sect. 4.4) points out four different meanings of 'goal function' which were used during the workshop and in the course of this volume. Only if the exact sense is clear we can discuss parallels between a natural system's goal functions and the social system's targets. The original meaning of the term goal function – a given mathematical trend – originates in ecological modeling as a methodological part of theoretical ecology. Ruth and Bossel, both with the background of theoretical ecology, use the term referring to modeling. Ruth (Sect. 4.5) shows on the base of the exergy principle (typical tendency of ecosystems to optimize the relation between work and energy), that it may be conducive to adopt ecological goal functions to the conception of industrial production-processes. On the assumption of a simple dynamic model of extraction, production and waste assimilation he describes paths to couple economic and environmental growth and development.

Bossel (Sect. 4.2) deals with orientors and with the transferability of ecosystem principles to human systems.

Dierssen (Sect. 4.7), Jüdes (Sect. 4.3), Marticke (Sect. 4.4) and Radermacher (Sect. 4.6) – though their arguments differ in detail – introduce the term goal function in the meaning of target, as any object aimed at. Their contributions deal with the central aspects of managing environment: "Eingriffsregelung", landscape planning, environmental impact assessment and conservation of species and biotopes. In these papers the target concept (target also means: total which is desired to be reached) of sustainable development stands in the limelight of a scientific audience, although there is no consensus concerning the concept. May the paradigm of sustainable development really lead to a globally developing 'culture of sustainability' (see Jüdes Sect. 4.3). Or is it a 'weasel word' as Marticke explains, but necessary as a useful regulative idea that shows the direction of societal development, nevertheless?

F. Müller M. Leupelt (Eds.)
Eco Targets, Goal Functions, and Orientors

Referring to Sect. 4.1 relevant targets have developed as consequence of utilization interests of special lobbies on an object-scale. On this scale concrete planning and planning application exists. Therefore, the present environmental state is a result of the prevalence of concrete utilization interests. In this context also the social institutions, such as administration, policy and law, did specialize. Defining the special targets of nature conservation (conservation of species, guilds and communities, habitats and sites, specific patterns, abiotic resources and ecosystem functions) Dierssen (Sect. 4.7) stresses the values of special "protective goods". He focuses on their individuality which is observable on the local scale. Demands of general applicability – e.g. from theoretical ecology – cannot replace the empirically based and subjective work of nature conservation. He also stresses the fact that the multitude of approaches in nature conservation needs to be connected to gain a "convincing synthesis" between the different approaches.

Also Marticke (Sect. 4.4) focuses on the variety of targets and values (intergenerational equity, protection of natural bases for the existence of life, intrinsic value of biodiversity,...), in national and international law. From the perspective of environmental law he assesses that the definition process of ecological standards is concentrated on a few targets, e.g. health care and conservation of species and biotopes. There are hardly existing examples where environmental policy already did implement instruments of complex targets such as functioning of landscape or the partial targets of sustainable development. It shows that environmental policy has not been able to integrate complex targets up to now. However, the question still remains: In how far do the existing instruments (local scale - national scale) suit to the realization of targets that concern the global scale, a long time horizon and the welfare of future generations? Marticke utters with regard to the dilemma of representing future generations: "This is the crucial point in the debate on 'rights' of future generations and nature: who shall have the right to speak 'in their name'? Our responsibility for future generations and the biosphere is necessarily *paternalistic* or *parental.*" In this context Jüdes (Sect. 4.3) who suggests that the environmental crisis is the consequence of a cultural crisis defines four levels of structuring, understanding and deriving measures due to sustainable development (problem analytical-ethical level, epistemological-conceptual level, organizational theoretical level, realizational level). Particularly, for the third and fourth level, organization and realization, there is hardly a concrete indication.

A far-reaching consensus can be identified, suggesting that the application of ecological principles can only proceed in the framework of democratic regulatives and according to the principle "to learn from nature". The sustainable development approach tries to connect social, economic and ecological targets. Analyzing the contributions of this Chapter in comparison with Chap. 2 illustrates the difference between social and natural sciences. In the field of social science, scientists did learn to work with diverging approaches. Social scientists manage to integrate this knowledge while formulating application targets. Taking this into consideration the authors warn against the claim of exlusiveness with respect to the goal function-concept.

Bossel (Sect. 4.2) indicates that "human system development is constrained by the laws of nature, but it is not limited to the slow and 'blind' processes of natural evolution." The principle of 'trial and error' connected to the principle of natural selection such as described by the theory of evolution, cannot apply to human beings for the reasons of ethics.

Human beings – as well as social groups – are provided with far more capabilities (conscious planning, choice, innovation, invention, and implementation) than other beings. The consequence is a high degree of flexibility and adaptability. In the course of human development instruments did advance according to the technical know-how and particularly, according to facilities of communication. This refers to the information term of the emergy-principle (see Jørgensen Sect. 2.4). The present technical stage of development is characterized with the concept of "globalization". The global social adaptation – may be up to the "sustainable society" demanded by Jüdes (Sect. 4.3) – is still missing. Therefore, the goal function concept may be very helpful to derive an indicator system for social and economic adaptation to sustainable development.

In comparison to the introduction (Sect. 4.1) the outcome may lead to disappointment. There are no contributions in this fourth section that present a comprehensive solution, how to integrate ecological targets into the principles of human nature utilization as a consequence of individual interests. There seem to be no utilization categories (see Sect. 4.1) to take human living conditions into account as a function of global, regional and local ecological conditions. However, besides this impression of disappointment the reader should see as an advantage that the goal function discussion indicates the differences between ecological orientors and human targets in a comprehensive way.

Chapter 5 gives an overview, to what extent landscape planning already integrates the ideas, that are discussed in Section 3 and 4, into its principles and methods.

Chapter 5

The Practical Consequences: Eco Targets as Goal Functions in Environmental Management

5.1 Introduction: Orientors and Goal Functions for Environmental Planning – Questions and Outlines

Reinhard Zölitz-Möller and Sylvia Herrmann

In today´s practice of environmental management and nature protection a great variety of orientors are in use. Criteria for the question, what should be protected and which goal should aimed at in landscape management, have been evaluated e.g. by Margules and Usher (1981), Roweck (1994) and Usher and Erz (1994). When determining eco targets for landscape development, one obviously has to take into account results from ecological research; it would be nonsense to aim at goals which cannot be reached because of physical constraints. Nevertheless, orientors for management of nature in anthropogenic landscapes seem to be determined to a high extent by societies decisions, perhaps more than by natural science. Hence, the goals in use are manifold, and sometimes they are mixed in a fairly arbitrary way. On the other hand, some nature protectionists and landscape planners may follow their own leading idea of management and protection, only.

This situation in landscape planning and management is unsatisfactory. It was one of the reasons, why to ask ecosystem researchers during the Salzau workshop in 1996, if there is another, perhaps more theoretically based way to define goals or orientors for the management of ecosystems and landscapes. The main idea was: Why not to ask for trends and rules of ecosystem development, if we have to manage ecosystems, and not have to deal just with species and biotopes?

The main questions during the workshop discussion have been, and for this chapter are:

- Which orientors are discussed in ecosystem theory? Which indicators for a quantification are proposed?
- If holistic orientors and indicators can be detected from the study of natural ecosystems - how to make them fruitful in societies discussion and decisions?
- Are orientors of ecosystem development possibly too general and perhaps meaningless, or are they precise enough for becoming significant for environmental management?
- Is the concept of goal functions applicable also for managed systems and landscapes, or exclusively for natural systems?
- To what extent is a holistic protection strategy already supported by recent environmental law? To what extent will such a strategy be accepted in envi-

F. Müller M. Leupelt (Eds.)
Eco Targets, Goal Functions, and Orientors

ronmental planning? Is it really beneficial or perhaps even obstructive to the solving of practical problems?

The papers in Chap. 5 try to give answers to these questions. The answers are, to a certain extent, summarized in the conclusion (Sect. 5.11). Others have been agreed upon – some quite quickly – during the workshop discussions, and shall be explained here as an introduction:

As a first result of work group dissussion, one should avoid the term *goal* to describe tendencies or trends in the development of ecosystems, because of its teleological meaning. This topic was already touched upon in Chap. 3 (see Sect. 3.1).

The term *goal function* sensu strictu should be applied only in the context of dynamic modeling: It is a mathematical expression, which may be coupled to a model via an algorithm; it is minimized or maximized in order to improve the performance of models. Nielsen and Marques (Sect. 5.3) and Gnauck (Sect. 5.6) give examples of this terminological understanding. Nevertheless, several other authors still use the term *goal function* in a weaker sense, discussing about orientors and aims of management and protection strategies in this chapter.

It is shown in the first Sections of this book, that recent ecosystem research provides a basket full of orientors for the development of natural ecosystems. Also holistic indicators could be found for these "tendencies", if not to call them "propensities" (Ulanowicz Sect. 2.10). It has been discussed during the workshop, if these emergent properties usually or even as a law of nature are optimized during the development of ecosystems. In this context, ecosystem properties like exergy storage and degradation, connectivity and cycling index, minimum entropy formation and losses have been touched upon. But the consensus of the discussion was found just on a very low level: Quite general tendencies have been talked about as orientors in the sense of laws of nature (*"Living systems tend to move away from thermodynamic equilibrium", "Water flows downstream")*. It is agreed upon, that such general orientors of ecosystem development are not accepted as eco targets for landscape management: They are too fuzzy or simply trivial.

This may be one of the reasons, why scepticists in ecology tend to state: Ecosystem research may not be able – even not after another 100 years of intensive research – to answer the simple questions of nature protectionists and landscape managers (see Roweck 1995). On the other hand, there are planners (see Roch Sect. 5.8), who do ask very general questions: What is the carrying capacity for a region? What are the threshold values for loads of landscapes? Sometimes they even really do expect quick answers from ecologists.

Fortunately, in the middle between these two extreme attitudes there is also some realism to be found: Its courageous representatives take the great risk of proposing practical solutions, e.g. for an operationalization of the sustainability concept (see Werner Sect. 5.10), which introduces the ecosystem view into the concepts of planners.

References

Margules CR and Usher MB (1981) Criteria used in assessing wildlife conservation potential: a review. Biological Conservation 21:79-109

Roweck H (1995) Landschaftsentwicklung über Leitbilder? Kritische Gedanken zur Suche nach Leitbildern für die Kulturlandschaft von morgen. LÖBF-Mitteilungen 4:25-34

Usher MB and Erz W (eds) (1994) Erfassen und Bewerten im Naturschutz: Probleme, Methoden, Beispiele. Quelle & Meyer, Heidelberg 1981, 340 pp

5.2 Quantifying the Interaction of Human and the Ecosphere: The Sustainable Process Index as a Measure for Co-existence

Christian Krotscheck

Abstract

It is argued that the most effective way to couple the evolution of humans and ecosphere sustainably is achieved by measuring the interaction of anthroposphere and ecosphere. Interactions link economy and ecology. Interactions are the bases of co-existence and are manifested in a very complex mass and energy flow network.

Based on mass and energy exchanges, the concept of the Sustainable Process Index (SPI) measures the potential impact (pressure) of anthropogenic activities on the ecosphere. The basic unit of SPI is area. It is the total area that is required to maintain the resources (e.g. food, water, minerals) as well as waste disposal demands (e.g. NOx emissions, waste water, by-products) per person, per service or per region (or city).

Due to the complex flow networks in the ecosphere the valuation of man-made effects on certain ecosystems is impossible. The chain of cause and effects vanishes. Thus, in contrast to any exergy-valuation-model, the SPI utilizes easy to access natural references in the valuation procedure, such as yields and concentrations. Even though the area in the SPI-model does not represent an existing field or forest, the pressure of human actions is sufficiently represented to provide a flexible and sensitive instrument for regional, project or individual planning.

5.2.1 Introduction

Ten years after the Brundtland-Report (WCED 1987) established sustainability as a moral imperative for politics, the concept of sustainable development has become prominent in scientific research. One of the main obstacles to operationalize a moral imperative, in practical terms for decision makers, is the prob-

F. Müller M. Leupelt (Eds.)
Eco Targets, Goal Functions, and Orientors

lem of measurement: how to distinguish sustainable from unsustainable activities.

Although the concept of sustainability also includes socio-economic development (e.g. Swaminathan and Saleth, 1992; Daly, 1991) this section will focus on eco-sustainability. Eco-sustainability is defined as the co-evolution of man and environment and thus, as the basis for co-existence of anthroposphere and ecosphere. This issue is (for cognitive reasons) separated from the co-evolution of man and man, socio-economic development, and from the co-evolution of ecosystem and ecosystem, ecological development. In contrast to socio-economic and ecological development, eco-sustainability describes the co-existence of two systems of different structures, different spatial and temporal dimensions. Therefore, the valuation of eco-sustainability is a priori a very delicate issue that needs to be addressed by a pragmatic approach of precaution (model of pressures on environment; see below).

As a corollary, a measure can only account for what it was designed to account for (e.g. temperature cannot be measured by watches). Indices for eco-compatibility are useless for purely economic decisions, issues concerning the socio-sphere or ecosystems development.

5.2.2 The Sustainable Process Index (SPI) as a Concept to Evaluate Eco-Sustainability

The Sustainable Process Index consists of a number of models to value the pressure of anthropogenic activities on the environment based on a common unit: area. It reflects a certain way to see and to evaluate interactions between humans and their environment. These interactions are perceived within artificially constructed borders, where the 'trade' of man and ecosphere can be balanced. The sustainability concept implies that human pressure on the ecosphere must be compatible such that the base for development is not endangered (see Bossel: the principle of partnership, this volume). The SPI is an efficient instrument to allocate the scarce resource 'area'.

Within the classification of the environmental space (Weterings and Opschoor 1992) the SPI clearly represents a pressure indicator. Any pressure indicator faces two challenges: *On the one hand it must be relevant within the system of human activities (the anthroposphere).* It does not make sense to gauge a property that may either not be influenced by human decisions or that is irrelevant to such decisions (socio-political dimension). *On the other hand a pressure indicator must evaluate a property that is relevant to the development of the system of natural processes (ecological dimension).* After all, it is the pressure on these natural processes that should be measured by such an indicator.

It is argued that mass and energy flows that are exchanged between anthroposphere and ecosphere are highly relevant 'pressures' on the environment and have a socio-political planning dimension as well as an ecological reference dimension (value; see Sect. 5.2.5). Although mass and energy flows are by no means the only influences exerted by man on the environment, they represent a very important class of environmental pressures. Almost all natural processes on our planet involve the exchange of mass and energy flows. Natural processes can therefore be seen as a complex network of mass flows that uses solar energy as the driving force.

Processes within the anthroposphere are also to a great extent defined by the exchange of mass and energy flows. They form a comparable network of mass flows, although of less complexity. These two networks of mass and energy flows are in constant contact and exchange. It is very important in this regard to note that these mass flows do not only exert an influence at the 'point' (the knot in the net) where they enter the other system. They cause a much more complex reaction, since all the knots within one mass flow system are interrelated by the flows they exchange *(also see B.C. Patten: ecosystems property).* Consequently mass flows exchanged with the ecosphere will eventually be responsible for a complex change in the way natural processes operate and develop. They therefore exert pressures on the environment that cannot be overestimated in their importance.

Mass flows are also at the heart of every human activity. They are essential elements in the system of industrial, or more generally societal, metabolism, that uses raw material extracted from the ecosphere, to provide services and products for human use and disposes of those flows that are no longer useful. The flows humans exchange with the ecosphere are defined by technologies and organizational systems. Therefore they depend directly on decisions made within the anthroposphere. As such, these mass flows are measurable and controlled by human decisions.

The factor 'area' has been selected as the basic unit for the computation of the SPI. This is based on the assumption that in a sustainable global system the only real input that counts in the long-term is solar energy. Its utilization per se is bound to the earth's surface. Furthermore, 'area' is a limited resource.

In the SPI concept (Krotscheck and Narodoslawsky 1996), area is not only seen as a recipient of solar radiation. All life and ecosystem's integrity are dependent on the quality of environmental compartments (most notably top soil, water and air) that are attached to the surface. So any given area of the planet's surface is not only defined by the amount of solar energy received, but also by the quality of the environmental compartments attached to it.

By taking into account the dual function of area, as both a recipient of solar energy and as a sustaining basis for ecosystems, the SPI measures the ecological pressure of an anthropogenic activity with respect to the quantity and the quality of the energy and mass flows it induces. Thus, activities needing more area for the same benefit are not suitable for sustainable development and may endanger the existence of ecosphere and anthroposphere.

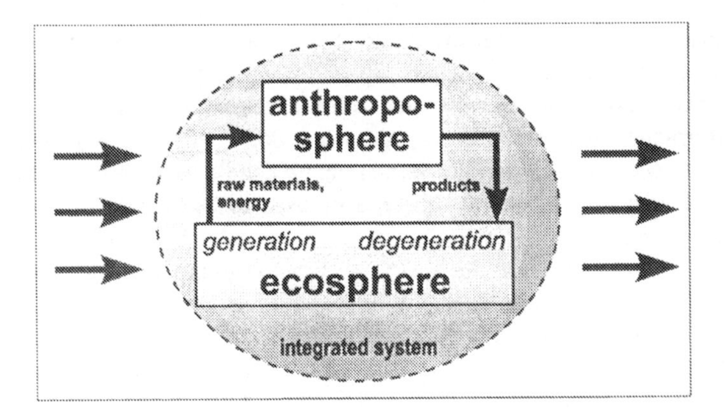

Fig. 5.2.1. Anthroposphere and ecosphere as co-evolving systems powered by solar energy

In coherence with the socio-economic understanding of 'process', the SPI concept includes models to transform the use of rawmaterials, energy, infrastructure, staff and the disposal or degeneration of products (including any emission or by-product) into the need for area. This characterization of a process can be extended to the total life-cycle of a product or to a region's economy. It describes the metabolism on the individual (e.g. Blum et al. 1995), the product/process (e.g. EC, 1996, Deistler et al., 1996) or the regional/global (e.g. Eder and Narodoslawsky 1996) scales (Fig. 5.2.1). Thus, the scale of application is a question of the target of the observation. There are different data-bases and models in each context, but the SPI models are flexibly designed to cope with different scales using the same principles. These principles have to be discussed to demystify the background of the SPI.

The first method to value raw materials is split into three categories: renewable raw material, fossil (organic) materials and non-renewable materials (e.g. minerals, ores). Fossil materials are seen as slowly developing renewable raw materials. The yield of the renewable resource directly relates a resource flow to an area (for fossil organic materials, the yield is about 0.002 kg m^{-2} yr^{-1}, referring to the sedimentation rate in the oceans). For non-renewable materials the definition of a yield is impossible. Here the pressure on the environment is mainly exerted by mining, extraction and refining. All these activities require a lot of energy and infrastructure, which is translated as pressure. Moreover, non-renewable materials induce significant pressures at the output side of the anthroposphere (see below).

The second principle is dedicated to energy supply: the utilization of energy is either done by direct conversion of solar energy (photovoltaics, solar collectors) or indirectly by burning energy vectors (natural gas, biomass, water, wind) which in turn have a rate of regeneration. Both, direct and indirect utilization of solar energy is bound to area and therefore can be aggregated in the SPI.

Analogous to economic valuation strategies, stocks such as infrastructure, equipment and staff, are calculated separately, although they also represent a kind of mass and energy network, yet with a different temporal scale. The area for infrastructure and personnel usually do not dominate the SPI. If no detailed analysis is necessary, the area for infrastructure can be easily represented in the SPI as the area of grey energy (the energy used to build infrastructure in the construction industry).

Last, but by far not least, the area to accommodate products and by-products (e.g. waste) should be discussed. Within the SPI, this area seems to be very important, considering that we are facing a waste disposal crunch rather than a resource depletion problem. Moreover, models for the input side to anthroposphere are rather commonly used in agriculture, forestry and fishery and need no detailed description.

5.2.3 Product Dissipation: the Problem of Operationalising the Degeneration Potential of Ecosphere

In the SPI concept the area for product degeneration is called the 'dissipation area'. It is discussed in detail for the environmental compartment of (continental) hydrosphere. The two basic principles behind the dissipation models are (as adapted from SUSTAIN 1994):

1. Anthropogenic pressures must not exceed local assimilation capacity.
2. Anthropogenic pressures must not alter quality and quantity of global cycles.

The keywords for these principles are: adapt locally and integrate globally. The local assimilation capacity is a measure of the rate with which ecosystems accept input and output streams without loosing their evolutionary potential. The global integration adds concerns of global cycles and the depletion of natural storage systems (e.g. fossil fuels as carbon storage), whose impacts on evolution are rather unknown. This requirement links the rate of exploitation to the rate of replenishment of these natural systems.

The quantification of these goals is done by relating man-made flows to natural flows. *It is argued that ecosystems are used to natural flows and their composition (e.g. geogenic flows, precipitation) and thus are able to accommodate such flows in their complex flow-network.* By combining rates of renewal of eco-compartments and local reference values for compositions, the spatial dimension of human activities to local ecosystems can be assessed. Global integration compares global cycles and the depletion/refill of storage systems with man-made flows to or from these cycles on a global scale. To exemplify these very abstract concepts, the quantification of anthropogenic emissions is discussed here.

Various case studies have shown that the flows *from* anthroposphere *to* ecosphere (e.g. waste, emissions, fertilizer) are especially crucial pressures of current activities. In the SPI concept, any stream leaving the anthroposphere is considered to be a product stream. It is furthermore supposed that these products are eventually dispersed into the environment. Their quantity and quality ought then to be considered as pressures to the ecosphere.

In order to estimate the area allocated to dissipation, the following reasoning is applied: *If there is a rate at which the content of a given environmental compartment (soil, water, air) is 'renewed', any product stream can be 'diluted' by the newly added mass, until this mass reaches qualities (e.g. concentrations) that are equal to the quality of the initial compartment.* Note that this is not a recipe to dispose wastes (moreover it would be economic insanity) but a principle to assess anthropogenic flows in accordance with ecological 'values'. It is neither a critical (acceptable) load that is described here, nor a limit-threshold value. Due to incomplete knowledge and synergy, critical loads are in principle undefineable.

Here, although the SPI has included the compartments 'soil' and 'air', this reasoning is best exemplified for 'substance release' to the compartment 'water' (i.e. the continental hydrosphere). In contrast to the definition of yields (input side of the economy), dissipation is linked to sinks in the SPI concept. These sinks (s) are described by the rate of renewal (R) and the current concentration of the dissipated substance 'm' in the compartment water.

$$s_m = R \cdot c_m \qquad kg_m\ m^{-2}\ yr^{-1} \tag{5.2.1}$$

The dissipation area can be calculated using the rate of renewal R in kg m^{-2} yr^{-1} of the environmental compartment 'water', the actual concentration of the substance c_m kg_m/kg in the compartment and the product flow $F_{P,m}$ kg_m/yr of the substance 'm' to this compartment. The area for dissipating a single component to the compartment 'water' is $A_{P,m}$:

$$A_{P,m} = \frac{F_{P,m}}{s_m} \qquad m^2 \tag{5.2.2}$$

In order to compare man-made and natural flows one must know the rate of renewal of the eco-compartment continental hydrosphere and the actual concentration of different components (e.g. heavy metals, sulfur, chloride etc.) in this compartment. The rate of renewal of the compartment water is calculated via precipitation P in kg m^{-2} yr^{-1} and the rate of water seeping through the ground/ precipitation 'r'.

$$R = P * r \qquad kg\ m^{-2}\ yr^{-1} \tag{5.2.3}$$

E.g. in Austria a (mean national) precipitation of 1200 kg m^{-2} yr^{-1} and a seeping ratio of about 30 % accounts for a rate of renewal of 360 kg m^{-2} yr^{-1} for compartment water. Assume that there is a concentration of 0.005 [mg] of cadmium per liter of ground water and we dissipate one kilogram of cadmium in Austria per year: an area of 0.56 km² (=1/360/0.005) for 'sustainable' dissipation is appropriated. Sustainable dissipation means that a natural, geogenic flow of e.g. 1 kg cadmium from volcanic activity is degenerated on the same area and finally leaves with the ground water to the sea (sediment).

5.2.4 The Concept of Product Dissipation – an International Comparison

By comparing the model to other internationally discussed quantitative measures for sustainable qualities, the reasons for the SPI's dissipation area for the quantification of anthropogenic output become clearer. Table 5.2.1 summarizes internationally known and highly aggregated valuation approaches which claim to assess pressures induced on the ecosphere. In addition to each concept's inventors, the basic dimension (unit of aggregation), the types of flows that can be aggregated and the reference of each approach are also stated. Obviously the dimension m² is the most prominent. Different flows induced on the ecosphere by man were characterized in input and output flows: only SPI is able to value both input and outputs.

Table 5.2.1. Characteristics of the different valuation strategies for eco-sustainability

concept[+)]	*unit*	*flow*	*reference*	*inventor (nationality)*
LCA	m²	output	technological, social	Hofstetter (Müller-Weng) (CH)
MIPS	kg	input	-	Schmidt-Bleek (GER)
SPI	m²	input, output	ecological	Krotscheck/ Narodoslawsky (AUT)
ACC	m²	input	agricultural	Rees/ Wackernagel (CAN)
WPE	$J\,kg^{-1}\,K^{-1}$	output	thermodynamic	Ayres (Prigogine) (F)
PCC	EURO	output	technological, social	Krozer (NL)

+) see below for short description of concepts

Another important aspect is the source from which the measures take their reference; (i.e. where do they derive their values or standards from?) It may be stated that only an ecological reference can guide towards eco-sustainability. Social standards (emission-threshold values, factor 4, reduction aims) are time dependent and shape ecosphere according to human values (see below). A short description follows to clarify the scale and possibilities of these measures in Fig. 5.2.2, (for a more detailed description, comparative case studies and a critical review, see Krotscheck 1997).

The Material Input per Unit-Service (MIPS). The Material Input per Unit-Service (MIPS) developed by Schmidt-Bleek (1993) lists the masses moved to produce a certain good regardless of the quality of these mass flows. One of the most distinguishing features of the MIPS is that it relates all impacts to one dimension (mass flows). Thus it gives a clear ranking of different competing activities and relates them to an aim of sustainability, namely, the reduction of mass flows and its pressure on the ecosphere. MIPS needs no reference value because it is argued that the more efficient a service can be reaped the better (the fewer kg of mass flow per service).

The Life-Cycle Balance Approach of Swiss BUWAL (LCA). The idea of this method (Müller-Weng et al. 1990) is to list all toxicologically (and to some extent ecologically) relevant flows caused by a process and to value their respective impacts on the environment (or on health). The pressure onto the ecosphere is described by 'critical volumes': The critical volume is the amount of a compartment (e.g. air) that is needed to dilute a (toxic) mass flow. The quality of the critical volume is 'allowed' by limit-threshold values (e.g. emissions). Hofstetter (1991) argued that the critical volumes can be aggregated on the base of area (per square meter of surface). Through aggregation to one unit, decisions concerning a shift of critical volumes between the compartments become possible.

The Appropriated Carrying Capacity (ACC). The basic idea of ACC (Rees and Wackernagel 1996), similar to the SPI concept, is that despite increasing technological sophistication, humankind remains in a state of obligate dependency on the utilities of ecosphere. The ACC therefore is the maximum persistently supportable load on the ecosphere to sustain development. The appropriated carrying capacity is proposed as the fundamental base for demographic accounting. The concept initially was designed to make the human 'footprint on ecosphere' of regions visible to their inhabitants. ACC, thus, is the area that a region uses to run its economy on the base of agriculture, forestry and fishery (renewable resources).

Waste Potential Entropy (WPE). The WPE is based on three ideas (Ayres and Martinás 1994):

1. all flows, from which a service for mankind can be reaped in the economy, can be characterized as flows of 'useful' embodied information;
2. the economy itself is an information processor;
3. the most general impact of pollution is the physical information lost in waste.

It is argued that the emission of material fluxes to the environment are more dangerous, the larger the difference is between the state of pollutants and the chemical equilibrium of the local environment.

Thus, entropy/exergy draws its reference from thermodynamics. Entropy/exergy has the advantage that it is a thermodynamic property that is derived from natural sciences. WPE measures the pressure exerted by a mass flow in relation to the quality of that flow and the standardized environment. In a way, entropy measures the potential change in the (chemical) driving forces of natural processes that may be caused by the mass flows which man emits to ecosphere.

Pollution Control Costs (PCC). The PCC concept (Krozer, 1995) tries to value material flows (output flows from the anthroposphere to the environment) via technological and structural standards in monetary units. PCC shows the costs for control measures (e.g. pollution control technology) to attain a specific environmental quality (e.g. carrying capacity). The underlying assumption is that the co-evolution of man and ecosphere requires efforts to prevent the degradation by residuals.

Note on the Material-Flow Based Entropy/Exergy-Concept. The sound theoretical base of entropy has a distinct advantage. It allows for a proven and tested method of calculations. Since all processes, natural and technological, are guided by thermodynamics, it makes these processes comparable and opens the discussion to rational scientific arguments.

But there are inherent problems this scientific concept must address: The first problem is the definition of the appropriate standardized reference values for the environment. Moreover, the three environmental compartments (or the nearly infinite number of ecosystems) are not in thermodynamic equilibrium and there are innumerable interactions between them (e.g. carbon, nitrogen cycle). Thus a consistent and realistic definition of the state of the 'standard-environment' is impossible.

Moreover, the natural systems (ecosphere) usually are far from the state of chemical equilibrium. Nature lives in a constant process of flows, reactions and interactions that are kinetically controlled. Many pressures of pollutants on the ecosphere are caused by accumulation of substances (e.g. highly chlorinated compounds), by catalytic processes (e.g. ozone depletion by CFC's), the effect of heavy metals on organic processes (e.g. cadmium, mercury) and by over-loading natural cycles (e.g. carbon dioxide in the atmosphere). All these pres-sures cannot be modeled with entropy. Thus, this scientific valuation concept might suggest safety where there is none. Three examples from Krotscheck (1997) emphasize this statement:

- WPE values the impact of ubiquitous substances relatively high. According to WPE, the impact of 52 kg of water is equivalent to 1 kg of carbon tetrachloride.

- WPE ranks carbon dioxide as a rather harmless substance. One kilogram of emitted CO_2 is valued equal to 0.12 kilogram of dissipated water.
- Catalysts are rather inert substances that stimulate other reactions (in the environment) by orders of magnitude. A high load of cadmium will change the behavior of natural systems considerably. This effect cannot be valued with WPE (the dissipation of 48 kg of water has an equivalent exergy loss as 1 kg of cadmium).

Therefore, it is doubtful whether the indirect valuation via entropy/exergy provides greater clarity in contrast to the more simple valuation via reference states alone (SPI).

5.2.5 Discussion of the SPI Dissipation-Concepts

The valuation of eco-sustainablity resembles a walk on the tightrope: on the one hand one must monitor carrying capacity of ecosystems to avoid catastrophic developments, and on the other hand one must know which human activity causes which effect in ecosphere, to enable policy interventions (actions) to avoid damage. How to deal with this problem?

In society this need for action may be accompanied by other changes (e.g. in income, environmental quality, security etc.). The knowledge of 'what matters' creates socio-political goals (e.g. sustainable development, if the state of environment or the state of the social system is insupportable). Notably any decision maker has to rely on data and their interpretation (valuation) of it. To reach a goal we therefore have to look for possible paths of development and try to figure out which effects a certain path causes in respect to that chosen goal. Mass and energy flows cause very important effects in the case of eco-sustainability. However, mass flows contain many different substances that cause very different effects in ecosphere. In addition, we have only limited knowledge of these effects or of synergy between various substances. Since there is such great variety of possible influences we need a normative, aggregated value that can relate possible influences: a value-basis to make effects comparable (e.g. money, SPI) in order to facilitate rational (policy) decisions.

Social discussion is an efficient tool to define preferences and to decide among varying, dynamic (and often contradictory) development pathways. Only intensive societal discussion within a democratic process can guarantee that a certain pathway will be implemented or accepted and progress towards a goal will be appropriately monitored. It is also a necessary precondition to orientate goals (and development pathways) to the preferences and requirements of the people involved.

However, the means of democratic societal discussion are ill-equipped to collect data on environmental pressures or to describe different pathways in an inter-subjective manner. Yet, this process is a necessary precondition for any societal decision making. Contrary to a democratic decision process, scientific discussion is well-equipped for this task (whereas science is unable to define social preferences). It therefore is the responsibility of scientific analyses to provide a relevant database (including a framework for valuation) as a foundation for societal discussion.

In this process, science cannot, and should not, 'make development decisions' nor can it 'tell the truth', given the inherently limited character of knowledge about environmental impacts of human activities. Science is only required to provide an inter-subjective description of the 'landscape' generated by a societal goal function and of the 'shape' of the analyzed pathways within this landscape. This simply combines two different kinds of human discourse in a way that is efficient to reach decisions.

This decision making process is especially important for ecological decisions, respecting landscape planning or regional (ecological) management. Scientific knowledge is implemented as a database as well as in aggregation concepts. Consensus in the scientific community (see de Jong), though not crucial, would strengthen the base for societal decision making.

The bottom line is the need for a scientific decision base with the least possible uncertainty. In the context of eco-sustainability, a principle characteristic of ecosystems and their processes is their unpredictability. The complicated flow structures and the complex networks are indescribable by science, at least not in the precise, mechanistic way we are used to. These complex dynamic systems show chaotic behavior: small changes might have large effects as a consequence. However, these indescribable and unpredictable aspects of ecosystems are not yet recognized by mechanistic hard-liners. But one can recognize a paradigm change occurring in this respect. For example Turner (1988) defined the deep-ecology approach by stating that ecosystems' behavior is unpredictable. Also, according to the system analyses of B.C. Patten, these natural networks cannot be really understood or measured. Thus, the precaution principle seems to be the only possible sustainable path for society: the human pressure on ecosphere has to be adapted to ecosystems in an appropriate way.

Let us now consider the formidable task for science to find or create an inter-subjective description or an aggregateable measure for eco-sustainability. In Fig. 5.2.2 the 'grid of uncertainty' shows where we should locate measures for eco-sustainability to avoid scientific pitfalls. On the abscissa we put possible databases (e.g. inventory analyses) and on the ordinate possible values to aggregate the database to one measure (e.g. inventory assessment using standards).

Looking at the axis there is fair knowledge of the input to processes. This knowledge is less precise for emissions and immissions (ground level concentration of pollutants in a certain region) from processes. Immissions are measurable with complicated, expensive technology. However, there is seldom a clear cut model that links individual emissions to measured immissions, with sufficient rigor.

The next step into detail is the question of the real impact. In recognition of the fact that all ecosystems are (more or less) interlinked, and that it is often the indirect conjunctions that dominate effects, science is locked in complexity (according to chaos-theory even small flows might end in big effects leaving open the question of where to draw the line for analyses). When it comes down to effects (the changes that occur in the development of ecosystem due to human pressures) science is unable to give any definitive answer about how causes and effects are connected. Even if we know the impact on environment, (e.g. the fading of lichen variety as bio-indication for human pressures) we often do not know the cause of the problem (the actual human activity involved) and which effects that source of pressure will have on the evolution of ecosystems.

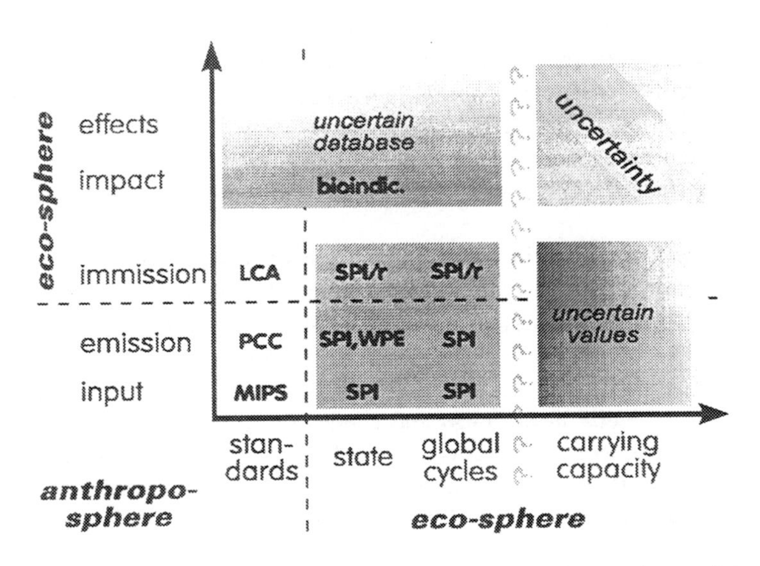

Fig. 5.2.2. Valuation-concepts in the grid of diminishing certainty (concerning results)

On the ordinate is the scale of how to value phenomena caused by anthroposphere. It questions of how much importance should be attributed to a single phenomenon. The values usually used are standards with a very certain value base which are defined by social discussion. Standards are fixed values for particular effects; e.g. the emission of cadmium is valued (weighted) with a defined number (e.g. emission limit-threshold value). Other, more uncertain ways to value effects, are to use environmental states or global cycles (see Sect. 5.2.3). It is assumed that ecosystems are used to current (natural) states (e.g. concentrations, flow rates) and that they can evolve on this basis.

The last valuation steps on the certainty hierarchy are the (local) flow networks (here we have to know about the quantity and quality of flows, that are exchanged between the system's constituent agents) and their carrying capacities. The carrying capacity would be the most appropriate value. However, it would be time and resource consuming to estimate carrying capacity to a useful precision for ecosystems that may have changed in the meantime. Furthermore, science faces the same problems in accounting for effects: the strong interconnections of systems and the complexity of the problem are causing unpredictable feedback and in turn make the value-base incorrect.

It is agreed that decisions have to be based on relatively certain facts. As carrying capacity, impacts and effects are fuzzy, measures have to rely on combining inputs, emissions or immissions with standards, states or global cycle values. The most obvious and accurate measure would be input/output of anthroposphere valued by standards (all well defined and accessible data). This would lead to the most secure decision base. However, another important aspect has to be considered besides the socio-economic dimension. We want to measure eco-sustainability – the pressure anthroposphere exerts on ecosphere. However, standards, brought about by social consensus (a rather anthropogenic quality), inherently cannot assess co-existence of two dissimilar systems. Measures based on this assumption are dependent on the social awareness of problems, scientific findings and the state of technology. As a corollary the ecological dimension has to be integrated.

Of the remaining space in Fig. 5.2.2; the shaded square shows the most potential location for indices, even if accuracy for values is declining. Within this square fair results for measuring eco-sustainability can be achieved. Within this square the SPI (SPI/r stands for a regionalized SPI) and, with some restrictions the exergy-concepts are to be found. The SPI is bridging anthroposphere and ecosphere on common bases: mass and energy flows. Mass and energy flows have both socio-economic (political) and ecological relevance. This results in a measure for co-existence on a macroscopic and easy to understand unit: area. Area is limited and a useful controlling instrument. Any effective instrument has to rely on a shortage of resources, since competitiveness is linked (almost by definition) to the shortage of a resources.

5.2.6 Conclusion

In the SPI concept (easy to access) ecosystem properties are used to link anthropogenic material and energy flows to the natural flow-scales. Anthropogenic flows are, thus, valued by natural properties and aggregated on the base of area. These ecosystem properties are on the one hand material and energy flows, among them biomass yields, transpiration and respiration. On the other hand serve material cycles as value-bases, e.g. the global cycle and storage systems of carbon, sulfur, nitrogen and water (e.g. the precipitation in combination with concentrations is used as simple measures for the diversity of flows or local diversity).

The weighing factors in the SPI dissipation-concepts do not reflect any known or proven impact or effect on the environment. They reflect a pressure on the environment against the value of (local) states and global cycles (ecological properties). However, this potential pressure is calculated with the direct reference to ecosystems, allowing the estimation of the pressure on the environment with respect to measurable qualities in the complex flow network.

The dissipation concept, values the (theoretical) area that is used for degeneration according to flows that ecosystems normally accommodate. In this way the regional characteristics of ecosphere are sustained. The concept allows the inclusion of the multiple use of area only, if the approach is applied to integrated local activities (regional planning). The SPI does also mirror the competition of different activities and, at least at an order of magnitude approach, the competition for land use.

In addition to the direct use of land (streets, buildings, arable land etc.), land use encompasses the indirect use of land (disperse pollution: emissions to all eco-compartments) which was experienced as the most important pressure. Thus, with the aim of land use management the indirect effects of measures for reorganizing area have to be considered as well, or even be given greater priority.

With respect to the co-evolution (co-existence) of man and ecosphere, the dissipation area as a precautionary concept is the best possible assessment from a scientific and political point of view. The reason for an applied precautionary principle is the inherently limited knowledge on ecosystemic processes, carrying capacity and on the effects on evolution. Due to system-theoretical considerations this lack of knowledge is a definitive fact and will not be overcome even by excessive research.

References

Ayres RU, Martinàs K (1996) Waste Potential Entropy - the ultimate ecotoxic? Centre for the Management of Environmental Resources, INSEAD, Fontainbleau, France

Blum WEH et al. (1995) Tallandwirtschaft Vorarlberg Endbericht. (Projekt i.A. der Vorarlberger Landesregierung und des Bundesministeriums für Umw, Jugend und Familie, Wien)

Deistler M, Krotscheck C, Burgholzer P (1996) Massenflußorientierte Kennzahlen zur Objektivierung des ÖKO-Audit. (Projekt im Auftrag des Ministeriums für wirtschaftliche Angelegenheiten, Wien)

European Community: Project No EV-5V-CT94-0374 (1996) Operational Indicators for Progress towards Sustainability. (Coordination: Krozer J, TME), European Commission, DG XII

Eder P, Narodoslawsky M (1996) Combined application of descriptive indicators of environmental pressure and input-output models on the evaluation of regional sustainability. (Proceedings of the Forth Biennial Meeting of the International Society for Ecological Economics, August 4-7, 1996)

Hofstetter P (1991) Bewertungsmodelle für Ökobilanzen. In: NBT-Solararchitektur, ETH Zürich (ed) Energie und Schadstoffbilanzen im Bauwesen. (Beiträge zur Tagung vom 7.3.1991, ETH Zürich, Zürich)

Krotscheck C (1997) How to Measure Sustainability? Comparison of flow based (mass and/or energy) highly aggregated indicators for eco-compatibility. EnvironMetrics, John Wiley and Sons, UK, forthcoming

Krotscheck C, Narodoslawsky M (1996) The Sustainable Process Index - A new Dimension in Ecological Evaluation. Ecological Engineering 6 (4):241-258

Krozer J (1995) Note on environmental-economic indicators for sustainability. Inst for Applied Environmental Economics (TME), The Hague, Netherlands

Müller-Wenk R, Abel S, Braunschweig A (1990) Methodik für Ökobilanzen auf der Basis ökologischer Optimierung. Bundesamt für Umwelt, Wald und Landschaft: Schriftenreihe Umwelt, vol. 133, Bern, Schweiz

Rees WE, Wackernagel M (1996) Our Ecological Footprint - Reducing Human Impact on the Earth. New Society Publishers, Philadelphia

Schmidt-Bleek F (1993) MIPS - A Universal Ecological Measure. Fresenius Environmental Bulletin 8:407-412

SUSTAIN (1994) Forschungs- und Entwicklungsbedarf für den Übergang zu einer nachhaltigen Wirtschaftsweise in Österreich. (Verein zur Koordination von Forschung über Nachhaltigkeit. Projekt im Auftrag der BBK, des BMWF und des BMUJF, Institut für Verfahrenstechnik der Technischen Universität Graz, Graz)

Swaminathan MS, Saleth RM (1992) Sustainable Livelihood Security Index: A Litmus Test for Sustainable Development. (Paper presented at the UNU Conference on the Concept and Measurement of Sustainability: The Biophysical Foundation. World Bank, Washington, DC, June 22-25 1992)

Turner RK (1988) Sustainability, resource conservation and pollution control: An Overview. In Turners RK (ed) Sustainable environmental management; Principles and practice. Belhaven Press, London

Weterings RAPM, Opschoor JB (1992) De milieugebruiksruimte als uitdaging voor technologieontwikkeling. RMNO, Rijkswijk, The Netherlands

World Commission on Environment and Development (WCED) (1987) Our Common Future. Oxford University Press, Oxford

5.3 Applying Thermodynamic Orientors: The Use of Exergy as an Indicator in Environmental Management

João C. Marques, Søren N. Nielsen and Sven E. Jørgensen

Abstract

The importance of thermodynamic laws at various hierarchical levels of living systems is discussed, and the same principles appear equally relevant to populations, societies and even ecosystems. Nevertheless, while at lower hierarchical levels interpretation appears relatively clear, it has been very difficult to transport the thermodynamic concepts out of their pure physicochemical context and to make evident their importance at other levels, e. g. ecosystem. Since thermodynamic relationships cannot easily be measured at the ecosystem level, theoretical assumptions have to be made in order to estimate thermodynamic balances of ecosystems. Moreover, an adequate data background must exist to make satisfactory calculations. The application of thermodynamic concepts in the analysis of ecosystems is exemplified, mainly focusing on the use of exergy, a well-known concept from energetic analysis of physical systems. The methodology to perform exergy estimations is described and difficulties discussed. Nevertheless, attempts made from other thermodynamic directions of interpreting ecosystems are also referred. The exergy concept is derived and its meaning in describing the condition of an ecosystem is discussed, comparing two exergy based ecological indicators, the Exergy Index and Specific Exergy, with several ecosystem properties, in the scope of ecosystem biological integrity assessment. Several examples of application of thermodynamic orientors are reported concerning ecological structures and balances of lakes, evaluation of terrestrial systems, and agro-ecosystems. Finally, the advantages of thermodynamically based holistic ecological indicators in comparison to other ecological indicators are discussed.

5.3.1 Thermodynamic Concepts: From Goal Functions to Ecological Indicators

Ever since Schrödinger's statement concerning living organisms "feeding on negentropy", the importance of thermodynamics to living organisms and living

F. Müller M. Leupelt (Eds.)
Eco Targets, Goal Functions, and Orientors

systems have been a subject of discussion. It is beyond any doubt that biological systems have to obey such basic physical laws as the first and second laws of thermodynamics. The importance of the first law seems to be easily captured by most biologists. Whereas, the destructive message of the second law, in many ways, appear to be incompatible with the very essence of life and therefore, more difficult to comprehend. Most studies carried out have been concentrated on how living systems go around this problem.

The importance of the thermodynamics laws can be discussed at various levels of hierarchical biological systems, such as biochemistry, cells, organisms, etc. The importance to lower levels can be illustrated by energetic analyses of biochemical processes (Eigen and Schuster 1977, 1978a, 1978b) or biophysical investigations of organisms. Within biophysics, energetic balances of organisms have been investigated ever since the early findings of Meyer and Helmholtz, clearly demonstrating the importance of dissipation to individual organisms. Indications also exist that the evolution of organisms through time, have performed in a way to develop organisms that are, also in the thermodynamic sense, more adjusted to their surroundings.

Certainly, at these levels the importance of thermodynamics can be understood and also to a certain extent be measured. Rationally, the same laws should be valid for assemblages of organisms. The idea arose that also the higher levels of biological hierarchy should have evolved in a way so as to optimize their thermodynamic relations. Therefore, the very same principles should also be relevant to populations, societies and even ecosystems.

While to lower levels the interpretation appears relatively clear, it has been very difficult to lift the thermodynamic concepts out of their pure physicochemical context and document their importance at other levels, such as the ecosystem level. There are two main obstacles. First, thermodynamic relationships cannot easily be measured at the ecosystem level. Therefore, theoretical assumptions have to be made in order to calculate thermodynamic balances of these systems. Second, in order to make adequate calculations a sufficient data background must exist. It is often difficult to obtain sufficient and relevant data since the calculations include variables and parameters, whose determination is not always automatically carried out by empirical researchers. Therefore, a tight connection between theorists and empiricists must be maintained in order to ensure sufficient quality and quantity of data for this type of calculation.

However, some examples of the application of thermodynamic analysis of ecosystems can be found in the current literature. Since most of the known case studies have been carried out on the basis of exergy, a well-known concept from energetic analysis of physical systems, the present paper focuses on the application of this concept. Moreover, attempts made from other thermodynamic directions of interpreting ecosystems will be included.

5.3.2 Derivation of the Exergy Concept

Exergy is a concept derived from thermodynamics that may be seen as energy with a built in measure of quality. Exergy is a measure of the distance between a given state of an ecosystem and what the system would be at thermodynamic equilibrium (Jørgensen and Mejer 1979). In other words, if an ecosystem was in equilibrium with the surrounding environment its exergy would be zero. This means that, during ecological succession, exergy is used to build up biomass, which in turn stores exergy.

In a trophic network, biomass and exergy will flow between ecosystem compartments, supporting different processes by which exergy is both degraded and stored in different forms of biomass belonging to different trophic levels. In thermodynamic terms, taking an ecosystem trophic structure as a whole, there will be a continuous evolution of the structure as a function of changes in the prevailing environmental conditions, during which the combination of the species that contribute the most to retain or even increase the built in exergy will be selected. At the level of the individual organism, survival and growth imply maintenance and increase of the biomass respectively. From an evolutionary point of view, it can be argued that adaptation is a typical self-organizing behavior of complex systems, which may explain why evolution apparently tends to develop more complex organisms. On the one hand, more complex organisms have more built in information and are further away from thermodynamic equilibrium than simpler organisms. In this sense, more complex organisms have also more built in exergy (thermodynamic information) in their biomass than the simpler ones. On the other hand, ecological succession is driven from more simple to more complex ecosystems, which seem at a given point, to reach a semi-balance between keeping a given structure, emerging for the optimal use of the available resources, and modifying the structure, adapting it to a permanently changing environment.

This system interpretation therefore takes into account the exergy storage in the organisms biomass and in the ecosystem structure, which appears as opposite to the exergy degradation approach proposed by Kay and Schneider (Kay and Schneider 1992, 1994; Schneider and Kay 1994a, b, c, 1995). The two approaches make sense if looked upon in the following way: The more complex an ecosystem structure is, more exergy it stores but, on the other hand, more exergy it must degrade to keep biomass and complexity. As a consequence, a decrease in the exergy available to be degraded by the system will inevitably determine a loss in biomass and complexity (stored exergy) of the ecosystem dissipative structure. Likewise, an increase in the exergy available will cause an opposite effect. In other words, through time, the ecosystem structure will always accommodate changes in environmental conditions, optimizing the use of the available resources.

A first formulation of exergy stored, which can be called Traditional Exergy, was given by (Jørgensen and Mejer 1979, 1981):

$$Ex = R \cdot T \cdot \sum_{i=0}^{n} [\, Ci \cdot \ln (Ci / Ci_{eq}) + (Ci - Ci_{eq})] \qquad (5.3.1)$$

Where R is the gas constant, T is the absolute temperature, Ci is the concentration in the system of component i, Ci_{eq} is the corresponding concentration of component i at thermodynamic equilibrium, and index 0 indicates the inorganic compounds of the considered element.

Part of the term, $C_i \cdot \ln (C_i / C_{i,eq})$, corresponds to exergy built in the structure of the system which may be called Internal Exergy, and $(C_i - C_{i,eq})$ represents a contribution from the external environmental conditions. Except for situations where abrupt changes are taking place, both traditional and internal exergy are expected to follow the same pattern of variation. This is an extension of thermodynamics to ecosystem level, where the derivation of the calculations was based on the classical potentials. The reference values, representing the hypothetical estimates of concentrations of biological systems and compounds at thermodynamic equilibrium, are expressed as probabilities, which are of course very low.

This formulation was, to a certain extent, successfully used as goal function in structural dynamic models (Jørgensen 1993; Nielsen 1992, 1994, 1995), that aimed to describe, applying the principles of optimization theory, the development of ecosystems through time as a response to changes in external factors, including the qualitative alterations of the trophic structure. Nevertheless, although successfully used in modeling, this algorithm was not suitable for wider application of the exergy concept as an ecological indicator. The main constraint was the problem in establishing a reference level for comparisons to be made.

More recently (Jørgensen et al. 1995), it has been derived from the previous formulation that an approximate estimation of exergy may be given by:

$$Ex = T \,.\, \sum_{i=1}^{i=n} \beta_I \cdot C_i \qquad (5.3.2)$$

Where T is the absolute temperature, C_i is the concentration in the ecosystem of component i (e. g. biomass of a given taxonomic group or functional group), β_i is a factor able to express roughly the quantity of information embedded in the genome of the organisms. Detritus was chosen as reference level, i.e. $\beta_i = 1$ and exergy in biomass of different types of organisms is expressed in detritus energy equivalents.

This new formulation, referred in first place as Modern Exergy (Jørgensen et al., 1995) does not correspond anymore to the strict thermodynamic definition, but provides a close approximation of exergy values. In this sense we propose to call it Exergy Index, instead. But independent from the name, this formulation allows to empirically estimate exergy from normal sets of ecological data, e. g. organism's biomass, provided that β_i value for the different types of organisms is known. Consequently, it appears suitable to be used complementary as goal function in modeling and as holistic ecological indicator.

Since the total exergy (Ex_{tot}) of an ecosystem depends on the temporal variations of the biomass and of the information contained in the biomass, if the total biomass ($Biom_{tot}$) in the system remains constant then exergy variations will rely only upon its structural complexity. In such a case, for a given instant, a measure of the exergy contained in each unit of biomass, previously referred to as Structural Exergy (Jørgensen 1994) but that we propose to call Specific Exergy (SpEx), is given by:

$$SpEx = Ex_{tot} / Biom_{tot} \quad (5.3.3)$$

Specific Exergy is as easy to estimate empirically as the Exergy Index, but its values seem to shift more drastically as a function of the temporal dynamics of ecosystems (Marques *et al.*, in press). Therefore, both formulations may be used as indicators in environmental management, but it might be advisable to use them complementary.

5.3.3 What is the Meaning of Exergy in Describing the Condition of an Ecosystem?

Using the Exergy Index and Specific Exergy as indicators it is possible to compare different ecosystems or different states of the same ecosystem at distinct moments in time. However, as indicators in environmental management, and consequently as goal functions in modeling, what is the intrinsic meaning of the Exergy Index and Specific Exergy values?

From the theory, it is reasonable to hypothesize that there is a relationship between the Exergy Index and Specific Exergy values and ecosystem characteristics such as biodiversity, community structure, or resilience. Table 5.3.1 indicated what trends should be expected in the variation of the exergy based ecological indicators and several ecosystem properties.

The following will consider some of the definitions of ecosystem integrity found in the literature, although an attempt to define the concept might be, in a certain extent, contradictory to the meaning of the concept itself:

- "Biological integrity is the maintenance of the community structure and function characteristic of a particular local or deemed satisfactory society" (Cairns 1977 in Kay 1993);
- "Biological integrity is the capability of supporting and maintaining a balanced, integrated, adaptive community of organisms having a species com-

position and functional organization comparable to that of the natural habitat of the region" (Karr and Dudley 1981 in Kay 1993);

- "If a system is able to maintain its organization in the face of changing environmental conditions then it is said to have integrity" (Kay 1989 in Kay 1993).

Table 5.3.1. Expected trends in the variation of the Exergy Index, Specific Exergy and several ecosystem properties

Lower values of the Exergy Index and Specific Exergy	Higher values of the Exergy Index and Specific Exergy
Lower biodiversity	Higher biodiversity
Lower functional redundancy	Higher functional redundancy
Lower buffer capacity and resilience	Higher buffer capacity and resilience
Less complex ecosystems	More complex ecosystems

These different definitions, formulated during more than a decade, evolved from a more motionless to a more dynamic notion of integrity, but the systems capability to maintain organization and function remain the central concept. If we relate Table 5.3.1 to any of these possible definitions, then the Exergy Index and Specific Exergy may be considered as holistic indicators of ecosystem integrity, acquiring a considerable interest in the context of systems ecology.

5.3.4 How to Perform Exergy Estimations in Practice

As a primary condition, in order to make the concept useful in environmental management and auditing, exergy estimations must obviously rely on adequate data background. Besides this elementary condition, various other aspects need, nevertheless, to be discussed.

As explained above, the Exergy Index and Specific Exergy may be empirically estimated from organism biomass, provided that β_i value for the different types of organisms is known (Marques et al., in press; see also Sect. 2.5). This might represent a main constraint. In fact, although partially operational, the method proposed by Jørgensen et al. (1995), which takes into account both biomass and the thermodynamic information due to genes, requires further improvements. Primarily, it is necessary to determine more accurate (discrete) weighting factors (β_i) compared to the ones used to date (Jørgensen et al. 1995; Marques et al., in press). In addition, using the organism's number of genes to express thermodynamic information (Jørgensen et al. 1995) is not practical. Actually, genetic mapping data of this kind available in published literature is very scarce because most of the organisms in biotic systems have not been characterized. The reasons for this are the absence of clear goals in such task and the long time procedures and high costs involved in the molecular work of gene analysis. A more practical approach must therefore, be found.

Genome dimension may be considered an efficient marker to characterize the complexity of organisms (Beaumont 1974; Cavalier-Smith 1985). Moreover, genome dimension is primarily a function of the required genetic information to build up an organism, and this genetic information is contained in the DNA. Therefore, it may be assumed that the dimension of the genome of different types of organisms is roughly proportional to their contents of DNA.

In actuality, several correlations are possible involving the quantity of DNA in the organisms and their level of complexity (e.g., cells dimension and DNA contents, histological complexity and DNA contents). Therefore, it is plausible to establish scales of numerical parameters, determined through such correlations, which will express the relative complexity of different taxonomic groups.

How to compare the contents of DNA of different organisms? In order to be compared, the DNA contents characteristic of each type of organism must be expressed as an absolute quantity. This may be achieved by determining the quantity of DNA in the cell nucleus from different representative organisms, after their isolation and purification. For the same type of cells (preferentially diploid), in the same stage of the cell cycle, which may be discriminated by flow cytometry analysis, the DNA quantity is constant, independently from the tissue where the cells proceed from. Data produced by this method is of course less exact than data obtained from genetic analysis. But this problem is not too relevant in this context, since the aim is to generate conceptual exergistic relationships between different kinds of organisms, accounting for their complexity as an indirect measure of the thermodynamic information built in their biomass.

A comprehensive database of weighting factors (β_i values), regarding representative organisms from different types of ecosystems (e. g. marine, riparian, forests, agricultural), would provide a simple methodology for the estimation of the Exergy Index and Specific Exergy from organisms biomass, and thus the use of these ecological indicators in the context of environmental management and audit.

5.3.5 Examples of Application of Thermodynamic Orientors

The number of applications is scarce and come from different thermodynamic directions of interpreting of ecosystems. Nevertheless, at least three kinds of examples of application of thermodynamic principles for functional analysis and evaluation of ecosystems have been reported in literature:

1. ecological structures and balances of lakes;
2. evaluation of terrestrial systems;
3. agro-ecosystems;

5.3.5.1 Aquatic Applications

The thermodynamical balances in terms of entropy have been established for a number of lakes and their importance in evaluating conditions in the lakes have been reported (Aoki 1988, 1989). In general, more eutrophic lakes produce more entropy. But it has also been demonstrated that the more eutrophic lakes showed an increase in negative entropy balance with the surroundings. What remains to be demonstrated in this type of study is that this negative entropy had also been used to build up structure.

The importance of several kinds of the above mentioned exergy calculations have been applied to a number of European lakes (see also chapter 2.12) mainly through the studies of Jørgensen (1993; 1994) and Nielsen (1992). The lakes have been investigated in connection with natural or human induced changes of the lake ecosystems, such as eutrophication and biomanipulation. In general exergy calculations based on the limiting nutrient have been used.

Summarizing the results, we may find clear indications that the changes induced in these systems are dictated by mechanisms that eventually lead to a raise in the exergy of the system. The mechanisms appear to act in a way that adaptation of organisms, succession of societies and ecosystem evolution are improving the thermodynamic balances of the system, expressed as exergy storage.

It has been shown that this continuous adaptation and adjustment of parameters not only leads to a higher exergy state but also brings the system closer to "the edge of chaos" (Jørgensen 1995 b) (see also chapter 2.3). In addition, two investigations show that intermediate disturbance leading to a raise in biodiversity is also reflected in a higher exergy (Jørgensen and Padisák 1996; Marques et al., in press; see also Sect. 2.5). Moreover, in this last example, the properties of the Exergy Index, Specific Exergy, and Biodiversity (species richness and heterogeneity) were studied along an estuarine gradient of eutrophication. A hypothesis was tested that would follow the same trends in space and time. The hypothesis was only partially validated, since exergy, specific exergy and species richness decreased as a function of increasing eutrophication while heterogeneity behaved predominantly in the opposite way. These results showed that Exergy and specific exergy appeared to be able to provide useful information regarding the state of the system.

5.3.5.2 Terrestrial Applications

Recently, several examples of using remote sensing to evaluate the thermodynamic function of terrestrial systems have been developed. The devices and equipment needed to do this are highly technical and therefore expensive. Meanwhile, there are some hopes for the future that this technology may be more accessible and also be applied at smaller scales. Description of more technical details may be found in Luvall and Holbo (1991).

The technique builds on an interpretation of the terrestrial ecosystems as dissipative structures that are degrading a gradient of exergy (Schneider and Kay

1994a, b, c; Kay and Schneider 1992, 1994). A simplified explanation of the arguments justifying this application are that the activity level of any vegetation is connected with photosynthesis. The photosynthetic level is in turn, also reflected in the evapotranspiration, which leads to a cooling of the system; the more active the system the more "cold" it will be. By manipulating the signals received it is possible to obtain a picture of this cooling of terrestrial systems. Not surprisingly, looking at the earth from space, one will find that the coldest places on the earth are in the rain forest areas of the tropics (Luvall, Schneider, and Kay, pers. comm.; see also Sect. 2.12).

5.3.5.3 Agro-Ecosystems

The use of goal functions as holistic ecological indicators in the management of agro-ecosystems is still at a preliminary stage, and there are not too many case studies using the Exergy Index or Specific Exergy, in this context. Nevertheless, an excellent example was provided using these two exergy based indicators and Ascendency in the scope of a pragmatic framework for describing, monitoring, modeling, analyzing, and evaluating quantitatively the productive performance and the ecological state of farms in the Philippines (Dalsgaard et al. 1995). An ecological performance evaluation was carried out through the application of these indicators for the development of agro-ecosystems. The results clearly suggest that an ecologically sound farming, with a diverse and integrated management of natural resources, can be more profitable and manageable. Moreover, results showed that the emulation of a mature ecosystem's characteristics might allow the design of more ecologically sustainable agro-ecosystems.

5.3.6 Advantages of Thermodynamically Based Holistic Ecological Indicators

The application of exergy and other goal functions has been compared and discussed in several papers. For instance, a good agreement was found between the properties of Exergy, Power, and Ascendency along an eutrophication gradient (Salomonsen 1992). Exergy (expressed as Internal Exergy), Specific Exergy (then referred to as, Structural Exergy), Emergy, Ascendency and the ratio indirect/direct effects, were also compared using several versions of three different lake models (Jørgensen 1994). Exergy was particularly highly correlated with Ascendency, but it has been found in general, that all the investigated goal functions were significantly correlated, although Emergy showed lower correlations with Ascendency and indirect/direct effects than the others. In another case study (Jørgensen 1995a), the Exergy Index (at that time simply referred to as Exergy), Specific Exergy (referred to as Structural Exergy), and Ecological Buffer Capacity were used to analyze ecosystem health, taking eutrophication of a lake as an example. The application of the three concepts was consistent, and general relations were found between the level of eutrophication and the biological structure of the lake. Also, as explained in Paragraph 5.3.5.1, the spatial and temporal variation of the Exergy Index, Specific Exergy, and Species Richness along an estuarine gradient of eutrophication showed a good agreement (Marques et al. in press; see also chapter 2.5).

Nevertheless, when compared to other holistic ecological indicators, the use of the Exergy Index and Specific Exergy for environmental management presents several advantages regarding its applicability. For instance, to estimate Emergy or Ascendency values one needs to have a good knowledge of the network. In other words, the organisms must be classified into functional groups, and the flows in the system quantified. This is a very difficult and time-consuming task. Alternatively, the Exergy Index and Specific Exergy estimations may be carried out simply by applying different weighting factors (βI values) to the biomass of all organisms. In most of the cases, it will be sufficient to take into consideration only the higher taxonomic levels (e. g. Class, Order, Family), and this will also represent a clear simplification when compared, for instance, with biodiversity estimations, which usually imply a much more puzzling taxonomic work. Besides, when defining biodiversity as the full range of biological diversity from intra-specific genetic variation to species richness, connectivity and the spatial arrangement of entire ecosystems at the landscape level (Solbrig 1991), then the Exergy Index and Specific Exergy values may, in a certain extent, comprehend biodiversity.

References

Aoki I (1988) Entropy Laws in Ecological Networks at Steady State. Ecol Model 42:289-303

Aoki I (1989) Ecological Study of Lakes From an Entropy Viewpoint. Ecol Modelling 49:81-87

Beaumont (1974) Genetics and Evolution of Organisms. Chapman & Hall, London

Cavalier-Smith T (1985) The Evolution of Genome Size. John Wiley & Sons, New York

Dalsgaard JPT, Lightfoot C, Christensen V (1995) Towards Quantification of Ecological Sustainability in Farming Systems Analysis. Ecological Engineering, 4:181-189

Eigen M, Shuster P (1977) The Hypercycle. A Principle of Natural Self-Organisation. Part A: Emergence of the Hypercycle. Naturwissenschaften 64:541-565

Eigen M, Shuster P (1978a) The Hypercycle. A Principle of Natural Self-Organisation. Part B: The Abstract Hypercycle. Naturwissenschaften 65:7-41

Eigen M, Shuster P (1978b) The Hypercycle. A Principle of Natural Self-Organisation. Part A: Emergence of the Hypercycle. Naturwissenschaften 65:341-369

Jørgensen SE (1992) Integration of Ecosystem Theories. Kluwer Academic Publishers, Dordrecht, The Netherlands, p 393

Jørgensen SE (1993) State of the Art of Ecological Modelling. In: McAleer M (ed) Proceedings of the International Congress on Modelling and Simulation, University of Western Australia, Perth, pp 455-481

Jørgensen SE (1994) Review and Comparison of Goal Functions in System Ecology. Vie Milieu 44(1):11-20

Jørgensen SE (1995a) Exergy and Ecological Buffer Capacities as Measures of Ecosystem Health. Ecosystem Health 1(3):150-160

Jørgensen SE (1995b) The Growth Rate of Zooplankton at the Edge of Chaos: Ecological models. J theor Biol 175:13-21

Jørgensen SE, Mejer H (1979) A Holistic Approach to Ecological Modelling. Ecol Model 7:169-189

Jørgensen SE, Mejer H (1981) Exergy as a Key Function in Ecological Models. In: Mitsch W, Bosserman RW, Klopatek JM (eds) Energy and Ecological Modelling. Developments in Environmental Modelling, Vol 1. Elsevier, Amsterdam, pp 587-590

Jorgensen SE, Padisak J (1996) Does the Intermediate Disturbance Hypothesis Comply with Thermodynamics? Hydrobiologia 323:9-21

Jørgensen SE, Nielsen SN, Mejer H (1995) Emergy, Environ, Exergy and Ecological Modelling. Ecol Model 77:99-109

Kay JJ (1993) On the Nature of Ecological Integrity. In: Woodley S, Kay J, Francis G (eds) Ecological Integrity and the Management of Ecosystems. St. Lucie Press, Canada, pp 201-213

Kay JJ, Schneider ED (1992) Thermodynamics and Measure of Ecosystems Integrity. In: Hyatt DE, McDonalds VJ (eds) Ecological Indicators, Vol 1. Proceedings of the International Symposium on Ecological Indicators, Fort Lauderdale, Elsevier, pp 159-182

Kay JJ, Schneider ED (1994) Embracing complexity, the Challenge of the Ecosystem Approach. Alternatives 20(3):32-38

Luvall JC, Holbo HR (1991) Thermal Remote Sensing Methods in Landscape Ecology. In: Turner M, Gardner RH (eds) Quantitative Methods in Landscape Ecology, Ecological Studies 82. Springer Verlag, New York, pp 127-152

Marques JC, Pardal MA, Nielsen SN, Jørgensen SE (in press) Analysis of the Properties of Exergy and Biodiversity Along an Estuarine Gradient of Eutrophication. Ecol Model

Nielsen SN (1992) Strategies for Structural Dynamic Modelling. Ecol Model 63:91-101

Nielsen SN (1994) Modelling Structural Dynamic Changes in a Danish Shallow Lake. Ecol Model 73:13-30

Nielsen SN (1995) Optimisation of exergy in a structural dynamic model. Ecol Model 77:111-122

Salomonsen J (1992) Examination of Properties of Exergy, Power, and Ascendency Along an Eutrophication Gradient. Ecol Model 62:171-183

Schneider ED, Kay JJ (1994a) Exergy Degradation, Thermodynamics, and the Development of Ecosystems. In: Szargut J, Kolenda Z, Tsatsaronis G, Ziebik A (eds) Energy Systems and Ecology, Vol 1. (Proceedings of the International Conference ENSEC 93, Cracow, Poland), pp 33-42

Schneider ED, Kay JJ (1994b) Complexity and Thermodynamics. Towards a New Ecology. Futures 26 (6):626-647

Schneider ED, Kay JJ (1994c) Life as a Manifestation of the Second Law of Thermodynamics. Mathl Comput Modelling 19(6-8):25-48

Schneider ED, Kay JJ (1995) Order from disorder: The thermodynamics of complexity in biology. In: Murphy MP, O'Neill LAJ (eds) What is Life: The next fifty years. Reflections on the future of Biology. Cambridge University Press, pp 161-173

Solbrig OT (ed) (1991) From genes to ecosystems: A research agenda for biodiversity, IUBS-SCOPE-UNESCO, Cambridge, Mass, p 124

5.4 Integrating Diverging Goal Functions: Time Scale Effects with Respect to Sustainability

Florian Jeltsch and Volker Grimm

Abstract

The conflict between the economic and the ecological point of view can be interpreted as a conflict between economic and ecological goal functions. Using the example of land-use in the southern Kalahari we apply the framework of goal functions to the problem of developing strategies of sustainable management. We use a grid-based simulation model to identify an appropriate ecological goal function. As a result, the widely used generic goal function of deterministic, continuous succession up to a climax state turns out to be inappropriate. Instead, the systems changes abruptly between distinct states which are separated by thresholds. Based on this experience and on a discussion of 'classical' goal functions such as equilibrium and stability, we argue that for developing strategies of sustainable management generic ecological goal functions should only be used with great caution. Management should instead be based on the identification and understanding of the goal function which is characteristic of the particular system in question. The application of these tailored goal functions is necessarily linked to an explicit acknowledgment of time horizons and inherent uncertainties, which are due to random fluctuations.

5.4.1 Introduction

On a worldwide scale, 70% of arid regions are endangered by degradation (Kruger and Woehl 1996). Particularly in the arid and semi-arid regions of Africa, overexploitation caused by a drastic increase in the population has continually degraded the environment (Kruger and Woehl 1996). Despite the numerous efforts to improve this situation, arid and semi-arid regions of Africa suffer a decrease in per capita production of 2% every year (Pieri and Steiner 1996). The resulting resource crisis and its link with population growth can be demonstrated with the example of Namibia. Around 1960, the population of Namibia exceeded the threshold of 1 Million people and since then, is consid-

F. Müller M. Leupelt (Eds.)
Eco Targets, Goal Functions, and Orientors

ered to be no longer self-sufficient in food production. The continuing exponential increase of the population led to the postulation that by approx. the year 2002, Namibia will no longer be self-sufficient in wood and by 2020 Namibia will no longer be self-sufficient in water (Ashley et al. 1995; Kruger and Woehl 1996). This threatening situation emphasizes the importance of a sustainable management of the scarce resources of arid and semi-arid Africa. But what does *sustainable management* mean in this context? *Sustainability* in its basic definition is the ability to maintain something undiminished over some period of time. From this basic definition it is clear that *sustainability* has many different aspects: what is to be sustained, at which scale and in which way, over which period of time, and with which level of certainty (cf. the 'checklist' for applying stability concepts in ecology; Grimm and Wissel 1997)?

The large number of existing definitions of sustainability basically reflect two different points of view: the economic and the ecological one. The economist looks for a sustained economic development, without compromising the existing resources for future generations. The economic goal is also the major goal for the applied biologist who wants to achieve a sustained yield from the exploitation of populations and ecosystems. The ecologist, on the other hand, aims at sustaining the biological diversity and the key processes (Gatto 1995) in ecosystems which are subject to human exploitation and intervention.

These two goals are also relevant to the discussion concerning land degradation in the southern Kalahari. Land degradation in this semi-arid region of southern Africa is mainly caused by livestock production which threatens the environment by a shift to less palatable plant species, a decrease in vegetation cover and litter, an increase in bare ground and soil and, most importantly, bush encroachment. *Bush encroachment* is a phenomenon widespread in southern Africa. The term describes the increase of unpalatable woody species to the cost of palatable herbaceous vegetation. From an ecological point of view "livestock industry has undermined the integrity of the Kalahari ecosystem by seriously damaging fundamental ecological processes" (Queiroz 1993). This is contrasted by the economic point of view, which states, that "and may be heavily degraded for purpose of nature conservation or wildlife utilization, but is unharmed for livestock production" (Abel 1993). Thus the question arises whether it is really possible that both judgments are simultaneously true. Can the land be ecologically damaged but be unharmed for livestock production? It may be useful to formulate the conflict between the economic and ecological point of view as a conflict of the economic and ecological *goal functions*. Economic goal functions are in principle easy to define because they simply reflect the economic goal that one wants to achieve – although this does not necessarily mean the economic goal functions are always *precisely* defined. For the southern Kalahari recommended stocking rates (potential carrying capacities; see below) are the economic goal function. Ecological goal functions are much harder to identify because they require an understanding of how and due to which processes ecological systems develop.

5.4.2 Ecological Goal Functions and Rangeland Management

A goal function describes a property of a system that, despite external disturbances or internal variation, shows a clear trend of approaching and maintaining a *goal*, i.e. a maximum, minimum, or a certain range of values. Classical examples from physics are energy and entropy. Goal functions are appealing because they refer to general system properties. Once a goal function for a class of systems had been found this would be the key to understanding and predicting the system's dynamics (Grimm et al. 1996). With goal functions one must not, overlook that *goal*, is, of course, only a metaphor: the systems behave *as if* they would, like human beings or animals, follow a certain goal. In fact, the goal emerges via self-organization (Müller et al. 1997).

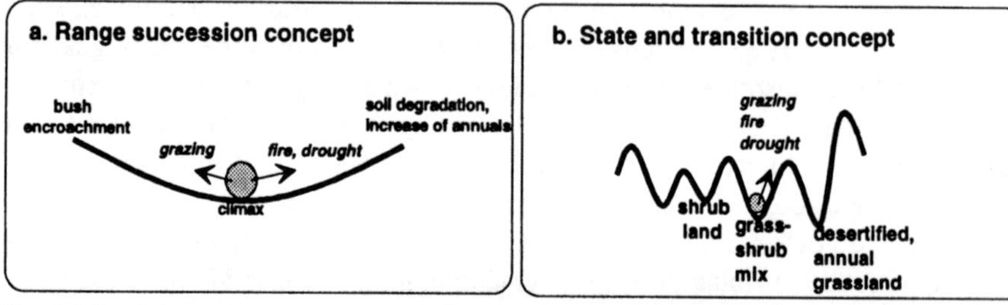

Fig. 5.4.1.a. Schematic description of the "range succession concept". The vegetation is represented by the gray ball which is shifted away from the climax state by various stresses such as grazing, fire and droughts. These induced changes are reversible, gradual and deterministic.
b. Schematic description of the "state and transition concept". Stresses cause "fast" transition of the vegetation towards other vegetational states, such as a bush encroached state or a state dominated by annual plants, once they exceed a certain threshold. These changes are only reversible by other stresses or external triggering factors.

Several general ecological goal functions have been proposed for ecological systems, for example *resilience* (sensu Holling 1973), *exergy* (Joergensen 1990), and *ascendancy* (Ulanowicz 1988), but have not yet been widely accepted and used in the ecological community. One reason for this may be that these functions have not yet been proven to aid understanding, and thus, improve prediction and management of real ecological systems. A more specific goal function assumed in vegetation dynamics is succession leading to a clearly defined end state or climax which is in equilibrium. This notion is the basis of the so-called *range succession concept* (Clements 1916) which is still widely accepted as a foundation for rangeland policy. With this concept, the process of achieving the goal, i.e. the climax, is thought to be a more or less deterministic, gradual and reversible change in vegetation (Fig. 5.4.1a). If this is correct, it would be possible to manage for an optimal yield of livestock without having to worry about ecological threats, since any temporary over-exploitation, which reduces the economic yield could easily be corrected by a reduction of the stocking rate.

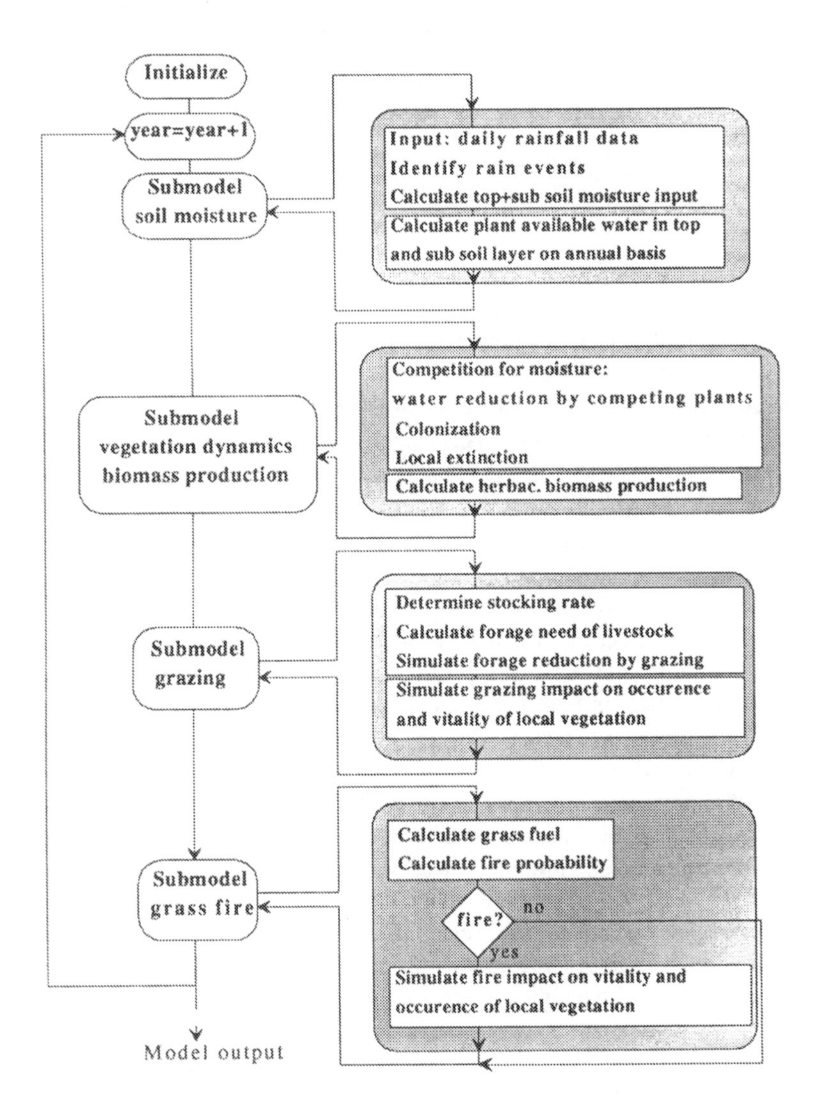

Fig. 5.4.2. Flow chart of the computer simulation model (see Jeltsch et al. 1997). Each simulation run starts with the *initialization* of parameters and distributions. Submodel *soil moisture*: In each time step (1 year) we initially consider the daily rainfall of the given year distinguishing between single rain events and amounts of precipitation. The calculations of the moisture input to top and sub soil layer result in the approximate plant available moisture in the top and sub soil layer. This moisture distribution is the same for all spatial subunits (cells). Submodel *vegetation dynamics and biomass production*: The dominant life forms in each cell compete for

the available moisture. The various life forms cause different degrees of moisture reduction. The available moisture determines the transition between different vitality states of the life forms. Colonization of empty space by different life forms is modeled to be stochastical. If moisture availability in a grid cell is not sufficient, extinction occurs within the cells with a certain probability. Based on the condition of life forms in the different cells the approximate herbaceous biomass production in the individual spatial subunit and in the total area can be calculated. Submodel *grazing*: For any given stocking rate the forage need of the livestock can be calculated. The local reduction of the biomass on the basis of individual cells simulates spatially explicit grazing by the cattle. The vitality of the herbaceous life forms deteriorates in proportion to the amount of remaining biomass, leading eventually to local extinction. Submodel *grass fire*: Finally, the amount of remaining grass fuel determines whether grass fires occur (with a certain probability), causing changes in the vitality and the local occurrence of the vegetation.

These ideas lead to the economic concept of a *potential carrying capacity* – a stocking rate that is to be recommended in all but very dry years (Fourie et al. 1985; Mace 1991).

An alternative ecological goal function is assumed by the *state and transition concept* (Westoby et al. 1989). This concept suggests that vegetation changes consist of fast transitions between states when the system is stressed beyond a threshold (Fig. 5.4.1b). In the given context, one has to distinguish between the *normal* state of a grass shrub mix, as it is typically found in the southern Kalahari, and *degraded* states such as a shrub encroached vegetation, or a *desertified* state with a lot of bare soil and mainly annual grasses. If such threshold dynamics exist in the context of livestock production in the southern Kalahari, a separation of economic and ecological goals and, in turn, sustainability would not be possible. The reason is that any stocking strategy that leads to a transition towards another state, e.g. a shrub dominated state, would decrease both the economic yield as well as the ecological value of the system, e.g. biodiversity.

Thus, it is obvious that as long as the goal functions of the ecological system are not fully understood and identified, any attempt to reconcile the economic and ecological point of view must necessarily fail. In the case of the southern Kalahari we need to know more about the stress response of the vegetation, about vegetation dynamics, time scales of changes and the impact and interaction of stresses such as grazing, grass fires and droughts.

A suitable tool to investigate questions like this on the relevant time scales of decades, are computer simulation models which allow us to focus on the relevant processes (Jeltsch and Wissel 1994; Wiegand et al. 1995; Jeltsch et al. 1996, 1997). Simulation experiments provide a valuable extension to empirical data in systems where time spans are too large and stochasticity too high to perform replicated field trials. But how can simulation models be constructed without assuming a general and not yet verified ecological goal function? Grid-based simulation models facilitate the replication of the synergistic spatial processes of ecological systems exclusively on the basis of local processes and interactions (Jeltsch and Wissel 1994; Grimm et al. 1996; Jeltsch et al. 1996, 1997).

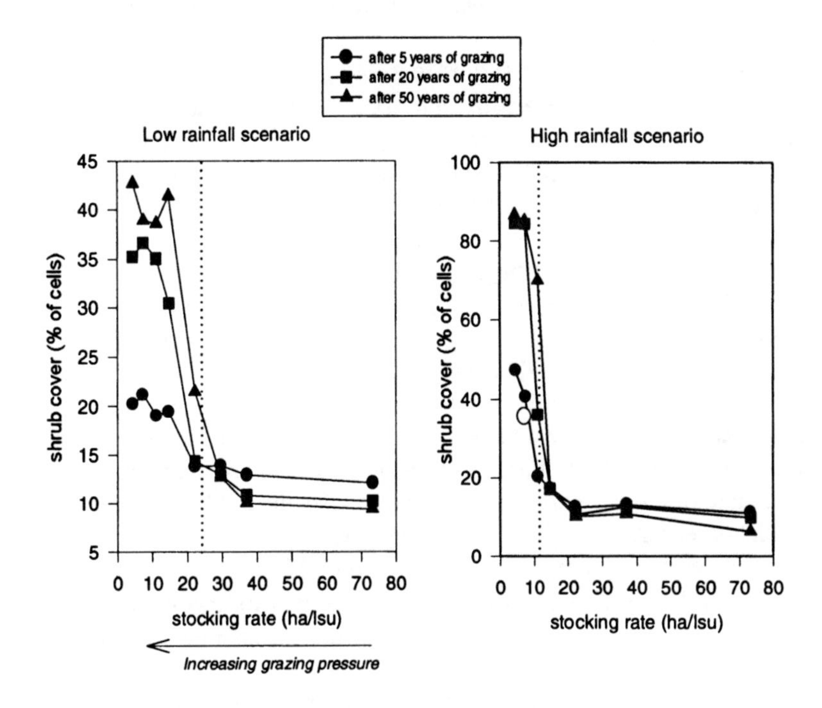

Fig. 5.4.3. Comparison of simulation results of a low and a high rainfall scenario – Twee Rivieren (220mm mean annual precipitation) and Pomfret (385mm mean annual precipitation): bush cover (percentage of shrub dominated grid cells) versus stocking rate (ha / lsu) after 5, 20, and 50 years of grazing. Note that the grazing pressure increases with decreasing values of the stocking rate (x-axis) which in arid regions is typically given in hectar (ha) per livestock unit (lsu). The unfilled data point in the Pomfret scenario shows results of a field study (Skarpe 1990), which corresponds very well with the simulation results. The dotted line gives the estimated potential carrying capacity (PCC) for the two sites (Fourie et al. 1985).

Thus, this bottom-up approach does not have to postulate an ecological goal function *a priori*, but the self-organization in the model which is designed to mimic the processes in the real system produces emergent properties which finally may be called goal functions.

5.4.3 Vegetation Dynamics in the Southern Kalahari: a Model Based Analysis

In the following example we will demonstrate how an ecological goal function of a particular system (southern Kalahari) is identified using a grid-based simulation model. The model is documented in Jeltsch et al. (1997); here we will only describe it briefly.

The modeled life forms are shrubs (e.g., *Rhigozum trichotomum*), perennial grasses and herbs (e.g., *Stipagrostis obtusa*) and annuals (e.g., *Schmidtia kalahariensis*). We modeled the most relevant spatial processes such as competition between life forms, colonization and extinction under the impact of realistic rainfall, grass fires and grazing scenarios (Fig. 5.4.2). The model is based on the available expert knowledge which is included in the model by using rules (for this *rule-based* approach, see Wiegand et al. 1995; Jeltsch et al. 1996, 1997).

The results of the simulation experiments prove the existence of a grazing threshold (Fig. 5.4.3). Increasing the grazing pressure at two different simulated localities with different rainfall leads, in both cases, to a sudden increase in shrub cover once a site specific threshold is exceeded. Below the threshold, i.e. at stocking rates which cause a lower grazing pressure, there is almost no visible effect on the shrub cover. Further investigations showed that the shrub encroached state of the vegetation persists for decades once it has appeared. Thus, the simulation experiments show that separate states of the system do exist and that these states are separated by a stress threshold. The *state and transition concept* seems therefore to be valid which means, that following the arguments given above, economic and ecological goal functions cannot be separated in this example.

However, the simulations show that the situation is more complex. Primarily one has to consider the time horizon in question. At stocking rates near the grazing threshold the amount of bush encroachment differs significantly depending on the time span of grazing, i.e., whether 5, 20 or 50 years are considered (Fig. 5.4.3). The differences are larger for the site with the lower average precipitation. These differences are important since the potential carrying capacity, i.e., the stocking rate which is recommended for these localities in all but very dry years (vertical line in Fig. 5.4.3), is, on both sites, close to the grazing threshold predicted by the simulations (Fig. 5.4.3) (Fourie et al. 1985). Hence, an economic goal which might cause an overexploitation in the long term does not necessarily lead to a change of the system state on a time scale of years. This is further complicated by the stochasticity of the system. Fig. 5.4.3 gives averages of fifty simulation runs for each data point. Between the individual runs the results may differ significantly depending on the stochasticity of the rainfall which, in the modeled area, is distributed in unpredictable rainfall events. The higher rainfall stochasticity at sites with a lower mean annual precipitation is reflected in a higher stochasticity of bush encroachment in these areas. Thus, with each stocking rate and under each grazing strategy, a certain risk of shrub encroachment exists as a consequence of the stochasticity. Consequently, any statement in this context can only be a probabilistic one.

The difficulty of reconciling economic and ecological goal functions is further documented in the comparison of the recommended potential carrying capacity with the simulated grazing thresholds for five localities along a rainfall gradient in the southern Kalahari (Fig. 5.4.4).

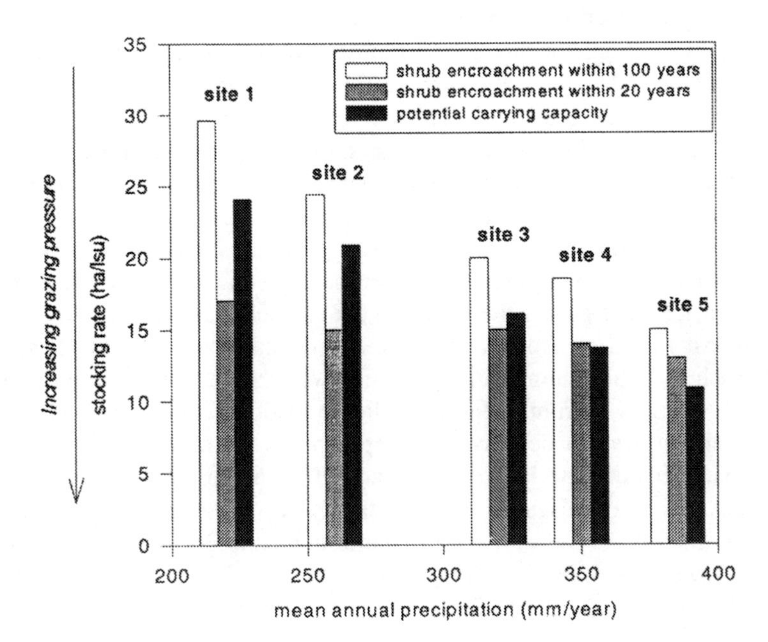

Fig. 5.4.4. The potential carrying capacity (PCC, black columns) for five modeled sites with different mean annual precipitation compared with simulation results of these sites. Stocking rates within the white columns show, on average, a doubling of shrub cover within a timespan of 100 years or less, whereas stocking rates within the gray columns lead, on average, to shrub cover doubling in timespans below 20 years. Notice, that the grazing pressure increases with decreasing values of the y-axis (compare Fig. 5.4.3).

Model results show that the recommended stocking rates (potential carrying capacity) does, on average, avoid shrub encroachment on a moderate time scale of twenty years but, with a high probability, leads to a drastic increase of the shrub cover in a time span of hundred years. Consequently, the estimations of a potential carrying capacity, which are based on the personal experience of land managers, underestimate the risk of shrub encroachment on longer time scales.

5.4.4 Discussion and Conclusions

Sustainable management of ecological systems is, *expressis verbis*, the attempt to reconcile the economic and ecological point of view. The possible conflict between these two points of view can be discussed without any reference to goal functions (Jeltsch et al. 1997). Thus the question arises whether the additional notion of goal functions really pays if it is to establish sustainable management strategies.

Talking of goal functions both in the economic and ecological context is appealing because it suggests that goal functions are a unifying framework for solving economic and ecological problems. On closer examination this becomes questionable. Economic goals may not easily be achieved, but in principle they can clearly be defined because they are anthropogenic, i.e. behind economic goals are the intentions of single persons, agencies, or companies. Ecological goal functions on the other hand are hard to identify for real, particular ecological systems. The term 'goal function' in ecology has the connotation of 'easily assessed' (such as in economy) and stands for the hope that there is some simple dynamic principle that guides the development of ecological systems. But this hope is almost as old as ecology (or even older) and was linked to classical concepts such as equilibrium (DeAngelis and Waterhouse 1987, Chesson and Case 1986), stability (Grimm 1996), resilience (Holling 1973), and climax. These concepts proved to be only partly appropriate for analyzing ecological systems which became especially obvious in the field of applied ecology, where management was often misguided by the classical concepts (see, for example, Wiegand et al. 1995 for a discussion of equilibrium concepts, and our example presented above concerning the climax concept). Since all of the classical concepts can be interpreted as goal functions, it appears that with goal functions there is a high risk of only 'retelling the story of the classical concepts'.

What can be done to link the notion of goal functions to the development of sustainable management strategies and to prevent the pitfalls of classical concepts? First of all, it should be acknowledged that goal functions are theoretical constructs which are superimposed on real ecological systems *a priori*. Thus, for each particular ecological system to be managed it has to be checked carefully if the goal function indeed has any predictive power. A complementary, and in our opinion more promising approach, is not to postulate a goal function *a priori* but to check if a goal function emerges via self-organization. A practical method for this are comparative analyses (Cole et al. 1991) where the system properties of similar or adjacent systems are compared in order to identify correlations between ingredients of the systems and their properties. Another important tool, that was presented here, are grid-based simulation models which account for three vital aspects of any ecological system which are largely ignored in the classical concepts (and in almost all goal functions that we know): spatial and temporal scales, disturbance and event-regimes, and effects of random variation. These models are still models in that they are not designed to map reality into a computer (which is impossible). Testing the model predictions with as many field observations as possible or even, if possible, with experiments is therefore an integral part of grid-based modeling.

The predictions of the Kalahari model have been discussed in detail in Jeltsch et al. (1996, 1997). Here it is emphasized that the model allowed to reject climax dynamics as a goal function and to identify distinct states, separated by thresholds, as a more appropriate concept. It should be noted that most goal functions which are discussed in ecology so far postulate a continuous development instead of thresholds and distinct states (but see Holling, 1973). The model accounts for disturbances (fire, grazing), event-regimes (rainfall), and stochasticity (rainfall scenarios, probabilistic rules for local interactions) and this is possibly the reason that the detected goal function is qualitatively different from more classical goal functions. A practical consequence of the detected goal function is that any management strategy which claims to be sustainable has to refer explicitly to time horizons and to uncertainties which are inherent in the system.

To summarize: generic goal functions should, in our opinion at the present state of ecological theory only with caution be involved in developing sustainable management. More case studies and goal functions of particular systems are required before we can properly assess the potential of goal functions for reconciling economic and ecological viewpoints.

References

Abel N (1993) Carrying Capacity, Rangeland Degradation and Livestock Development Policy for the Communal Rangelands of Botswana. Pastoral Development Network 35c:1-9. Overseas Development Institute, London, UK

Ashley C, Mueller H, Harris M (1995) Population Dynamics, the Environment and Demand for Water and Energy in Namibia. DEA Research Discussion Paper No. 7. Namibia

Chesson PL, Case TJ (1986) Overview: Nonequilibrium Community Theories: Chance, Variability, History and Coexistence. In: Diamond JM, Case TJ (eds) Community Ecology. Harper and Row, New York, pp 229-239

Clements FE (1916) Plant Succession: an Analysis of the Development of Vegetation. Carnegie Institute Publication 242:1-512, Washington DC, USA

Cole J, Lovett G, Findlay S (1991) Comparative Analyses of Ecosystems. Springer, Berlin

DeAngelis DL, Waterhouse JC (1987) Equilibrium and Non-Equilibrium Concepts in Ecological Models. Ecol Monogr 57:1-21

Fourie JH, van Niekerk JW, Fouché HJ (1985) Weidingskapasiteitsnorme in die vrystaatstreek. Glen Agric 14:4-7

Gatto M (1995) Sustainability: is it a well defined concept? Ecol Appl 5:1181-1183

Grimm V (1996) A Down-to-Earth Assessment of Stability Concepts in Ecology: Dreams, Demands, and the Real Problems. Senckenbergiana Maritima 27:215-226

Grimm V, Wissel C (1997) Babel, or the Ecological Stability Discussions: An Inventory of Terminology and a Guide for Avoiding Confusion. Oecologia 109:323-334

Grimm V, Frank K, Jeltsch F, Brandl R, Uchmanski J, Wissel C (1996) Pattern-Oriented Modelling in Population Ecology. Sci Tot Environm 183:151-166

Holling CS (1973) Resilience and Stability of Ecological Systems. Ann Rev Ecol Syst 4:1-23

Jeltsch F, Wissel C (1994) Modelling Dieback Phenomena in Natural Forests. Ecological Modelling 75/76:111-121

Jeltsch F, Milton SJ, Dean WRJ, van Rooyen N (1996) Tree Spacing and Coexistence in Semi-Arid Savannas. J Ecol 84:583-595

Jeltsch F, Milton SJ, Dean WRJ, van Rooyen N (1997) Simulated Pattern Formation Around Artificial Waterholes in the Southern Kalahari. J Veg Sc 8(2):177-189

Joergenson SE (1990) Ecoystem Theory, Ecological Buffer Capacity, Uncertainty and Complexity. Ecol Model 52:115-122

Kruger AS, Woehl H (1996) The Challenge for Namibia´s Future: Sustainable Land-Use under Arid and Semi-Arid Conditions. Entwicklung und ländlicher Raum 4:16-20

Mace R (1991) Overgrazing Overstated. Nature 349:280-281

Müller F, Bredemeier M, Breckling B, Grimm V, Malchow H, Nielsen SN, Reiche EW (1997) Ökosystemare Selbstorganisation. In: Fränzle O, Müller F, Schröder W (eds) Handbuch der Umweltwissenschaften. Ecomed, Landsberg

Pierei C, Steiner, KG (1996) The Role of Soil Fertility in Sustainable Agriculture With Special Reference to Sub-Saharan Africa. Entwicklung und ländlicher Raum 4:3-6

Queiroz JS (1993) Range Degradation in Botswana: Myth or Reality? Overseas Development Network 35b:1-17. Overseas Development Institute, London, UK

Ulanowicz RE (1988) On the importance of higher-level models in ecology. Ecolog Modelling 43:45-56

Westoby M, Walker BH, Noy-Meir I (1989) Opportunistic management for rangelands not at equilibrium. Journal of Range Management 42:266-274

Wiegand T, Milton SJ, Wissel C (1995) A simulation model for a shrub-ecosystem in the semi-arid Karoo, South Africa. Ecology 76:2205-2221

5.5 Deriving EcoTargets from Ecological Orientors: How to Realize Ecological Targets at the Landscape Scale

Hartmut Roweck

Abstract

To envisage the cultural landscape of tomorrow different scenarios can be perceived. The development of landscape depends predominantly on socio-economic demands and technical possibilities – counterbalanced by measures of nature conservation and environmental protection. However, the economic prerequisites and political structures determine the realization of landscape development far more than the results of ecosystem research. The integration of diverging sectoral planning results – deriving data from basic research – may lead to ideas of cultural landscape that considerably limit the degree of freedom for land-use utilization. This applies in particular to the conservation of species, since in this field the definition of standards is hardly possible. In this situation, it is useful to describe the sensitivity of landscape compartments as a measure to assess impact factors on protective goods due to land-use systems. The 'sensitivity'-approach may facilitate a cautious embedding of land-use systems without neglecting the protection and development of abiotic and biotic landscape potentials.

5.5.1 Introduction

When preparing a paper on goal functions, the most obvious step is to either, restrict oneself to those goals whose realization from the outset may be related to a whole landscape, or one attempts to show that all "ecological targets" may only be achieved in landscapes, even if they are at first formulated without reference to any particular area. However, it is not so much scaling-problems, which sometimes make preparations for realization difficult, but the frequently observed overrating of the relevance of detailed results of eco-system research for

F. Müller M. Leupelt (Eds.)
Eco Targets, Goal Functions, and Orientors

the tasks of restitution and regeneration of disturbed systems or for the development of environmentally tolerable land use systems.

The current support by administration of strongly application-oriented research in the geo- and life science areas certainly plays a role here, where the classical sciences have to increasingly make an effort to deliver practical results which, as far as possible, can be implemented directly.

It is not these results which should be criticized but their all too often very simplistic evaluation and merging with popular ecosystem management concepts by experts from other fields. Such books often include graphic pairs of pictures: "the disturbed system" – "the intact system" (e.g. Duvigneaud 1984) in which "intactness" is usually defined by means of structural diversity and "completeness" in terms of the text book descriptions (e.g. Blab 1993). An orientation, for instance, towards regional examples is still a rare exception. This warning about simplification and standardization is also, however, relevant to experts within the subject area in view of the immense complexity and individual variability of ecosystems.

Our analyses, regardless of how comprehensive they may be, can only cover a small section of the system under investigation, and even in the best possible case, only record short phases in the development process of single ecosystem individuals. The application of the results to other individuals of the same type is often as difficult as prognoses concerning development over a period of time. Furthermore, the value of even data gathered over e.g. twenty years is already considerably lessened by the fact that episodic events (mostly extreme variations in climate) which are so important for the coenosis composition are only registered by chance and at most, once, so that they are likely to be regarded as disturbances in the analysis procedure.

The widespread belief that we are suffering from a lack of data that has practical applications is without a doubt, correct in relation to certain disciplines. However, a wide-ranging and thorough analysis of ecosystem complexes that would, for example, be able to give indispensable coenosis sets for certain systems in the sense of a standardization on the "type-level" is, to my understanding, probably not possible for fundamental reasons (Roweck 1996).

5.5.2 Distinction between Basic Research and Application

Today, even with complicated modelling processes we do not understand the most species-poor systems well enough to reliably predict the effects of particular impacts in detail. Especially with regard to zonal, more mature systems, "bottom up"-approaches certainly do not enable to be derived justifiable standards for coenoses, critical loads etc. from the present data. What can easily be derived is the necessity for further research, but once most of the main disruptive factors in coenosis and ecosystems have been uncovered, differentiation should again be more clearly defined between the tasks of basic research and a more application-oriented gathering of data in the future (Roweck 1997).

The lack of distinction in this area can be seen, for example, in the way the environmental protection evaluation criteria that are most often found in the literature today are used (Table 5.5.1). These criteria have their main reference mostly in the area of ecosystem analysis. An evaluation is indeed possible when target-oriented within the framework of procedures for political agreement (or as preparation for this), but the valency of particular systems or conditions should not be simply the result of an adding up of values or other mathematical operations (Dierssen 1994, Scherner 1995). In connection with such judgements even many of these criteria lose their meaning .

Table 5.5.1. Small selection of the most often cited (literature) and used (planning) nature conservation evaluation criteria, organized according to their relationship with organisms and ecosystems

organism-related	ecosystem-related
- rareness	- rareness
- endangerment	- endangerment
- number of species	- age
- species diversity, eveness	- presence of indicators
- completeness (of coenosis)	- habitat size
- population size (of target species)	- density, location
- population size (of indicators)	- diversity of structures
- type of population growth	- gradient density
- strategytype, lifeform	- completeness (of systems)
- minimum viable population size	- heterogenity, homogenity, complexity of landscapes
- "ornithological value"	- maturity
- ...	- indicator potential
	- habitat network, buffering effects, ...
	- naturalness
	- stability
	- resilience
	- regeneration potential
	- potential for restitution
	- potential for substitution
	- land use capability
	- ecological carrying capacity
	- hemerobic level
	- ...

If "diversity" as a value-determining criterion has been decided upon, the idea of "increasing diversity" up to a certain level may be suggested itself. In a similar manner, demands to "increase the number of species" or to "restore naturalness" arise. However, as far as both quantitative and qualitative aspects are concerned, all those items that would indicate the course we should take are missing.

Approaching the subject area "quality", one might imagine a "Parliament of Animals" where the individual parties express their demands. We could easily see that no landuse system would be able to survive the fulfilment even of a small section of possible demands. This is particularly true if quantitive aspects are included, e.g. if we wanted to attempt to provide many species with sufficient space for viable populations (Roweck 1993).

The example touches on a problem which is often encountered when we attempting to define, by means of the "bottom-up approach", which organisms, and which functions are indispensible for the continued existence of particular ecosystems. The clue of many expert investigations consists strictly speaking of the question "How much nature do we have to save?" (instead of asking "How much nature are we able to save?")

If one attempts to sum up all available ecosystem analyses data results, concerning cultural landscapes, the rather modest sounding answer to this question would be "as much as possible". To get "as much as possible" is not just a terse demand by unreasonable nature lovers but concerns its realization, the only professionally justifiable result of all these carefully conducted analyses.

In systems not (or almost not) influenced by man, keeping the natural dynamics of the systems in mind, an attempt to describe their carrying capacity or resilience can be made. However, when faced with our modern landuse systems, such limits have been exceeded in the majority of cases. For this reason, from an ecological point of view, it is only possible to plead for the maximum protection of biotic and abiotic resources, which lead to; the realization of a "concept of minimal impact", to the employment of in this sence best available technologies, which have to be embedded as carefully as possible in landscape contexts.

5.5.3 Problems of Quantification

Returning to quantification, one might envisage a traditionally-used harmonious cultural landscape, primarily without, and then with, an increasing number of energy-generating wind turbines - in the last picture the landscape is just a background to a "wind farm". In this example, politicians require a maximum amount of energy produced in an environmentally friendly manner and therefore, ask ecologists the following questions: how many wind turbines does the landscape tolerate

1. from an aesthetic point of view, and
2. as an ecosystem complex?

The answer is dependent upon how far ahead one looks: if external effects of current energy generation are taken into account, the answer tends more in the direction of "local wind farms" than if we solely concentrate on the increased risk for migratory birds or on the increasingly unrecognizable landscape. In this case, as few wind turbines as possible (or as necessary) would be advocated. This "as few as possible" or, the other way round, "as many as possible" is in many cases, the only justifiable answer from an ecological perspective to questions of "how many". This is true for questions relating to organisms and habitats as well as for the frequently required limitations in the abiotic area. It is necessary to take part in round table discussions where limits are set. Not because these limits can be derived from the results of ecological research, but rather to be in a position to illustrate the consequences which can be expected, and thus, to make our contribution to a "maximum protection of biotic and abiotic resources".

5.5.4 Combining "Top down-" and "Bottom up"-Approaches

In brief: the "bottom up-approach" (to which the majority of the evaluation criteria shown before may be assigned), our causal-analytical attempt to understand ecosystems as far as their functions are concerned so that their reactions to particular changes may be calculated, throws up fascinating questions again and again, but results to just a limited extent in useful tools for practical nature conservation work. Ecosystem analysis is, however, acquainted with another approach, in which we could intensify our activities. It is the "top down-approach" so to speak, the attempt to transfer information to individual systems that are as similar as possible to those investigated. A considerable part of what we currently know about and do in nature conservation is at any rate, derived from this type of system comparison. Ideally, both approaches complement each other. The approach coming from above is especially important for the "step from the spot into the landscape".

The interesting aspect in this context is, that here criteria become important that refer to surface-system characteristics which describe the "intensity of impact", and usefully complement fundamental criteria such as population size, number of species etc. Such criteria are useful because they allow direct interpretations with regard to impacts; with their help the amount of scope for various landuses may be more clearly stated, once the sensitivity of systems towards specific influence-factors can be given – influences which occur in connection with impacts, uses and also the decrease of uses (Fig. 5.5.1).

The fact that the development of our cultural landscape will certainly also be determined by socio-economic demands and the technical resources of the time, is a further argument for intensifying the top down-approach. Adam (1996) recently emphasized the "primacy of economic matters" in the development, endangerment and protection of our cultural landscape. We have been trying to influence the development of future landscapes with considerations of nature conservation and environmental protection only for a comparatively short time. With increasing knowledge of ecological relationships, it is possible to make more and more precise statements about the chances to support landscape potentials and lessening impacts; in the best case, however, this knowledge only represents a flexible counterweight. The extent of realization is determined above all by economic requirements and political structures.

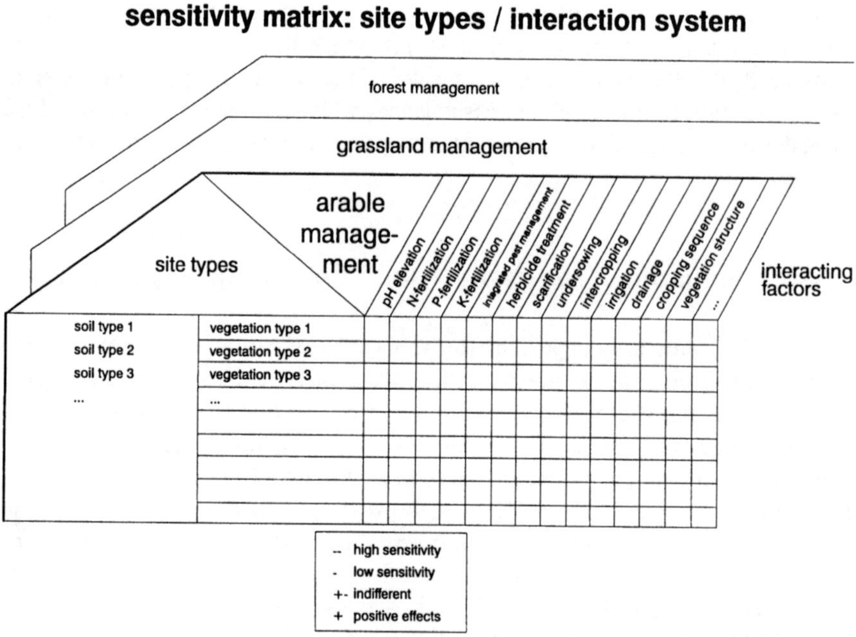

sensitivity matrix: site types / interaction system

forest management

grassland management

arable management

site types

pH elevation, N-fertilization, P-fertilization, K-fertilization, integrated pest management, herbicide treatment, scarification, undersowing, intercropping, irrigation, drainage, cropping sequence, vegetation structure, ...

interacting factors

soil type 1 | vegetation type 1
soil type 2 | vegetation type 2
soil type 3 | vegetation type 3
... | ...

-- high sensitivity
- low sensitivity
+- indifferent
+ positive effects

Fig. 5.5.1. Site sensitivities in confrontation with selected impact factors of landuse systems (explanation in the text)

So, if the extent of what is feasible does not depend so much on the results of ecological research anyway, it makes sense to orientate our evaluation systems towards the maximal exploitation of the opportunities which arise through changing economic conditions. If nature conservation acts essentially as something to fill a gap, then this should at least happen in such a way as to optimally exploit the available possibilities.

In Central-Europe, these possibilities concern in particular, the land in intensive arable use – so this approach is less relevant for areas and systems with a high biotic potential for development and those which are less intensively used. To focus on areas in which, from a scientific point of view, there are great difficulties in defining biotic protection goals, anyway.

If the sensitivity of systems to specified changes (impacts, cultivation etc.) is known, or can be derived from comparisons of systems without too much effort, then it should be possible to depict the relative levels of such sensitivities (Fig. 5.5.1), which reflect the current state of knowledge (Ringler 1978). They can be used in a matrix to qualitatively and quantitatively estimate the risks that may occur in connection with actual or even conceivable landuses (i.e. land uses which do not exist at the moment). If we divide the types of cultivation to be tested, are derived into impact-relevant factors, it can be easily seen that it is not, for example, the cultivation of corn as a whole in particular locations which is problematic, but rather specific elements of this cultivation.

From the opposition of relevant groups of factors with differing degrees of sensitivity, emerge (Fig. 5.5.2), among other things, starting points in the search for types of land use which are less harmful to the ecosystems involved. This results in a high degree of flexibility in relation to the embedding of land uses in landscape contexts.

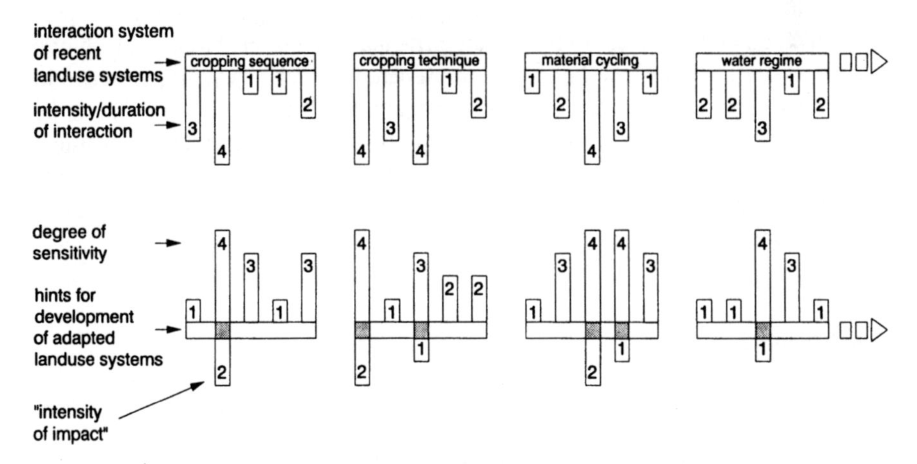

Fig. 5.5.2. Useful indications for the most careful possible embedding of unalterable landuses in concrete landscape situations emerge from the opposition of weighted interacting factors with sensitivity levels for the landscape elements to be evaluated.

In the same way, it is possible to prepare a sensible lessening of impacts on systems with a higher degree of naturalness, as the factors with negative effects in our cultural ecosystems are mostly relevant to negative trends in systems closer to nature too - and (this is the point) they often come from the same sources.

So firstly, if (whether we like or not) it is necessary to accept the "primacy of economic matters", and secondly, hesitate for good reasons when defining ecological standards, at least on the biotic level – then we have to look for goal functions, more than is currently done, on the impact side. This may sound very simple, but has many implications for the realization level. There is a basic difference whether society spends money on programs that try to keep or reestablish a certain level of biotic diversity in cultural ecosystems, or on backing the reduction of factors, that hinder the development of a site-specific diversity, even if the connection is not very obvious.

5.5.5 Conclusions

According to the needs of a particular time, a confrontation of sensitivities to interacting factors, as described, offers the possibility to formulate in a procedure for political agreement, to what extent support or restrict for must be given to what type and intensity of particular landuses, or to put it the other way round, what degree of resource protection is necessary and affordable (Roweck 1995).

Approaches such as this allow, in addition, a landscape-related representation of the current state of expert knowledge. The updating of this knowledge, usually only leads to gradual changes in sensitivity levels, assuming that the most important interacting factors as well as their principal effects on biotic and abiotic resources are known. Data that has been prepared in such a way, for instance in the form of subject maps, also has the advantage that it can, to a large extent, be kept free of moral statements (Lehnes 1994, Höffe1980). Later, in the political agreement process, the data can be related to such statements. And finally, data prepared in this way give a suitable basis for the preparations needed to realize the "as much as possible" scenario, mentioned above.

References

Adam T (1996) Mensch und Natur: das Primat des Ökonomischen. Natur und Landschaft 71(4):155-159

Blab J (1993) Grundlagen des Biotopschutzes für Tiere. Kilda-Verlag, Bonn

Dierssen K (1994) Was ist Erfolg im Naturschutz? Schriftr f Landschaftspfl u Natursch 40:9-23

Duvigneaud P (1984) La synthese ecologique. Doin editeur, Paris

Höffe O (1980) Naturrecht - ohne naturalistischen Fehlschluß, ein rechtsphilosophisches Programm. Klagenfurter Beitr z Philosophie, Reihe 2, Wien

Lehnes P (1994) Zur Problematik von Bewertungen und Werturteilen auf Ökologischer Grundlage. Verhandl Ges f Ökologie 23:421-426

Ringler A (1978) Nutzungsspezifische Empfindlichkeitskarten in der Landschaftsplanung. Natur u. Landschaft 53:90-95

Roweck H (1993) Zur Naturverträglichkeit von Naturschutzmaßnahmen. Verhandl Ges f Ökologie 22:15-20

Roweck H (1995) Landschaftsentwicklung über Leitbilder? LÖBF-Mitt 4:25-34

Roweck H (1996) Möglichkeiten der Einbeziehung von Landnutzungssystemen in naturschutzfachliche Bewertungsverfahren. Akad F Naturschutz Bad-Württ 23:129-142

Roweck H (1997) Angewandte Aspekte bei der Auswertung des Bornhöved-Projektes. UFZ-Berichte 5:169-178

Scherner R (1995) Realität oder Realsatire der "Bewertung" von Organismen und Flächen. Schriftenreihe f Landschaftspfl u Natursch 43:377-410

5.6 Applying Ecological Goal Functions: Tools for Orientor Optimization as a Basis for Decision Making Processes

Albrecht Gnauck

Abstract

The control of complex dynamic systems requires decisions which are based on multiple contradictory goals, on different evaluation procedures and on distinguished valuation scales. Particularly for ecosystems, these goals are influenced by a high number of uncertain state variables, by the dynamic characterization of system components and their interrelationships, by the kind of anthropogenic actions and by restricted information structures. Furthermore, decision making processes in ecology are coupled with difficulties in formulating ecological goal functions or orientors, where the economic consequences of management alternatives and the social benefits play an important role. Applying multicriteria decision making methods in ecology, optimal compromise solutions between ecological and man-made control operations will be obtained. Particularly methods which are related to Pareto optimality allow valuations of contradictory ecological goal functions. For example, the investigation of a river basin by water quality simulation models allow statements for single important variables of water quality and their ecological significance only. Implementing a water quality model into a decision support system, allows (Pareto-) optimal values of the goal functions to then be computed. An example is presented on how to formulate goal functions for environmental protection and water quality management. Dissolved oxygen, biochemical oxygen demand, the amount of sewage water quantity and costs of enlargement of sewage treatment plants are taken into consideration. As a result, Pareto optimal solutions are computed for different ecological control strategies.

5.6.1 Introduction

Ecological systems are considered as complex dynamic control systems with many of biotic (living) and abiotic (non-living) components and interrelation-

F. Müller M. Leupelt (Eds.)
Eco Targets, Goal Functions, and Orientors

ships. A classification of these types of thermodynamic open systems can be given by the number of components and interrelationships and by the degree of computability of the system behavior (Gnauck 1987). Specific information on the results of control processes within the system and the system response to environmental disturbances can only be obtained by the complete computability of ecosystem states. With regard to the interval [0,1] the system states are characterized by probability density functions which contain information on the changes of the systems behavior, resulting from internal and/or external disturbances (Straškraba and Gnauck 1985). Adapting a definition of complex dynamic systems given by Šiljak (1978), ecosystems are characterized by a high degree of dimension, by uncertainty and by restricted information structures. The first characteristic leads to decomposition of the ecosystem structure and to an analysis of subsystems. Uncertainty is connected with the dynamic characteristics of ecosystems: observability, controllability, perturbability, reachability, robustness and also stability and sensitivity. The restricted information structure of ecosystems provides aggregated information on local, regional or global scales in time and space which are used for ecosystem modeling. This classification allows for the following consequences:

- Ecosystems may be characterized by a high number of components and interrelationships with medium computability for small changes of an ecosystem state.
- The complexity of an ecosystem is expressed by a low computability of the system behavior (strongly organized complexity) as well as by a high number of components and interrelationships (weakly organized complexity).

The application of optimality principles to ecosystems is burdened with difficulties of fundamental nature. In most cases, ecosystem analysis is essentially a conflict analysis characterized by socio-political, socio-economical and environmental value judgments (French 1988; Saaty 1980; Lootsma 1993). The regulations of ecosystem evaluation are open to methods of environmental law (Brandt 1994). As such, many of different methods can be applied to ecosystem analysis and evaluation. Plachter (1991) as well as Bastian and Schreiber (1994) distinguish three levels to formulate ecological goal functions:

1. Characterization of ecosystems by means of natural sciences.
2. Comparison of the functioning of ecosystems and their socio-economic environment.
3. Design and application of ecological standards for environmental planning.

Kinnebrock (1994) stated that optimization processes are natural. This implies that nature uses optimization procedures for its development. In addition, man optimizes most of the activities in their individual life. In ecology it is difficult to arrive at the „best“ solution. This implies that a multi-related ecological process or system will always be understood as a search for acceptable compromise solutions which requires an adequate evaluation methodology (Strassert 1995). Multiple-criteria decision making techniques aim at providing such a set of tools, where Pareto-optimization (Pareto 1896) seems to be an adequate method to evaluate multi-goal functions of ecosystems, if contradictory or conflicting goal functions are taken into consideration. Pareto optimality implies that the better fulfillment of one goal function can have a negative consequence on other goal functions (Gnauck 1987). In this sense, various statements for decision making in ecology will be given and an outline of how to formulate goal functions for ecosystem management. Another aim of this contribution, is to stimulate ideas on developments of optimal control strategies for environmental protection. This is not only a practical question but a theoretical one.

5.6.2 Optimization as a Control Task in Ecology

The investigations of ecosystems allow statements for single important state variables while the control of ecosystems implies situations of decision making which are characterized by multiple states, concurrent ecological goal functions and also, by different evaluation scales and appreciation. In analyzing ecological systems, two control problems arise (Gnauck 1987):

1. Theoretical analysis, mathematical description and simulation of the changing ecosystem behavior.
2. Elaboration of optimal control methods to realize active, goal oriented operations referred to the state and behavior of ecosystems.

The first statement covers both the biocoenotic-structural and functional analysis of an ecosystem and the modeling procedures used to predict the change of ecosystem states. The second statement is directed to the utilization of ecosystems by man and to ecosystem protection. The operating scales and ecological goals are defined by the biotic components of ecosystems and by anthropogenic activities related to ecosystems. As an example, in Table 5.6.1, two concepts of controlling aquatic systems are given, depending on the goals of water quality management. The first concept is directed to obtain optimal system states, while the second, focuses on obtaining a constant operating behavior of the system. An environmental manager knows that the most important quality of an ecosystem functioning correctly, is by exhibiting certain functional, or qualitative properties decided upon as being important (Hokkanen 1997). Once assured that the ecosystem behaves correctly, it is also important that its operating costs are minimal or that it works in optimum time or of whatever performance measure is chosen. While functioning is taken for granted, the latter quantitative properties will often decide the best operation, or otherwise, of the system. In principle, such problems are solved by decision models which consist of simulation models embedded in optimization procedures, resulting in non-unique solutions of the problem. The common goal of such procedures is to recognize all values of control variables which are elements of a specific control domain. The non-uniqueness of the solutions is primarily provided by the differences between control variables and goal variables. Consequently, for ecosystems (and also for landscape ecology and environmental planning purposes) the analysis of a control system is a summary of model building, simulation and scenario techniques in connection with optimization procedures, game theoretical methods and knowledge based expert systems.

Table 5.6.1. Concepts of control for water quality management

Control option	Goal of control	Performance of control	Examples
Optimal control	Extreme values of state variables	Integral criteria	Water quality of surface waters, ground waters, natural ecosystems, landscapes
Constant systems behavior	Fixed dynamic systems behavior	Standards, normative values, rules of operation	Water treatment plants, artificial ecosystems

5.6.2.1 Remarks on Pareto-Optimality

Decision making and search for a compromise is one of the main ecological concepts of organisms and living systems. On a wider scale, these problems are the basis of systems analysis in ecology which are directed to the development of strategies for survival. Pareto-optimality in ecology is characterized by concurrent goal functions of the living components, by alternatives of action, by consideration of multiple goals, by selection of one element of the set of all alternatives and by the individual weighting of one single goal, related to the subjective preference structure of the decision maker. The decision making process is characterized by choice of one objective source of action related to a subjectively valued compromise, on partial satisfaction of the concurrent and conflicting ecological goals. This implies that, decision making in ecology, is always searching for a compromise between contradictory goal functions. Optimization models in ecology deal mainly with some form of cost minimization designed to meet environmental quality and/or ecological standards. Decisions are based on scenarios resulting from simulation models. The individual weighting of the goal functions often depend on their actual reachable system state.

5.6.2.2 The Meaning of Decision Support

Decision support in ecology implies that a final state can be reached by different forms of action starting from an initial state. The final state is described by any number of conflicting criteria. Thus, a decision of an individual means a selection of one (subjective, preferred) vectorial evaluated alternative of action chosen from a set of all objective allowable alternatives to reach the final state (Habenicht 1990). Decision making processes in ecology are mostly based on interactive decision support systems (DSS) with hierarchical structures (Gnauck 1987). The hierarchy is given by different working levels: Preparation level, simulation level, testing level and optimization level. Lewandowski and Wierzbicki (1988) classified DSS by simple software tools (e.g. data bases), by expert systems and knowledge based systems (recognition of special ecological situations) and by model based DSS. From the last classification type the decision maker is supported through the use of vector optimization procedures or by multicriterial selection procedures to generate and/or to select alternative ways of action (Habenicht 1990). In principle, DSS consists of three main components:

1. Component of communication (communication between the DSS and an user (e.g., decision maker)).
2. Component of model and/or knowledge base (formulation of the actual decision problem).
3. Component of problem solving (solving procedures of vector optimization problems).

The computer-aided decision making process in ecology can be divided into three steps. Two, are subjective components and one, an objective component of the decision making process:

1. *Formulation of the decision problem*
   In this subjective component, the decision variables and the contradictory goal functional are defined.
2. *Computation of proposals for decision making*
   In this objective component of decision making a set of efficient alternatives will be generated by multi-objective optimization. The dominant alternatives are selected and stored in the result domain.
3. *Actual decision making process*
   This subjective component of the decision making process covers the computer-aided selection of a compromise by individual weighting of the goal functions by the decision maker.

5.6.3 Multi-Objective Optimization and Multi-Criteria Decision Analysis

In order to solve a decision problem, three requirements have to be fulfilled (Hwang and Masud 1979, Chankong and Haimes 1983):

1. Each solution of the decision problem is an efficient alternative (problem of validity).
2. Each efficient alternative may be a solution of the optimization procedure (problem of non-discrimination).
3. A solution of the decision problem is an efficient alternative, if it exists (problem of identification).

Ester (1987) distinguishes between multiple attribute decision making (MADM) problems and multiple objective decision (MODM) problems. In the first case the alternatives can be explicitly evaluated by discrete values of the goal functions. In the latter case the evaluation of alternatives is implicitly given by functionals and can change continuously in the domain of admissible solutions. In such cases an important feedback between optimization and modeling procedures arise. Global goals of how to reach efficient compromise strategies and algorithms for determining efficient compromise solutions for MODM-problems, are formulated by Peschel (1980).

5.6.3.1 Multi-Criteria Decision Making

A multi-criteria decision making process is given in four different phases (Simon 1960, Steuer 1986):

1. During the first phase (search phase) a system will be analyzed for changes in the structure and the function. If the results of analysis are positive, one or more decision making problems arise as a result.
2. The second phase (modeling phase) is characterized by formulation of the goal functions, by the directions of alternatives, by the modeling process and by the embedding procedure of the simulation model, into the optimization procedure.
3. The third phase (selection phase) is represented by choosing one process of action from the total set of alternatives. This chosen alternative has to be implemented into the system under consideration.
4. The fourth phase is characterized by the decision making process. The decision making problem is represented by a set of alternatives which contribute to reaching the different goals. For a given decision problem the set of allowable alternatives is defined by external (real) conditions.

5.6.3.2 Orientor Optimization

In order to evaluate an ecosystem behavior, two general assumptions have to be fulfilled:

1. The ecosystem state is known for all relevant aspects, in time and space.
2. For all relevant aspects, criteria of evaluation have to be known.

The different criteria are called orientors by Bossel (1992). When evaluating ecosystem development it is necessary to map the ecosystem state using orientors. These are expressed by summarizing indicators describing the system state by set values of the orientor space. Indicators are represented by constraints, by quality criteria and by weighting factors. There are several reasons why multi-criteria or orientor optimization can provide a feasible set of tools in helping ecological and environmental decision making. According to ecosystem dynamics several different points of view have to be taken into consideration. The ecosystem information are uncertain. They are based on past events and on the history of the ecosystem. Therefore, assumptions on the ecosystem behavior in time and space have to be made. The number of decision makers, representing several properties of an ecosystem, is large, as well as the number of different interest groups, involved directly or indirectly in the planning and decision making processes. In this sense, a single objectively derived best solution does not generally exist, and problem solving without any computational support may distort the final results, as one cannot control all criteria simultaneously. Orientor optimization is used as a decision aid in environmental planning processes in order to clarify the planning processes necessary for decision making, to avoid diverse distortions, and in order to control all the information, criteria, uncertainties, and degree of importance. Solving ecological or environmental problems is possible to identify ecologists engaged in some or all of the following activities:

- Defining the problem for environmental policy making.
- Analyzing and modeling different ecological situations (scenarios).
- Designing possible solutions in the form of alternatives.
- Carrying out a detailed evaluation of the proposed alternative solutions in terms of their ecological feasibility, cost effectiveness, probable ecological effects on different populations, and political acceptability.

5.6.4 Pareto-Optimization for River Water Quality Management

The management of freshwater ecosystems requires mathematical models for different time horizons which allow a process control related to special goals of water resources management. Water quality management strategies mainly deal with some form of cost minimization designed to meet river water quality and effluent standards. To manage the water quality sub-optimal rules (water quality standards, effluent conditions, conditions of pollutant input) are set up, where simple optimization procedures are mostly related to only one goal function. Hence, management decisions were derived from by scenarios resulting from simulation models. By way of contrast, decision support models may be seen as developments in water quality modeling. These which give out Pareto-optimal control designs for conflicting goal functions of the water uses in catchment areas.

5.6.4.1 Structure of the DSS CADS

One of the broadly described DSS is CADS (Computer Aided Decision System) developed by Straubel and Wittmüss (1983). It summarizes the following structure (Model et al. 1995):

Preparation level. General arrangements for computing, input of external process model parameters, input of parameters for optimization runs.

Learning level. Analysis of the time behavior of the process model, investigation of the influence of different management strategies on the process model time behavior in dependence of changing external parameters and of the performance of the goal functions, learning how to choose a suitable management strategy by game theoretic methods.

Testing level. Check for reachability of control targets related to the goal functions, search for a special control (management) strategy to reach given targets within a chosen time horizon, computation of the individual optima of the goal functions.

Optimization level. Computation of Pareto-optimal solutions by ranking the goal functions, by relaxation methods or by computation of the set of all compromise points with decreasing a-priori knowledge of the decision maker.

The following optimization procedures are implemented in the computer package:

- Numerical gradient approximation procedure (for concave or convex continuous problems).
- Simulation of Darwinian evolution strategy (for non-concave or non-convex continuous and/or integer problems).
- Mixed optimization procedures (for non-concave or non-convex continuous and/or integer problems).
- A thermodynamic method (simulated annealing).
- Branch and bound procedures (to solve special problems).
- Vector optimization on graphs (for search of routes on state-dependent vectorial evaluated arcs).

The DSS CADS works in an interactive dialogue form where no special requirements are necessary for the type of process equations and for the mathematical formulations of the goal functions. Time-dependent restrictions of the management variables in the form of lower and upper bounds and other implicit formulated restrictions between state and management variables, are also taken into account.

5.6.4.2 Decision Making for Water Quality Management

In contrast to the use of water quality simulation models only a few applications of decision support systems for solving water quality problems exist (Hahn and Cembrowicz 1981; Krawczak and Mizukami 1985; Cembrowicz 1988; Model, Wittmüß and Gnauck 1995, Ivanov et al. 1995). Most of the river water quality problems are related to the interactions between the discharged matter and river organisms. The primary mechanisms for conservative non-degradable wastes are transport and dilution while degradable waste concentrations are described mathematically by considering the natural transport and decay processes. Because a complex relationship exists between the biochemical oxygen demand (BOD) of the waste discharged and the dissolved oxygen (DO) concentration at points downstream of the discharge, this relationship has become one of the main indicators governing river water quality. The DSS CADS was used to find out Pareto-optimal solutions for water quality management options. For this reason, a modified Streeter-Phelps type model (Streeter and Phelps 1925) SPROX describing parts of the River Spree (Gnauck 1984) was embedded into the DSS. The River Spree is characterized by a high level of eutrophication and by banked-up water levels with different water utilization in the catchment area. The model takes into account the input of constituents into the river from the drainage area, the transport of constituents along the river, physical, biological, chemical and biochemical reactions within the aquatic ecosystem, and also the basic load of constituents caused by the underlying geology of the river bed and the drainage area.

The DO - BOD interaction model equations are formulated as follows:

$$\begin{aligned} dDO^{(i)}(t)/dt = \; & -(K_2(TW) + Q^{(i)}/V^{(i)})DO^{(i)}(t) - K_1(TW)BOD^{(i)}(t) \\ & + Q^{(i)}/V^{(i)}\, DO^{(i-1)}(t-t_F) + K_2(TW)\, CS(TW) \\ & + DOE^{(i)}(t)\, QE^{(i)}/V^{(i)}\, u^{(i)}(t) \end{aligned} \qquad (5.6.1)$$

$$\begin{aligned} dBOD^{(i)}(t)/dt = \; & -(K_1(TW) + Q^{(i)}/V^{(i)})BOD^{(i)}(t) \\ & + Q^{(i)}/V^{(i)}\, BOD^{(i-1)}(t-t_F) \\ & + BODE^{(i)}(t)QE^{(i)}/V^{(i)}u^{(i)}(t), \end{aligned} \qquad (5.6.2)$$

where t - time variable, DO - dissolved oxygen concentration (mg/l), BOD - concentration of biological oxygen demand (mg/l), K_1 - BOD decay rate constant, K_2 - reaeration rate constant for DO, Q - mean volumetric flow rate, V -mean volume of a river segment, TW - water temperature (^oC), CS - saturation concentration of DO, i - number of segment, E - input of DO or BOD, t_F - time of flow.

DO saturation concentration within a segment:

$$CS(t) = 14.65 - 0.41022TW + 0.007991TW^2 - 0.0000474TW^3. \qquad (5.6.3)$$

Annual time course of water temperature TW(t):

$$TW(t) = 13.16 + 10.23\cos(2\pi\, t-213)/365). \qquad (5.6.4)$$

Because a one-dimensional model was used segmentation was done in longitudinal direction by 11 segments, where segmentation cuts are performed in such a way, that the ensuing segments represent river stretches with close to homogeneous characteristics. The global parameters describing DO-producing and DO-consuming reactions as well as pertinent hydraulic parameters (flow rate, flow velocity etc.) considered as constant for each segment. Segments are also determined by tributaries and essential waste water inputs. Inputs of organic load are located at the beginning of a segment (if any), where a segment is considered as a continuous stirred tank reactor with complete mixing approximated. Water quality variables observed, are given at the beginning and at the end of each segment. Then the output of the i-th segment will be the input of segment (i+1).

According to the chosen model structure, the goal functions were formulated as extreme values, dependent on the ecological and socio-economical standards. The optimization results presented are valid for mean flow conditions.

On the basis of DSS CADS two management alternatives are considered.. The individual optima of the goal functions were computed to obtain information on the decision space. The results are presented in Table 5.6.2.

The goal functions are symbolized by $f_i(t)$:

1. Efficient control of waste discharges from point sources.
   The goal functions are given by
   f_1 - mean value of dissolved oxygen (DO) ⇒ Max
   f_2 - mean value of biochemical oxygen demand (BOD) ⇒ Min
   f_3 - total amount of waste discharge from point sources (QE) ⇒ Max
2. Cost analysis for reconstruction of existing or construction of new treatment plants.
   The goal functions are given by
   f_1 - mean value of dissolved oxygen (DO) ⇒ Max
   f_2 - mean value of biochemical oxygen demand (BOD) ⇒ Min
   f_3 - total investment costs ⇒ Min

For the first management strategy, a maximum of waste water input will be reached for a minimum value of DO while the amount of degradable waste is medium. The enlargement of sewage treatment plants show a low DO value connected with a high BOD value as a zero cost effect.

Table 5.6.2. Domain of Pareto-optimal solutions for two river water quality control options (The underlined values represent the individual optima of each goal function)

Control option	Goal function F(1) DO (mg/l)	Goal function F(2) BOD (mg/l)	Goal function F(3) QE (m³/a), Mio DM
Control of purified wastewater input	<u>10,4</u>	32,8	$50,1*10^6$
	10,4	<u>32,8</u>	$50,1*10^6$
	9,7	56,5	<u>$55,1*10^6$</u>
Enlargement or reconstruction of sewage treatment plants	<u>10,5</u>	40,8	87,6
	10,5	<u>40,8</u>	87,6
	8,7	100,0	<u>0,0</u>

Table 5.6.3. Optimal solutions for river water quality management

Management alternative	DO (mg/l)	BOD (mg/l)	Costs (Mio. DM)/QE (m³/a)
Waste water input	9,85	52,3	54,00
	10,00	46,8	53,35
	10,14	42,1	52,37
	10,32	36,6	50,72
	10,42	32,8	50,06
Enlargement or recon-	8,78	101,0	0,00
struction of sewage	8,78	100,7	2,00
treatment plants	8,79	100,6	3,20
	8,79	100,5	4,00
	8,79	100,3	5,20
	8,80	100,3	6,00
	8,80	100,1	7,20
	8,84	98,9	8,50
	8,84	98,7	10,50
	8,85	98,5	11,70
	8,88	97,2	14,20
	8,89	*97,0*	*16,20*
	9,49	*76,6*	*16,50*
	9,50	76,3	18,50
	9,51	75,9	21,70
	9,55	74,7	25,00
	9,56	74,1	28,20
	9,60	*72,8*	*31,50*
	10,20	*52,2*	*33,00*
	10,22	51,7	37,00
	10,26	50,2	41,50
	10,31	48,5	47,20
	10,32	48,3	54,00
	10,34	47,1	79,60

In each case the computed values span the space of Pareto-optimization solutions. In Table 5.6.3, Pareto-optimal solutions for the investigated problems of the whole river are presented. Only five optimal solutions could be computed for the development of an optimal management strategy of waste water input from point sources to the river. For a low amount of purified waste water input an acceptable yearly mean concentration of DO is obtained in accordance with a low BOD concentration. Because of the small differences in DO, the decision criterion should be the yearly mean BOD concentration. A broader basis for decision making is given for the enlargement or reconstruction of waste water treatment plants. The yearly mean DO concentration increases with increasing costs while the BOD concentration decreases. This is very clear and was expected. However, two places indicated in the list are significant for decision making. The first place is characterized by DO=8.89, BOD=97.0 and costs=16.20 Mio. DM. For a slightly higher budget (costs=16.50 Mio. DM) a decrease of BOD concentration of about 20 mg/l and an increase of DO concentration to 9.49 mg)l will be reached. An analogous cost effect will be obtained for the second place. An increasing budget from 31.50 Mio. DM to 33.00 Mio. DM results in a diminishing BOD concentration of about 20 mg/l. In both cases the effect on DO concentration gives an increase of DO of 0.6 mg/l.

Table 5.6.4. Pareto-optimal solutions for important pollution inputs

Number of segment	Capacity (m^3/d)	Capacity (%)	BOD (mg/l)	BOD (kg/d)	Costs (Mio. DM)
3	0	0	140	3013	0,00
	10750	50	82	1765	16,50
	21500	100	24	516	33,00
7	0	0	580	4009	0,00
	1037	15	561	3878	2,00
	2074	30	542	3746	4,00
	4500	65	426	2945	8,50
	4835	70	348	2405	9,00
	5875	85	329	2274	11,00
	6912	100	310	2143	13,00
9	0	0	94	2282	0,00
	1250	5	83	2015	2,00
	2500	10	78	1894	2,00
	3750	15	73	1772	6,00
	5000	20	66	1600	8,00
	7500	30	59	1491	11,20
	18750	75	45	1122	30,40
	22000	90	28	702	33,60
	23750	95	20	514	39,60
	25200	100	13	333	41,60

The suitable Pareto-optimal solution will then be chosen by the subjective dominated preference structure of the decision maker. The above described decision making process was carried out in detail for all river segments to obtain Pareto-optimal solutions. Predominantly polluted river reaches are; segments 3, 7 and 9.

Table 5.6.4 shows the Pareto-optimal results for different capacities of the treatment plants for a 85%-BOD-removal. According to the available budget a combination of different enlargements and/or reconstruction of sewage treatment plants can be taken into consideration for environmental planning processes. Table 5.6.5 shows various scenarios for Pareto-optimal decision making. Depending on planning constraints, monetary constraints and regional environmental planning objectives, a subjective decision is made on an objective computational basis.

Table 5.6.5. Scenario analysis for Pareto-optimal decision making

Number of scenario	Number of River segment	Enlargement of sewage treatment plant (%)	Capacity (m^3/d)	Costs (Mio. DM)	Sum (Mio. DM)
1	3	100	21.500	33,00	33,00
2	9	90	22.000	33,60	33,60
3	3	100	21.500	33,00	36,20
	9	10	2.500	3,20	
4	3	100	21.500	33,00	63,40
	9	75	18.750	30,40	
5	3	100	21.500	33,00	74,60
	9	100	25.000	41,60	
6	3	50	10.750	16,50	64,60
	7	65	4.500	8,50	
	9	95	23.750	39,60	
7	3	50	10.750	16,50	40,70
	7	100	6.912	13,00	
	9	30	7.500	11,20	
8	3	50	10.750	16,50	32,20
	7	65	4.500	8,50	
	9	30	7.500	11,20	
9	3	100	21.500	33,00	87,60
	7	100	6.912	13,00	
	9	100	25.000	41,60	
10	3	100	21.500	33,00	48,00
	7	70	4.835	9,00	
	9	15	3.750	6,00	

5.6.5 Conclusions

The use of control theory in ecology is an approach which promises greater theoretical understanding of complicated natural processes at the complex ecosystem level. Empirical studies, classical biological investigations and mathematical models have been mainly directed, to superficially describe the structure and/or behavior of ecosystems. The shift of interest towards the forces decisive for the macroscopic, holistic features of ecosystems is fairly similar to that of the thermodynamic approach. The study of goal functions can be seen as a study of mechanisms of change in ecosystems, rather than a study of ecosystems as they are observable at a given instant of time, or at a given site and situation. Such study is far from simple, because there is no direct method to observe a goal function of a whole ecosystem or its subsystems. Until now, it has not been distinguished which formulations of optimality are most appropriate for ecosystems, and how far different goal functions are equivalent. This is also a question of practical concern. The method of solution of the optimal problem depends on the formulation. In this context multi-objective optimization, especially Pareto-optimization is of growing interest to solve ecological problems. Game theoretic formulations of goal functions appears to correspond most closely to what is occurring in nature. However, no solutions are available at present for situations remotely close to what is of interest from an ecosystems point of view. Therefore, orientor optimization could provide an outline for the utilization of ecosystems and for ecotechnological purposes.

References

Bastian O, Schreiber KF (1994) Analyse und ökologische Bewertung der Landschaft. Fischer, Jena Stuttgart

Bossel H (1992) Modellbildung und Simulation. Vieweg, Braunschweig Wiesbaden

Brandt E (1994) Umweltaufklärung und Verfassungsrecht. Blottner, Taunusstein

Chankong V, Haimes YY (1983) Multiobjective decision making – theory and methodology. North-Holland, Amsterdam

Cembrowicz RG (1988) Siedlungswasserwirtschaftliche Planungsmodelle. Springer, Berlin Heidelberg New York

Ester J (1987) Systemanalyse und mehrkriterielle Entscheidung. Verlag technik, Berlin

French S (1988) Decision theory – an introduction to the mathematics of rationality. Ellis Horwood, Chichester

Gnauck A (1984) Das Selbstreinigungsmodell SPROX. F/E-Bericht Inst für Wasserwirtschaft, Berlin

Gnauck A (1987) Kybernetische Beschreibung limnischer Ökosysteme. Thesis B, Fak Bau-, Wasser- und Forstwesen, Techn Univ Dresden, Dresden

Hokkanen J (1997) Aiding public environmental decision making by multicriteria analysis. Univ of Jyväskyla, Jyväskyla

Habenicht W (1990) Neuere Entwicklungen auf dem Gebiet der Vektoroptimierung. Arbeitspapier 1/90. Univ Hohenheim, Stuttgart

Hahn HH, Cembrowicz RG (1981) Model of the Neckar River, Federal republic of Germany. In: Biswas AK (ed) Models for water quality management. McGraw-Hill, New York, pp 158-221

Hwang CL, Masud ASM (1979) Multiple objective decision making – methods and applications. Springer, Berlin Heidelberg New York

Ivanov P, Masliev I, Kularathna M, Kuzmin A, Somlyody L (1995) DESERT - Decison support system for evaluating river basin strategies. WP-95-23, IIASA, Laxenburg

Kinnebrock W (1994) Optimierung mit genetischen und selektiven Algorithmen. Oldenbourg, München Wien

Krawczak M, Mizukami K (1985) River pollution control as a conflict. In: Malanowski K, Mizukami K (eds) Constructive aspects of optimization. Polska Akad Nauk, Inst Badan System Ser, Tom 6, Warsaw, pp 219-239

Lewandowski A, Wierzbicki AP (eds) (1988) Theory, software and testing examples in decision support systems. WP-88-071, IIASA, Laxenburg

Lootsma FA (1993) Scale sensitivity in the multiplicative AHP and SMART. J Multi-Criteria Decision Anal 2:87-110

Model N, Wittmüß A, Gnauck A (1995) Berechnung von Sanierungsstrategien für Fließgewässer. In: Gnauck A, Frischmuth A, Kraft A (eds) Ökosysteme - Modellierung und Simulation. Blottner, Taunusstein, pp 174-204

Pareto V (1896) Course of economic politique. Pouge, Lausanne

Peschel M (1980) Ingenieurtechnische Entscheidungen. Verlag Technik, Berlin

Plachter H (1991) Naturschutz. Fischer, Stuttgart

Saaty TL (1980) The analytic hierarchy process, planning, priority setting, resource allocation. McGraw-Hill, New York

Šiljak DD (1978) Large scale dynamic systems: stability and structure. Elsevier-North Holland, New York

Simon H (1960) The new science of management decision. Harper & Row, New York

Steuer RE (1986) Multiple criteria optimization – theory, computation, and application. Wiley, New York

Straškraba M, Gnauck A (1985) Freshwater ecosystems – modelling and simulation. Elsevier, Amsterdam

Strassert G (1995) Das Abwägungsproblem bei multikriteriellen Entscheidungen. Peter Lang, Frankfurt/Main

Straubel R, Wittmüß A (1983) Ein interaktives Entscheidungsmodell zur optimalen Steuerung eines Systems mit mehreren Zielfunktionen. Messen-Steuern-Regeln 26:2-4

Streeter HW, Phelps B (1925) A study of the pollution and natural purification of the Ohio River. Publ Health Bull 146:1-25

5.7 Marine Ecological Quality Objectives: Science and Management Aspects

Folkert De Jong

Abstract

An overview is presented of the discussion concerning the political need for, and scientific possibilities of developing ecological targets for the Wadden Sea and the North Sea. A review is given of various definitions of ecological objectives. Existing ecosystem goals are evaluated for their management usefulness and scientific credibility with specific emphasis to the Wadden Sea eco target concept. The desire to develop clear and quantified ecological targets is discussed within the framework of a rational management concept. It is concluded that specific targets do not fit the emerging notion that ecosystems are dynamic and unpredictable and that social and economical factors play an important part in environmental management. The concept of adaptive management is discussed as a possible approach to holistic environmental management.

5.7.1 Introduction

Public policies for the prevention and mitigation of negative effects of human activities on the environment have changed substantially since the end of the 1960s. Since the early days of this development the need for and feasibility of quality objectives for the environment have been discussed. Environmental objectives applied in practice were of a sectoral character, for example critical limits of hazardous substances in sea water. With the increasing need for integrative environmental policies, the demand for indicators of ecosystem quality became apparent.

Ecological quality objectives are not a new phenomenon. There exists a host of objectives in national and international treaties, laws, conventions, agreements and declarations. They contain expressions like 'ecosystem health', 'a sustainable ecosystem', 'a natural ecosystem' and are generally of a qualitative nature.

The wish to develop specific quantified objectives for the quality of ecosystems is of a more recent nature. Since the Brundtland Report and the Earth Summit, the awareness of the need for preserving natural resources for future

F. Müller M. Leupelt (Eds.)
Eco Targets, Goal Functions, and Orientors

generations has increased. The main rationale is an anthropocentric one: the interest of the natural resource capital must satisfy the needs of future generations. History has shown that the conservation of nature for the sake of Man is the predominant option (Wildes 1995). In the development of natural resource management strategies there will therefore be an on going and probably increasing demand for indicators of which nature can cope with.

On many occasions politicians have called for the development of Ecological Quality Objectives based upon sound scientific research. This is understandable since it would be more acceptable to have precise indicators of the limits of human use of ecosystems.

In practice this has proven to be not so easy. Firstly there is the problem of choosing indicators of ecosystem health, and after they have been defined, the indicators have to be validated. The more precise the choice, the more vulnerable to scientific critique the set of indicators are. Consequently, the lower its credibility and the chance of reaching a consensus. These two factors are essential to the effect science may have on politics (Underdal 1989; Boehmer-Christiansen 1995).

Mainly because of its inherent uncertainties, scientific information is not a sufficient basis for decision making and must be supplemented by discretion (Jasanoff 1990). For this reason, precautionary action has been a central element in US environmental policies since the mid 1970s. The precautionary approach entered the political arena in Europe in the second half of the 1980s. An important impetus for the adoption of this political principle is seen in the controversy between the United Kingdom and most continental states concerning the application of an emission or immission approach to environmental policies (Gündling 1989). The emission approach was practiced by the UK and based upon the principle of environmental capacity which was defended on scientific grounds (Boehmer-Christiansen 1989).

The precautionary approach as adopted at the 2nd International North Sea Conference (London 1987) reads as follows: 'in order to protect the North Sea from possibly damaging effects of the most dangerous substances, a precautionary approach is necessary which may require action to control inputs of such substances even before a causal link has been established by absolutely clear scientific evidence' (London Declaration §VII).

Since the London Conference, the precautionary approach has been the cornerstone of North Sea environmental policies, but in recent years some disadvantages of this policy have become clear. At the last (4th) North Sea conference (Esbjerg 1995) it was decided to phase out discharges of dangerous substances within one generation (25 years). This decision can be valued as the ultimate victory of the precautionary and, consequently, the emission approach. It has, on the other hand, not been possible to reach agreement on a further tightening of the measures for the reduction of nutrient inputs, nor on the designation of a fishing-free area, not even for scientific purposes. It is unlikely that this will change at future North Sea Conferences. Regarding nutrients even an alleviation of past agreements may be possible: It has been argued that reductions of nutrient inputs negatively affect fish stocks (see amongst others Boddeke 1993) and it has been proposed to investigate options for fertilizing the sea (MacKenzie 1996).

The precautionary approach is implemented mainly through uniform emission standards. It is this uniformness that is currently being criticized, for example by Gray (1996). He argues that the uniform application of emission standards for nutrients in the North Sea has little or no effect in the open sea areas. Gray poses the question whether 'It is ethical to spend vast amounts of money in preventing nutrients entering the Skagerrak where the effects will not be observed, when spending the same amount of money, in say Poland, would lead to enormous improvements in the environment' (Gray 1996). Also here, supplementing the precautionary approach with quality objectives for the ecosystem, would probably improve environmental policies.

This article sets out to describe and analyze the development of ecological quality objectives, with specific emphasis on developments in the framework of the trilateral Wadden Sea cooperation. The experiences with the trilateral eco-target concept will be discussed in relation to the role of ecological knowledge in environmental policy making.

5.7.2 The Wadden Sea and the North Sea

The Wadden Sea is a shallow sea extending along the North Sea coasts of The Netherlands, Germany and Denmark. It is a highly dynamic ecosystem with tidal channels, sands, mudflats, saltmarshes, beaches, dunes, river mouths and a transition zone to the North Sea; the offshore zone.

During low tides large parts of the Wadden Sea emerge. These so-called tidal flats cover about two-thirds of the tidal area and are one of its most characteristic features. With a total length of 500 km it is the largest unbroken stretch of mudflats in the world. They account for 60 % of all tidal areas in Europe and North Africa. Like most coastal areas in the world, the Wadden Sea is under intensive human use. In the eleventh century, drainage of peatlands for agricultural purposes began, followed by large scale building of dikes (Knottnerus 1996).

To date the most important human activities are tourism and recreation, shellfish fisheries, exploitation of natural gas, military training and commercial shipping, most of which have increased since World War II as a result of economic growth. The relatively high level of pollution of the Wadden Sea is caused by its location at the rim of the highly populated northwestern Europe. Three major rivers, the Ems, the Weser and the Elbe, exit into the Wadden Sea. Their catchment areas add up to some 231.000 km^2 and extend up to the Czech Republic-Austrian border. Moreover, pollutants are emitted into the Wadden Sea by air and via North Sea currents.

A comprehensive overview of the quality status of the Wadden Sea ecosystem and the impact of human activities is given by De Jong et al. (1993).

In the early 1970s environmental scientists stated that the Wadden Sea ecosystem cannot be divided according to national borders. The Wadden Sea is, from an ecological point of view, one system. The politicians from the three Wadden Sea countries were called upon to work together in the protection and conservation of the area. The first trilateral governmental conference on the protection of the Wadden Sea was held in 1978 in The Hague, The Netherlands. The second Wadden Sea Conference took place two years later in Bonn, Germany. In 1982, at the third Conference in Copenhagen, Denmark, a Joint Declaration was agreed upon by the three countries. According to the Joint Declaration, the Wadden Sea countries declared their intention to coordinate their activities and measures in order to implement a number of international legal instruments in the field of natural environmental protection, amongst others the Ramsar Convention and the EC Bird Directive. Since 1982 four more Governmental Wadden Sea Conferences were held and the trilateral cooperation strengthened and intensified.

Trilateral Wadden Sea policies are mainly conservation orientated. To a large extent, Wadden Sea water quality policies have been determined by developments in the Oslo and Paris Conventions and the North Sea Conferences. The three Wadden Sea countries, The Netherlands, Germany and Denmark have participated in the North Sea Conferences and the Oslo and Paris Conventions.

The Oslo Convention of 1972 regulates sea based discharges, such as dumping. The Paris Convention of 1974 aimed at reducing land based discharges into the Convention Area defined as the northeast Atlantic. In 1992, the two Conventions were merged. In 1984 the first ministerial North Sea Conference was held in Bremen, Germany. The North Sea Conferences, the fourth of which was held in 1995 in Esbjerg, Denmark, have been an important impetus for tightening the North Sea environmental protection regime.

Since the beginning of the 1990s, North Sea and Wadden Sea policies have developed into more integrated ecosystem policies (De Jong 1994). This was reflected by political decisions to develop ecological quality objectives for both systems.

5.7.3 Ecological Quality Objectives: Pros and Cons

At the sixth trilateral Governmental Conference in Esbjerg in 1991 the trilateral policy was embedded in a common structure consisting of a Guiding Principle, a number of Management Principles and a set of common objectives for human use. The Guiding Principle of the trilateral Wadden Sea policy, as agreed at the Esbjerg Conference in 1991 is 'to achieve, as far as possible, a natural and sustainable ecosystem in which natural processes proceed in an undisturbed way'. It was also agreed that this Guiding Principle be specified by means of common Ecological Targets and measures to reach those targets.

The decision to develop Ecological Targets for the Wadden Sea ecosystem was not an isolated case (de Jong 1992a): At the 3rd North Sea Conference, which had been held one year earlier in The Hague, the ministers of the eight North Sea riparian states had already decided 'to elaborate techniques for the development of ecological objectives for the North Sea and its coastal waters' (§35.1.iii). European Community work on the development of a Directive for the Quality of Water started in the early 1990s.

In addition, at the national level, initiatives had been taken to extend sectoral policies, which were mainly based upon the setting of uniform emission standards, to integrate ecosystem policies with the aid of ecological objectives for the ecosystem. In The Netherlands the so-called Amoeba concept had already been adopted in 1990 as the future objective for the status of the Dutch North Sea. A comprehensive description of the concept is given by Ten Brink et al. (1991).

The development of Ecological Quality Objectives was accompanied by intensive discussions concerning the possibility of reaching its requirements. In The Netherlands the discussion focused mainly on suitable methodologies and principles and not on political aspects. The Amoeba concept was questioned because of its ecological starting points; the representation of the North Sea ecosystem by a set of indicator species (Dankers et al. 1991; Dankers & de Vlas 1994). Dankers et al. (1991) promoted the use of processes as ecological objectives. A first attempt was made to integrate the different physical and biological processes and to propose ecological reference states. The result of this critique found that there was insufficient support for applying the Amoeba concept to the environmental policies of the Dutch Wadden Sea. In Germany both the political rationale and the ecological feasibility of Eco-logical Quality Objectives were hotly debated (excellent overview SDN 1994).

Those in favor of quality objectives argued that the establishment of objectives is inherent to policy and management, because politics and policy are essentially dealing with the balance of conflicting interests. A lack of clear objectives is a disadvantage to political negotiations and in the day to day nature management. The inventor of the Amoeba concept, Ben Ten Brink, explained that the development of quantitative conservation goals were mainly motivated by the urge to 'have a weapon in the battle' against nature deterioration (Van der Windt 1995). In the introduction to the above mentioned SDN Report, the environment minister of the German state of Schleswig-Holstein, Müller, stated clearly that: 'Only with such Quality Objectives can future administrative interventions and additional expenditures for North Sea protection be made understandable to citizens and politicians' (Müller 1994).

The opponents argued that the use of ecological quality objectives may easily lead to a 'fill-up' policy. Some even feared that an immission oriented policy would undermine the precautionary principle (Lutter 1991; Hoppenheit 1992). A policy of precaution combined with the application of best available techniques and best environmental practices is sufficient in environmental protection (Hoppenheit 1992; Reise 1994). Negative effects of fisheries, for which the precautionary principle is no straightforward solution, could be solved by the designation of fishing-free zones (Reise 1994).

Secondly, debates have arisen about the feasibility of the development of ecological quality objectives. Opponents stressed the variability of the marine ecosystem (Hoppenheit 1992; Reise 1994), questioned the representative of so-called key or indicator species in terms of the quality of the eco-system (Dankers et al. 1991), and emphasized the scientific problems attached to so-called emergent properties of ecosystems such as stability and resilience (Reise 1994).

Advocates of marine ecological quality objectives acknowledged this critique but highlighted the need for pragmatism and simplicity (especially, in dialogue with politicians) (Colijn 1994), or stated that a complex ecosystem requires a complex, integrated assessment (Borchardt 1994).

5.7.4 The Many Faces of Ecological Quality Objectives

The complex nature of the ecological quality objectives discussion has evolved primarily because scientific and political arguments have become mixed up (De Jong 1992b). In addition, the definitions used were very different, causing much confusion. Friedrich and Claussen (1994) reported that the German Working Group for the development of marine ecological quality objectives (AG UQZ) started with the definition agreed in the North Sea Conference framework: "the ecological quality objective is the desired level of the ecological quality relative to the reference level".

In the search for suitable parameters, the following criteria were applied by AG UQZ:

- indicator of certain anthropogenic impacts for which reduction measures can be given;
- the range of natural variation of the indicator must be known;
- natural and anthropogenic causes must be distinguishable;
- a causal relationship between anthropogenic factors and effects must exist;
- the quality objective must be inferred by scientific method.

Knauer (1994) defined ecological quality objectives as direct goals for ecosystems, landscapes etc. from an ecological perspective. Examples are historical reference situations, surface areas, minimum areas and species composition.

Borchardt (1994) defined ecological quality objectives as long-term goals which must guarantee that natural processes can proceed undisturbed, that chemical substances occur in natural concentrations and that the diversity of landscape elements is maintained.

Reise (1994), however, considered the wish to develop ecological quality objectives as the result of a mechanomorphic vision of nature, as it could also be found in ecological research. He took into consideration concepts such as functional mechanisms and stability properties and interpreted the objective behind ecological quality objectives as the wish to return to a stable situation after disturbance (by human activities).

The objective of the Dutch approach, the Amoeba concept, is a sustainable ecosystem. This objective is operationalized by a number of indicator species which were selected on the basis of representativity, balanced distribution for different compartments, political appeal, available knowledge and human impact (Colijn 1994). It can easily be concluded that there are substantial differences in both the notions of what ecological quality objectives are (should be) and what purpose they should serve. What makes it even more complicated is that both are interrelated. Table 5.7.1 lists the qualities which ecological quality objectives must possess, should all definitions be applied at the same time. It will be very difficult, if not impossible, to find parameters which meet all these conditions.

Table 5.7.1. Cumulative list of conditions for Ecological Quality Objectives, proposed by different authors (see text)

Ecological Quality Objectives: Conditions
• Indicative of Human Impact
• Quantitative
• Causal Relationships must be known
• Scientifically Deducible
• Realistic Monitoring Requirements
• Attractive to Politicians
• Representative Ecosystem Compartments

5.7.5 Scientific Credibility and Management Usefulness

Ecological quality objectives should be adaptable in practice and backed by (sufficient) scientific knowledge. The role of science in policy-making is predominantly determined by its credibility. Uncertainty and lack of consensus are the two main factors which reduce scientific credibility (Underdal 1988; Miles 1988). Management usefulness and scientific credibility often seem to have an inverse relationship. In the following diagram (Fig. 5.7.1) a number of ecological quality objectives is presented according to an estimation of their scientific credibility and their (potential) usefulness in management.

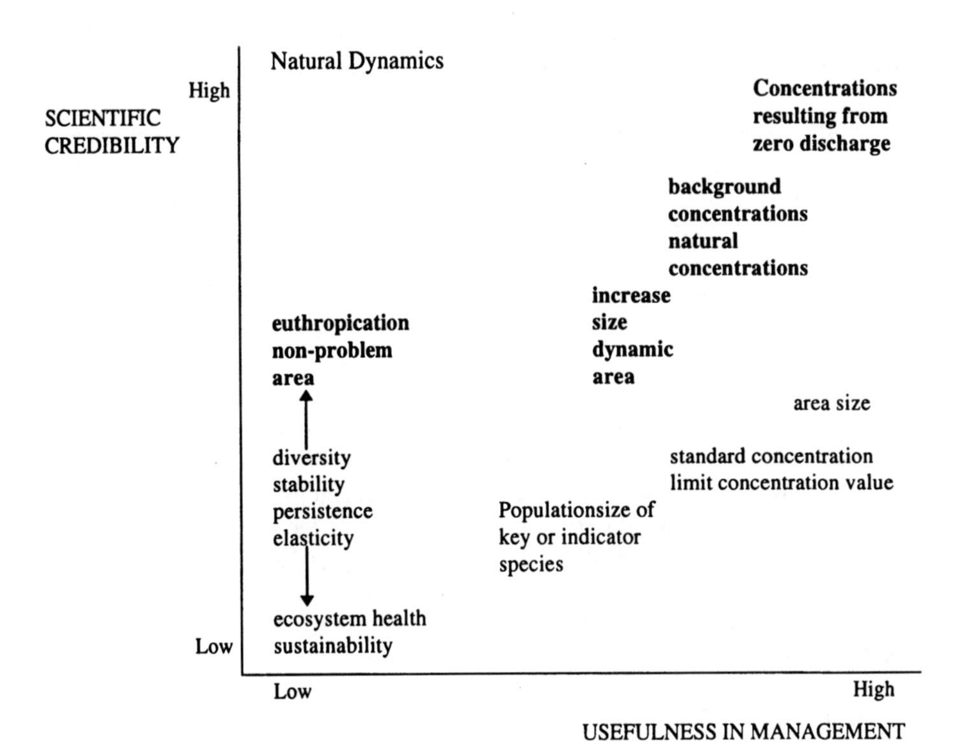

Fig. 5.7.1. Relative scoring of various ecological objectives for their scientific credibility and usefulness in management

The examples were taken from very different sources, ranging from general political agreements to specific concepts such as the Amoeba and the eco target concepts. When interpreting the diagram it should be kept in mind that:

- the scales are qualitative and not necessarily linear;
- the items listed in the diagram are of a very different nature; their only common denominator is their status as ecological quality objectives. The adjective 'ecological' is interpreted to include quality objectives covering the whole ecosystem as well as biological, physical and chemical elements of the ecosystem. According to some classifications, not all entries have been defined as ecological quality objectives but as guiding principles, standards etc.. In the diagram such hierarchical differentiations have not been made;
- the entries in bold are from the trilateral eco target concept.

The entries in the diagram are briefly elucidated below:

Natural Dynamics. The goal 'to achieve a natural and dynamic ecosystem' can be found in various policy documents and laws. Because of its generality it is of little use in management. The scientific credibility or, in this case the lack of scientific controversy concerning this goal, will be relatively high.

Diversity, Stability, Persistence, Elasticity. These emergenting ecosystem properties have become standard expressions in policy. These concepts have led to controversy amongst ecologists. As a result, the position in the diagram may differ depending on the school of ecological thought. Cramer (1988) has classified ecology as a pre-paradigmatic science, which is characterized by an open discussion about its theoretical basis. Attempts have been made to unify theories, but these have not found general support and are subject to increasing critique (Zonneveld 1989; Bröring and Wiegleb 1990; Peters 1991; Bradbury 1996).

Despite the scientific critique, a number of concepts such as stability, diversity, ecosystem health, persistence and resilience, have found their way into environmental policies. Dekker and Knaapen (1986), for example, describe the profound impact of the stability-diversity hypothesis and the island biogeography theory on Dutch nature conservation policies despite increasing scientific critique. They suspect that the ethical desire for integrity and stability of nature has been an important factor in the popularity of this hypotheses and their success in nature conservation. Dekker and Knaapen (1986) use the term 'political career' for such successful influences of ecological concepts. Biodiversity is a recent example of a 'political career' manifested by the adoption of the Biodiversity Convention in 1992. Generally however, the influence of such abstract theories and concepts does not go beyond the political level. From a management perspective they are not very useful, as long as they have not been operationalized.

Ecosystem Health. This approach has already been integrated in policy for a period of time. The assessment of the 'health' of the marine ecosystem will only be possible after the concept has been made operational. According to Sherman (1994) 'a single precise scientific definition is problematical'. This author has listed 'variables' which may be used for indexing changes in ecosystem health. These include stability 'variables' such as resilience, resistance and sustainability and complexity 'variables' such as species richness and connectivity (Sherman 1994).

Sustainability. The sustainability concept has been applied to economy, i.e., in the sense of sustainable development, and as an ecosystem property i.e. a sustainable ecosystem. The concept seems of little value in nature management without further specification (compare Dankers and De Vlas 1994). In terms of ecology, the definition of sustainability is uncertain. Higler et al. (1993) state that there are many fundamentally different ideas concerning sustainability and that the concept can only be described in more general terms, such as natural processes and biodiversity parameters. Furthermore it should be clear what the overall management goal is: nature reserve or human use. In terms of the latter, it follows that the ecosystem property 'sustainability' is the result of management. It is clearly illustrated by Zonneveld (1989) and De Vlas (1992), that sustainability may also mean maximizing (sustained) yield by recurrent perturbation of the system.

Population Size of Keystone Species or Indicator Species. This type of objective is, amongst others, applied in the Amoeba concept as an operationalization of the sustainability concept. The applicability in management has been estimated as "medium", mainly because in most cases the steerability of the objectives (cause-effect relationship) is limited. From an ecological point of view, the use of selected species as indicators of ecological quality is questionable (Bröring and Wiegleb 1990; Dankers and De Vlas 1994; Macgarvin 1995).

Standard Concentrations. Until now, standard or limit concentration values have most frequently been applied in management, for example as environmental quality objectives in the United Kingdom. These objectives are useful from a management point of view. Ecological credibility is questioned because it is unknown how representative their in vitro ecotoxicological testing is for the ecosystem.

Area Size. Objectives of the area size-type are mainly applied in the management of terrestrial ecosystems. They are useful from a management point of view because the objective can be precisely specified: number of hectares of a certain habitat type to be reached within a certain time period. If the overall goal is to achieve natural development, the fixing of a specified area size is questionable from an ecological point of view. If, on the other hand, the overall goal is, for example, to increase the population size of certain species, it may be ecologically appropriate to aim for a certain area size.

5.7.6 The Wadden Sea Eco-Target Concept

The entries in the diagram in bold (see Fig. 5.7.2), are derived from the eco target concept, developed by the trilateral Ecological Target Group (ETG), according to the political decision in 1991, and laid down in the Final Report of this Group (CWSS 1993). A slightly amended version was adopted at the 7th Governmental Wadden Sea Conference (1994).

The starting point for the ETG was the Guiding Principle, adopted in Esbjerg in 1991, which aims at achieving a natural and sustainable ecosystem, in which natural processes proceed without being disturbed. For reasons given above, the concept of sustainability was not further considered in the elaboration of the Guiding Principle. Already existing approaches, such as the Amoeba concept, were the basic starting material for discussions (CWSS 1993).

As all members of the ETG were aware of the problems with regard to biological parameters, particularly numbers of individuals of certain species, it was decided to instead, focus on *habitat conditions*. The general reasoning behind this was that, if the habitat quality is good then it follows that species can develop naturally and without disturbance.

One of the terms of Reference of the ETG was to develop a Reference Description for the Wadden Sea ecosystem. The ETG encountered most problems in fixing reference values for the dynamic parameters. According to the Guiding Principle, the Wadden Sea ecosystem must develop without disturbance. This, however, precludes fixed reference values or even reference ranges for the dynamic parameters. Most experts agree that the Wadden Sea is a system in development. This implies that developing reference values on the basis of historical records – for example the period around 1930 as in the Amoeba concept – is an arbitrary questionable action. How much value has a reference point which is based on a period of time that is probably unique in a continuously developing Wadden Sea ecosystem?

There are more ways of developing reference values, for example through modeling or the use of reference areas. Modeling does not solve the problem of arbitrariness, however. It would be a fair option to select undisturbed reference areas in the Wadden Sea. Of course, these are not ideal references because many anthropogenic influences, especially inputs of chemicals, cannot be excluded. But for the evaluation of developments with regard to species numbers and diversity such reference areas are possibly the closest one can get.

The main element of the eco target concept is the presence of all typical Wadden Sea habitats in their natural state. "Natural" means in this case free of disturbance from human influences (i.e., physical and chemical). For all habitats, targets were adopted along the line of '*increase the area which is natural, undisturbed or dynamic*'. The rationale for this open-end type of formulation was threefold (compare also Roweck, this volume): first, there is substantive evidence for drastic declines in the quality status and area size of most of the habitats in the past (see amongst others De Jong et al. 1993); second, the lack of clear ecological arguments for the optimum size and quality of the different habitats and, third, the social and economic claims on the area. In the diagram, the target has been positioned a bit left of the 'area size' goal, because the management applicability is less (no precise area size is given in the eco target) but, for the same reason, the ecological credibility is higher.

In addition to these habitat-related parameters, it was agreed to aim for *background concentrations* of naturally occurring hazardous substances (heavy metals) and *concentrations resulting from zero discharges* for xenobiotic substances. The latter objective is positioned in the diagram in a slightly better position than 'background concentrations', because its management applicability and scientific credibility are higher. The eco target on eutrophication '*a Wadden Sea which can be regarded as a eutrophication non-problem area*' has a low management score because the concept 'eutrophication non-problem area' has not yet been specified. The scientific backing is also problematic. From an ecological point of view: is a differentiation between eutrophication problem and non-problem areas possible?

The eco target concept contains also species and community parameters. For species it covers functional parameters, such as birth rate (common seal) or escape distance (selected bird species). The community targets comprise the presence of mature intertidal mussel beds, seagrass stands and *Sabellaria* reefs. These targets have indicative function, supplementary to habitat and chemical targets. For reasons of simplicity they have not been included in the diagram.

Eco targets have a good scoring in terms of both scientific credibility and management usefulness. Dankers and De Vlas (1994) concluded that the ecotarget concept is the result of an evolutionary process, during which some developments, like the Amoeba concept, turned out to be 'dead end' or 'side branches' on the evolutionary tree. Although it is certainly true that the elaboration of eco-targets is part of an ongoing development I would prefer to call it a synthesis of different lines of thinking, which has resulted in what I consider for the moment the maximum in terms of practicability for management, ecological credibility and political feasibility.

This does certainly not mean that the concept is ideal. Especially the operationalization of the eutrophication target and further specifications of the acceptable level of disturbance (in other words what does 'undisturbed' really mean in practice) must be addressed in the years to come.

5.7.6 Ecological Goal Functions as an Holistic Management Tool?

Both from the scientific and political perspective the trilateral eco targets do not seem very satisfactory. They are still general and of a descriptive nature. From several sides the wish has been expressed to specify the targets further. In this section the possibilities of and needs for such a further precision will be analyzed in relation to the use of scientific knowledge in public policies in general.

5.7.6.1 Rational Management

The wish to develop scientifically based goals for the management of ecosystems is rooted in the general conception of *rational management*. According to Underdal (1989) there is consensus among decision makers and scientists that theoretical understanding of cause-effect relationships, as well as relevant descriptive information is a necessary condition for *rational management*. Knowledge is both, a tool for the diagnosis of environmental problems and for prescribing remedial action. Two kinds of knowledge are needed, namely, information for describing the problem and a theory for prediction and explanation. Science is not the only provider of knowledge but, "the more technical the measurement required, the more complex and less transparent the cause-effect relationships, and the more stable the dynamics of the system studied, the greater seems to be the comparative advantage of systematic research over more impressionistic modes of generating knowledge" (Underdal 1989).

Regulatory agencies in the United States formed in the early 20th century and were, according to Jasanoff (1990), mainly dealing with fact finding. The tasks of these agencies became increasingly complex. From the beginning of the 1970s new scientific duties emerge, such as sponsoring basic research, conducting inspections, performing risk assessments and developing analytical methodologies. At first the agencies could not cope with these tasks, because the scientific basis for public policy was not developed: public science or regulatory science, Nelkin (1987) attributes the vast growth of scientists employed by the administration to the increasing complexity of policy decisions and the use of science as a source of authority, by which consensus in public affairs can be reached. "Scientific standards have a universal appeal as an authoritative basis of rational decision making" (p 289).

Küppers et al. (1978) discuss the relevance of scientific information for the administration. They state that an increasing number of problems have a cross section character, and do not fit in the existing monitoring and administrative structures. Administrations must be reflexive: They must adapt to the changing (intersectoral) nature of (environmental) problems. This means that the administrations can not rely upon 'accidentally applicable knowledge', but instead needs of a systematic, longer term knowledge relevant to eliminated problems. This requirement is stressed by elongating action claim and the concomitantly enhanced consequences of decision, be they social, economic or fiscal.

Ecological science has, with the rise of environmental problems in the past 35 years, become indispensable in environmental policies (Nelkin 1975; Küppers et al. 1978; Cramer 1987; Jasanoff 1990). A recurrent question is whether ecology can fulfill the expectations of politicians and the promises of scientist. In other words: can ecology provide answers to (the management of) present and future environmental problems?

Rational management has become subject to increasing critique in recent years, especially with regard of the application of ecological knowledge in environmental management (Van Leussen 1996; Grumbine 1997). Practical experience has shown that complex environmental problems can not, or to a very limited extent, be managed on the basis of science alone.

Mangel et al. (1996), who consulted 380 key scientists and managers from all around the world about conservation strategies, concluded that an important shift in the understanding of conservation has occurred. It used to be regarded as 'primarily a scientific issue, with the key components to be considered including the ecological and most narrowly biological information about the species or stocks involved, their ecosystems, and the scientific analysis of those data to develop management regimes' (p 355). One important change they observed was the recognition that ecosystems are no longer seen as 'stable, closed, internally regulated and behaving in a deterministic manner, but as open systems in constant flux and usually without long-term stability, and affected by a series of human and other, often stochastic, factors, many originating outside the ecosystem itself' (p 356).

Mangel et al. (1996) also observed a second important shift in thinking, namely the recognition of the 'fundamental role of social and economic factors in determining what the goals of management will be and what management actions will be taken' (Mangel et al. 1996, p 356).

5.7.6.2 Adaptive Management

Adaptive management or management through "learning by doing" is an evolving discipline, which attempts to find answers to the management of increasingly complex environmental problems, caused by the paradigm shift in the notion of ecosystems, the fundamental role of social and economic factors in environmental management and the involvement of additional interested parties. Van Leussen (1996) envisages adaptive management as a process trying "to integrate new and existing knowledge with the observed developments in practice, in correspondence with a long-term vision". It is a continuous process, in which new insights may lead to adjustments of the management program. Variability and unpredictability in ecological and social systems are taken as the rule and the management program should be designed accordingly. Reduction of uncertainty will come from practical experiences rather than basic research and the development of ecological theory (Van Leussen 1996).

There is little practical experience with adaptive management but the experiences so far with the development of ecological targets for the Wadden Sea ecosystem fit into the theoretical picture. In the flow scheme in Fig. 5.7.2 the iterative processes are shown, by which objectives are developed, implemented, evaluated and amended in the case of international Wadden Sea policies in a three year cycle according to the tri-annual Governmental Conferences.

The ecological targets reflect both, the need for a recovery of the natural values of the Wadden Sea ecosystem and the persistence of human activities in the area. The targets make clear that an increase of natural and undisturbed habitats all over the Wadden Sea is a necessary condition for the restoration of the ecosystem. At the same time, the targets have been formulated in an open-end way: no fixed claims for the desired situation are given. This means that there is room for negotiation in each tri-annual period, both, from the user and the nature protection sides. Such negotiations, in which (new) ecological insights may play an essential role, can best be carried out at the local and regional level. The recent adoption of a scheme for mussel fisheries in the coming decade in the German federal state Schleswig-Holstein is a good example of the regional elaboration of general ecosystem targets. The scheme, which was developed by both fishermen and administrators inter alia guarantees the exploitation of certain areas for mussel culture and forbids fishery for seed mussels on the tidal flats. This practice will help maintain and recover mature intertidal mussel beds. Scientific knowledge about the development of mature mussel beds and the importance of these structures for the sediment balance, biodiversity and food supply in the tidal area has played an important role. First it helped prioritizing the problem of declining mature mussel beds by the adoption of an ecological target and second it facilitated the practical implementation of this target.

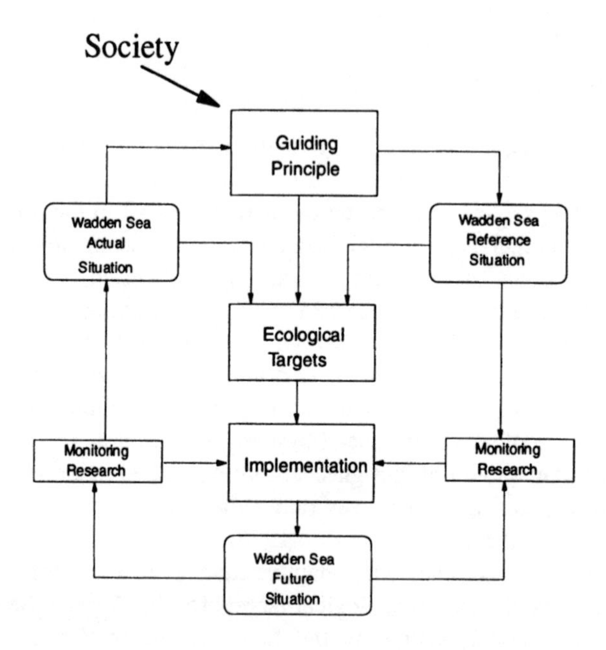

Fig. 5.7.2. Flow scheme of development, implementation and amendment of ecological targets in current trilateral Wadden Sea nature conservation policies

As illustrated in the diagram, political, social and economic factors, as well as the results of monitoring and research, may influence the overall policy goals and consequently the ecological targets. It is stressed here that also basic ecological knowledge, including theory building, can, in the long-term, have a profound influence on overall principles for environmental policies.

5.7.7 Consequences for Holistic Ecological Goal Setting

The classical, rational management based approach to the development of ecological quality objectives is best illustrated by the detailed hierarchy of ecological objectives as elaborated in Germany (see for example Scholles 1990) starting with a Guiding Principle (Leitbild) at the top level and ending at the basis with quantified standards. Such a hierarchy finds its basis in ecology: from emergent ecosystem properties to parameters at the species level (Müller et al. 1997). However, as was illustrated in Sect. 5.7.5, practical experiences with the development of ecological targets show that scientific discord increases with increasing level of detail of ecological goals. A lack of scientific consensus is easily (mis)used by opponents.

There is another disadvantage of precise goals, derived from strict scientific principles: it will take much negotiation and consequently much time to achieve consensus. Often political decisions can only be achieved and implemented within a certain time window, depending on specific socio-economic and political constellations. The development of ecological objectives for the North Sea may serve as an example. In 1991, at the 3rd North Sea Conference, it was decided that techniques for the development of ecological objectives will be developed. To date, six years and a number of expert workshops later, the discussion is still at a very fundamental level. In the meantime, North Sea environmental issues disappeared from the political agenda.

Brunner and Clark (1997) question clear and detailed goals, because, "given the ambiguities, uncertainties and questions of practical prudence, it may be unclear whether a satisfactory course of action exists in a particular ecosystem". The setting of clear goals would interfere with complex decision making processes in a pluralistic society.

The paramount role of social, economic and political factors in environmental decision making has since long time been recognized in the social studies of science (see amongst others Boehmer-Christiansen 1989; Boehmer-Christiansen 1994; Metzner 1996; Metzner Sect. 3.3). Since the beginning of the 1990s ecologists and ecosystem managers appeal for more involvement of the social sciences in dealing with environmental problems (Machlis 1992; Gandraß et al. 1996; Mangel et al. 1996; Roe 1996; von Storch et al. 1997). Some of these appeals seem to aim at extending the rational management conception beyond the natural sciences and to strive for integrating these with the social sciences in order to be better prepared for the irrational aspects of society. The desire to develop holistic goals in ecosystem management, taking into account not only ecological conditions but also social problems and needs, must be seen in the same light. However, the limited predictive power of the social sciences, a quality which it shares with ecology, makes such expectations very unrealistic. As was argued earlier, one of the main obstacles for rational ecosystem management is the complexity and unpredictability of ecosystems. Holistic management will face even bigger problems by including additional unpredictable factors.

Another major obstacle for the development and application of specific holistic targets is the sectoral structure of society. In 1977 Küppers et al. concluded that this had been the main reason for the difficulties in basing the first German environment program on ecological principles. In the past decades the notion of the need for interdisciplinarity has become almost universal, but this hardly had practical consequences, although there are substantial differences between countries. The practice of involving interest groups (stakeholders) in the management of natural resources, which is rapidly developing nowadays, is an additional "sectoral structure" factor.

5.7.8 Conclusions

With increasing anthropogenic impact on natural resources, the call for indicators of 'what nature can cope with' has become louder. Politicians have expressed a desire for science based, quantified objectives. Experiences so far with developing and applying ecological targets have revealed that the level of specification and quantification of ecological goals is limited by the following obstacles:

- the lack of consensus about basic ecological concepts results in increasing scientific discord in the operationalization of such concepts;
- ecological quality objectives are constructs at the interface of science and policy. They contain an important normative aspect;
- negotiations about specified goals are time consuming and may extend beyond the period a certain issue is on the political agenda;
- clear and specific goals limit the participation possibilities of interest groups.

This has two practical implications: politicians and policy makers must realize and accept that there are no ideal indicators of ecosystem quality like in chemistry litmus. Ecologists must realize that a scientific solution in itself will never be sufficient in political debates and that the proposed ecological solutions will be 'watered down' to more general goals that have appeal to a broader audience.

The poor results of the application of ecology in environmental management have led to a rethinking of the role of ecology. In addition to the paradigmatic shift in the ecological conceptualization of ecosystems, which has changed from stable and deterministic to dynamic and unpredictable, the fundamental role of social and economic factors is also beginning to be realized by scientists and administrators.

This development has caused an increasing appeal for including the social sciences in the development of natural resource policies. Although this may lead to more clarity about the social and economic effects of certain management strategies, it is still questionable whether it will alleviate the problems of the acceptance of certain policies by regional and local inhabitants and user groups.

The consequences for holistic ecosystem management strategies can be summarized as follows:

- General management goals should be developed on the basis of ecological and social needs, preferably in consultation with all interest groups.
- Specification and implementation of the general goals can best be done at the regional and local level, together with involved interest groups;

- The general and, consequently, the specified goals, are adapted on the basis of practical experiences, new scientific findings and changing social circumstances.

References

Boddeke R (1993) Phosphat und Fisch in der Nordsee. In: SDN-Kolloquium Eutrophierung und Landwirtschaft. SDN, Wilhelmshaven, pp 58-75

Boehmer-Christiansen S (1989) The Role of Science in the International Regulation of Pollution. In: Andresen S, Østreng W (eds) International Resource Management: the Role of Science and Politics. Belhaven Press, London/New York, pp 143-167

Boehmer-Christiansen S (1994) Reflections on Scientific Advice and EC Transboundary Pollution Policy. Paper prepared for the Conference "Scientific Expertise in the European Public Policy Debate". London School of Economics and Political Science, 14/15th of Sep 1994

Borchardt T (1994) Ökologische Qualitätsziele für Wattenmeer Nationalparke. In: SDN-Kolloquium Eutrophierung und Landwirtschaft. SDN, Wilhelmshaven, pp 64-76

Bradbury RH, Van der Laan JD, Green DG (1996) The Idea of Complexity in Ecology. Senckenbergiana maritima 27(3/6):89-96

Bröring U, Wiegleb G (1990) Wissenschaftlicher Naturschutz oder ökologische Grundlagenforschung? Natur und Landschaft 65 (6):283-292

Colijn F (1994) Ökologische Qualitätsziele in den Niederlanden. In: SDN-Kolloquium Eutrophierung und Landwirtschaft. SDN, Wilhelmshaven, pp 46-63

CWSS (1993) Final Report of the Eco-Target Group. CWSS, Wilhelmshaven

Dankers N , Dijkema KS, Reijnders PJH, Smit CJ (1991) The Wadden Sea in the Future, Why and How to Reach. RIN contributions to research on management of natural resources 1991-1. Research Institute for Nature Management, Texel

Dankers N, De Vlas J (1994) Ecological Targets in the Wadden Sea. Development of Ecological Targets as a Political Instrument; Previous Attempts and the Present State in the Wadden Sea. Ophelia Suppl 6:69-77

De Jong F (1992a) Ecological Quality Objectives for Marine Coastal Waters: The Wadden Sea Experience. International Journal of Estuarine and Coastal Law 7(4):255-276

De Jong F (1992b) Ökologische Qualitätsziele für den marinen Bereich: politisch unerwünscht oder wissenschaftlich unmöglich? DGM-Mitt 1:30-32

De Jong F (1994) International Environmental Protection of the Wadden Sea and the North Sea and the Integration of Pollution and Conservation Policies. Ophelia Suppl 6:37-45

De Jong F, Bakker JF, Dankers N, Dahl K, Farke H, Jäppelt W, Madsen PB, Kossmagk-Stephan K (1993) Quality Status Report of the North Sea: Subregion 10, The Wadden Sea. CWSS, Wilhelmshaven

De Vlas J (1992) The Wadden Sea in Maximum, Sustainable Use. In: Dankers N, Smit CJ, Scholl M (eds) Present and Future Conservation of the Wadden Sea. Neth Inst Sea Res Publ Ser 20: 49-53

Friedrich A, Claussen U (1994) Grenzen mariner Qualitätsziele. In: SDN-Kolloquium Eutrophierung und Landwirtschaft. SDN, Wilhelmshaven, pp 9-20

Gandraß J, Kurz J, Salomons W (eds) (1996) Monitoring Strategies for the Next Century. Workshop report Eurobasin workshop. GKSS, Geesthacht

Gray JS (1996) Environmental Science and a Precautionary Approach Revisited. Mar Poll Bull 32(7):532-534

Grumbine RE (1996) Reflections on "What is Ecosystem Management?" Cons Biol 11(1):41-47

Gündling L (1990) The Status in International Law of the Principle of Precautionary Action. In: Freestone D, IJlstra T (eds) The North Sea: Perspectives on Regional Environmental Cooperation. Graham and Trotman/Martinus Nijhoff, London Dordrecht Boston, pp 23-30

Hastings A, Hom CL, Ellner S, Turchin P, Godfray HCJ (1993) Chaos in Ecology. Is Mother Nature a Strange Attractor? Ann Rev Ecol Syst 24:1-33
Higler LWG, Dankers N, Koop HGJM, Opdam PFM (1993) Sustainability of Ecosystems in northwest Europe. Study in Commission of the Advisory Council for Research on Nature and Environment (RMNO) and the Biological Council of the Royal Dutch Academy of Sciences (KNAW). Institute for Forestry and Nature Research (IBN-DLO), Wageningen
Hoppenheit M (1992) Qualitätsziele für die Nordsee? Pro und Kontra. Deutsche Hydrograph Zeitschrift 44(5/6):289-292
Jasanoff S (1990) The Fifth Branch. Science Advisors as Policy Makers. Harvard University Press, Cambridge London
Knauer P (1994) Umwelqualitätsziele für das Wattenmeer-Notwendigkeit und Widerstände. In: SDN-Kolloquium Eutrophierung und Landwirtschaft. SDN, Wilhelmshaven, pp 21-37
Knottnerus OS (1996) Structural Characteristics of Coastal Societies: Some Considerations on the History of the North Sea Coastal Marshes. In: Roding J, Heerma van Voss L (eds) The North Sea and Culture. Verloren, Hilversum, pp 41-63
Lutter S (1991) Alle reden vom Vorsorgeprinzip, niemand wendet es an. DGM Mitt (4):12-16
Macgarvin M (1995) The Implications of the Precautionary Principle for Biological Monitoring. Helgoländer Meeresunters 49:647-662
Machlis GE (1992) The Contribution of Sociology to Biodiversity Research and Management. Biological Conservation 62:161-170
MacKenzie D (1996) Norway's Fish Plan 'a Recipe for Disaster' New Scientist 1/1996:4
Mangel M et al. (1966) Principles for the Conservation of Wild Living Resources. Ecological Applications 6(2):338-362
Metzner A (1996) Social Constructions of Environmental Issues in Scientific and Public Discourse. Cottbuser Beitr z Sozialwiss Umweltforschung, BTUC-AR 7:74-127
Müller E (1994) Vorwort. In: SDN-Kolloquium Ökologische Qualitätsziel für das Meer. SDN, Wilhelmshaven, p 8
Müller F, Breckling B, Bredemeier M, Grimm V, Malchow H, Nielsen SN, Reiche EW (1997) Emergente Ökosystemeigenschaften. Handbuch der Umweltwissenschaften. ecomed, Landsberg
Peters RH (1991) A Critique for Ecology. Cambridge University Press, Cambridge
Reise K (1994) Ökologische Qualitätsziele für eine ziellose Natur? In: SDN-Kolloquium Ökologische Qualitätsziele für das Meer. SDN, Wilhelmshaven, pp 38-45
Roe E (1996) Why Ecosystem Management can't work without Social Science: an Example from the California Northern Spotted Owl Controversy. Environm Management 20-5:667-674
Scholles F (1990) Umweltqualitätsziele und -standards: Begriffsdefinitionen. UVP Rep 3:35-37
SDN (1994) Ökologische Qualitätsziele für das Meer. Schriftenr d Schutzgemeinschaft Deutsche Nordseeküste e.V., Wilhelmshaven
Sherman K (1994) Sustainability, Biomass Yields, and Health of Coastal Ecosystems: an Ecological Perspective. Mar Ecol Prog Ser 112:277-301
Ten Brink BJE, Hosper SH, Colijn F (1991) A Quantitative Method for Description and Assessment of Ecosystems: The AMOEBA approach. Mar Poll Bull 23:265-271
Underdal A (1989) The Role of Science in International Resource Management: a Summary. In: Andresen S, Østreng W (eds) International Resource Management: the Role of Science and Politics. Belhaven Press, London New York, pp 253-267
Van der Windt H (1995) En dan, wat is natuur nog in dit land. Ph D Thesis Univ Groningen
Van Leussen W (1996) Public Policy Aspects of Integrated Water Management. Ministry of Transport, Public Works and Water Management, The Hague
Von Storch M, Stehr N (1997) Climate Research: the Case for the Social Sciences. Ambio 26(1):66-71
Wildes FT (1995) Recent Themes in Conservation Philosophy and Policy in the United States. Environmental Conservation 22(2):143-150
Zonneveld IS (1989) Theorieen en concepten: een tussentijdse balans. Landschap 1:65-76

5.8 Landscape Planning: On its Role in Developing Sustainable Land Use Concepts

Isolde Roch

Abstract

Categorizing, securing and developing natural resources and systems are presented here as tasks of landscape planning of direct relevance to the preservation and sustained economic utilization of the environment. Various objectives pursued in landscape planning, particularly those derived from coping with former East German practices of careless resource exploitation, display an interdisciplinary and "integrative" character and are thus seen to facilitate the consideration of settlement structures and economic activity in environmental planning. These objectives will be contrasted with the actual responsibilities presently defined by German planning legislation and how these effect the ability of landscape planning to pursue the goal of environmentally sensitive regional development. Following a critical discussion of environmental planning procedures, suggestions for model landscape planning approaches that flexibly address the land-use potentials of different ecological systems will be presented. Finally, research questions posed by new planning approaches with regard to ecosystems studies, interdisciplinary concepts and public policy will be addressed.

5.8.1 Objectives of Landscape Planning as Derived from East German Experience

Traditionally, landscape planning has focused on environmental protection, the preservation of natural areas and landscape development. The latter involves a cautious and sparing use of resources, addressing problems of damaged and blighted areas and reconciling traditional elements of local landscapes with modern land use forms – all of which are to be carried in an environmentally sustainable, aesthetically pleasing and contextual appropriate manner. Because of its broad approach to both basic natural and structural aspects of cultural landscapes, landscape planning has acquired not only the capability but also the responsibility to participate in developing basic conditions for an improved spatial planning process.

F. Müller M. Leupelt (Eds.)
Eco Targets, Goal Functions, and Orientors

Landscape planning in Eastern Germany can boast of a long tradition of integrative practice, combining a wide spectrum of planning and planning-related issues as well as a variety of technical approaches. The "General Landscape Plan" of 1989/90 for the former administrative region of Dresden (Bezirk), for example, covered an area of 8000 square kilometers and represented in 1:25000 scale maps specific spatial elements such as "Protected Areas", "Economic Structure" and "Areas of Landscape Damage" (see Table 5.8.1). These various elements were superimposed on a general 1:100000 scale map. In the Map entitled "Environmental Characteristics", dominant vegetation forms were illustrated according to the quality of their natural states - represented by 11 categories indicating the relative degree of exposure to human activities. These categories included, coal-mining areas, urban settlements, basic agricultural zones, orchard and special crop areas, differentiated agricultural areas, mixed zones of forests, pastures and croplands, mixed zones of pastures, ponds/marshlands, and croplands, predominantly forested areas, areas of forest and ponds/marshlands and, finally, forested areas with limited urban development and agricultural activity. The map "Protected Areas/Land Use Restrictions" differentiates between areas of inherent ecological importance (natural protection areas and watersheds), industrial activity preserves (e.g. mining areas, areas reserved for freshwater fisheries), other restricted-use areas (storm damage potential, zones for nuclear power stations, public monuments) and reserved-use areas (e.g. for the analysis of groundwater resources). A further map, "Landscape Damage", documents natural resource areas where ameliorative action is required as well as the environmental quality of industrial and settlement zones.

Table 5.8.1. Contents of the General Landscape Plan for the District of Dresden

Land-use restrictions	*Economic structure*	*Landscape damage*
Natural protection areas	Industrial firms (physical plant size, number of employees)	Smoke and other emissions
Landscape protection areas	Location of industrialized agriculture	Air pollution (SO_2 and NO_2)
Mining preserves	Location of mining enterprises	Water contamination
Watersheds	Surface transportation network	Erosion
Water reservoirs	Residential areas and surroundings	Devastated areas
Weather-sensitive areas	Locations of importance to tourism	Uncollected trash; illegal rubbish dumps

These maps were the result of objective data-gathering and an interdisciplinary synthesis of information. When made official as part of the General Landscape Plan in 1989 (before the political changes of November), they prompted a sharp political reaction; members of the district council expressed shock and outrage over the manner in which valuable natural resources were being exploited in the Democratic Republic. The maps, however, also provided positive recommendations for differentiated land-use restrictions. One vital element of the General Landscape Plan, for example, was the representation of activity zones and structural economic areas (see Table 5.8.1). This allowed for the expression of cause- and-effect relationships with regard to physical space and thus signaled restrictions in development potential under the then given conditions. One of the most important results of the plan was a preliminary concept for establishing a comprehensive biotope system, based on existing environmental resources and the need to create "bio-corridors" connecting forested areas (Bezirk Dresden 1989, Roch 1990). Another was the identification and clear visual depiction of serious land-use conflicts in the Dresden region (see Fig. 5.8.1).

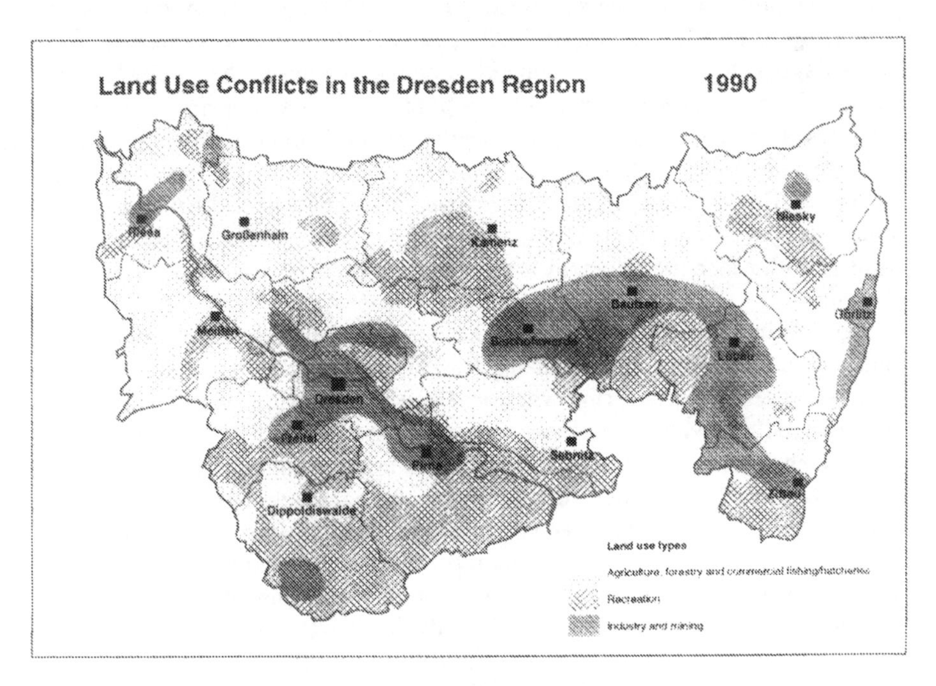

Fig. 5.8.1 Land Use Conflicts in the Region of Dresden (1990)

Areas of conflicting land uses thus located and described were recognized to be symptomatic of an overall problem of short-sighted exploitation of irreplaceable natural resources with little regard for protection. Concrete development recommendations were formulated by representatives of city, county and district agencies using a 1:25,000 map and submitted as a basis for establishing the aforementioned biotope system, as well as for defining renaturation and restricted development areas. The restriction of urban development and/or other construction activity had the prime objective of protecting the most sensitive natural areas primarily through defining an urban growth boundary. This, and the other maps discussed here were basically conceived as preliminary studies for formal regional plans. However, the government of Saxony could only give informal recognition of the concepts and proved unable to further the realization of development goals contained therein; indeed, the state government could do little to prevent uncontrolled development within the legal vacuum that existed after the end of GDR planning practice but before the establishment of new regional planning legislation.

The General Plan was a belated product of attempts by landscape planners to establish a sectoral planning framework, starting with the "Landscape Diagnosis" of the early 1950s (Wübbe 1994) and continuing on to the debate over cultural heritage legislation in the late 1960s. As objectives inviting general consensus, landscape protection and development were given high priority. Another generally accepted goal was the rational utilization of natural resources necessary for industrial development and the support of human settlement (Roch 1997a). The General Plan was primarily concerned with the collection of environmental data and, based on this, a comprehensive ecological evaluation in order to classify areas regarding their suitability for various uses and their exploitation as future resources. This objective was to be integrated into economic and structural planning concepts in recognizance of the necessity to negotiate planning and development with economic actors, as well as to address social requirements of communities.

Starting in 1990, attempts to integrate landscape and spatial planning were undertaken as a means of minimizing land use conflicts and re-evaluating general land use categories. This was partially a consequence of the "cause and effect" logic of traditional territorial planning that – heretofore with scant concern for environmental issues – was orientated towards optimizing the economic region factors. Through this comprehensive and extensive evaluation of landscape conditions, landscape planning presented itself as an instrument allowing wide-reaching controls on the development of different land-uses if allowed to operate on equal footing and in co-operation with economic and settlement development planning. Concrete steps towards this integrative planning approach were initiated after the political transformations of 1989 and 1990; "holistic" development concepts for counties (with an area of roughly 155 square kilometers) on a scale of 1:25,000 were elaborated by urban and regional planning, and environmental-affairs authorities of the administrative region of Dresden. This ad hoc exercise, which could have provided a valuable foundation for the proactive and cautious management of spatial processes, never acquired legally binding status as German Federal law fragments planning authority ultimately giving municipal governments planning "sovereignty" over basic land-use issues. As such, comprehensive and sectorally integrated plans, such as the one developed for the counties of Saxony, can, at best, fulfill an advisory role.

5.8.2 Options for Landscape Planning in Promoting Sustainable Regional Development

Nevertheless, the experiences and arguments of landscape planners were acknowledged and subsequently reflected in the new Planning Act of the State of Saxony, carried by an overwhelming parliamentary majority conscious of the grave planning problems that the incipient process of political and economic transformation would bring. Saxonian planning law designates that landscape planning must provide basic data and concepts for the elaboration of state development plans (Landesentwicklungspläne) and regional plans (Regionalpläne). Although clear guidelines have been established with regards to the specific instruments of open space development (Table 5.8.2), this basic task of landscape planning is carried out in several different ways; a fact that owes as much to the lack of uniform methodological guidelines as to the very different natural characteristics of Saxony's regions (see list below).

Development of Open Space Areas in Saxony (Germany):

1. Conservation and protection
   Designation of priority and potential preservation areas for:
   - natural protection and landscape conservation
   - agriculture and soil-protection
   - forestry and soil-protection
   - climate-protection
   - water-protection
   - the securing of subsoil natural resources
   - recreational purposes
2. Promotion and development
   Designating regional ecological networks
   - by regarding landscape structures
   - by securing important landscape areas through interregional coordinated system of protection areas
   - by creating regional landscape networks in densely populated areas (urban agglomerations)
3. Care of Open Spaces
   (Planning Region "Oberes Elbtal/Osterzgebirge" only)
   Designation of areas requiring special landscape regeneration measures, including:
   - areas of extension and/or regeneration and reforestation
   - areas of high erosion risk
   - zones designated as future forest or wooded areas

Instruments of Regional planning. Despite unanimity among landscape planners that their role is to secure, promote and protect biotopes and natural resources, as well as to offer prospects for the sustainable development of resources, the individual goals, instruments, and results emanating from practical application of these principles are quite different. This fact is corroborated by Finke (1993) and Hoppenstedt et al. (1996) who indicate that the character and procedural aspects of framework landscape planning in western German states differ considerably and that this is mostly due to the specifics of the respective state planning laws. Framework landscape plans differ considerably, for example, with respect to contents (Kistenmacher et al. 1993); this generates confusion over the inherent function of formal planning documents and often leads to misunderstandings between the states themselves.

Table 5.8.2 Structural Development Objectives for Open Spaces

regional plan "Chemnitz-Erzgebirge"	*regional plan "Oberes Elbtal/Osterzgebirge"*
the promotion of characteristic biotopes, plant- and animal-species and landscape systems	in case of further settlement, taking into account historical forms of settlement, of cultural monuments and of the characteristic landscape forms
the establishment of networks of protected areas	regional development in harmony with sustainable uses
securing and improving soil, water and air quality	using land sparingly, concentrating infrastructure, avoiding further truncation of interconnected landscape areas
increasing forest/wooded areas and the development to mature natural communities	the avoidance of settlement and infrastructure development in environmentally sensitive areas
the promotion of extensive and structurally diverse forms of agriculture	Compensation of damage inflicted on the landscape
the protection and/or regeneration of near-natural water (brooks, rivers) characteristic "landscape image"	
the protection of space-saving and open ground-retaining (e g asphalt minimizing) urban development	

The "variations on a planning theme" that we see in the German state of Saxony are characteristic of attempts to tailor land-use concepts to the specifics of locality and region – despite the fact that cultural and natural areas are rarely coterminous with administrative regions. Regional land-use patterns are, to a large extent, determined by specific natural resource characteristics that require special consideration if development strategies appropriate to local conditions are to be defined and implemented. The basic ecological condition and natural potentials of a region are of prime importance; these guide the definition of specific environmental quality objectives within different periods of time. Reclamation, reforestation and other measures, for example, can help achieve short-term results that can be extended within a wider regional context. Obviously, this process mandates regular evaluation and re-evaluation of objectives and measures to be implemented, based on the results previously achieved.

Experience shows, that present environmental quality objectives do indeed assist in improving the environment. However, we need to go beyond generalities and develop means for tailoring environmental quality objectives to specific regional situations. This requires knowledge about the self-regenerative properties of natural regions as well as the land-uses they are conducive to and the potential dangers they face. This is, in fact, being pursued through the definition of possible protective measures obtainable through landscape planning. Based on a long-term strategy of development while maintaining the "material equilibrium" of different natural zones and landscape types, it has been understood that environmentally-oriented planning functions must consider the multiplicity of land-use demands in order to assure future viability of natural areas. It is, for example, conceivable that the renaturation of agricultural land could be combined with the protection of endangered species and less intensive forms of cultivation. A failure to realize this and a failure to link natural areas and biotopes together as regional systems could result in limited conservation measures, preserving smaller, parcel-like areas more or less as "natural museums".

Potential for regional action based on integrative approaches lies in communicating and co-operating with technical and physical planning authorities. With respect to land-use goals formulated by these sectoral planning bodies as well as other public agencies, we can envisage the following:

1. use of watersheds and dam catchment areas through private/public water utilities
2. definition of protected or restricted forest areas with the collaboration of public/private forestry interests
3. development of protective strips along rivers and other waters; the establishment of flood water retention areas through public agencies
4. promoting extensive cultivation methods as well as renaturation through cessation of cultivation
5. recultivating former mining and deforested areas defining tourism development concepts in conjunction with communities and industry representatives.

The constant improvement of inter-agency co-operation is aimed towards establishing a compromise between various development and planning interests; it is perhaps the best way to promote objectives of natural area and landscape protection while avoiding the need for compensatory financial measures.

Above and beyond this, however, landscape planning has the essential and formal role of incorporating conservation and sustainable development goals within the framework of regional and municipal master plans. Here it would be important to modify procedural methods in order to establish conservation and development criteria and to prepare actual development schemes (Bebauungspläne) as a joint exercise involving public agencies and developers. Co-operation between various planning agencies must, nevertheless, be supplemented by a consolidation of available development grants and subsidy programs in order to facilitate inter-agency (and integrative) projects. This approach would, furthermore, apply fully to the restoration of landscapes and natural areas. Because renaturation and redevelopment measures take effect on very different time scales, it is necessary to monitor the progress of landscape restoration and its affects on surrounding areas. This requires the periodic re-evaluation of environmental quality objectives.

Exemplary procedures such as the ones discussed above are elements of the landscape planning process that could be incorporated into integrated and comprehensive concepts of sustainable regional development. However, the success of comprehensive approaches can only be guaranteed if public and private stakeholders in regional development (e.g. firms, farming and forestry interests, ministerial agencies supervising urban development) are included in defining phases of the process at the regional and local levels, allowing for different concerns to be voiced and recognized in the implementation of strategic plans. With regard to the latter, projects or concrete measures should reflect different time-scales, different spatial levels (the spheres of influence" of individual measures) and the need to integrate projects hierarchically and in a mutually re-enforcing manner so that they may constitute a cohesive policy. At the same time, however, this approach demands considerable effort on the part of the actors involved to reach a degree of consensus that allows for effective project management and goal coordination. Of course, regional and local implementation must always be pursued with overall environmental quality objectives in mind. Successful strategic approaches in managing, for instance, local renaturation or tourism development projects, cannot, by themselves, assure sustainability in spatial development and do not eliminate the need for inter-local and inter-regional coordination (Roch 1997b). In fact, both strategic management and comprehensive planning are possible within the present administrative framework even if the environmental quality objectives that have been established do not always correspond to the higher goals of sustainable development.

5.8.3 Approaches to Ecologically-Based Land-Use as an Element of Sustainability

Realistic approaches towards an ecological sensitive and sustainable land-use planning process bring us back to the roots of ecosystem research. We must be able to answer questions as to the flexibility, resistance and vulnerability of different types of ecosystems in order to determine potential dangers as well as land-use patterns that best suit their specific nature and character. Thus, apart from the above-mentioned links to economic and urban development, landscape planning must re-establish close ties to ecosystem investigation. This will represent a new set of responsibilities for the profession and one that will also require an integrated approach with other planning disciplines. This will, of course, entail the elimination of serious information deficits caused by previous separation of ecological research and the planning process.

The objectives of an *Integrated Planning Approach* would include:

- the elaboration of land-use recommendations and advisory reports on the ecological limits of various land-use combinations,
- the determination of environmental quality objectives for different land-use forms and with regard to prevailing ecological conditions,
- the determination of priorities for the protection and the securing of natural areas and their potentials,
- the determination of conditions necessary for "sustainable connectivity" between natural areas, urban settlements and economic activities,
- establishment of a process of monitoring and evaluating the effectiveness of integrated planning as well as determining appropriate future measures

In order to implement each of these it will, however, be necessary to find common ground in communicating ideas and in establishing the appropriate spatial level. Map scales and the detail with which ecosystems are to be evaluated must be negotiated between the involved professionals. One might expect that the traditional scales of landscape and regional planners of 1:25,000/1:50,000 with a maximum generalization of up to a scale of 1:100,000 will remain appropriate for the treatment of settlement patterns and natural areas. However, these scales are quite different to those used in ecosystems research. Thus, guidelines regarding the generalization of spatial representations must be agreed upon (see Fig. 5.8.2 for an example illustration).

Results of joint efforts should lead to differentiated recommendations regarding allowable land-uses for different landscape types and that must be defined in concrete terms by the authorities and agencies whose actions directly affect urban and regional development. In order to translate recommendations into concrete action a variety of public measures must be formulated; these could, for example, involve tax holidays and other benefits to enterprises whose activities are deemed to be ecologically appropriate for specific areas or location disincentives for firms likely to cause serious land-use conflicts. The elaboration of such measures will require new forms of communication and cooperation between planners, economists and public agencies.

In order to complete this panorama of differentiated and 'tailor-made' ecological and spatial development approaches, we must keep in mind that the various expectations and requirements of those who live in the regions must be considered. The quality of life can be improved in all regions.

However, the means by which regions define and achieve *their* high quality of life will differ considerably; integrated planning approaches must distance themselves from the traditional and uniformly egalitarian practices. If the quality of life can be freely defined, then we will be provided with an incentive to promote greater choice in the development paths and in the economic and social scenarios regions can follow. Furthermore, the process of interregional exchange might be dynamized, lessening the impact of transfer payments and economic centralization. Economic growth pressures could be kept at manageable levels through the exchange of complementary services, products and resources.

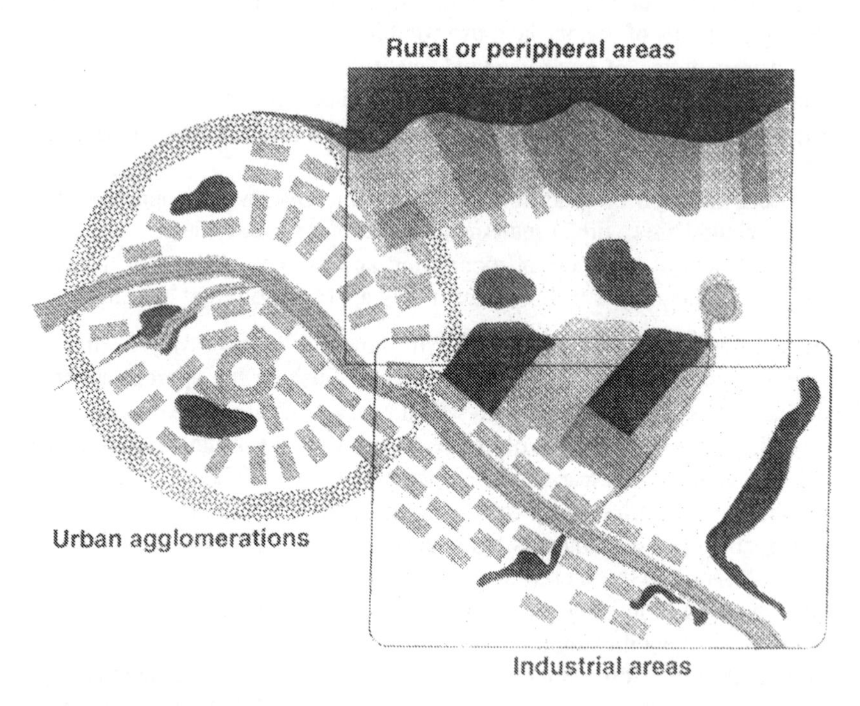

Fig. 5.8.2. Land-Use Types

With this brief sketch of some potential results of an integrated approach to landscape planning we emphasize the central importance attributed to ecological considerations in attempting to reconcile economic and social objectives and, in the long run, to develop planning goals along the lines of sustainability. Unquestionably, the ideas presented here require continual work and flexible and pragmatic strategies. Areas in which uncertainties and/or information deficits still remain are:

- insufficient knowledge of ecological processes,
- lack of clarity regarding the division of coordinating responsibilities between public actors,
- deficiencies in the evaluation of planning measures
- insufficient co-operation between the relevant disciplines of Ecology, Spatial Planning, Economics and Policy Sciences

5.8.4 Conclusion

The approach outlined here, with its objective of producing "model" land use patterns that accommodate specific situations, is one that will probably require a generation to fulfill. Because of the scope of environmental problems and land-use conflicts that must be dealt with, however, it is necessary that the groundwork for this approach be initiated as quickly as possible. Existing conditions allow for the elaboration of concepts and objectives of regional, sub-regional, landscape and site development, as well as for the realization of protection and rehabilitation measures for specific problem areas. This 'room for maneuver' opens possibilities for differentiated and site-sensitive planning as an alternative to uniformity and standardized procedures. This would also allow for a more effective use of endogenous potentials in stabilizing local population dynamics and enhancing the global attractiveness of unique cultural and economic regions. Regions that require urgent attention include metropolitan areas and regions experiencing environmental crisis and rapid industrial transformation, particularly in Central Europe.

Landscape planning, with its wide range of tasks and competencies, would be a likely candidate for assuming an integrative planning function. In any case, the landscape planning profession is called upon to represent in a conscientious and creative manner the issues of environmental stabilization and landscape protection in the decisive process of plan execution.

References

Bezirk Dresden (1989) Generallandschaftsplan für den Bezirk Dresden: Beschluß Nr. 158/89 des Rates des Bezirkes Dresden, Dresden

Finke L (ed) (1993) Berücksichtigung ökologischer Belange in der Regionalplanung der BRD. ARL-Band 124, Verlag d ARL, Hannover, p 164

Haber W (1994) Sustainability und Sustainable Development – ökologisch kommentiert. ARL Arbeitsmaterial 212:156-187

Hoppenstedt A, Müller B, Erbgut W (1996) F+E Vorhaben "Weiterentwicklung der Landschaftsrahmenplanung in Problemregionen und ihre Integration in die Regionalplanung". Endbericht der Vorstudie. Arbeitsgemeinschaft Planungsgruppe Ökologie + Umwelt Hannover; TU Dresden, Universität Rostock, Hannover Dresden Rostock

Hübler K-H, Roch I, Mundil R, Hana W (1995) Leitlinien und Entwicklungsziele zur umweltschonenden Raumentwicklung des sächsisch-böhmischen Erzgebirges – ein Beitrag zur Regionalplanung. Endbericht des F+E-Vorhabens, Band B, Leitlinien, Entwicklungsziele und Maßnahmeempfehlungen. Instit f Stadtforschung u Strukturpolitik (IfS) 736/EB, Berlin

Kistenmacher H et al. (1993) Planinhalte für den Freiraumbereich. ARL-Band 126, Verlag d ARL, Hannover

Roch I (1990) Zwischenergebnisse bei der Erarbeitung des Generallandschaftsplanes im Bezirk Dresden. Landschaftsarchitektur 3:24-32

Roch I (1997a) Integration der Landschaftsrahmenplanung in die Regionalplanung – Erfahrungen und Wege aus Sachsen und Thüringen. Landschaftsarchitektur 7:1-7

Roch I (1997b) Zielkonzepte umweltgerechter Entwicklung von Grenzregionen in der Umsetzung - Projekterfahrungen und Schlußfolgerungen. In: Graute U (ed) Sustainable Development for Central and Eastern Europe. Springer-Verlag, Berlin New York, forthcoming

Sächsiche Staatsregierung (1994) Sächsisches Landesplanungsgesetz. Sächsisches Gesetz- und Verordnungsblatt 59, 17th of Nov 1994, Dresden

Wolf J (1997) Nachhaltigkeit und Gleichwertigkeit – zur Notwendigkeit eines neuen Ansatzes für das Leitbild der Raumordnung, Beitrag zum FRU-Förderpreis 1997, Berlin

Wübbe I (1994) Landschaftsplanung in der DDR. Diplomarbeit Studiengang Landschaftsplanung, Technische Universität Berlin, Berlin

5.9 Deriving Eco Targets from Ecological Orientors: Ecological Orientors for Landscape Planning?

Astrid Berg and Wolfgang Riedel

Abstract

The fact that ecosystem research focuses on "ecological goal functions" and aims at "ecosystem integrity", leads to the discussion about the integration of these aspects into planning processes. It has to be clarified to what extend ecological goal functions can be applied to landscape planning and if goal functions can help to develop integer management strategies to operationalize the concept of sustainable landuse ?

5.9.1 The Role of Landscape Planning

Landscape planning is an essential instrument for realizing the goals of nature conservation and landscape maintenance; it has to share its interest in the landscape with other discplines. The Federal Minister of Environment, Nature Conservation and Reactor Safety (1993) (Bundesminister für Umwelt, Naturschutz und Reaktorsicherheit) formulates its following functions:

Landscape planning defines the capacity of given natural systems by assessing their potentials and functions. The fact that interactions between soil, water, air, climate, plants and animals are considered, indicates its multidisciplinary character. The influence of all present and planned land-use issues on this interactive system as well as its feed-back to land use is illustrated. Consequently, this way of planning - the overlapping of various sectors – leads to nature and landscapes being the starting point for determining any future land use.

Landscape planning aims to integrate all interests of nature conservation and landscape maintenance into the process of physical planning considerations. Moreover it elucidates the limited capacity of natural resources.

Landscape planning defines the ecological design which is necessary to maintain the capacity of natural systems and of scenic quality.

F. Müller M. Leupelt (Eds.)
Eco Targets, Goal Functions, and Orientors

Landscape planning supplies standards for the assessment of environmental effects, caused by activities of other planning disciplines. Thus, as a basis for all planning processes, it is essential.

To fulfill the latter function, ecologically and socio-culturally well-elaborated, expertise is needed together with comprehensive information about the pros and cons of development versus conservation issues. This step is incorporated into the general planning process (Bundesminister für Umwelt, Naturschutz und Reaktorsicherheit 1993):

1. problem setting: definition of the problems, planning scope, goals
2. mapping and assessment: assessment of the natural potentials: populations, impact, protection, vitality
3. concept: development of goals of nature conservation and landscape maintenance, alternatives for the solution of conflicts
4. procedure: definition of requirements and activities for the realization of the goals.

The data should support an interactive way of planning, which aims at minimizing conflicting situations that arise from conflicts between planned activities and the existing environment, i.e. between development and conservation interests. This assures the consideration of all natural resources and resource interactions as well as the influence of the other planning disciplines, that are dependent on natural resources or land-use issues. Finally, land-use proposals have to be formulated which should minimize the impact on environmental systems.

5.9.2 Environmental Goals, Environmental Quality Objectives and Environmental Standards/Critical Loads in Landscape Planning

As being the authorized planning discipline for nature conservation and landscape maintenance, landscape planning must formulate environmental goals, which are the basis for assessing the present and possible spatial changes and impacts in a given landscape or landscape entity (Marzelli 1994). Environmental goals should orient towards the natural potential and the specific character of a certain area, which originate from the natural site conditions and its cultural and historical development (Finck et al. 1993). Moreover it should unify the protection of abiotic, biotic and esthetical resources. As demonstrated in Fig. 5.9.1, environmental goals should lead to the definition of more detailed environmental quality objectives for all ecosystem components and further on to corresponding environmental standards/ critical loads.

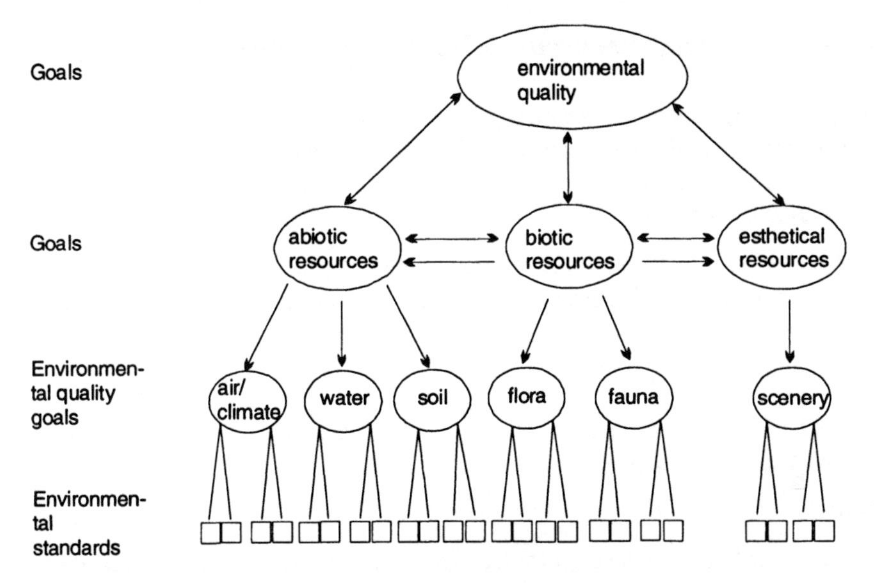

Fig. 5.9.1. Hierarchies of goals and environmental fields (Marzelli 1994; modified)

Environmental goals have to be formulated on different scales; they exist on the global scale as well as on landscape scales. Being derived from environmental goals, environmental quality objectives – "specific, factual, spatial and if necessary temporal defined qualities of resources, potentials or functions" (Kriedemann 1996, p 8) – refer to the ecosystem scale and define the environmental quality, which is aimed in sense of provision. They can be seen as dynamic instruments for environmental policies, which constantly have to adapt towards new findings, e.g. in natural sciences (UVP-Förderverein 1995). The environmental quality often cannot be reached in the reference time, in which case remote goals indicate the long-range aimed environmental quality, requiring intervening steps (UVP-Förderverein 1995).

Environmental standards/critical loads, "quantified environmental quality objectives" (Marzelli 1994, p 12) are definite assessment standards, making environmental quality objectives to become operative; they determine the impact and formulate the required quality and degree of protection by defining the characteristics, the measuring technique and the general conditions for a certain parameter and indicator, respectively. Generally, environmental standards refer to separate ecosystem compartments. Therefore, the existing federal standards do not master the regional characteristics (UVP-Förderverein 1995).

The formulation of those goals requires interaction between scientific knowledge and social values. Eco-sciences are able to tell how a defined environmental goal can be reached, but society has to decide which kinds of ecosystems and how many should be protected (Marzelli 1994).

5.9.3 Ecosystem Research – a Basis for Landscape Planning ?

Talking about scientific knowledge being required for the definition of environmental goals and consequently for decision-making in ecological planning, particularly in landscape planning and environmental policy, the following questions arise:

- Which kind of information is needed in planning practice ?
- Does any of the eco-sciences, particularly ecosystem research, supply these data ?

or the other way round

- Is the knowledge about ecosystems and their characteristics considered during landscape planning processes ?

If not

- can this information be used for the formulation of (new) environmental goals and are the data applicable to planning practice ?

As mentioned above, landscape planning should include a comprehensive analysis and a constant monitoring of a given landscape or landscape entity, focusing on the interactions between the different ecosystem compartments water, soil, air, climate, flora and fauna and on impacts. A crucial aspect, which has to be considered in planning processes, are the dynamics of natural development in ecosystems and therefore the factor 'time'. Time solely can lead to e.g. a compensation action, which has been planned by using static assessment standards, becoming an impact on the landscape (Haemisch and Kehmann 1992).

Bearing the natural dynamics of ecosystems in mind, the following aspects have to be considered during planning processes (Kleyer et al. 1992):

1. How far do land-use changes affect the site under investigation ?
2. Are there any additional areas being affected ?
3. Which kind of species, structure, land-use intensity, trophy, dynamics, diversty, productivity are present/ would develop ?
4. How do land-use changes effect material-fluxes, nutrient-pools, -cycling and nutrient-outputs ?
5. Which species are affected? Which species have to be protected ?

Finally, environmental standards and critical loads have to be defined for the landscape under investigation.

Because landscape planning still being action-oriented and not goal-oriented, at present landscape analyses refer to certain planning objects. It follows that effects of other planned objects in the region under investigation are not considered during the assessment process. This situation is worsened by the fact that data gained in the field are often not complete. Some ecological aspects, which are essential for a scientifically correct assessment, are not asserted, because the required investigations are defined as "extraordinary contributions" to landscape planning and thus are too expensive e.g. for a small municipality being planning authority. This situation would improve if landscape planning and eco-sciences, particularly ecosystem research, were co-operating on their common scale of ecosystem complexes in a more effective way.

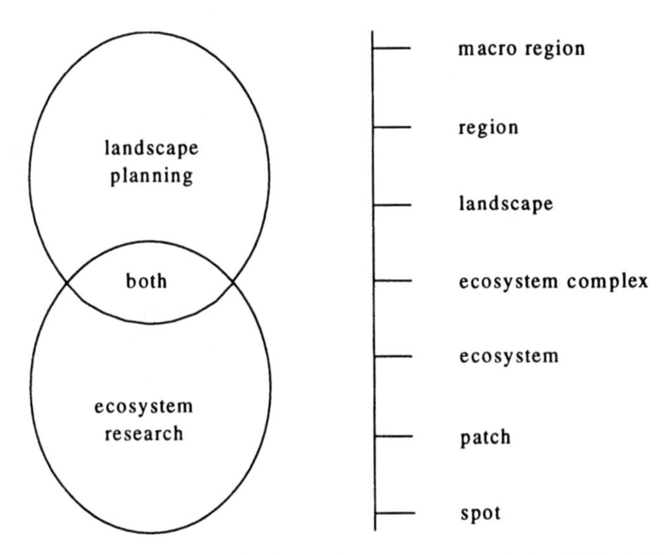

Fig. 5.9.2. Scales of landscape planning and ecosystem research (see Zölitz-Möller et al. 1997)

Ecosystem research focuses on the interactions between the different ecosystem compartments water, soil, air, climate, flora and fauna and on the concept of "ecological goal functions". The latter bases on "thermodynamics of irreversible processes and criteria for the self-organization of dissipative structures" (Schneider and Kay 1994, p50) and states that being affected by the given site conditions, which define the capacity of the system, the different ecosystem parameters grow for a certain period of time until they reach a constant level. They are optimized during the process of succession. Referring to Schneider and Kay (1994), the following ecological goal functions, describing the natural orientation of ecological developments on different scales, can be listed:

- the use of solar energy/radiation in productive processes rises,
- the density of chemical fluxes and energy rises,
- the amount and length of chemical- and energy-cycles rise,
- the storage capacity for chemicals and energy rises,
- the loss of chemicals decreases,
- respiration- and transpiration-losses rise,
- biodiversity and the degree of organization of the systems rise and

- hierarchical structures and signal filtering grow.

Referring to ecological integrity and functionality, respectively, these characteristics are found in mature systems. It therefore can be concluded, that integer ecosystems are also sustainable. Consequently, ecological integrity can be seen as the ecological part of the 'sustainability' definition (Müller 1996).

Ecological assessment basing on this theory would be comprehensive and process-oriented, because it refers to the balance of nature as being a functional entity, considers the natural potential of a given site for development and represents the environmental goal of a 'sustainable development'. In its expertise, published in 1994, the German advisory council of environmental issues (*Sachverständigenrat für Umweltfragen, SRU)* stresses this environmental goal. It discusses the practicability of "[... eine(r) dauerhaft-umweltgerechte(n) Entwicklung] a long-term environmental-suitable development" and pays attention particularly on ecology, which is often regarded as a model of a new normative and integrative science, following an integer approach. But in the opinion of the SRU ecology cannot fulfill these requirements, showing a purely descriptive character. Working on the problem of maintaining stable structures over long periods without impacts on systems leading to irreparable damages (Projektzentrum Ökosystemforschung 1996), ecosystem research is addressed to supply data to landscape planning, which refers to expected interaction changes in a given landscape after a certain impact.

5.9.4 Concept for Unifying the Interests of Landscape Planning and Ecosystem Research

Environmental assessment can be regarded as the link between descriptive ecosystem analyses and the subsequent actions in landscape planning and environmental policy (Heinig 1997). But in practice environmental goals and standards – once formulated during planning processes – tend to make themselves independent. Too strict strategies lead to landscape planning becoming a restrictive discipline. Therefore it has to be discussed, how precise environmental goals should be and how many are used/wanted to create a concept ranging between strict formulation on the one hand and dynamic flexibility and ability of projection on the other (Jessel 1994). The use of environmental goals in nature conservation and environmental protection requires process-oriented thinking and not the fulfillment of fixed aims. Progressively, goals have to be elaborated, which are accepted by all partners and which are practicable.

Not all aspects of the concept "ecological goal functions" are of interest for landscape planning, because many cannot be applied to planning practice, e.g. the use of solar energy/radiation in productive, ecosystem processes and the amount and length of chemical- and energy cycles of ecosystems. Assessment standards/ecological indicators should give complex information, which enables an assessment of a landscape in respect to the environmental goal of 'sustainable development'. For planning practice indicators should express

- the potential of ecological systems to develop,
- the duration of effects of environmental management actions,
- the development of self-organization-capability of ecosystems,
- the long-term characteristics of the natural basis for the quality of life of future human generations,
- the maintenance of structures and process-networks of ecosystems,
- the buffering capacity, resilience and loading capacity of ecological systems,
- the realization of ecosystem principles for integral management strategies and
- the determination of the stage of development, the trend of development and the entirety of influences of man upon ecosystems (Schneider and Kay 1994).

The very complex and interwoven character of the required information leads to the question of implementation. How can the theoretical concept of 'ecological integrity' be realized methodically and be applied to practicable measuring techniques?

Environmental quality objectives have to consider the discussed concept. Because generally these goals hardly being weighed with other concerns, they would be elaborated best in the scope of a newly defined and extended landscape planning. At present this planning discipline already shows a partially integrative character and thus good promise: in practice and in charge of the law it orients towards land-use interests and the environmental goods soil, water, climate, vegetation, fauna and landscape. The issues rarely being treated are mankind, air and cultural heritage and additionally water, which is not sufficiently considered (drinking water) (UVP-Förderverein 1995). Beside the integration of these aspects, future landscape planning must be enabled to treat and assess material input into ecosystems in more detail to formulate integer goals aiming at a 'sustainable development'.

5.9.5 Conclusions

At present hardly any collaboration between ecosystem research and landscape planning exists. This is not the result solely of lacking data transfer and the data hardly being interpretable by planners, but mainly of the differing concepts and strategies, existing for landscape development and nature protection, respectively. On the one hand the causal analyses, carried out in ecosystem research studies, are too specific and complex (Heinig 1997) and thus not all of the resulting goal functions are applicable to landscape planning. On the other hand landscape planning is not process-oriented and does not integrate those ecological goal functions, which are crucial for the development of sustainable management strategies.

All in all, both disciplines have to change their concepts, at least in parts, to guarantee a conceived and effective nature conservation.

References

Der Bundesminister für Umwelt, Naturschutz und Reaktorsicherheit (1993) Landschaftsplanung - Inhalte und Verfahrensweisen. Bonn

Finck P, Hauke U, Schröder E (1993) Zur Problematik der Formulierung regionaler Landschafts-Leitbilder aus naturschutzfachlicher Sicht. Natur und Landschaft 68/12:603-607

Haemisch M, Kehmann L (1992) Naturschutzbilanzen - Definierte Umweltqualitätsziele und quantitative Umweltstandards im Naturschutz. Natur und Landschaft 4:143-148

Heinig S (1997) Ökosystemare Umweltbewertung. Naturschutz und Landschaftsplanung 29:53-56

Jessel B (1994) Leitbilder - Umweltqualitätsziele - Umweltstandards. Laufener Seminarbeiträge 4:5-10

Kleyer M, Kaule G, Henle K (1992) Landschaftsbezogene Ökosystemforschung für die Umwelt- und Landschaftsplanung. Zeitschrift für Ökologie und Naturschutz 1:35-50

Kriedemann K (1996) Grundlagen für ein Konzept zur Festlegung von Naturschutz-Qualitätszielen im Land Mecklenburg-Vorpommern einschließlich eines Indikatorensystems zur Raumüberwachung

Marzelli S (1994) Zur Relevanz von Leitbildern und Standards für die Ökologische Planung. Laufener Seminarbeiträge 4:11-23

Müller F (1996) Ableitung von integrativen Indikatoren zur Bewertung von Ökosystemzuständen für die Umweltökonomischen Gesamtrechnungen. Interne Mitteilung, Statistisches Bundesamt, Wiesbaden

Projektzentrum Ökosystemforschung (1996) Ökosystemforschung im Bereich der Bornhöveder Seenkette. Interne Mitteilungen des Projektzentrum für Ökosystemforschung, Kiel

Sachverständigenrat für Umweltfragen (1994) Umweltgutachten 1994. Für eine dauerhaft-umweltgerechte Entwicklung. Metzler-Poeschel, Stuttgart

Schneider ED, Kay JJ (1994) Life as a manifestation of the second law of thermodynamics. Math. Comput. Modelling 19:25-48

UVP-Förderverein: AG Umweltqualitätsziele (ed) (1995) Aufstellung kommunaler Umweltqualitätsziele. Anforderungen und Empfehlungen zu Inhalten und Verfahrensweise. Dortmunder Vertrieb für Bau- und Planungsliteratur, Dortmund

Zölitz-Möller R, Heinrich U, Nachbar M (1997) Environmental Planning with Help of Digital Geographical Information and Ecosystem Approach. Geo-Informations-Systeme 10, in print.

5.10 Integrating Diverging Orientors: Sustainable Agriculture: Ecological Targets and Future Land Use Changes

Armin Werner and Hans-Rudolf Bork

Abstract

A *sustainability of land use systems* can only be discussed, if biotic aspects of the landscape as well as regional, social, and economic aspects are integrated. Therefore, agriculture in the classical borders of fields or farms cannot be sustainable The goals for sustainability in a region can be defined, but there is not yet sufficient knowledge concerning the correct indicators that allow evaluation of sustainability in actual or suggested land use systems. The goal function of sustainability for a region has to be defined in a process together with the relevant groups. In this process, complex methods are necessary to evaluate the impact of land use or agriculture. For this purpose, the usage of landscape simulation models is suggested. A study on land use changes in a large region of north-eastern Germany is the basis for a discussion of some results and the limitations of this technique. To define the intended goals for sustainability, and to develop the necessary evaluation tools as well as the adapted land use systems it is necessary to undertake interdisciplinary and integrative research. For this, a new culture in science is required that enhances joint and discipline cross-discipline work.

5.10.1 Introduction

A major challenge for politics, administration, science and technology is to develop strategies of sustainability in their field of work. The basic idea of the *sustainability concept* is to have the same chances for future generations as we have today. The intention of this chapter is to discuss the necessary steps to achieve such a sustainability of chances in the area of agricultural land use or ecosystems, in general. In this chapter `ecosystems´ cover all aspects of ecological systems, including social systems.

F. Müller M. Leupelt (Eds.)
Eco Targets, Goal Functions, and Orientors

Firstly, the relevant targets that should be addressed and the goals that should be reached have to be defined (see table 5.10.1). This also implies looking for indicators with which it is possible to estimate or measure the development of sustainability in the land use systems. In addition, it is necessary to have suitable tools that allow the impact of land use or agricultural measures on the defined

Table 5.10.1. Definitions and explanations of goal related terms according to the sense used in this chapter

Term	Sense	Examples
ecological target	*a distinct objective in the protection of abiotic or biotic resources that is in the focus of adapted measures in land use or agricultural management* ecological targets are certain changes in land use management or specific measures in land use during the process of improving the ecological performance of land use systems	• reduce nutrient load to the environment by using site and crop adjusted fertilizing • improve biodiversity by reducing negative disturbances due to agrochemicals
ecological goal	*an objective that should be reached as part of the overall quality of a landscape situation*; a desired situation of abiotic or biotic resources in a landscape, ecological goals are less specific than ecological targets concerning the necessary measures to reach the goals	• low nitrate leaching • low or zero soil erosion • high habitat quality for field birds • secured biotic integrity
indicator	*a variable that is used to indicate the distinct physical state of a certain object or situation in the ecosystem, on the farm or in the rural area, it describes an ecosystem property;* indicators can be either measured directly or are based on calculations or estimations; indicators are used to evaluate the landscape situation and the degree of fulfillment of ecological goals	• nitrogen balance of a farm • amount of soil sediment being eroded or deposited • number of bird species in a biotope
goal function	*a set of goals that should be fulfilled in a specific land use situation*; the goals should be achieved all at the same time (the goal function equals an objective function in the terms of the optimization technologies from operations research science)	minimize risk of nitrate pollution *and* increase biotic quality *and* increase crop yield *and* maintain number of jobs in a rural area
orientor	a generalized characteristic for the goals of an ecosystem; a system-orientor can be understood as an intended general goal of the ecosystem or the land use system, orientors typify the complete set of ecological goals for a land use system or an ecosystem	• energy should be dissipated thoroughly on a regional scale • the level of excess nutrients in the ecosystems should be kept below the critical load level

indicators to be determined and to evaluate the impact on the possibility of reaching the goals. Research for suitable evaluation tools has to analyze, whether the involved sciences are capable of achieving suitable progress in defining the relevant goals and are able to develop sustainability in land using; current scientific approaches, present scientific education and the dominant types of joint research activities. All these aspects will be addressed in this Chapter. The terms in Table 5.10.1 will be used to distinguish the different levels of discussion.

5.10.2 The Goals of Sustainability in Land Use

The term *sustainable development* emerged in the 1980's to describe a specific type of desired human activity. It is concerned with economic goals as well as ecological and social objectives and their strategies (Dabbert 1993; Christen 1996). Dealing with sustainability in land use, the ecological goals are defined and aimed towards reaching an environmentally sound situation and to maintain the biological integrity of agricultural and forest property landscapes. Integrating social goals in these strategies should lead to viable communities in the rural areas, sufficient employment capacities and a functioning infrastructure. Therefore, the goal function of sustainability in land use consists of conflicting goals (high economic yield versus low emission of pollutants) but to some extent also of corresponding goals (viable farms and employment). Solutions for this problem can either have a high proportion of one of the three goal-sets (economy, ecology, social complex) or they can treat all three goal-sets equally (Lélé 1991).

When the dominance of economy increases in these solutions, the sustainability of the corresponding (land use) system becomes weaker (Turner 1988). A very strong sustainability is only achievable (Eblinghaus and Stickler 1996: p. 92), when an extreme protection of natural resources is maintained, and the economic impact is reduced. This shows, that the intentions of the involved persons (actors) determine the degree of sustainability that should be obtained (Gatto 1995). Thus, the degree of sustainability can only be a result of negotiations between the relevant groups in society.

The concept of sustainability is also criticized (Goodland 1996, Trainer 1990). However, there are great advantages in defining participatory methods as the future approach to solving solutions for conflicts between economy and ecology.

5.10.3 Can Agriculture be Sustainable?

Such a question seems rather strange for agronomists and forestry staff from Central Europe. For them the answer is obvious: Yes, it can. The term *sustainability* has a literal correspondence with the German word *Nachhaltigkeit*. It was first used to describe a type of planning and production in forestry: Only as much wood grown in a certain period of time, is harvested. Thinking in such patterns was a necessary reaction to large and severe deforestation in Central Europe during the Middle Ages, which clearly did not sustain the forests. Later, with knowledge of plant nutrition, agriculture adapted this type of thinking through managing humus and nutrients in the soil to prevent losses in productivity from nutrient depletion or humus destruction. Only specific amounts of plant nutrients or humus should be extracted or degraded which can ultimately be replaced or rebuilt in the same amount of time. The corresponding term is soil fertility (German: *Bodenfruchtbarkeit*).

The idea of soil fertility is to secure or increase the ability of the land to allow crops to grow and produce sufficient yields, as well as to carry agricultural production. In Central Europe, the actual soil fertility of arable fields, in most areas, is higher than it has ever been before since the beginning of man-related agriculture (Finck 1979; Heyland 1990). This statement can be proved with the high nutrient levels (Hamm 1989; Köster et al. 1988) or the level of organic carbon (humus) (Rogasik et al. 1994) in most soils, the steadily increasing crop yields of agricultural areas in Central Europe (Diercks et al. 1990) including the very high and increasing crop yields of organic farming under these conditions (Heß et al. 1992). This implies, that soil fertility is sustained on many fields in Central Europe; it has even increased. Therefore, to many farmers in Europe there is no question whether agriculture (or forestry) can be sustainable. To describe sustainability in this sense, is acceptable when reducing sustainability to just soil fertility. Nevertheless, this encompasses only a small proportion of the idea of sustainability.

When the concept of sustainable development is applied to any type of land use, the economic and social targets always cross the borders of single fields, farms or even communities. For example, biological integrity of a landscape can not be evaluated or managed just on a single field. In most cases, it is necessary to consider a whole landscape or at least parts of it. This is true for almost all biological targets and some abiotic targets. Furthermore, social sciences define and observe their indicators on levels of communities or within other administrative system borders. This leads to a situation where agriculture can only be judged on its sustainability together with other functions and activities in the rural areas. Therefore, sustainability of agriculture can only be assessed, if all surrounding factors (economic and social) are considered in conjunction. This implies that, agriculture, within the classical system borders of production, cannot be sustainable. Linking agriculture with nature and environmental protection measures, and with the social and economic system of a region, results in complex land use systems. It is possible to evaluate these land use systems on their sustainable development. Thus, there is no sustainable agriculture, but there can be sustainable land use systems within a region.

5.10.4 Sustainability-Concept in Land Use and Ecological Targets

Primarily, system borders where defined in terms of the suitable application of the sustainability concept on land use. As a next step, it is necessary to identify those indicators with which it is possible to detect or measure sustainability of a land use system. It is then possible to define those targets that should be addressed when certain measures of land use should be changed.

Table 5.10.2. Single goals and indicators for different landscape compartments as a base for sustainable land use systems – selected examples – [source: Werner 1997 b, adapted]

landscape compartment	*goals* *maintain or develop*	*protect against*	*possible indicators*	*remarks*
single farm	economic viability	violently changing frame conditions	profit per unit area, work or product	cannot be planned on a long term basis
soil	• soil surface • soil structure • soil fertility	• erosion • compaction • loss of humus	• soil depth • bulk density • organic matter content	• no threshold available • depends on site and usage • depends on site and usage
hydro-sphere	drinking water production	• reducing amount • contamination	• amount of drainage • amount of leachate	• forest decreases • zero-emission not feasible
atmo-sphere	thermal balance	contamination with greenhouse gases	amount of emitted gases	zero-emission not feasible
biosphere	• habitat diversity • habitat quality	• loss of habitats • loss of species	• landscape structure • species diversity	• what type • how much?

A set of system indicators is necessary, which is consistent with a diverse set of goals from the concept of sustainability. Only if these indicators are defined and measurable can the sustainability of different land use systems be analyzed (Hansen 1996). Table 5.10.1 groups a number of selected indicators according to their relevant compartment in the landscape. Each indicator can also be the object in a goal function. When using these indicators, problems arise. For many it is not possible to define reasonable thresholds or even relevant frame conditions. In addition, it is impossible to derive the degree of sustainability of a land use system with a single indicator. The sustainability of a land use system can only be analyzed by using the pattern from a set of indicator values. However, there is no definition as yet, in which indicators are relevant and what such a pattern should look like.

A major characteristic of the sustainability idea is the long perspective of most objectives. The land use systems should continue with a desired quality into the future. Until now, there are no criteria or even defined indicators that allow the prediction of long term effects on a reliable basis (Hansen 1996). All state variables in landscapes have high dynamics, particularly when looking at longer time periods.

As an example the `sustainable´ content of organic matter in the soil can be used. Organic matter is relevant as an indicator for humus in the soil and thus, soil fertility. Soil type or climatic site conditions alone cannot define the correct value of that indicator. The type of land use (forest, arable cropping, grassland), the management system (for example, with or without organic fertilizer, dominance of specific crops) and the levels of input, determine the steady state level of the humus content in the soil (Bachinger 1995). This dependency on the land use system applies to most indicators of agro-ecology and forest-ecology (Table 5.10.2). When determining the sustainability of the production system from an ecological point of view, particularly the spatial (lateral) interactions of the agro-ecosystem with the neighboring or distant ecosystems, have to be regarded. These are off-site effects caused by eroded sediments, by leached or washed out nutrients or pesticides, by depletion of natural non-renewable resources and further disturbances to other ecosystems.

Due to the flow of nutrients, energy, products, information and labor to and from farms, the current agricultural structures are *open (eco-)systems* (Haber 1986). Such open systems can have detrimental effects within other ecosystems if they are managed without realizing the impact on other systems. The actual situation of these open systems can hardly be defined as sustainable in the basic sense of the sustainability concept. Therefore, it can be assumed, again, (see Sect. 5.10.3) that it is not possible to define states or processes of sustainability just within the system borders of agriculture (fields, farms).

Only those indicators which can be influenced through land use are valid indicators for the development of sustainable land use systems. Therefore, those measures of landscape structuring and agricultural management have to be identified which can alter the performance of the indicators. In this way, the indicators become sections of ecological targets. These are targets of an ecologically adapted land use or a specific agricultural management measure. A typical *ecological target* in the development of sustainable land use systems is the reduction of a nutrient load to the environment (target), through adaptation of practical farming, to site and crop-adjusted fertilizing (measure).

5.10.5 Goal Function Construction and Orientor Deduction for Sustainable Land Use

The summary of the above reasoning is, that constructing and implementing the concept of sustainable land use systems largely depends on the existence and usage of multiple ecological targets. They will be part of a set of goals, a goal function, that includes economic and ecological objectives. The essential characteristics of the resulting ecosystem can be summarized as the general 'intention' of that ecosystem (e.g. maximum energy dissipation in a region). These typified characteristics are described by the term *orientor*.

Abiotic goals can be easily defined because existing knowledge concerning the effects of certain abiotic variables or landscape situations on the environment are quite substantial. In most cases, the abiotic goals are set as environmental standards for measurable or indirect environmental indicators. Currently most of these standards are derived by looking at the detrimental effects on human beings. Therefore, mainly the minimum-acceptable-effect or no-effect concept for public health is used (for example the maximum concentrations of nitrate or pesticides in drinking water are used for addressing the quality of groundwater). In the future, the allowable effects on ecosystems will also have to be included (for example, using the *critical load concept*).

The numerous possibilities of managing the interaction between land use and abiotic parts of the environment help to include abiotic goals in the general object function more easily. The same is true with the definition and management of economic and most social goals.

It is more difficult to define goals of nature conservation (species and habitat protection) for agricultural areas. The difficulty lies in deriving the necessary goals when looking at the biotic quality of a land use system. Which species or what type of habitats should be developed or maintained depends largely on the regional specific characteristics and the type of land use. Particularly with organisms, there is the problem of conflicting goals (for example, developing a landscape for birds from open landscapes, and at the same time, for butterflies from hedges). In addition, most of the interaction of land use management and biotic quality of the landscape ecosystems is not identified, and the relevant processes are, in most cases, not yet understood. To just demand higher *biodiversity* in a landscape, is too simple. As such no general solution for biotic goals can be derived. A rather practical approach is to identify the typical originality of a landscape (*orientor approach*). This has to be done in context with a specific land use. Additionally, the maximum achievable biotic goals, according to typical species or habitats in such landscapes, have to be identified (Plachter 1995).

It is also difficult to evaluate completely, a biotic situation from its performance according to a set of predefined biotic goals. For this Heidt and Plachter (1996) propose to define the range of acceptable output for each biotic indicator and to evaluate the pattern of degrees in the fulfillment of goals.

5.10.6 Evaluating the Effects of Land Use Changes and the Influence of the Human Factor

The development of a regional land use system in a region is the result of several decision making processes by land owners and land users. The decision is derived from economic and legal, but also other, mainly personal arguments. Therefore, it is not trivial to determine the development of land use systems (Kächele and Dabbert 1995). To achieve this, it is necessary to determine the reaction of farmers to different frame conditions (O´Callaghan 1992). In addition, the available technologies, their costs and their results have to be known and quantified. If these processes and relations are formalized in a practical way, then the changes of the land use systems over time or the effects of changing frame conditions can be estimated. An elegant and effective way to describe the internal structure of farms and to include the economically-minded farmer is to build economic simulation models (Claramunt et al. 1994; Mothes and Lutze 1997). With such a model, Kächele and Dabbert (1995) evaluated the impact of changing agricultural politics on farm structure and land usage within a region of about 10.000 sq.km in Northeast Germany. The changing economic general set-up for the farmers from 1992 to 1996, were analyzed. These changes were due to completion in the reform of the common EU-agricultural politics (CAP-reform). When the frame conditions change, effects on the land use, largely depend on the farmers' skill to organize his business, as efficiently as possible. Therefore, two abilities of the farmers were analyzed as scenarios. In the first scenario, the farmers were highly skilled and had a high adaptability to the new (economic) situation. In the second scenario, the farmers are less skilled and therefore, have higher fixed costs of production than in the first scenario. When Kächele and Dabbert (1995) applied their regional model, they determined the type of production that the farmers would probably follow under such conditions. In addition, they calculated, if, and how much of a land use area would be abandoned. This abandoned land (not used for production) should either be kept open or could develop into forest. In the first case, the open landscape can be kept under the climatic conditions of Central Europe only when applying special maintenance measures (cutting, grazing) which are not part of the regular agricultural management. The new forests can either be planted or can result from succession processes of the vegetation after several decades.

This leads to new areas of arable land, grassland and forests (see Table 5.10.3). Furthermore, the new land use differs in management. These new land use systems change the effects of agriculture on different ecological and social indicators. By using process oriented models or balancing algorithms, the changes of these indicators were analyzed (Bork et al. 1995; Dalchow et al. 1995) (see Fig. 5.10.1).

The adaptation of highly skilled farmers to the new economic situation leads to a rather small reduction in the land use area (mainly due to abandoned grassland) and changes in the management. This changed land use causes a reduction of negative ecological effects especially on abiotic indicators (Fig. 5.10.1). There are only minor effects on biotic indicators of land use.

If farmers have problems in adapting well to the new situation (less skilled farmers), a higher proportion of their formerly used land (32%) is given up. The resulting positive effects for abiotic indicators, are considerably large. The land use changes also lead to substantial improvements in some, but not all biotic indicators (Fig. 5.10.1). Unfortunately, the social and economic indicators (employment and animals) develop very negatively, in both scenarios. This shows the rather difficult situation of goal fulfillment in a landscape. It is not possible to improve all relevant indicators or orientors at the same time.

high adaptability of farmers

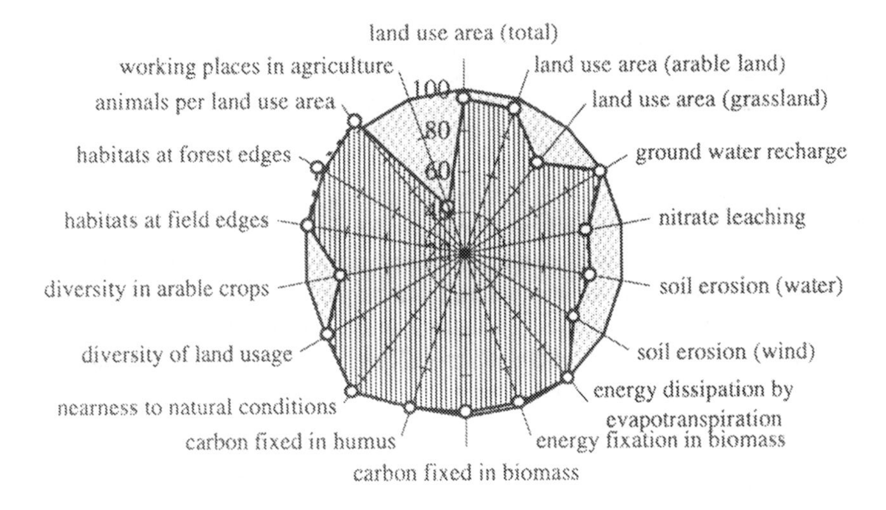

low adaptability of farmers

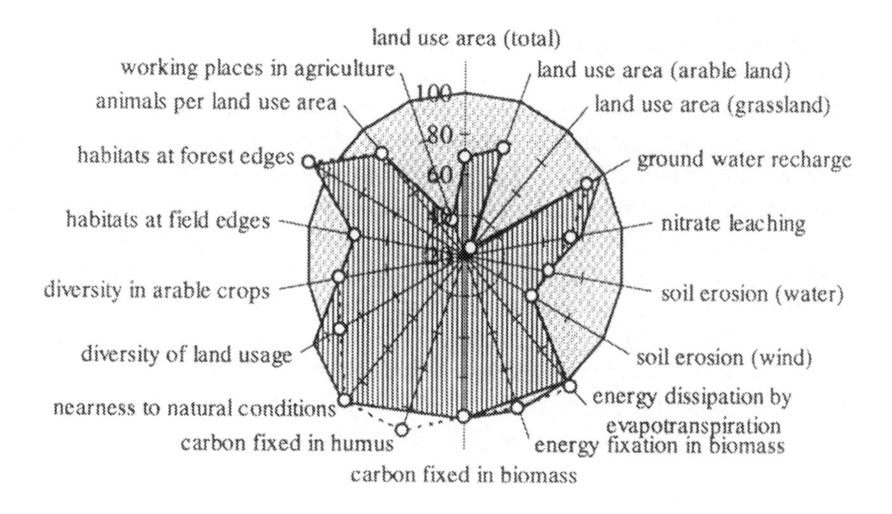

Fig. 5.10.1. Simulated impact of land use changes onto a set of orientors for land use quality within an area of 10000 sq.km of Northeast Germany in comparison to the reference situation (1992 = 100 %). Land use changes are caused by adaptation of farms to EU-agricultural policies from 1992 until 1996 with farmers able to adapt well (top) or less well (bottom) to the new situation. Abandoned land is either afforested or kept open through landscape husbandry. [Data from Bork et al. 1995: pp. 125, 153, 193, 213, 229, 248, 251, 270, 271, 331, 344, 345, 347; own sketch]

The described scenario-technique allows to take different frame conditions to be taken, to search for the best situation or to analyze given frame conditions. These tools can thus be used for evaluating technologies as well as politics (Bryant 1995). Therefore, such studies are helpful in discussing with relevant decision making groups for land use planning as well as for the policies in the process of developing sustainable land use systems.

5.10.7 Multiple Goal Optimization

The very intrinsic feature in the sustainability concept of land use, is the existence of achieving several goals. All goals are summarized in a *goal function.* Solutions to such a goal function can only be found by comparing the result of one (or several) goal(s) to the results of the others. Improving the performance of one goal in the land use system can have negative, positive or no effects on the others. To find optimal solutions within the economic, ecological and social constraints makes it necessary to describe the relationship between the land use management and the achievement of these goals. For this, the use of simulation models is the appropriate method (Kächele and Dabbert 1995; Rotter et al. 1995). The ecological quality of a land use system is primarily based on the activities on farms and in other types of land use. These activities are results of the decision making process in the land use systems. Therefore, those models that describe the land use systems have a strong background in the economic aspects of land use, but they also have to include the ecological effects of the activities in their structure. With such models, it is possible to calculate the economic and ecological results when certain management decisions are followed. They also allow management options that fulfill a set of predefined goals to be looked for (Kächele and Zander 1997).

An important prerequisite for developing such economic-ecological land use models is the ability to describe and model the ecological effects of all measures that can (or should) be taken in practical land use. Therefore, it is necessary to formalize the technologies that are feasible in actual or future land use and to estimate the effects on the relevant ecological goals. In the decision support system ADAMS (Agricultural Data, Analysis, and Modeling System) (Kächele and Zander 1997) this is done for a large set of different field crops and a number of site specific management strategies for each crop (Werner et al. 1997). The evaluation algorithms are based on simple process oriented models (e.g. estimating the risk of nitrate leaching) as well as on expert judgment (e.g. estimating the impact of management activities on single organisms) (Meyer-Aurich et al. 1997). For a typical region in Northeast Germany, crops which are currently grown and the feasible management practices are analyzed using these algorithms (table 5.10.3). Different crops show different levels of efficiency in fulfilling the desired goals. Unsatisfactory results point to a need for improvement of the checked management technologies (e.g. the corn crop in table 5.10.3).

With these simulation models, it is possible to analyze changes in ecological effects when different management strategies are followed. The loss (or gain) for the other goals when improving the performance of one specific goal, can be described mathematically in *trade-off functions* (Kächele and Zander 1997). Using ADAMS the scientist or the planner are able to look for the best management strategy and the resulting economic costs to fulfill ecological goals (Kächele and Zander 1997; Meyer-Aurich et al. 1997; Werner et al. 1997).

5.10.8 Practical Approaches to Sustainable Land Use Systems

The definition of sustainable land use systems and thus, the analysis of whether a land use system is more sustainable than the other, is not yet possible. Because

Table 5.10.3. Differences between corn and terms of other, typical crops of an agricultural region in the ability to comply with four abiotic and five biotic goals. Evaluation algorithms based on simple process models or expert judgment, evaluation scale: 1 (good) up to 3 (bad); site: South Uckermark, Brandenburg. [Source: Werner 1997 b, adapted]

	ecological goals (min.: reduce, max.: improve)								
	max.	max.	min.	min.	max.	max.	max.	max.	max.
	ground-water protection	ground-water recharge	soil erosion (wind)	soil erosion (water)	oligotrophic plant communities	typical field weeds	habitat for partridge	habitat for amphibians	habitat for crane
	evaluation value								
AVERAGE from 17 arable field crops [1]	1,5	1,4	1,8	1,6	1,8	1,8	1,5	2,4	2,7
	difference to the AVERAGE (negative value: less good than AVERAGE, positive value: better than AVERAGE)								
with corn for silage [2]	-0,5	0,1	-0,9	-1,0	-0,6	-0,6	-0,5	0,1	1,1
with corn [3]	-1,3	-0,6	-1,2	-1,4	-0,4	-0,4	-0,4	-0,2	0,7

[1] for four site types and 4 to 52 management systems for each crop
[2] for four site types and 12 management systems
[3] for four site types and 8 management systems

of this, it is necessary to develop a strategy that tries to consider the relevant elements of sustainability in the current land use research and in the practical land use management. A major problem is, that it is not possible to define sustainable land use systems yet. However, there is already a need in research, development and even politics to look for pathways of sustainability. The most practical way to solve this puzzle is to increase the efficiency in resource use (Wit 1992) and to reduce risks of pollution, erosion or detrimental soil compaction. In addition the hospitality of the agriculturally used landscape for typical or rare organisms has to be increased (table 5.10.4). In most cases this needs further research and the development of adapted agricultural management systems or even new technologies.

5.10.9 Consequences for Integrated Rural Planning

Three major characteristics of applying the sustainability concept to land use can be identified. First, the goal function (activities of land use should focus on it) is not a set of fixed objectives. It is a result of continuing negotiations between the relevant actor groups in the rural areas. Therefore, this goal function has to be flexible according to the actual or future demands. Scientific standards cannot define the goal function nor can this be done by scientific deduction. Secondly, the necessary negotiations will only be successful if they are done participatory, from the very beginning. All relevant groups, including the sciences have to be involved. Thirdly, the way of evaluating and planning sustainable land use systems is a typical process of decision making with uncertainties. No strict rules can be applied and a continual process of learning from experience and scientific progress is necessary.

Table 5.10.4. Criteria that land use systems should fulfill as a minimum prerequisite to develop sustainability of land use in the future. [Source: Werner 1997 b, adapted]

abiotic perspective
♦ high efficiency in the use of nutrients and energy
♦ no relevant surplus or deficits in the nutrient balances (fields or regions)
♦ low risk of soil erosion and detrimental soil compaction
♦ low risk of emission of nutrients or pesticides
biotic perspective
♦ high proportions of areas within the landscape left undisturbed (5% - 15% of the land use area)
♦ adaptation of land use management systems according to biotic potentials or sensibilities, locally or in the region
economic perspective
♦ sufficient income to build up reserves for future investments
♦ minimum profit for the farmer or entrepreneur
♦ positive returns on investments
social perspective
♦ participatory and iterative definition of goals for landscape development and rural planning
♦ joint efforts in approaching sets of diverging goals

All activities for rural planning have to be adapted to these characteristics. This also leads to a new situation for planning organizations, for public administration and in politics. Decision making in rural areas should be done in crossing all disciplinary borders.

Particularly, the definition of relevant problems and the solutions (Table 5.10.5) to solve them, should be examined together. A feasible way of handling goals of protecting environment and nature could be through organizing regional co-operatives for the environment (Table 5.10.5). Members of these groups should be farmers, communities, the public administration and any group interested in the development of that rural area.

Table 5.10.5. Proposed features for rural planning with the objective of developing sustainability in land use.

relevant problems
• contamination of water bodies with plant nutrients or pesticide residues
• average surplus of nutrient supply to fields
• soil erosion and detrimental soil compaction
• loss of species and habitats in agriculturally used rural areas
• reduction in economic profitability for farmers
• loss of job possibilities in rural areas
• abandoning of agricultural sites with low natural productivity
• marginalisation of rural areas
future groups and organizations to handle aspects of sustainability
• agriculture, forestry and communities
• regional co-operatives for the environment
• urban and rural areas
• environmental and nature protection groups
spatial scale to plan for sustainable development
• water catchment-areas as minimum regions
• nature regions or complete landscapes
• administrative regions (single or several communities)
important goals in the process of increasing sustainability of rural areas
• reduce flow of plant nutrients into neighboring ecosystems
• improve general and special habitat quality of fields
• enlarge area left undisturbed within agriculturally used landscapes
• increase spatial heterogeneity of site conditions within landscapes
• reduce import of plant nutrients into the system regions
• increase flow of plant nutrients and organic carbon from urban to rural areas
• increase and stabilize the economic and social functions of rural areas
negotiation in the process of planning for sustainability of rural areas
• necessity to increase knowledge about relevant processes for all participants
• joint definition of achievable goals and expected success
• application of conflict solving methods (for example mediation, moderation)
• searching for compromises
• iterative
• continuing and repeating

Primarily, the achievable regional goals should be determined, sensitive areas within the region defined and a set of possible management solutions developed. Tax money that is spent on protecting the environment or on ecological services should be given to such co-operatives (Bahner 1996). Payments could be better distributed within a region than any general program, that is responsible for all farmers or protection groups in the area. These organizations could also include more and more other goals for the rural area and therefore would be the base for public *sustainability developers* of their own region.

Public administration will play an important role in this process, being responsible for planning and managing the rural areas. All relevant sub-organizations that remain separate (farming, forestry, environment, rural planning, etc.) have to link their activities. A very integrative approach of their work is necessary. Additionally, all relevant disciplines will have to be viewed at the same time and not separately as is done at present, in most cases. In the future, these organizations have to expand their current activities into negotiations, moderation and conflict solving. For most of the administrations this will result in new tasks and specific demands in skills and knowledge (Werner et al. 1997). A restructuring process of these organizations should be initiated very soon. Examples of relevant problems, of goals and possibilities of achieving solutions are listed in table 5.10.5.

A major problem in developing sustainable land use will be the regional fluxes of plant nutrients or organic matter into fields, but also from the settlements into the region. The nutrients that are exported with the crop products from the fields into the cities or towns do not return to the original sources in most cases. The same is true with most plant nutrients (phosphorous and potassium) coming from distant mining sources into the rural areas. The fields are fertilized with these nutrients, but the limited sources are continuously and irreversibly exploited. A major management rule of (global) sustainable development is violated: to exploit non-renewable resources, only if they can be replaced by alternatives (Pearce and Turner 1990).

All plant nutrients and organic carbon should be kept in an almost short cut regional cycle. Such regional cycling can only be reached if settlements and rural areas plan and work together in a co-operative way. To reach this goal, the current infrastructure for waste management has to be changed and new technologies in crop production and food processing are necessary. Also aspects of trading and the behavior of consumers and their expectations of food have to de altered.

5.10.10 Implementation of Ecological Targets in Sustainable Land Use and the Science Issue

With the main characteristics of sustainability pathway in mind (flexible goal functions, participial approaches, decision making with uncertainties), the attitude of most natural scientists is understandable. The sustainability debate is put into the domain of politics, or pushed into the area of social sciences. Unfortunately, such an attitude prevents the formation of any part of sustainable development, because the first step to sustainability cannot wait. The work towards sustainable land use is dominated by the general problem of the sustainability concept, not knowing what the correct goals or orientors are and how systems perform on a long term basis. For this reason, the permanent interaction with scientists is necessary, to continuously find solutions for sustainable development.

Therefore, it is necessary, that research and development are flexible and adapt their activities according to the demands of the sustainability concept. The main point is, that an approach of sustainable research and development has also to be participatory. In interaction with relevant groups in rural areas, the scientists related to land use systems and rural areas have to solve several questions. Especially those questions, that are relevant to continue the permanent completion of the sustainability idea. They are also asked to identify the processes and problems relevant to sustainability.

Due to the multiple goal approach of sustainable land use, all relevant disciplines have to be covered. These have to work together in an integrative way to solve the scientific problems of sustainable development. For most scientists, this leads to a new challenge: working in interdisciplinary groups and linking the success of their work to the results of other scientists. In addition, when working beyond all borders between disciplines, the involved sciences have also to include the users in the scientific activities. Thus, the process of sciences, working on sustainability aspects has not only to be interdisciplinary. It can be characterized as `transdisciplinary´ (Ganzert 1994).

To address the very complex systems made from interaction of land use and landscape ecology, the system science approach seems to be the most appropriate. With this procedure the system borders, the internal structure and the relevant variables have to be defined (or at least clarified) in a joint effort involving all disciplines. The system science approach also includes models as decision support tools. Such decision support methods will be the major link between all disciplines working on the sustainability problem of land use. To develop such tools, it is necessary to first define the relevant questions or tasks for this decision-making process. After these questions and tasks have been addressed in an interdisciplinary way, it is possible to determine the steps that are necessary to develop the decision-making tool and the relevant research. Using such models makes it necessary to define common interfaces for knowledge, data and rules in the decision-making process. Therefore, the development and application of decision making models or tools are a very integrative approach.

It will not be easy to meet all these demands in the interaction between science and the development of sustainable land use. Working on interdisciplinary, integrative and applied research particularly will not be easy. At present, the evaluation of scientific activities is based on the judgment of how well a discipline is covered and how many publications a scientist produces. This prohibits to a large extent the co-operation of interdisciplinary teams. It is much easier to produce many publications with work done by just by one author, than to run through the time consuming process of solving communication problems due to different definitions of terms and theories. Until now, it has not been proven, which approach is more effective for progressing in solving the scientific challenges concerning sustainability. Either to profit from the excellent performance of individual scientists or to enhance the synergy effects that can emerge when different disciplines work together as teams. If the scientific community wants to find out which of the two pathways is the most appropriate, then we need a new culture of science (Werner and Meyer-Aurich 1997). Scientists should be encouraged to do interdisciplinary and integrative work, to work in groups and to use system science as an analytical tool. In addition, the evaluation of scientific work should avoid forcing the scientists to work just on the minimum publishable piece of a problem. An integrative approach that offers complete solutions for problems is more helpful for the advancement of science when one has the sustainability perspective in mind.

5.10.11 Conclusions

From the topics and discussions in this chapter the following conclusions can be drawn:

- The system borders of agriculture (fields, farms) do not allow inclusion of all relevant objects of sustainability (especially biotic and social aspects). Therefore it is not possible to address *a sustainability* of agriculture. Only when extending agriculture with regional aspects to complex *land use systems*, the sustainability concept can be applied.
- It is possible to characterize the goals of sustainability in land use (closed nutrient cycles, intact biotic integrity, sufficient income and jobs in the rural area ...). But it is not yet possible to define those indicators with which it is possible to detect or control sustainability in practical farming or land use.
- Therefore, indicators of sustainability in land use have to be developed (especially for long term effects). For this, it is necessary to analyze the impact of different land use strategies or agricultural measures on ecological goals. Suitable methods are described in the text.

- Tools to evaluate the effects of different or changing land use systems can be found in landscape- or system modeling. These tools can also be used to find optimal solutions when a set of goals and a goal function is defined for a specific landscape or region. When such a tool was applied to a region of 10.000 sq.km in Northeast Germany, it could be shown that larger changes in the rate of used land do not cause strong changes in the ecological effects, but do cause strong changes for social indicators.
- Defining the correct goals for sustainability in a practical situation cannot be achieved through science alone. This is a social process, that has to involve all relevant groups (mainly the regional actors). Therefore, the process of finding the goal function (set of ecological and socio-economic goals) that should be fulfilled in a region, has to be done in a participatory and iterative way.
- Landscape models can support the decision making process of these groups. The application of such models offers the opportunity to analyze different pathways when developing strategies for future land use systems as well as to check the impact of several agricultural systems or measures onto ecological goals.
- The correct usage of the terms *indicator, ecological goal, goal function* and *orientor* can help to support the development of sustainable land use systems. Particularly the discussion between actors in the process of defining the intentions and possible pathways of sustainability in a region can be clarified.
- To enhance the scientific work for the development of sustainable (land use) systems, a new scientific culture is suggested. Scientists have to be trained to do more interdisciplinary and integrative work. A common bracket for all disciplines involved could be system analysis and system modeling. In addition, the procedures of how the scientific work of researchers is evaluated, have to be adapted. It is more important to favor joint and complex research and its publication, than to favor single activities and small, easily publishable pieces of work.

References

Bachinger J (1995) Effects of organic and mineral fertilizers on chemical and microbial parameters of C- and N-dynamics and root parameters. In: Mäder P, Raupp J (eds) Effects of low and high external input agriculture on soil microbial biomass and activities in view of sustainable agriculture. Proceedings of the second meeting in Oberwil (Switzerland), 15th to 16th Sept 1995, pp 52-58

Bahner T (1996) Landwirtschaft und Naturschutz - vom Konflikt zur Kooperation: eine institutenökonomische Analyse. (Europäische Hochschulschriften: Reihe 5, Volks- und Betriebswirtschaft, Vol. 2005), Lang, Frankfurt/Main

Bork H-R, Dalchow C, Kächele H, Piorr H-P, Wenkel K-O (1995) Agrarlandschaftswandel in Nordost-Deutschland unter veränderten Rahmenbedingungen: ökologische und ökonomische Konsequenzen. Verlag Ernst & Sohn, Berlin

Bryant CR (1995) Interest groups, policy makers and the challenge of modeled or perceived futures for the rural environment. In: Schoute JFTh, Finke PA, Veeneklaas FR, Wolfert HP (eds) Scenario Studies for the Rural Environment. Kluwer Academic Publishers, Dordrecht, pp 599-607

Claramunt C, De Sède MH, Prélaz-Droux R, Vidale L (1994) GERMINAL: A spatially referenced information system for rural environmental management. In: Sustainable Land Use Planning, Proceedings of the 1992 CIGR workshop, Wageningen. Elsevier Science Publishers, pp 257-270

Christen O (1996) Nachhaltige Landwirtschaft („Sustainable agriculture"). Ideengeschichte, Inhalte und Konsequenzen für Forschung, Lehre und Beratung. Berichte über Landwirtschaft 74:66-86

Dabbert S (1993) Mayer contra Liebig: zur Aktualität einer historischen agrarwissenschaftlichen Auseinandersetzung für das Konzept der „Nachhaltigen Nutzung von Agrarlandschaften". ZALF-Bericht 6. ZALF, Müncheberg

Dalchow K, Bork H-R, Kersebaum KC, Piorr H-P, Wenkel K-O (1995) Agro-Landscape Changes in North-East Germany, Ecological and Socioeconomic Consequences. Arch Nature Protection and Landscape Ecology 34:1-15

Diercks R, Heitefuß R (1990) Integrierter Landbau: Systeme umweltbewußter Pflanzenproduktion. Grundlagen-Praxiserfahrungen-Entwicklungen. BLV-Verlag, München , p 420

Eblinghaus H, Stickler A (1996) Nachhaltigkeit und Macht: zur Kritik von Sustainable Development. IKO-Verlag, Köln

Finck A (1991) Düngung, ertragssteigernd, qualitätsverbessernd, umweltgerecht. Ulmer, Stuttgart

Ganzert C (1994) Umweltgerechte Landwirtschaft. Nachhaltige Wege für Europa. Economia Verlag, Bonn

Gatto M (1995) Sustainability: is it a well defined concept? Ecol Applications 5:1181-1183

Goodland R (1996) Environmental sustainability: universal and non-negotiable. Ecol Applications 6:1002-1017

Haber W (1986) Über die menschliche Nutzung von Ökosystemen - unter besonderer Berücksichtigung von Agrarökosystemen. Verhandlungen der GfÖ XIV:13-24, Hohenheim

Hamm A (1989) Entwicklung der P-Bilanz in der Bundesrepublik Deutschland. In: Aktuelle Probleme des Gewässerschutzes: Nährstoffbelastung und -elimination. Verlag Oldenbourg, München Wien, pp 99-109

Hansen J W (1996) Is Agricultural Sustainability a Useful Concept? Agricultural Systems 50:117-143

Heidt E, Plachter H (1996) Bewerten im Naturschutz: Probleme und Wege zu ihrer Lösung. Beitr Akad Natur- und Umweltschutz Baden-Württemberg 23:193-252

Heß J, Piorr A, Schmidtke K (1992) Grundwasserschonende Landbewirtschaftung durch Ökologischen Landbau? Dortmunder Beiträge zur Wasserforschung, Veröffentlichungen des Institutes für Wasserforschung GmbH Dortmund und der Dortmunder Stadtwerke AG 45

Heyland KU (1990) Integrierte Pflanzenproduktion. System und Organisation. Ulmer, Stuttgart

Kächele H, Dabbert S (1995) Ökonomisches Regionalmodell. In: Bork H-R, Dalchow C, Kächele H, Piorr H-P, Wenkel K-O (eds) (1995) Agrarlandschaftswandel in Nordost-Deutschland unter veränderten Rahmenbedingungen: ökologische und ökonomische Konsequenzen. Verlag Ernst & Sohn, Berlin, pp 70-84

Kächele H, Zander P (1997) Modeling Multiple Objectives of Land Use for Sustainable Development. Agricultural Systems, submitted

Köster W, Severin K, Möhring D, Ziebel HD (1988) Stickstoff-, Phosphor- und Kaliumbilanzen landwirtschaftlich genutzter Böden der BRD von 1950-1986. LUFA, Hameln

Lélé S M (1991) Sustainable Development: A critical Review. World Development 6:607-621

Mothes V, Lutze G (1997) Reflexionen sozioökonomischer Probleme in der Landschaftsmodellierung. Arch. für Naturschutz und Landschaftsforschung 35:239-253

Meyer-Aurich A, Zander P, Werner A, Roth R (1997) Developing land use strategies appropriate to nature conservation goals and environmental protection. Landscape and Urban Planning, in print

O'Callaghan (1992) Decision Making in Land Use. In: MC Whitby (ed) (1992) Land Use Change: the Causes and Consequences. ITE Symposium no 27, HMSO London, pp 79-87

Pearce DW, Turner RK (1990) Economics of natural resources and the environment. New York. Cited in Hohmeyer O (1996) Mögliche Klimaveränderungen und CO_2-Vermeidungsstrategien in Europa. In: Eichhorn P (ed) Ökologie und Marktwirtschaft. Probleme, Ursachen und Lösungen. Gabler Verlag, Wiesbaden, pp79-120

Plachter H (1995) Functional criteria for the evaluation of cultural landscapes. In: Droste B von, Plachter H, Rössler M (eds) Cultural landscape of universal value. Fischer, Jena, pp 393-404

Rogasik J, Dämmgen U, Obenauf S, Lüttich M (1994) Wirkungen physikalischer und chemischer Klimaparameter auf Bodeneigenschaften und Bodenprozesse. In: Brunnert H, Dämmgen U (eds) Klimaveränderungen und Landbewirtschaftung, Teil II, Landbauforschung Völkenrode 148:107-139

Rotter R-P, Veeneklaas F-R, Diepen CA van (1995) Impacts of changes in climate and socio-economic factors on land use in the Rhine basin: projections for the decade 2040-49. Studies on environmental sciences, Elsevier Scientific Publishing, Amsterdam, pp 947-950

Trainer T (1990) A Rejection of the Brundtland-Report. IFDA-Dossier 77:72-84

Turner RK (1988) Sustainable Environmental Management. Principles and Practise, London

Werner A (1997a) Bedeutung des Maisanbaues im Konzept von Landnutzungssystemen und einer nachhaltigen Landwirtschaft. In: Umweltgerechter und ertragsorientierter Maisanbau. Proceeding der Fachtagung Umweltgerechter und ertragsorientierter Maisanbau. Soest 8.-9. Juli 1997. Universität-Gesamthochschule Paderborn

Werner A (1997b) Nachhaltige Landwirtschaft - ein Leitbild zur Beurteilung von landwirtschaftlichen Produktionsformen? Mitteilungen der Agrarsozialen Gesellschaft, Göttingen, submitted

Werner A, Meyer-Aurich A (1997) Agrarforschung - Quo Vadis? Anpassung einzelner Forschungsbereiche der Agrarwissenschaften an veränderte Rahmenbedingungen und neue Herausforderungen. Perspektiven im Forschungsbereich Pflanzenproduktion. In: Dachverband Agrarforschung (eds) Schriftenreihe agrarspectrum, Band 26, Frankfurt, pp 46-65

Werner A, Müller K, Wenkel K-O, Bork H-R (1997) Partizipative und iterative Planung als Voraussetzung für die Integration ökologischer Ziele in die Landschaftsplanung des ländlichen Raumes. Zeitschrift für Kulturtechnik und Landentwicklung, in print

Werner A, Roth R, Meyer-Aurich, A, Zander, P (1998) Estimating the ecological effects of crop management strategies. German Journal of Agronomy, in preparation

Wit CT de (1992) Resource use efficiency in agriculture. Agricultural Systems 40:125-151

5.11 Conclusions: Potentials and Limitations of a Practical Application of the Eco Target and Orientor Concept

Sylvia Herrmann and Reinhard Zölitz-Möller

Introduction

The inclusion of ecosystem development goals would be an important aspect for environmental planning and management, particularly, with respect to the aspect of sustainability. However, eco targets and orientors such as self-organization principles, energy-dissipation, catastrophe topologies, exergy and Holling cycles compiled by analysts to determine the inherent principles of ecosystem development, may provide "intriguing perspectives for further study and some intuitive images of what a healthy environment is all about. But they have not been made practically operational for routine application" (Regier 1993). These impressions had also arisen by discussing the possibilities of using orientors as a tool for the planning process according to the contributed papers in this chapter. Orientors appear to be more of an idealistic picture of the desirable state of ecosystems, being in harmony and without any human influence. They are either designed within more theoretical (modeled) worlds of ideas, e.g. as background for dynamic modeling based on physical laws mostly or derived from measurements and surveys of natural ecosystems. That means, they try to describe a more or less optimal state or functioning of ecosystems. If orientors should be used as tools in environmental planning and management, various requirements have to be met.

- The proposed final states have to be transferable to the actual situation of the planning area. Particularly in European landscapes, these areas are often more or less strongly man-influenced.
- Orientors must fit for different types of ecosystems and at different scales (from ecosystem to landscapes). Therefore, not only aquatic but also terrestrial ecosystems have to be included and according to the consideration of landscapes or regions a "complex of complexes" has to be described and considered.

F. Müller M. Leupelt (Eds.)
Eco Targets, Goal Functions, and Orientors

- Adequate parameters for the quantification of the quality of the actual state of the ecosystems or landscapes (concerning the relevant orientors) have to be provided.
- The measurement of these parameters must be easily adaptable in practice and for different scales.
- Depending on the strong cultural influences, natural cycles and development are often overlaid by the overwhelming effects of economic factors. This implies that, for developing goals and describing orientors for the future state of ecosystems in cultural landscapes a compromise between ecosystem inherent and external factors, such as, economy and culture has to be made. Otherwise, the practical application will fail.
- The acceptance of orientor based strategies in environmental planning has still to be elaborated, because until now, more anthropogenic or economic reasons had been focused on.

Demands of Planners and Proposals of Researchers

The papers provided, reflect the expectations of the planners on the one side and the whole range of possible answers of the researchers on the other.

There exist expectations on the side of planners to obtain "environmental quality objectives for different land use forms and with regard to prevailing ecological conditions" (Roch Sect. 5.8) from the ecosystem research, for the orientation of future management actions. Even if the application of orientors within the planning process will "require a generation to fulfill" (Roch Sect. 5.8), the intention to base planning decisions on a better ecological basis by using these tools is visible. Even if an agreement to this proposal exists, it is stated, that the communication between ecosystem researchers and planners is not very easy because of the differing concepts for landscape development and nature protection (Berg and Riedel Sect. 5.9). Planners seem to think too statically and researchers look too closely at specific objects and forget the necessity of transferable statements.

Furthermore, there exists a skepticism concerning the possibility of finding general principles for the development of ecosystems (Roweck Sect. 5.5). Beside this, the power of the socio-economic demands and the political structures seems to dominate over the results of the ecosystem research and lead to their underestimation within the actual planning process. Nevertheless, the practical management has also to deal with certain transferable parameters to reduce the work that is necessary for every new planning situation.

Different answers to the questions of planners are given by the scientists and even within the scientific community there are a lot of alternative views of which eco targets and orientors should be used, are appropriate and how close they are to being applied in practice. The more theoretical considerations of exergy as an indicator for certain system states (Marques et al. Sect. 5.3) show an interesting approach. The authors admit, that the question of practical application of the indicator is still open. Much research work has to be done to provide a simple methodology for the estimation of the Exergy Index. Particularly for terrestrial systems, only a few, even promising examples exist. The approach refers purely to the ecological aspects.

The Sustainable Process Index (SPI) as a measure for obtaining information about the quality of the co-existence of man and ecosphere is a very general indicator (Krotscheck Sect. 5.2). These interactions are proposed to show the link between economy and ecology, including the potential impact of anthropogenic activities on the ecosphere. This index refers to the factor "area" as a central unit for the measurement of these effects. Because of the limited knowledge about ecosystem processes, the carrying capacity and the effects of evolution, the author proposes to use precaution values as orientors. In this concept, the demand to include economic aspects is also fulfilled. However, the considerations are restricted to production activities and show only a balance of area consumption on the one hand and production height on the other. Therefore, this index seems more appropriate for the general political discussion than as a decision support tool in the usual planning process.

Jeltsch and Grimm (Sect. 5.4) propose to use modeling techniques to derive sustainable management concepts. In addition they suggest that "the conflict between the economic and ecological point of view can be interpreted as a conflict between economic and ecological goal functions". Through their modeling it was detected, that certain common goal functions such as the succession to a climax state could be inappropriate. Their recommendation is to be very careful with "classical" goal functions ("generic ecological goal functions"). They state, that management should be based on tailored goal functions, that refer to the time horizons and inherent uncertainties, that are characteristic for the particular system in question. In this, they reveal one of the problems of the pure eco-centric goal functions, if used for planning and management purposes in man-influenced systems.

Very strongly relating to application in practice, is the approach of Gnauck (Sect. 5.6), particularly in terms of economic aspects. He proposed an optimization tool to find the best decisions in the discussion of the performance of environmental management measures. In his approach, ecological factors are included, but stand vis-à-vis to the pure economic values. The decisions of politicians would be oriented very strongly to the amount of money that is available to perform the measures. Therefore, the optimization of planning will mostly take place as the effective use of the available amount of money with regard to ecological constraints. According to the financial situation of most of the communities, in many cases only minimum standards of ecological factors can be expected to be reached by this procedure.

De Jong (Sect. 5.7) shows an approach for nature protection management. He states that, consideration of the public and politics to reach planning objectives is necessary. Therefore, the best fit between scientific credibility and usefulness in management has to be found. This implies a bridging of the gap between the necessary reductionism of the practical measurements (e.g. chemical concentrations) and the claims of a completely scientific view of the problem. His paper also shows the falling-apart of more general eco targets or orientors such as stability, diversity, persistence, ecosystem health and sustainability on the one side and the concrete aspects, such as, chemical concentrations, area size, population size and indicator species.

Werner and Bork (Sect. 5.10) went further on, to combine ecological, economic and, in addition, social aspects within their proposed orientor concept for rural areas. The goal of sustainable land use is, in their opinion, only to reach within a framework of integrated rural development, that includes the actors of land use (farmers, etc.) in a participating process. Because of the lack of scientific knowledge to define sustainable land use-systems, they propose at least to "look for pathways of sustainability". That means, to increase the efficiency of resource use and to reduce risks of pollution, erosion and soil compaction in agricultural areas. Overall the eco targets have to be flexible, according to actual or future demands and must be worked out through negotiations of all relevant groups involved. This implies, we have to cross all disciplinary borders. This also requires a new type of researcher, who is more willing to work interdisciplinary and to provide a problem adapted and actor oriented scientific knowledge. Therefore, the authors propose, that the system science approach seems to be the most appropriate, including models as decision support tools. Before reaching a high efficiency of such a research agenda, the actual system of scientific evaluation, based on sectoral judgment, should be altered and the synergy effects of cooperative work should be included.

Conclusions

Firstly, the papers presented, demonstrate the wide range of different aspects that the eco target and orientor concept could have. It is a relatively new scientific concept and therefore, still in development. Aspects of practical application had been included only very scarcely until now. So, e.g. the meaning of the terms "goal function" or "orientor" is often not clearly stated and used in differing ways within the papers.

The, partly, high expectations of planners face the problem that until now no formulation of orientors in sensu strictu for the application within the planning practice is possible. Only the development of some specific, particularly aquatic systems (e.g. lakes) can be described. One of the reasons could be, that many orientor concepts contain the idea of cycles and cycling, that should be closed or in a certain exchange. These criteria are much more easily obtained in an aquatic system, because of the more homogeneous distribution of matters and reaction of the system to changes, compared with terrestrial systems. The proof, that such

goals are transferable to other situations or represent common descriptions is hitherto given in only a few examples.
The same problem is, that most of the proposed goals or orientors refer to single ecosystem behavior. In the planning process landscapes or regions often have to be considered. Furthermore, strongly integrating and complex indicators such as ecosystem health, require a lot of defining factors, that must be determined each time by measurement. But for planning purposes, easily measurable parameters are needed to monitor the desirable development and resulting state of the concerned ecosystem and quantification of the proposed orientors must be possible. The overwhelming influence of economic and cultural factors on the planning process (except in very large natural areas or regions) makes it necessary to include the orientor concept in a holistic concept, where ecological, economic and cultural factors should be taken into consideration.

Goal functions and orientors are, on the other hand, usable for the modeling of ideal states of ecosystems or can be used for scenario techniques to visualize possible paths of development. The combination of these models with Geographical Information Systems can lead to expert systems, that e.g. support the decisions in environmental policy. Particularly for monitoring purposes, the use of orientors seems to be a promising procedure, if the necessary indicators can be provided. In general, the eco-centric approach of the eco target and orientor concept is considered more suitable for the upper levels of environmental policy than for a practical environmental management with its lower time scale. As some of the orientors can be determined by indicators comparable to environmental quality objectives ("Umweltqualitätsziele"), they can be used operationally in the planning process, e.g. as orientation for the environmental policy. But we have to be aware, that orientors have a modeling background and therefore, represent the point of view of their designer. For this reason, they cannot reflect the whole reality and planners have to use them carefully, considering their specific range of validity.

Recommendations

Some recommendations and therefore, objects of further scientific considerations had arisen.

- To make eco targets and orientors more suitable for planning purposes, at least a larger range of ecosystem types must be included, for which they are valuable and tested.
- As the proposed orientors combine complex facts, many variables are necessary for their description. Thus, to use the orientor concept for planning and environmental management purposes, a list of indicators must be provided.
- To avoid mapping and measuring indicators for every new project, a regular inquiry such as agricultural statistics should be performed, that every planner

can use. Müller (1996) proposed an indicator system for the detection of the actual ecological state and its change. If these indicators were brought together in an integrating concept and updated officially in certain time steps, a more intensive use within the planning process would be possible. However, some doubt remain: Is there hope for monitoring all these parameters on more than a few sites (mainly biosphere reserves)? If not, it would not really be helpful for everyday decisions in regional planning.

- If it would be possible to use orientors as integrating "slogans" in advertising, one could possibly promote within the planning process complex facts to the public. Therefore, the meaning of the proposed orientors has to be integrating but also succinct.

As the multitude of targets will remain, integrative concepts for planning and management should combine ecological orientors with economic and social factors. This means an interdisciplinary work of scientists and planners is necessary to provide such a concept. Therefore, planners have to include more dynamic eco-systemic aspects and give space to these arguments beside the necessities of economic and cultural demands. Ecosystem researchers, on the other hand, have to consider, that purely eco-centric point of views are not feasible for planning purposes. That means, the formulation of optimal states of ecosystems should be the starting point for discussions, but for the negotiations within the planning process, room to move must be provided. For this reason, it is also necessary to derive a minimum quality level of ecosystems ("Mindeststandard", SRU 1996; Reck et al. 1994). Both values will give the freedom to negotiate with the economic and cultural aspects and to find a compromise, that gives more significance to the ecosystem aspects within planning than is actually done. As man is able (and often willing) to exceed the carrying capacity of nature with his management actions, the reception of eco targets and orientors could be one of the possibilities to come to a more sustainable development in the future.

References

Müller F (1996) Ableitung von integrativen Indikatoren zur Bewertung von Ökosystemzuständen für die Umweltökonomische Gesamtrechnungen. Statistisches Bundesamt, Wiesbaden

Rat von Sachverständigen für Umweltfragen (SRU) (1996) Konzepte einer dauerhaft-umweltgerechten Nutzung ländlicher Räume. Metzler-Poeschel, Stuttgart

Reck H, Walter R, Osinski E, Kaule G, Heinl T, Kick U and Weiß M (1994) Ziele und Standards für die Belange des Arten- und Biotopschutzes: Das "Zielartenkonzept" als Beitrag zur Fortschreibung des Landschaftsrahmenprogrammes Baden-Württemberg. Laufener Seminarbeitr 4:65-94

Regier HA (1993) The Notion of Natural and Cultural Integrity. In: Woodley S, Kay J, Francis G (eds) Ecological Integrity and the Management of Ecosystems. St. Lucy, Ottawa, pp 3-18

Chapter 6

Conclusion: Targets, Goals, and Orientors

6 Targets, Goals, and Orientors: Concluding and Re-Initializing the Discussion

Felix Müller, Jan Barkmann, Broder Breckling, Maren Leupelt, Ernst-Walter Reiche and Reinhard Zölitz-Möller

6.1 Change as a Resultant of Conflicting Targets

This volume turns out to be a multilevel documentation of change. It includes reports on the unavoidability, the necessity and the objectives of change as well as the challenges and problems of the corresponding alterations. Change is discussed in a multitude of details at several levels of organization, ranging from single organisms to the interacting systems of man and nature. Parallel to the contents of the single papers, the whole line of argumentation in this volume symbolizes a dynamic change, which also includes an exemplary presentation of the associated problems. These difficulties, the related delays and the sequential time lags emerge since there are huge distinctions between the world views of the companions in this game of concept development, such as theorists ("dreamers"), applied scientists ("realists") and planners ("pragmatists"). Similar groupings can be found with respect to eco-centric (deep ecological) and anthropo-centric (human purpose focused) ideas, or to economic (profit optimizing), social (welfare improving) and ecological (nature conserving) philosophies. Other divergences obviously exist between top-down (holistic) and bottom-up (reductionistic) approaches, between abstract (systemic) and single-case oriented (individualistic) accesses, between long term (intergenerational) and short term (intragenerational) views, between the implication of broad (interregional) and small (intraregional) spatial extents, between the pre-eminent contemplation of direct (short-term and accountable) and indirect (long-term but priceless) effects, or between conservative (museal) and more "progressive" (developmental) strategies of environmental protection. Each of these adverse features can be attributed by commendable, specifically elaborated concepts. Each of these concepts is directed by individual, societal acknowledged targets. The resultant direction of change, thus, will be determined by the power of persuasion of the single target, and theoretically the resultant should coincide with the social consensus. The demonstrated diversity of goals does not only elucidate the enor-

F. Müller M. Leupelt (Eds.)
Eco Targets, Goal Functions, and Orientors

mous creativity of socio-economic and scientific conceptions. The goal variety also demonstrates the helplessness that might arise due to the expectable unlimited duration of discussions. They will be necessary to find "the one applicable solution" to the questions, what concepts will be the optimal results of the imperative adaptation of targets (see Chap. 1) and how the related environmental change will be evaluated and directed with respect to the modified orientation.

6.2 Environmental and Social Change toward the Target of Sustainable Development

The sustainable development paradigm offers a sufficiently unified view of social, economic and ecological change as it sets a universally acclaimed target for human action. Contrary to many proposed conceptions, the sustainability concept has the potential to integrate the demands of ecological, economic and sociological arguments on a multi-scale fundament of argumentation (e.g. WCED 1987; Daly et al. 1990; MISEREOR and BUND 1996; Moldan 1997). However, to approximate the necessary interdisciplinary and pluralistic, sustainable concept the corresponding targets have to be adjusted. Theoretically, most companions, even the political representatives have agreed to this idea in principle; but the conversion into practical action seems to be extremely difficult. Again, one of the focal problems in this situation originates in the multitude of targets. Therefore, an intensive discussion on the developmental directions of economies, societies and environments has to continue in the future.

One central and integrative part of the resulting sustainability concept should be a hierarchical categorization of functional scales, as outlined in Fig. 6.1. In such hierarchies the subsystems develop on the constraints that are determined by the superior levels (Allen and Starr 1980; O'Neill et al. 1986; Allen and Hoekstra 1992). In many cases this context of constraints emerges as a result of the interactions in the constrained subsystems (emergent properties) (Müller 1996; Müller et al. 1997a,b). Thus, there is self-organized top down control as well as bottom up regulation. However, during phases of steady states, the lower levels can only develop within the constraints of the higher niveaus. If we apply this model to the interactions of man and nature (Fig. 6.1), it becomes obvious, that human developmental degrees of freedom, in fact, are constrained by natural hierarchical levels. Disregard of this contextual hierarchy leads to a de-coupling of the whole system and will direct towards extraordinary phase transitions. So, humans have to adapt their activities to the carrying capacities of the environment. Within this rather simple cognition, the main interests are focused: A positive long-term development of economies, ecosystems and social systems will only be possible if humans considers the natural hierarchies adequately. In this situation, humans advantage is an intellectually guided freedom of action, a talent for rationality, and the potential speed of change that human systems are capable to attain in comparison with natural systems (Bossel Sect. 4.2).

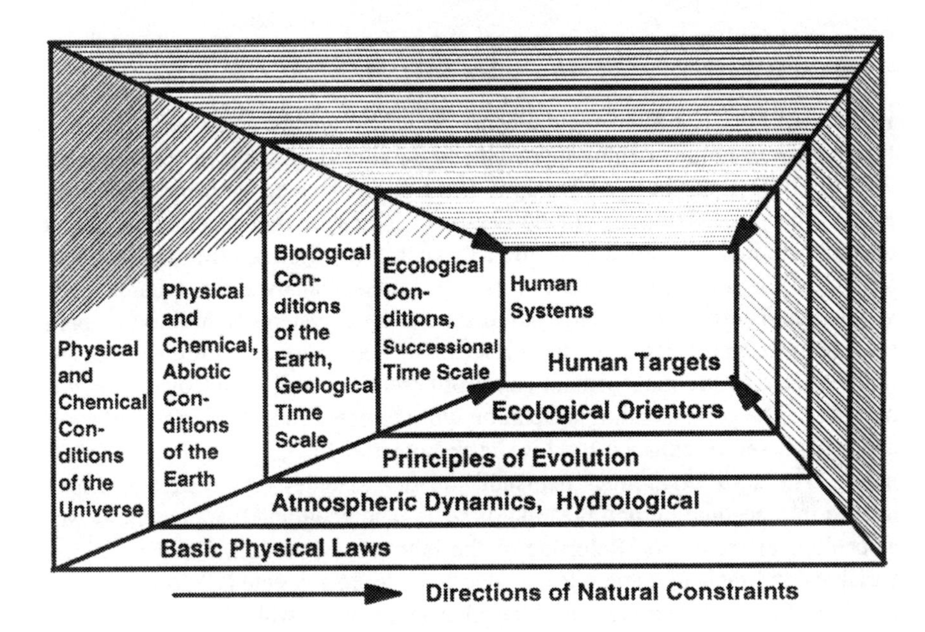

Fig. 6.1 A hierarchical sketch on the operative system of constraints which act as boundary conditions to human target setting. Many environmental problems have arisen because human decisions have not taken into account the long term, indirect and broad scale changes they effect within the natural system of constraints that limits their own actions.

Fig. 6.1 suggests to interpret sustainability as a basically ecological concept because the fundamental constraints of the human systems' activities are environmental limitations, such as scarce environmental resources, reduced ecological buffer capacities concerning the multiple forms of human entropy production, and a restricted capability for short-term, evolutionary adaptation. Therefore, a concept of sustainable landscape management has to integrate

- long term strategies, focusing on systems' succession and evolution,
- hierarchical strategies including scale transitions and the spatio-temporal integration of processes that are operating on at least three scales (O'Neill et al. 1989),
- interdisciplinary strategies which are integrating ecological, social and economic requirements,
- realistic strategies that incorporate, quantify and nominate uncertainties,
- holistic strategies that take into account the potential causes of indirect effects,
- nature based strategies that reflect how natural processes can be used as a source of inspiration for technical application,

- theory based strategies that support the self-organization capacities of ecological systems.

Particularly, the last three requirements document a high conceptual interrelationship between the orientor approach and the idea of sustainable development. First of all, as it has been described in Chap. 2, the concept of ecological orientors is theory-based, aiming at a derivation of indicators that describe the degree of and potential for self-organization in ecological systems (Müller and Fath Sect. 2.16). It is the general idea of its application to analyze the dynamic pathways of undisturbed ecosystems and to set disturbed systems right into that path, thus, optimizing the degree of self-organization capability on multiple scales. By this way, landscape management can profit from the well-established strategies of nature. The corresponding adaptation will include a reduction of disturbances and will therefore enhance the long-term potential for human appropriation of nature. The third requirement demands holistic strategies. These strategies have to take into account an optimization of social and economic parameters as well as ecological indicators. Referring to the latter, both, structural and functional variables and their potentials for change have to be respected. And as the propensities for a structurally enriching change increase with the availability of potential actors, the conservation of biodiversity must be one of the key targets in sustainable management strategies. The connected conservation activities should be performed as parts of an improved ecological integrity on the regional scale, on the one hand. On the other, the utilization of special endangered areas has to be restricted in dependence of single conservation targets.

Besides the protection of structural subsystems, the functioning of ecosystems has to be preserved and (re)developed with a high priority. Obviously, this also requires nature protection strategies, that try to preserve specific sets of ecosystem elements. In addition to these static features, the multiple functions that natural systems perform for the human utilization have to be taken into account (see Chap. 1 and 4). These characteristics refer to the capability of ecosystems to proceed their energy and matter degrading processes and to evolve (and optimize these processes) in an adaptable manner. To follow this target, the development of the respective ecosystems – change – plays a vital role. The courses of development, thus, comprise a higher significance than the potential final states and structures. The orientation should not be directed toward potentially brittle, overconnected, senescent systems states, but toward the way and the processes which undisturbed systems use while getting there.

This objective is essentially determined ecologically. However, if we review the diversity of the specific targets that are described as competing or complementary world views in the first paragraph of this Chapter, a common aiming point, toward which all companions can align sights, may be perceived with the holistic environmental approaches of the sustainability concept. In this aspect, ecological orientors could play an important role as indicators and as models.

Thus, the preceding papers suggest some interesting inputs from many different points of view. These ideas should be used to outline an ecosystem based impetus to the integrated developmental strategies of economies and societies. However, the central question of this book remains: What can we learn from the regularities of change in ecosystem structures and functions? We will arrange some points of view and answers to this focal question on the following pages. Doing so, we will refer to the structure of the volume and to the questions which have been asked in Chap. 1.

6.3 Recapitulating the Points of View and Questions

The general idea of argumentation in this book is the following: In the beginning, the theoretical concept of ecosystem development (orientors and goal functions) is compiled, discussed from multiple aspects, and exemplified by ecosystemic case studies (Chap. 2). As this concept contains some ideas that have to be analyzed critically from the point of view of epistemology (e.g. teleology), from the basic ideas of evolution (e.g. conceptual parallels) from social, political and ethical aspects (e.g. applicability in human systems), the goal function concept is discussed philosophically in Chap. 3. Based on these considerations, ecological targets are put in correspondence with human aims, and the propensity for a conceptual integration of these different goals (sustainability conceptions) is demonstrated (Chap. 4). Finally, the realized methods and applications of the orientor concepts are described, and the potential for a future utilization in landscape management and planning is investigated (Chap. 5).

The Scientific Origin and the System Analytical Evidence of Ecological Orientors

The orientor principles as well as the goal function concept originate from general systems analysis and synergetics, that investigate the fundamental processes of self-organization. The ideas have been introduced to ecosystem theory by Jørgensen (1992) and Bossel (1992). They have been chosen to describe the general development of open systems and to optimize the results of ecological, structural dynamic models (Nielsen 1992). In this sense, goal functions are mathematical functions, utilized in ecological modeling. They can be employed to describe the direction of an ecosystemic development as a consequence of the self-organizing ability of these systems. Ecological orientors are the state variables which denote the directed behavior and development of ecosystems. Their dynamics are regularly directed towards specific, optimized states. This occurs as an emergent result of the involved interactions. These states are the attractors of the system. Their values depend on internal and specific conditions of the observed site.

Working with these terms, we have to be aware of the modeling character that both notions comprise: Whenever a system is defined, the observer only selects a small segment of reality. He constructs a model that is used as a metaphor of the reality he is interested in. That model and its functional implications are hypotheses for the behavior of the ecological processes it refers to. Even if the modeling results are validated we may address them only as the "best guess" of the modeler about ecosystemic tendencies; goal functions, thus, should not be understood as properties of "objective reality". On the other hand, as already the system is defined as a model, the only access to complex systems are complex models. And of course, all propositions about the systems are propositions about the utilized models. Therefore, the empirical validation and the exact definition of a model's domain of validity play a focal role for the credibility of systems analytical statements. These paradigmatic implications of system science are generally known to most modelers of complex systems. Therefore, they have to be respected as basic prerequisites of the orientor principle (see Barkmann et al. Sect. 3.1). It is still important to point at these 'intellectual and philosophical finesses' because the pragmatic purposes that induced a certain system definition or modeling strategy are crucial when the adequateness of a model for application has to be considered.

Which are these orientors and goal functions? The introductory questions in Chap. 1, referring to this point are the following: What are the general principles of ecosystem development? Can we define certain, regularly appearing attractors and system-based orientors throughout the development? How are ecological goal functions correlated with the general features of self-organization, emergent properties, thermodynamics, gradient degradation, and general ecosystem dynamics? What are the corresponding indicators and how can they be quantified, implemented, and evaluated?

In Sect.s 2.1 to 2.16 a variety of such variables is discussed, some of them are illustrated and some orientors are demonstrated in case studies. The respective answers are summarized in Sect. 2.16. In this Section, different orientors are traced back to fundamental prerequisites and features of dissipative self-organization. These basic characteristics of self-organized entities are the following: openness of systems, input of exergy, internal exergy degradation, entropy production, increasing distance from thermodynamic equilibrium, nonlinearity, phase transition, internal regulation, hierarchy, (meta)stability, and irreversiblity (see Müller and Fath Sect. 2.16). Every single item is linked with the idea of thermodynamic gradient dynamics: dissipative self-organization leads to structures that comprise more and more internal gradients situated between the subsystems and system elements (Schneider and Kay 1994a,b). These gradients provide the "energy source" (exergy) of all ecological processes. They operate as processual units, effecting a more and more efficient dissipation of the massive external gradients, the system is exposed to. To erect internal gradients of spatial and functional concentrations, more and more exergy is utilized. Thus, parts of the captured energy is transformed into structure and information in an increasing quantity (Jørgensen 1992). The maintenance of these internal gradient hierarchies, requires increasing amounts of exergy, which are transformed into entropy. These processes display non-linear dynamics and result in a growing distance of the entire system of hierarchical gradients from thermodynamic equilibrium. According to the increasing complexity, more and more functional and structural hierarchical levels are introduced, throughout these dissipative dynamics displaying more and more emergent properties. These general developmental schemes exist in many different dissipative systems. The resulting changes of single variables are summarized in seven classes in Sect. 2.16. These orientor groups are:

- thermodynamic orientors (e.g., exergy storage, exergy consumption, entropy production)
- information-theoretical orientors (e.g., information, heterogeneity, ascendency)
- structural orientors (e.g., mutualism and symbioses, fecundity, niche diversity)
- network theoretical orientors (e.g., indirect effects, cycling index, utility)
- ecophysiological / functional orientors (e.g., loss reduction, respiration, resource utilization)
- system dynamics orientors (e.g., resilience, buffer capacity)
- system organization orientors (e.g., feedback control, complexity, connectedness).

Within these classes many different indicators are defined and different schemes of aggregation for these variables are proposed. Typically, all holistic descriptions of systems have to be built up by a set of variables; there is no single variable which might describe ecosystem states alone. Thus, multidimensional evaluation models seem to be the only suitable tool for a holistic ecosystem evaluation (see, e.g., Kutsch et al. Sect. 2.12).

Advanced models of ecosystems development take into account that development does not base upon linear dynamics. In correlation with the maturity of a developing ecosystem, the risk of breakdown increases as well, because in "climax stages" all of the captured exergy is necessary to maintain the attained complex structure. There is no energetic reserve that could be invested into the resilient buffering of disturbances. Thus, the adaptability of highly mature systems is reduced enormously, and the high connectivity may turn into overconnectedness. Therefore, also the chances of creative destruction (Holling 1986; see also Bass Sect. 2.11), that are important parts of self-organized dynamics, have to be integrated into management plans.

The Philosophical Barriers and the Social Implication of the Orientor Concept

According to the guiding ideas of this volume, inputs and proposals from philosophy, politics, and sociology, are documented in Chap. 3 to define the general prerequisites and limitations of the orientor, and particularly the related goal function approach and its applications. The corresponding question in Chap. 1 is: "What are the suppositions for a valid derivation of ecological orientors in self-organized systems?"

One central problem of the orientor concept arises from the temptation to argue in a teleological manner. Therefore, it has to be stressed that nature in scientific perspective does not and can not possess any *intended* goal that might be derivable from the observation of biogeochemical or thermodynamic processes. Consequently, the usage of the term "goal" should be restricted to well-defined circumstances (e.g., modeling procedures) to avoid teleological misunderstandings. These mistakes could easily lead to the impression that social or political issues are improperly mingled with scientific affairs. Therefore, it is necessary to distinguish between intentional human goals for environmental development as a result of societal interactions on the one hand and trends or tendencies in ecosystem development as a result of physicochemical and organismic interactions on the other. A second distinction, which also has been touched upon in the preceding paragraph, has to be made between the properties of ecological models as mathematical or systems analytical constructions and the assumed properties of "real world systems".

From hierarchy theory (Allen and Starr 1982; O'Neill et al. 1986) it can be derived that each level of integration in ecological systems has its own specific properties. They are emerging on the particular observational level as an outcome of the interactions of lower levels (Wiegleb and Bröring 1996; Müller 1992). Within these hierarchies there are different qualities of interaction which should be distinguished: The term "goal function" can address natural ecological interactions as well as social phenomena. Both can be understood in terms of self-organization. The involved types of interactions are different, however: Natural systems display emergent behavior by spontaneous organization. The self-organization depends on the internal interactions of the system elements within their environmental context of external impacts and other constraining site conditions. Social systems have a self-organizing structure that is intellectually reflected and modified due to intentional reasoning. This means that goals in an entirely strict sense can be imposed as consciously defined management objectives within social systems only.

Looking at the characteristics of natural ecological processes, we can find another important conceptual distinction: Theoretical models on the one side which are set up in order to investigate the formal implications of certain mathematical assumptions, describing an ecological background, and on the other side models of "real world systems", where an abstract formalism is used to depict the considered essence of underlying ecological processes. In the first case it is justified to use the term "goal function" as a description of model properties. In the second case, however, it will be more adequate to talk about developmental tendencies of orientors or ecological development trends. Between these types of models there is a certain feedback of mutual impacts: Theoretical considerations provide a repertoire of potential descriptions, and on the other side receive inspirations and challenges from practical problems. Thus, there is also an implicit influence of social images on the theoretical research activities and paradigms. Vice versa, the understanding of ecological processes influences social goal setting processes profoundly. This is most obvious in attempts to design sound ecosystem management strategies, which is impossible without at least a fragmentary understanding of the underlying ecological processes. From this perspective, ecosystem management and ecosystem analysis represent an iterative cycle. Understanding the specific feedback processes requires to distinguish properties of natural and social systems, their "goal seeking behavior", and be aware of their qualitative differences in order to improve our knowledge about how both work together as one entity.

The Natural Orientation of Self-Organized Systems and The Goal Setting of Human Systems

The dispute of philosophical and societal ideas (Chap. 3) continues in the 4th Chapter, which discusses the interactions between societal targets and ecosystem use in the framework of the sustainability concept. The central questions are: What are the orientors in human systems, how can they be quantified, and how can they be combined with the ecological orientors? How can natural trends of ecosystem development be coupled with the goals of society? These problems are introduced on the base of conflicting landscape utilization strategies and the principle differences between ecological, cultural, and economic goals. The discussion identifies the scales of the respective processes and of the decision making procedures as important variables. The variables differ enormously if local, regional, scientific, or social systems are involved. An adjustment of the corresponding scales seems to be a suitable strategy to improve the functional order and to realize the constraints in processes of environmental politics and decision making.

In the papers of Chap. 4, the authors emphasize the idea of sustainability as a focal concept for the interdisciplinary integration of ecological, social, and economic strategies and targets. In this context, the functional appreciation of nature, the anthropo-centric view of nature (nature providing services for humans) as well as the scientific understanding (patterns of flows and dynamics), seem to be suitable approaches to achieve the aspired combination of disciplinary world views and methodologies. Thus, the ecologically determined environmental goods and the respective environmental quality objectives should base on a functional approach, in any way. Function in this case includes structure, as the latter notion represents the pattern of elements that take an acting part in the systems' functional processes. Basing on this assumption, the orientor approach, as it is explained and illustrated in Chap. 2, turns out to be an optimal strategy to integrate the multitude of evaluating benchmarks that significantly influence ecological decision processes.

The contributions in Chap. 4 also focus on the limitations of the transferability of ecosystemic principles to human systems. Human and cultural systems can in fact learn from natural processes and dynamics to a certain extent. However, in a number of cases, this will not be possible. This is another reason why the terms "ecological orientors" or "goal functions" should be handled with care. Particularly, a direct, non-reflected ethical translation of ecosystem processes to human behavior would imply to be actively caught in the trap of the naturalistic fallacy (see Chap. 3 and Bröring and Wiegleb Sect. 2.3).

Despite the differences between human and natural systems, we can derive brilliant techniques from natural processes—studying the products of natural evolution. Environmental policies focus on many different fields: species and biotopes protection, ecosystem protection, environmentally compatible production, protection of resources. Consequently, a further target of providing environmental policy has to be the production of material and products that are compatible with the material flows of nature. During evolution nature realizes innovations continuously by small steps of limited interference. That's how nature keeps her chemical, physical, and structural composition over long periods of time. Each level of synthesis with its special storage of energy and molecular order is used as long as possible before the selective reorganization on the next level, and, thus, nature introduced the principle of scarcity. So, on the one hand, there is a principle-inventing natural system and on the other hand, human systems that can react with a high flexibility and adaptability and that are equipped with prerequisites to find solutions for the imperative ecological problems. Humans can improve their industrial processes according to natural processes (see Sturm 1997; Ruth Sect. 4.5, Leupelt and Reiche Sect. 4.1). They may even be able to use functional approaches for the construction of land use strategies (Bork et al. 1995; Hatfield et al. 1994), environmental target systems and ecological evaluation schemes. Ecological orientors could be focal elements of such modern changes. Thus, a focal question is, what the recent practice of environmental management looks like.

The Orientor Principles in Practical Application

Practical applications are illustrated in Chap. 5. The central questions for this part of the book (see Chap. 1) are: Can we use orientor principles in environmental management? Could orientors be applied to define new sustainable strategies for landscape management? Will it be possible to develop a hierarchy of system-oriented, holistic goals for the realization of principles such as sustainability, health, and integrity, based upon an integration of eco-centric and anthropo-centric orientors? Each question can clearly be answered "yes", particularly, as far as they are directed toward future developments in landscape management. It will become more and more important to derive scientifically based, convincing and evident eco targets, environmental quality objectives and environmental standards that can compete with the popular economic scales and indicators of monetary functions. Such eco targets should be built on a systemic fundament which takes into account

- the complex character of ecosystems, their intra- and inter-systemic connectedness and the significance of indirect effects,
- the holistic character of ecosystems that are integrating units of transfer and storage, connecting all environmental media and sectors,
- the interactions of processes which operate on different spatial and temporal scales,
- long-term effects, chronic interactions and regional (not only local) processes,
- the correlated significance of an integrative analyses of structures and functions,
- the dynamic aspects of ecosystem change,
- the significance of the ecosystem's capacities for self-organization,
- the potentially constructive role of destruction,
- the system of natural constraints that human systems are exposed to and that they depend on.

For the derivation of ecosystemic indicators that are capable to cope with these demands, the orientor principles will be very helpful. As the papers in Chap. 5 illustrate, the fundamentals are existing, the theoretical background and the empirical evidence are satisfactory, and the step of application as an ecological instrument for the development of the sustainablity concept should be the logically following consequence. In this framework, it will be feasible to arrange the different orientors in a functional order which will also ease the coupling of the sustainability concept with the ideas of ecosystem health, ecosystem integrity and ecosystem protection.

However, if we take a look at the status quo in environmental management, as done in Chap. 5, no direct use of orientors – taking into account ecosystem developments – can be found in recent landscape or regional environmental planning procedures yet. Although applied models (e.g., Marques and Nielsen Sect. 5.3 or Jeltsch and Grimm Sect. 5.4), optimization tools (e.g., Gnauck Sect. 5.6) and intelligent accounting methods (e.g. Krotscheck Sect. 5.2 or Radermacher Sect. 4.6) are available, these concepts have not yet been applied in a considerable extent. The responsible attitude of the practitioners can be characterized by the following sentences: From a pragmatic point of view, the evidence and the advantages of holistic indicators are not obvious because they are extremely difficult to calculate, too complex (complicated) to be operated in every-day-work, and an application would therefore be too expensive. The exergy approach, particularly, has been tested in aquatic ecosystems, while most management decisions have to be made concerning terrestrial systems. The necessary data, for example, the data for an application of the Sustainable Process Index (SPI) (Krotscheck Sect. 5.2) or for the ecological footprint ideas, are not available in most cases. Furthermore, planners are used to work with threshold values. Where will the required holistic data come from in future environmental management? On the base of these and similar questions, asked by pragmatists, the optimism which has been visible from a purely scientific point of view above, might be reduced.

Nevertheless, the construction of a set of indicators for monitoring purposes represents a rather rapidly applicable method to use the knowledge on ecosystem development for the evaluation of the state of an ecosystem or a landscape with respect to sustainable land use systems from the functional perspective. This kind of monitoring might be an expensive challenge, but it could reach a reference status for scenario techniques and thus serve as a paradigmatic prototype that demonstrates the potential of holistic evaluation procedures. Furthermore, a utilization of the orientor approach in monitoring would deliver excellent fundamentals for political decision making processes. However, even this approach does not seem to meet the needs of decision makers on the lower level. The application is restricted to longer time scales and to projects with relatively large spatio-temporal extents. The goal function approach, thus, recently seems to be more suitable for environmental policy or regional environmental planning than for landscape planning on a local level, where most of the planning work is done in many countries. Consequently, orientors can preferably be used to determine the ecological constraints for local planning activities. On the other hand, also for the lowest level of environmental politics, the complex tools for an integrative planning procedure are available. There are many good and practical simulation models, there are experienced techniques to combine these models with geographic information systems, there are elaborated instruments to operate these units with optimization procedures, and there is successful experience in applying scenario strategies.

Therefore, the pessimistic opinions are not convincing from a theoretical point of view (the "dreamer", see above). Actually, the application of the described principles and techniques will be an important improvement for applied ecological activities. However, as the concepts still are new, as they might sound confusingly new and perhaps unusual, nobody should wonder about the declining reactions of practitioners: The thermodynamic equilibrium is hard to determine for any landscape plan that has been based on the idea of a structurally indicated ecological balance. Additionally, the theoretical concepts seem to be correlated with an increasing complication and the economical border conditions of environmental practice might not support the idea, that more expensive procedures are welcome. In this situation, the reflections from the first Section of this Chapter become crucial again: Change is always accompanied with problems, delays and partial resistance. Thus, we have to hope that Bossel (Sect. 4.2) is right, pronouncing that the advantage of humans is their rationalism, flexibility and the speed they can attain compared with natural systems.

6.4 A Variety of Questions - and an Optimistic Position

There still remain two questions, we will focus on for the final remarks: What are the potential consequences of ecosystem evolution and development for the definition of environmental management goals? What can we learn from the regularities of change in ecosystem structures and functions?

These questions still have to be discussed, and again the answer will be a risky prognosis: The developmental tendencies of ecological systems fit into the general trends that dissipative, self-organized systems exhibit. The tendencies can be observed by means of specific orientors, which elucidate specific areas of the developmental performance, and which should therefore be used as indicators. These indicators can be arranged in a pattern that describes the system as a whole, avoiding the usual restrictions to structural units, alone, but integrating the important functional ecosystem elements, as well. In this way, indirect, chronic and delayed effects, and intersymestic impacts can be included into evaluation procedures. The application of the indicators will demonstrate to which degree the observed system behaves similar to natural entities. As these systems are related with a very high capability for self-organization and adaptation, and as the corresponding parameter combinations and features have been evolving for billions of years, we should utilize the ecosystem experience, use the orientors' signals to manage nature in a more nature-near manner, thus improving the integrity and the health of the respective ecosystems. This management plan will support the energetic and materialistic functionality of the ecosystems as well as their structural diversity. Additionally, it will also improve the long-term functionality of natural or nature-near systems for human utilization. Thus, a landscape management which adopts these ecosystem tendencies is a profound and promising strategy and contributes to the ecological goals of sustainable development.

To reach this state, further investigations and applications will be necessary, particularly, concerning the derivation of indicators and the methods of handling multiple, multidimensional sets of ecosystem features. Therefore, the focus of future research should address the question, how to transfer the theoretical concepts into environmental practice. On the preceding pages most prerequisites for this step have been fulfilled. Now, the more urgent step is the transfer into every-day-application.

We resume these views, in reference to the three initializing theses of this book in three concluding propositions:

This volume documents that the theoretical principles of dissipative self-organization can be applied to ecosystems, and that the corresponding theoretical concepts are suitable to describe the general tendencies of ecological development, predominantly its functional changes. These directed modifications can be described by emergent and collective state variables – orientors or goal functions in the modeling context, which are generally directed toward certain attractors. A combination of the variables will amount to valuable, theory-based indicators of environmental states.

This volume stresses the eminence of dynamic features as targets and indicators in nature protection. The support of the self-organizing capacity of ecosystems, the promotion of ecosystem functions, and the permission of change – even if the change seems to be destructive at first glance – turned out to be important guidelines for holistic, sustainable environmental management strategies. These goals are capable to unify important views from ecology, economy and sociology. Therefore, the orientor based sustainability research is a good point of departure for an interdisciplinary enhancement of recent, system-based environmental endeavors such as the investigation of ecosystem health or ecological integrity.

This volume points out an enormous need for further interdisciplinary communication and collaboration. The semantic and conceptual problems that impede dialogue have turned out to be much more serious than expected during the conceptual phase of the goal-function-project. These difficulties lead to a considerable time lag between the development of scientific ideas and their practical application. Additionally, all proposals that are too indirect and complex for mechanical application by planners or ecologists, run the risk to be refused due to economic reasons—without considering the potential advantages. Therefore, all companions of the environmental object have to improve their motivations for a firmer collaboration. The theorists have to adapt their ideas and visions to operable concepts, and the practitioners should open up to the application of new strategies, although they still might seem to be rather complicated. Consequently, an enhanced exchange of information, methods, and problems is one of the basic necessities that we will need for the implementation and the scientific improvisation of the sustainable development concept.

References

Allen TFH and Hoekstra TW (1992) Toward a Unified Ecology. Columbia University Press, New York, p 384

Allen TFH and Starr TB (1982) Hierarchy – Perspectives for Ecological Complexity. Chicago University Press, Chicago, p 310

Bork H-R, Dalchow D, Kersebaum KC and Stachow U (1995) Regionalanalyse der Auswirkungen veränderter agrarökonomischer Rahmenbedingungen auf Agrarlandschaftsnutzung und Umweltqualitätsziele. Zeitschrift für Kulturtechnik und Landentwicklung 36(4):194-201

Bossel H (1992) Real-structure process description as the basis of understanding ecosystems and their development. Ecol Modelling 63:261-276

Hatfield JL and Karlen DL (eds) (1994) Sustainable agriculture systems. Lewis Publishers, Boca Raton, p 316

Holling CS (1986) The resilience of terrestrial ecosystems: local surprise and global change. In: Clark WM and Munn RE (eds) Sustainable Development of the Biospere. Oxford, pp 292-320
Jørgensen SE (1992) Integration of Ecosystem Theories: a Pattern. Kluwer Academic Press, Dordrecht, p 383
MISEREOR and BUND (eds) (1996) Zukunftsfähiges Deutschland. Ein Beitrag zu einer global nachhaltigen Entwicklung. Studie des Wuppertal Instituts für Klima, Umwelt und Energie. Birkhäuser Verlag, Basel Boston Berlin
Müller F (1992) Hierarchical approaches to ecosystem theory. Ecol Model 63:215-242
Müller F (1996) Emergent properties of ecosystems – Consequences of self-organizing processes? Senckenbergiana maritima 27:151-168
Müller F, Breckling B, Grimm V, Malchow H, Nielsen SN and Reiche EW (1997a) Ökosystemare Selbstorganisation. In: Fränzle O, Müller F and Schröder W (eds) Handbuch zur Ökosystemforschung. Ecomed, Landsberg
Müller F, Breckling B, Grimm V, Malchow H, Nielsen SN and Reiche EW (1997b) Emergente Ökosystemeigenschaften. In: Fränzle O, Müller F and Schröder W (eds) Handbuch zur Ökosystemforschung. Ecomed, Landsberg
Nielsen SN (1992) Application of Maximum Exergy in Structural Dynamic Models. Ph.D. thesis. National Environmental Research Institute, Copenhagen
O'Neill RV, De Angelis DL, Waide JB and Allen TFH (1986) A hierarchical concept of ecosystems. Monogr Popul Biol 23. Princeton University Press, Princeton, p 254
O'Neill RV, Johnson AR and King AW (1989) A hierarchical framework for the analysis of scale. Landscape ecology 3(2/4):193-206
Schneider ED and Kay JJ (1994a,b) Life as a manifestation of the second law of thermodynamics. Math Comput Model 19:25-48
WCED (Brundtland SH) (1987) Our Common Future. Oxford University Press, Oxford
Wiegleb G and Bröring U (1996) The Position of Epistemological Emergentism in Ecology. In: Albers B, Dittmann S, Krönicke I, Liebezeit G (eds) The Concept of Ecosystems. Senckenbergiana Maritima 27 (3/6):179-193

Subject Index

Springer and the environment

At Springer we firmly believe that an international science publisher has a special obligation to the environment, and our corporate policies consistently reflect this conviction.

We also expect our business partners – paper mills, printers, packaging manufacturers, etc. – to commit themselves to using materials and production processes that do not harm the environment. The paper in this book is made from low- or no-chlorine pulp and is acid free, in conformance with international standards for paper permanency.

Printing: Mercedesdruck, Berlin
Binding: Buchbinderei Lüderitz & Bauer, Berlin

Lightning Source UK Ltd.
Milton Keynes UK
02 March 2010

150826UK00008B/25/A